D1233742

Volume 38

Exact constants in approximation theory

ENCYCLOPEDIA OF MATHEMATICS AND ITS APPLICATIONS

1 Luis A. Santalo *Integral geometry and geometric probability*
2 George E. Andrews *The theory of partitions*
3 Robert J. McEliece *The theory of information and coding: a mathematical framework for communication*
4 Willard Miller, Jr. *Symmetry and separation of variables*
5 David Ruelle *Thermodynamic formalism: the mathematical structures of classical equilibrium statistical mechanics*
6 Henryk Minc *Permanents*
7 Fred S. Roberts *Measurement theory with applications to decisionmaking, utility, and the social sciences*
8 L. C. Biedenham and J. D. Louck *Angular momentum in quantum physics: theory and application*
9 L. C. Biedenham and J. D. Louck *The Racah–Wigner algebra in quantum theory*
10 W. Dollard and Charles N. Friedman *Product integration with application to differential equations*
11 William B. Jones and W. J. Thron *Continued fractions: analytic theory and applications*
12 Nathaniel F. G. Martin and James W. England *Mathematical theory of entropy*
13 George A. Baker, Jr and Peter R. Graves-Morris *Padé approximants, Part I: Basic theory*
14 George A. Baker, Jr and Peter R. Graves-Morris *Padé approximants, Part II: Extensions and applications*
15 E. C. Beltrametti and G. Cassinelli *The logic of quantum mechanics*
16 G. D. James and A. Kerber *The representation theory of the symmetric group*
17 M. Lothaire *Combinatorics on words*
18 H. O. Fattorini *The Cauchy problem*
19 G. G. Lorentz, K. Jetter, and S. D. Riemenschneider *Birkhoff interpolation*
20 Rudolf Lidl and Harald Niederreiter *Finite fields*
21 William T. Tutte *Graph theory*
22 Julio R. Bastida *Field extensions and Galois theory*
23 John R. Cannon *The one-dimensional heat equation*
24 Stan Wagon *The Banach–Tarski paradox*
25 Arto Salomaa *Computation and automata*
26 Neil White (ed) *Theory of matroids*
27 N. H. Bingham, C. M. Goldie and J. L. Teugels *Regular variation*
28 P. P. Petrushev and V. A. Popov *Rational approximation of real functions*
29 Neil White (ed) *Combinatorial geometries*
30 M. Pohst and H. Zassenhaus *Algorithmic algebraic number theory*
31 J. Aczel and J. Dhombres *Functional equations containing several variables*
32 Marek Kuczma, Bogden Chozewski and Roman Ger *Iterative functional equations*
33 R. V. Ambartzumian *Factorization calculus and geometric probability*
34 G. Gripenberg, S.-O. Londen and O. Staffans *Volterra integral and functional equations*
35 George Gasper and Mizan Rahman *Basic hypergeometric series*
36 Erik Torgersen *Comparison of statistical experiments*
37 Arnold Neumaier *Interval methods for systems of equations*
38 N. Korneichuk *Exact constants in approximation theory*

ENCYCLOPEDIA OF MATHEMATICS AND ITS APPLICATIONS

Exact constants in approximation theory

N. KORNEICHUK
Institute of Mathematics
Ukrainian Academy of Sciences, Kiev

Translated from the Russian by
K. IVANOV
Bulgarian Academy of Sciences, Sofia

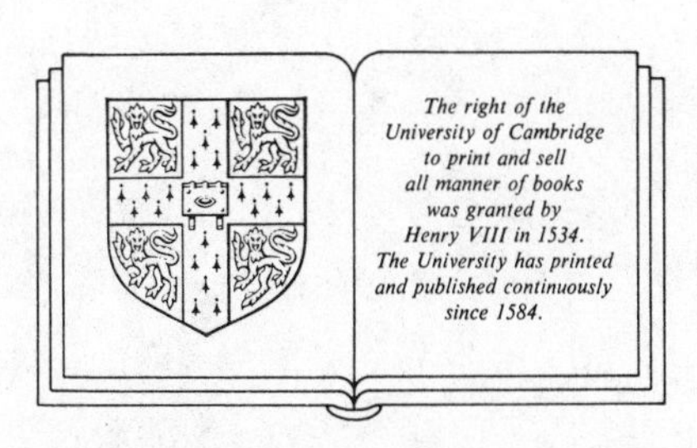

CAMBRIDGE UNIVERSITY PRESS
Cambridge
New York Port Chester
Melbourne Sydney

Published by the Press Syndicate of the University of Cambridge
The Pitt Building, Trumpington Street, Cambridge CB2 1RP
40 West 20th Street, New York NY 10011, USA
10 Stamford Road, Oakleigh, Melbourne 3166, Australia

First published 1991

Printed in Great Britain at the University Press, Cambridge

British Library cataloguing in publication data

Korneichuk, N.
Exact constants in approximation theory.
1. Mathematics. Approximation
I. Title II. Series
511.4

Library of Congress cataloguing in publication data

Korneĭchuk, Nikolaĭ Pavlovich.
[Tochnye konstanty v teorii priblizheniia. English]
Exact constants in approximation theory / N. Korneichuk:
translated from the Russian by K. Ivanov.
p. cm. – (Encyclopedia of mathematics and its applications)
Translation of: Tochnye konstanty v teorii priblizheniia.
Includes bibliographical references.
ISBN 0 521 38234 3
1. Approximation theory. I. Title. II. Series.
QA221.K6613 1991
511'.4–dc20 90-1443 CIP

ISBN 0 521 38234 3 hardback

PN

CONTENTS

PREFACE

This book was written for the series Encyclopedia of Mathematics and Its Applications published in the USA and United Kingdom since 1974. A few years ago I was asked to write a monograph *Exact Constants in Approximation Theory* by the editor of the series Professor G.-C. Rota.

The purpose and objectives of the series were presented by the editorial board as their credo in the following statement:

> It was not too long ago that the view that in present-day mathematics one can find the basic concepts of tomorrow's science was widely spread. Nowadays, however, mathematical results, proudly secured behind the barrier of complicated terminology with uncompromising strictness, are very often far removed from their potential users . . . The ample results along with sophisticated modern mathematical presentation have resulted in the mathematician's reluctance to view his achievements from aside, with the eyes of an interested applied scientist. This fact and our conviction that every science should sooner or later test itself by becoming mathematical are the motivating force of the Encyclopedia.

In their letter to authors the editorial board gives them a free choice of material and style of presentation imposing only two requirements that the monograph should meet, namely: (1) the content should not be of transient significance; and (2) the form of presentation should make the subject matter comprehensible for a wide circle of reades, including non-specialists who may be dealing with other branches of mathematics.

In our view a sign of the healthy development of every theory is the fact that at a given stage it is used in adjoining or applied fields of science. Approximation theory like some other theoretical branches of mathematics, having originated from practical problems, was for a while mainly with its own internal problems. At present, however, it can be observed that the ideas and methods of approximation theory have spread to various fields of the natural sciences to a much higher extent than, let us say, twenty or fifty years ago. The same is true for the successful application of its achievements in applied research. The major successes in the solution of extremal problems (finding N-widths, in particular) which have served as a theoretical foundation in investigating many practical optimization problems were of high importance.

The choice of material was predetermined by the title of the book, that is it had to include results in which the exact constants in approximation problems are found. Thus, the question of the subject matter was more or less clear from the very beginning, whereas the question of the form of presentation proved to be much more difficult for the author. This was not only because of the necessity of making the monograph comprehensible for non-specialists. It would not have been that difficult to present the results for exact constants in a systematic way supplementing them with some clarifications. But that was not what I wanted. To get to the essence of the method of solution is much more important than to know the final results, not only for the pure scientist but also for the applied scientist who is looking for mathematical approaches to the solution of practical problems for which the standard schemes do not work. Furthermore quoting from the editors' statement again, they say 'what is most important in a theory is not always contained in the formulations of the theorems, it can be found in the proof's arguments, in algorithms, examples and even in drawings'. This applies to the present book because in obtaining the exact constants in an approximation problem a new approach usually emerges based on entirely new ingredients (e.g. a new exact inequality) often bearing a simple geometrical meaning. It is frequently found that this new approach also paves the way to the solution of other often quite different problems.

Moreover, not every mathematically precise proof should be included in a book written for a wide circle of readers. Sometimes it takes time to determine both the exact place of the new result in the general system of mathematical facts and the chain of arguments leading to it in the shortest and most natural way. In the long run both the result and its proof find their places. It is such proofs which combine depth and simplicity that bring aesthetic pleasure and allow us to speak of beauty in mathematics. In some sense they are exact constants of a kind because they cannot be further improved. I do not mean to claim that the proofs in the present book meet those high requirements though I am not hiding my aspirations towards such an ideal.

The most important key results as well as the statements clarifying the essence of the method are supplemented by detailed proofs. We would like to point out that the methods for solving the extremal problems used in this book are based, as a rule, on geometrical considerations and quite often a fact, whose analytical description is complicated, becomes clear when a drawing is made.

One can judge the contents of each chapter by its title. I shall say only a few words about the general principles of my presentation. The exact constants results in approximation problems, as we presently understand it, belong, in most cases, to approximation by polynomials (trigonometric or algebraic) and by polynomial splines; Chapters 4 to 7 are devoted to these questions. Some sections (e.g. 4.3 and 4.4) and results can be studied independently of the preceding material. The majority of the exact constants related, in particular, to the best approximation are based on deep facts of analysis and function theory, therefore, I considered it expedient to present this foundation material in Chapter 1 (duality relations) and Chapter 3 (comparison theorems). As a matter of fact, exact constants inequalities proved in Chapter 3 belong to the general subject matter of the book by their very nature. Chapter 2 expresses both my intention for independent presentation and my wish to avoid the explanation and substantiation of some general facts from approximation theory (actually well known to specialists) later on. Chapter 8 gives the logical completion of the main lines of solving extremal problems by determining the values of N-widths of some classes of functions of finite smoothness. Furthermore, it is found that the exactness of the constants in most of the results in Chapters 4 to 7 can be interpreted in a much wider sense.

In attempting to achieve an optimal volume for the book we have, of course, not been able to give all aspects of modern approximation theory even from the exact constants point of view. Some material from optimal quadratures theory, although very rich in exact results, remains beyond the scope of the present book. Not enough light has been shed on optimal recovery of functions and linear functionals, approximation of convolution functional classes either.

The references are given in three lists. Lists A and B contain monographs in approximation theory and books of a more general or other nature, respectively. We quote them as [1A], [1B], etc. The third list is of articles and other original publications in approximation theory which are directly related to the subject of the book.

The author would like to thank the referee V. M. Tikhomirov for his support in publishing the book and his recommendations for its improvement, to V. F. Babenko whose remarks on the manuscript enable me to improve some parts of the presentation and to Zh. E. Myrzanov for his assistance in preparing the manuscript for printing.

N. Korneichuk

LIST OF MOST IMPORTANT NOTATION

Notation of general nature

$\forall$	generality quantor: 'for all'
$\exists$	existence quantor: 'there is'
$\mathbb{R}$	the set of real numbers
$\varnothing$	the empty set
$x \in A$	the element x belongs to the set A
$x \notin A$	the element x does not belong to the set A
$A \cup B$	the union of sets A and B
$A \cap B$	the intersection of sets A and B
$A \backslash B$	the difference of sets A and B
$A \oplus B$	the direct sum of sets A and B
$A \subset B$	the set A is contained in the set B
$\underline{A}$	the closure of set A
$\{x : P_x\}$	the set of all elements x possessing property P
$\sup\limits_{x \in A} f(x)$ (or $\sup \{f(x): x \in A\}$	the supremum of the values of functional f on set A
$\inf\limits_{x \in A} f(x)$ (or $\inf \{f(x): x \in A\}$	the infimum of the values of functional f on set A
$=:$	equal by definition
$\operatorname{sgn} \alpha$	a quantity equal to 1 if $\alpha > 0$, equal to -1 if $\alpha < 0$, and zero if $\alpha = 0$
ess sup	essential supremum
meas E	the Lebesgue measure of set E
dim X	the dimension of linear space X

$f \perp \varphi$ for $f \in L_1[a, b]$ and $\varphi \in L_\infty[a, b]$ means $\int_a^b f(t)\varphi(t)\,\mathrm{d}t = 0$

$f \perp \mathcal{N}$ for $f \in L_1[a, b]$ and $\mathcal{N} \subset L_\infty[a, b]$ means $f \perp \varphi \; \forall \, \varphi \in \mathcal{N}$

$[\alpha]$	the integral part of real number α
δ_{ki}	Kronecker symbol: $\delta_{ii} = 1$, $\delta_{ki} = 0$ $(k \neq i)$

1

Best approximation and duality in extremal problems

The notion of best approximation was introduced into mathematical analysis by the work of P. L. Chebyshev, who in the 1850s considered some of the properties of polynomials with least deviation from a fixed continuous function (see e.g. [2B]). Since then the development of approximation theory has been closely connected with this notion. In contrast to the early investigations which concentrated on the best approximation of individual functions, since the 1930s more effort has been put into the approximation of classes of functions with prescribed differential or difference properties. A variety of extremal problems naturally arises in this field and the solution of these problems led to the concept of exact constants.

The most powerful methods for solving the extremal problems for the best approximation of functional classes are based on the duality relationships in convex analysis. In this chapter we deal mainly with such relationships. The theorems proved in Sections 1.3–1.5 connect different extremal problems. Their solutions are of independent importance, but we shall also use them as a starting point for obtaining exact solutions in particular cases. General results connected with the best approximation of a fixed element from a metric (in particular from a normed) space and the formulation of extremal problems for approximation of a fixed set are given in Sections 1.1–1.2.

1.1 Best approximation

1.1.1 The functional of best approximation Let X be a metric space and let $\mu(x, y) = \mu(x, y)_X$ denote the distance between the elements x and y. In the case when X is a normed linear space we have $\mu(x, y) = \|x - y\| = \|x - y\|_X$. The deviation of $x \in X$ to the non-empty subset $\mathcal{N}$ of X, i.e. the quantity

$$E(x, \mathcal{N}) = E(x, \mathcal{N})_X := \inf_{u \in \mathcal{N}} \mu(x, u)_X, \tag{1.1}$$

is said to be the best approximation of x from the set $\mathcal{N}$. For any fixed $\mathcal{N}$ the equation (1.1) defines a functional on X (a non-negative number corresponds to any $x \in X$). This functional is called the best approximation functional and is denoted by the same notation $E(x, \mathcal{N})$. Here are some of its properties.

Proposition 1.1.1 The best approximation functional is uniformly continuous on X for any subset $\mathcal{N}$. If X is a normed space and $\mathcal{N}$ is a linear manifold of X then $E(x, \mathcal{N})$ is subadditive:

$$E(x_1 + x_2, \mathcal{N}) \le E(x_1, \mathcal{N}) + E(x_2, \mathcal{N}) \tag{1.2}$$

and positively homogeneous:

$$E(\alpha x, \mathcal{N}) = |\alpha| E(x, \mathcal{N}) \qquad \forall\, \alpha \in \mathbb{R}. \tag{1.3}$$

Proof Let $x_1, x_2 \in X$. For any $u \in \mathcal{N}$ we have

$$E(x_1, \mathcal{N}) \le \mu(x_1, u) \le \mu(x_1, x_2) + \mu(x_2, u).$$

Taking the infimum on u on the right-hand side of this inequality we get

$$E(x_1, \mathcal{N}) \le \mu(x_1, x_2) + E(x_2, \mathcal{N}),$$

i.e.

$$E(x_1, \mathcal{N}) - E(x_2, \mathcal{N}) \le \mu(x_1, x_2). \tag{1.4}$$

Interchanging x_1 and x_2 in (1.4) we obtain the inequality

$$|E(x_1, \mathcal{N}) - E(x_2, \mathcal{N})| \le \mu(x_1, x_2),$$

which implies the uniform continuity on X of the functional $E(x, \mathcal{N})$.

Now let X be a normed space and let $\mathcal{N}$ be a linear manifold of X. For any $u_1, u_2 \in \mathcal{N}$ we have

$$E(x_1 + x_2, \mathcal{N}) \le \|(x_1 + x_2) - (u_1 + u_2)\| \le \|x_1 - u_1\| + \|x_2 - u_2\|,$$

which proves (1.2) because u_1 and u_2 are arbitrary. Also for any $x \in X$, $\alpha \in \mathbb{R}$, $\alpha \ne 0$, we have

$$E(\alpha x, \mathcal{N}) = \inf_{u \in \mathcal{N}} \|\alpha x - u\| = |\alpha| \inf_{u \in \mathcal{N}} \|x - u/\alpha\| = |\alpha| \inf_{u \in \mathcal{N}} \|x - u\| = |\alpha| E(x, \mathcal{N}),$$

which proves (1.3). Property (1.3) also gives (under the assumptions for X and $\mathcal{N}$)

$$E(-x, \mathcal{N}) = E(x, \mathcal{N}). \qquad \square$$

The simple example $X = \mathbb{R}$, $\mathcal{N} = [0, 1]$, $x_1 = 1$, $x_2 = 2$ shows that (1.2) and (1.3) may not be true if we drop the condition that $\mathcal{N}$ has to be a linear manifold. In addition for some x_1 and x_2 one may have a strict inequality in (1.2), i.e. the best approximation functional is not additive.

The subset Q of a linear space is said to be *convex* if $\alpha x_1 + (1-\alpha)x_2 \in Q$ whenever $x_1, x_2 \in Q$ and $0 \le \alpha \le 1$.

Proposition 1.1.2 If $\mathcal{N}$ is a convex subset of X, then the functional of best approximation $E(x, \mathcal{N})$ is also convex, i.e. for any $x_1, x_2 \in X$ and $0 \le \alpha \le 1$ we have

$$E(\alpha x_1 + (1-\alpha)x_2, \mathcal{N}) \le \alpha E(x_1, \mathcal{N}) + (1-\alpha)E(x_2, \mathcal{N}). \tag{1.5}$$

Proof Let $x_1, x_2 \in X$. For every $\varepsilon > 0$ there are $u_1, u_2 \in \mathcal{N}$ such that

$$E(x_1, \mathcal{N}) > \|x_1 - u_1\| - \varepsilon, \qquad E(x_2, \mathcal{N}) > \|x_2 - u_2\| - \varepsilon.$$

Taking into account that $\mathcal{N}$ is convex and $0 \le \alpha \le 1$ we get

$$\begin{aligned} E(\alpha x_1 + (1-\alpha)x_2, \mathcal{N}) &\le \|\alpha x_1 + (1-\alpha)x_2 - [\alpha u_1 + (1-\alpha)u_2]\| \\ &\le \alpha\|x_1 - u_1\| + (1-\alpha)\|x_1 - u_1\| \\ &\le \alpha E(x_1, \mathcal{N}) + (1-\alpha)E(x_2, \mathcal{N}) + 2\varepsilon. \end{aligned}$$

These inequalities prove (1.5), because ε can be arbitrarily small. □

If the infimum in (1.1) is attained for $u_0 \in \mathcal{N}$, i.e. $\mu(x, u_0) = E(x, \mathcal{N})$, then u_0 is called *an element of best approximation for x in* $\mathcal{N}$. The very important problems of existence, uniqueness and characterization of the element of best approximation arise in this setting. We shall investigate them as far as is necessary for the proofs of our main results.

1.1.2 Existence and uniqueness of the element of best approximation The set $\mathcal{N}$ in the metric space X is called an *existence set* if for every $x \in X$ there is an element of best approximation in $\mathcal{N}$. Every existence set is necessarily closed, but there are closed sets which are not existence sets (see Exercises 1.1 and 1.2). Only additional restrictions on $\mathcal{N}$ will provide this property. We shall consider this problem only in normed spaces.

Proposition 1.1.3 Every closed locally compact subset $\mathcal{N}$ of the normed space X is an existence set.

Proof Let us fix $x \in X \backslash \mathcal{N}$ (*for* $x \in \mathcal{N}$ the existence is obvious). Then

$$\inf_{u \in \mathcal{N}} \|x - u\| =: d > 0,$$

because $\mathcal{N}$ is closed and, by the definition of an infimum, for every $m = 1, 2, \ldots$, there is $u_m \in \mathcal{N}$ such that

$$\|x - u_m\| < d + 1/m.$$

The sequence $\{u_m\}$ is bounded:

$$\|u_m\| \le \|x\| + \|u_m - x\| < \|x\| + d + 1/m, \qquad m = 1, 2, \ldots$$

and, because of the local compactness of $\mathcal{N}$, there is a subsequence $u_{m_j} \to u_0 \in X$. Moreover, $u_0 \in \mathcal{N}$ because $\mathcal{N}$ is closed. Now from the inequalities

$$d \le \|x - u_{m_j}\| < d + 1/m_j, \qquad j = 1, 2, \ldots$$

we get $\|x - u_0\| = d$ if we let $j \to \infty$. This means that u_0 is an element of best approximation for x in $\mathcal{N}$ and proves the proposition. □

Corollary 1.1.4 Every finite dimensional linear subspace $\mathcal{N}$ of the normed space X is an existence set.

The dependence of the uniqueness of the element of best approximation on both the structure of the approximating set and the properties of the metric in X is more delicate. The norm in X is called *strictly convex* if the unit sphere $\|x\| = 1$ does not contain any segment, i.e. from $\|x\| = \|y\| = 1$ and $0 < \alpha < 1$ it follows that $\|\alpha x + (1 - \alpha)y\| < 1$. A space with a strictly convex norm is called strictly normalized. Such space can be defined by the following equivalent condition: in the inequality $\|x + y\| \le \|x\| + \|y\|$ one has equality only in the case $y = cx, c \ge 0$. Obviously, in a strictly normalized space, there is no sphere $\|x - x_0\| = d$ containing a segment.

Proposition 1.1.5 Let X be a strictly normalized space and $\mathcal{N}$ be a convex subset of X. If there is an element of best approximation in $\mathcal{N}$ for $x_0 \in X$, then this element is unique.

Proof Let $u_1, u_2 (u_1 \ne u_2)$ be two elements of best approximation for x_0 in $\mathcal{N}$:

$$\|x_0 - u_1\| = \|x_0 - u_2\| = E(x_0, \mathcal{N}) =: d.$$

If $0 \le \alpha \le 1$ and $u_\alpha = \alpha u_1 + (1 - \alpha)u_2$, then $u_\alpha \in \mathcal{N}$ and

$$d \le \|x_0 - u_\alpha\| \le \alpha\|x_0 - u_1\| + (1 - \alpha)\|x_0 - u_2\| = d.$$

Therefore $d = \|x_0 - u_\alpha\|$, which means that the sphere $\|x_0 - x\| = d$ contains the segment $[u_1, u_2]$ – a contradiction with the strict convexity of the norm in X. □

In a general normed space X the uniqueness of the element of best approximation can be implied by the structure of the approximating set $\mathcal{N}$. The set $\mathcal{N}$ in a normed space X with the property that for every $x \in X$ there exists a unique element of best approximation is called a *Chebyshev set*.

1.1.3 Operator of best approximation In the case when $\mathcal{N}$ is a Chebyshev set in the normed space X the equality

$$E(x, \mathcal{N}) = \|x - Px\|, \qquad Px \in \mathcal{N}, \tag{1.6}$$

defines an operator P on X. We shall call it *the operator of best approximation*.

Proposition 1.1.6 If $\mathcal{N}$ is a locally compact Chebyshev set in a normed space X, then operator P given by the equality (1.6) is continuous. If $\mathcal{N}$ is a Chebyshev subspace, then operator P is homogeneous and, in particular, odd: $P(-x) = -P(x)$.

Proof Let $x_m \to x_0$, $u_m = Px_m$, $u_0 = Px_0$. The sequence $\{u_m\}$ of elements of $\mathcal{N}$ is bounded, because of the inequalities

$$\|u_m - x_0\| \leq \|u_m - x_m\| + \|x_m - x_0\| = E(x_m, \mathcal{N}) + \|x_m - x_0\|$$

and because of the convergence of the sequence $\{E(x_m, \mathcal{N})\}$ (see Proposition 1.1.1). Let us assume that $u_m \not\to u_0$. From the local compactness of $\mathcal{N}$ we conclude the existence of a subsequence $\{u_{m_j}\}$ such that $\lim_{j\to\infty} u_{m_j} = u_* \neq u_0$. Moreover $u_* \in \mathcal{N}$ because $\mathcal{N}$ is a Chebyshev set and therefore closed. Taking a limit when $j \to \infty$ in the inequalities

$$\|x_{m_j} - u_{m_j}\| = E(x_{m_j}, \mathcal{N}) \leq \|x_{m_j} - u_0\|, \qquad j = 1, 2, \ldots$$

we get $\|x_0 - u_*\| \leq \|x_0 - u_0\|$, i.e. u_* is an element of best approximation for x_0 in $\mathcal{N}$ which is different from u_0. This contradicts the assumption that $\mathcal{N}$ is a Chebyshev set. Hence $u_m \to u_0$, i.e. $Px_m \to Px_0$.

In the case when $\mathcal{N}$ is a Chebyshev subspace, using (1.3) for any $\alpha \in \mathbb{R}$ we can write

$$\|\alpha x - \alpha Px\| = |\alpha|\,\|x - Px\| = |\alpha| E(x, \mathcal{N}) = E(\alpha x, \mathcal{N}) = \|\alpha x - P(\alpha x)\|$$

and hence $P(\alpha x) = \alpha Px$. □

It is important to note that, in general, the operator of best approximation is not additive – see Exercise 1.4.

Let $\mathcal{N} = \mathcal{N}_N$ be an N-dimensional Chebyshev subspace of the normed space X and let $\{x_1, \ldots, x_N\}$ be a basis in $\mathcal{N}$. Then the operator of best approximation can be represented in the form

$$Px = \sum_{k=1}^{N} c_k(x) x_k, \tag{1.7}$$

where c_k are functionals defined on X. From Proposition 1.1.6 we get

Corollary 1.1.7 The functionals c_k are homogeneous and continuous.

Indeed, the equality $P(\alpha x) = \alpha Px$ in this case is equivalent to

$$\sum_{k=1}^{N} c_k(\alpha x)x_k = \sum_{k=1}^{N} \alpha c_k(x)x_k,$$

which implies $c_k(\alpha x) = \alpha c_k(x)$ because of the uniqueness of the representation (1.7). The continuity of the functionals c_k follows from the continuity of the operator (1.7) and the well known fact that the convergence in a finite dimensional space is equivalent to componentwise convergence.

1.1.4 Best approximations and linear methods The operator P of best approximation is given implicitly by the equality (1.6). The explicit indication of the element $u = Px$ is possible only in some exceptional cases – for example when X is a Hilbert space and $\mathcal{N}$ is its subspace. Moreover, in general, P is not an additive operator and this makes its construction and investigation very difficult.

Hence, when one considers approximations of the elements $x \in X$ by a fixed set $\mathcal{N}$, it is often preferable to replace the operator P by another operator A: $X \to \mathcal{N}$ which allows us effectively to find in $\mathcal{N}$ the element $u = Ax$ closed to Px. The most important and common is the case when X is a normed space, $\mathcal{N}$ is its subspace and A is a *linear* operator. In this case one says that the operator A is given by *a linear method of approximation*.

Proposition 1.1.8 Every linear operator A mapping the normed space X into an N-dimensional subspace $\mathcal{N}_N$ with a basis $\{x_1, x_2, \ldots, x_N\}$ can be represented as

$$Ax = \sum^{N} \beta_k(x)x_k, \tag{1.8}$$

where β_k are bounded linear functionals on X.

The linearity of the functionals β_k follows from the linearity of the operator A and the uniqueness of the representation (1.8). The boundness of β_k follows from its continuity which is implied by the continuity of A.

1.2 Formulation of extremal problems

1.2.1 Approximation by a fixed set The problem of the best approximation for the element $x \in X$ by a fixed subset $\mathcal{N}$ of X consists not only of finding the element of best approximation to x in $\mathcal{N}$ (if it exists) but also of determining the quantity $E(x, \mathcal{N})$, i.e. expressing it by different characteristics depending both on the element x and the approximating set $\mathcal{N}$. Let us note immediately that for individual elements x this task can only be fulfilled in rare cases. In particular, when X is a Hilbert space and $\mathcal{N} = \mathcal{N}_N$ is an N-dimensional subspace, then the operator of best approximation is the orthogonal projection onto n and

$$E(x, \mathcal{N}_N)^2 = \|x\|^2 - \sum_{k=1}^{N} (x, u_k)^2 \qquad \forall\, x \in X,$$

where $\{u_1, u_2, \ldots, u_N\}$ is an orthonormal basis in $\mathcal{N}$.

In non-Hilbert spaces – except for the simplest cases, e.g. the approximation of functions by constants – it is exceptional to be able to solve the problem of finding the best approximation of an individual element. In view of the above remarks, the problem of best approximation is usually formulated in a more general sense: to evaluate the quantity $E(x, \mathcal{N})$ according to the characteristics of some set $\mathcal{M}$ containing the element x. Obviously the most interesting problem is to find a bound on $E(x, \mathcal{N})$ from above which cannot be improved as long as $x \in \mathcal{M}$. This means that we want to find the quantity

$$E(\mathcal{M}, \mathcal{N}) = E(\mathcal{M}, \mathcal{N})_X := \sup_{x \in \mathcal{M}} E(x, \mathcal{N})_X = \sup_{x \in \mathcal{M}} \inf_{u \in \mathcal{N}} \mu(x, u), \qquad (2.1)$$

which characterizes the deviation of the set $\mathcal{M}$ from the set $\mathcal{N}$ in the metric of the set X. In other words, it is necessary to estimate $E(x, \mathcal{N})$ in the best possible way under the constraint that x belongs to $\mathcal{M}$.

It is possible to solve problem (2.1) exactly in many important and meaningful cases of approximation in linear normed spaces, in particular when $\mathcal{N}$ is a linear manifold of trigonometric polynomials or splines of fixed order. However even when we know the quantity $E(\mathcal{M}, \mathcal{N})$ we still cannot guarantee the possibility of finding $u \in \mathcal{N}$ for any $x \in \mathcal{M}$ such that $\mu(x, u) \leq E(\mathcal{M}, \mathcal{N})$. Hence, in practice, one usually uses a method of approximation effectively defined by an operator A ($Ax \in \mathcal{N}$), whose error on the whole class $\mathcal{M}$ is obviously bounded from above by the quantity $\sup_{x \in \mathcal{M}} \mu(x, Ax)_X$.

Let X be a normed space, $\mathcal{N}$ be a linear manifold and $\mathcal{L}(X, \mathcal{N})$ denote the set of all linear operators A from X to $\mathcal{N}$. One can formulate the following problem: find the quantity

$$\mathscr{E}(\mathcal{M}, \mathcal{N}) = \mathscr{E}(\mathcal{M}, \mathcal{N})_X = \inf_{A \in \mathcal{L}(x, \mathcal{N})} \sup_{x \in \mathcal{M}} \|x - Ax\|_X \qquad (2.2)$$

and determine the operator $A_* \in \mathcal{L}(X, \mathcal{N})$ (if it exists) supplying in (2.2) the infimum in (2.2), i.e. $\mathscr{E}(\mathcal{M}, \mathcal{N}) = \sup\{\|x - A_* x\|_X : x \in \mathcal{M}\}$. The operator A_* determines the best linear method of approximation by the elements of $\mathcal{N}$ for the set $\mathcal{M}$.

For every $A \in \mathscr{L}(X, \mathscr{N})$ we have

$$\|x - Ax\| \geq E(x, \mathscr{N}), \qquad \forall\, x \in X$$

and hence

$$\sup_{x \in \mathscr{M}} \|x - Ax\| \geq E(\mathscr{M}, \mathscr{N}) \tag{2.3}$$

and

$$\mathscr{E}(\mathscr{M}, \mathscr{N}) \geq E(\mathscr{M}, \mathscr{N}). \tag{2.4}$$

Inequalities (2.3) and (2.4) clarify the practical value of knowing the quantity $E(\mathscr{M}, \mathscr{N})$ in the case of normed spaces: this quantity gives a lower bound for the approximation of the elements of $\mathscr{M}$ by linear methods and can be used as a guide for evaluating the approximation properties of a particular method or the possibility of linear approximation in general.

Equation (2.2) can be formulated in a restricted sense: the infimum is taken on a proper subspace of the whole class $\mathscr{L}(\mathscr{M}, \mathscr{N})$ of linear operators $A: X \to \mathscr{N}$. For example one can consider $\mathscr{L}_{\perp}(\mathscr{M}, \mathscr{N})$ – the class of all linear projections A from X to $\mathscr{N}$, i.e. $A \in \mathscr{L}(X, \mathscr{N})$ and $Ax = x$ if $x \in \mathscr{N}$. We then consider the quantity

$$\mathscr{E}_{\perp}(\mathscr{M}, \mathscr{N}) = \inf_{A \in \mathscr{L}_{\perp}(X, \mathscr{N})} \sup_{x \in \mathscr{M}} \|x - Ax\| \tag{2.5}$$

for which we obviously have

$$E(\mathscr{M}, \mathscr{N}) \leq \mathscr{E}(\mathscr{M}, \mathscr{N}) \leq \mathscr{E}_{\perp}(\mathscr{M}, \mathscr{N}). \tag{2.6}$$

In the case of the Hilbert space X the inequality signs in (2.6) are to be replaced by equality signs. It is natural to look for other meaningful cases when the inequalities in (2.6) can be replaced by equalities.

1.2.2 Problems for exact asymptotics There are many problems in approximation theory which fit into the following scheme. In the metric space X we fixed a sequence of subsets $\mathscr{N}_k$ $(k = 1, 2, \ldots)$ such that

$$\lim_{k \to \infty} E(x, \mathscr{N}_k) = 0$$

for every $x \in X$. We want to find for a fixed $\mathscr{M} \in X$ the exact asymptotic of the sequence $E(\mathscr{M}, \mathscr{N}_k)$ $(k = 1, 2, \ldots)$.

The index k is connected with the dimension of $\mathscr{N}_k$. Usually $\{\mathscr{N}_k\}$ is an expanding sequence: $\mathscr{N}_1 \subset \mathscr{N}_2 \subset \ldots$; in this case the sequences $E(x, \mathscr{N}_k)$ and $E(\mathscr{M}, \mathscr{N}_k)$ are decreasing.

If $\{\mathscr{N}_k\}$ is an expanding sequence of sets with a similar structure (e.g. a sequence of polynomial spaces with increasing degrees), then one can define by a common rule the sequence of operators $A_k \in \mathscr{L}(X, \mathscr{N}_k)$ $(k = 1, 2, \ldots,)$

and can investigate the asymptotic behaviour of the quantities $\|x - A_k x\|$ $(x \in X)$ when $k \to \infty$ and

$$\sup_{x \in \mathcal{M}} \|x - A_k x\|; \qquad k = 1, 2, \ldots; \qquad \mathcal{M} \subset X. \tag{2.7}$$

It is interesting to compare the sequences $\{\mathscr{E}(\mathcal{M}, \mathcal{N}_k)\}$ and $\{E(\mathcal{M}, \mathcal{N}_k)\}$ and to discover when their asymptotics coincide.

If for every $k = 1, 2, \ldots$, we can determine the examined quantities $E(\mathcal{M}, \mathcal{N}_k)$, $\mathscr{E}(\mathcal{M}, \mathcal{N}_k)$ or (2.7) exactly using some numeric characteristics of the sets $\mathcal{M}$ and $\mathcal{N}_k$, then the problems formulated in this section usually have a trivial solution; the asymptotic expansions are used when the exact determination of these quantities does not seem possible.

1.2.3 Problems for N-widths An essentially new class of extremal problems arises if we fix the set $\mathcal{M} \subset X$ and want to minimize the quantity $E(\mathcal{M}, \mathcal{N})$ or $\mathscr{E}(\mathcal{M}, \mathcal{N})$ over all approximating sets $\mathcal{N}$ which share certain properties, e.g. over all linear manifolds $\mathcal{N}_N$ of dimension N. We shall restrict ourselves to considering only this last case, which is the most interesting and important case from both the practical and the theoretical point of view.

Let X be a normed space and let $\mathcal{M}_N$ be the class of all subspaces of X of dimension not greater than N. If $\mathcal{M}$ is a central symmetric set in X then the quantity

$$d_N(\mathcal{M}, X) = \inf_{\mathcal{N} \in \mathcal{M}_N} E(\mathcal{M}, \mathcal{N})_X \tag{2.8}$$

is called the *Kolmogorov N-width* of the set $\mathcal{M}$ in the space X; it is equal to the half of the smallest 'width' of the set $\mathcal{M}$. For an arbitrary (not necessarily centrally symmetric) set $\mathcal{M} \subset X$ the N-width $d_N(\mathcal{M}, X)$ can be defined as the infimum of $E(\mathcal{M}, \mathcal{N})$ over all possible translates $\mathcal{N} + a$ of the elements $\mathcal{N}$ of $\mathcal{M}_N$.

Starting with the best linear approximation $\mathscr{E}(\mathcal{M}, \mathcal{N})$ of the centrally symmetric set $\mathcal{M}$, one comes to the problem of finding the quantity

$$\lambda_N(\mathcal{M}, X) = \inf_{\mathcal{N} \in \mathcal{M}_N} \mathscr{E}(\mathcal{M}, \mathcal{N})_X, \tag{2.9}$$

which is called the *linear N-width* of the set $\mathcal{M}$ in X. One also considers the *projection N-width*

$$\pi_N(\mathcal{M}, X) = \inf_{\mathcal{N} \in \mathcal{M}_N} \mathscr{E}_{\perp}(\mathcal{M}, \mathcal{N})_X, \tag{2.10}$$

where the quantity $\mathscr{E}_{\perp}(\mathcal{M}, \mathcal{N})_X$ is defined in (2.5). It will be convenient to set

$$d_0(\mathcal{M}, X) = \lambda_0(\mathcal{M}, X) = \pi_0(\mathcal{M}, X) = \sup_{x \in \mathcal{M}} \|x\|_X.$$

Let us note the following fact (see e.g. [25A, p. 218]). If $\mathcal{M}$ is an algebraic sum of a compact and finite dimensional subspace then

$$d_N(\mathcal{M}, X) = \inf_{\mathcal{N} \in \mathcal{M}_N} \inf_{A \in C(X,\mathcal{N})} \sup_{x \in \mathcal{M}} \|x - Ax\|, \tag{2.11}$$

where $C(X, \mathcal{N})$ is the class of all continuous mappings of X into $\mathcal{N}$. Clearly this is connected with the continuity of the operator of best approximation.

Obviously the N-widths d_N, λ_N, π_N decrease when N increases and we always have

$$d_N(\mathcal{M}, X) \leq \lambda_N(\mathcal{M}, X) \leq \pi_N(\mathcal{M}, X). \tag{2.12}$$

Together with estimating of the quantities (2.8) to (2.10) it is natural to look for such subspaces in $\mathcal{M}_N$ which supply the corresponding infimums; these subspaces are called *extremal*. In the linear N-width problem it is very important (especially when one looks for possible applications) to find a subspace $\mathcal{N}_N^* \in \mathcal{M}_N$ such that

$$\mathscr{E}(\mathcal{M}, \mathcal{N}_N^*)_X = \lambda_N(\mathcal{M}, X)$$

and to construct effectively a linear operator $A_* \in \mathscr{L}(X, \mathcal{N}_N^*)$ such that

$$\sup_{x \in \mathcal{M}} \|x - A_* x\|_X = \lambda_N(\mathcal{M}, X).$$

The operator A_* together with the subspace $\mathcal{N}_N^*$ is the best linear method for approximating elements of the set $\mathcal{M}$ among the approximations by linear combinations with not more than N elements of X.

In Sections 1.2.1–3 we have formulated the extremal problems by starting with the functional of best approximation $E(x, \mathcal{N})$. In some functional spaces one can consider other functionals instead of $E(x, \mathcal{N})$, taking into account some special features of x or $\mathcal{N}$, e.g. non-symmetric and, in particular, one-sided approximations. We are not going to state here the general formulations of these problems; particular cases will be considered in the next sections and chapters.

1.3 Duality of extremal problems in linear spaces

1.3.1 Main theorems Duality theorems, which reflect the basic relationships in convex analysis, allows us to reduce the problem of finding $E(x,\mathcal{N})$ when $\mathcal{N}$ is convex to a more visible extremal problem in the conjugate space and when $\mathcal{N}$ is a linear manifold to the problem of finding the extremum of an explicitly given linear functional.

In the following we only consider *real* linear spaces.

Theorem 1.3.1 Let X be a linear normed space, let X^* be the conjugate space of X (i.e. the set of all linear bounded functionals defined on X) and let $\mathcal{N}$ be a convex set in X. Then we have for every $x \in X$

$$E(x, \mathcal{N}) := \inf_{u \in \mathcal{N}} \|x - u\| = \sup_{f \in X^*, \|f\| \leq 1} [f(x) - \sup_{u \in \mathcal{N}} f(u)]. \tag{3.1}$$

If $x \in X \backslash \bar{\mathcal{N}}$ then there is a functional $f_0 \in X^*$ with norm $\|f_0\| = 1$ for which the supremum on the right-hand side of (3.1) is attained.

We shall prove (without involving any additional ideas) a more general variant of this duality theorem which will be used when we consider non-symmetric and one-sided approximations in functional spaces. Let X be a linear space. Let p be a finite non-negative real function defined on X everywhere with the properties

$$p(0) = 0, \qquad p(x + y) \leq p(x) + p(y),$$
$$p(\alpha x) = \alpha p(x), \qquad \alpha \geq 0.$$

Such a p is called a *non-symmetric seminorm*. If $\mathcal{N} \subset X$ then we set

$$p(x, \mathcal{N}) = \inf_{u \in \mathcal{N}} p(x - u).$$

Let X' denote the set of all linear functionals on X and for any $f \in X'$ set

$$p^*(f) = \sup \{f(x) : x \in X, p(x) \leq 1\}. \tag{3.2}$$

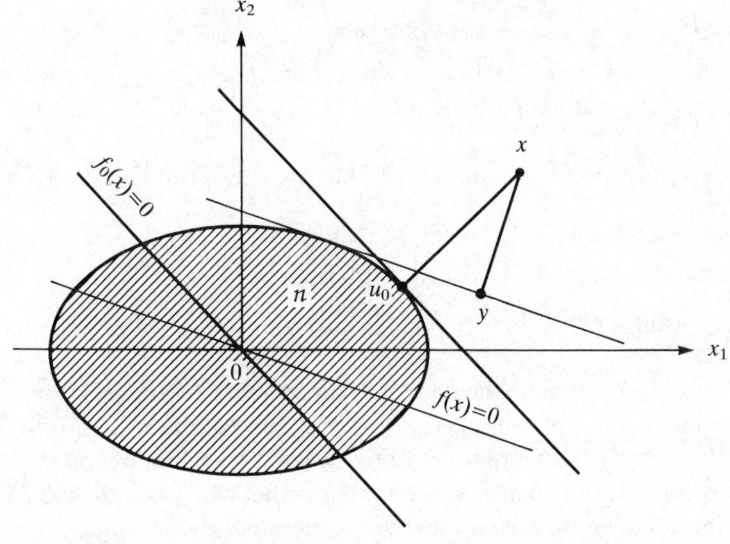

Fig. 1.1 Geometrical sense of Theorem 1.3.1 by $X = R_2$ ($\mathcal{N}$ is convex set in R_2).

$$xu_0 = \inf_{u \in \mathcal{N}} \|x - u\|, xy = f(x) - \sup_{u \in \mathcal{N}} f(u), f \in R_2^*, \|f\| = 1.$$

From this definition we immediately get

$$f(x) \le p^*(f)p(x). \tag{3.3}$$

Obviously, when X is a normed space and $p(x) = \|x\|$, then $p(x, \mathcal{N}) = E(x, \mathcal{N})$ and $p^*(f) = \|f\|$ for $f \in X^*$. In general, the quantity $p^*(f)$ may not be finite. It is important that, as in the case $p(x) = \|x\|$, the condition $p(x) \le 1$ in (3.2) can be replaced by $p(x) < 1$ and therefore for any $f \in X'$ and $\eta > 0$ we have

$$\sup \{f(x) : x \in X, p(x) < \eta\} = \eta p^*(f). \tag{3.4}$$

Let us prove this. Set $\sup \{f(x) : x \in X, p(x) < \eta\} = A$ for fixed $f \in X'$ and $\eta > 0$. In view of (3.2) one has $0 \le A \le \eta p^*(f)$ and so we have to consider only the case $p^*(f) > 0$. First suppose that $p^*(f) < \infty$. Assume that $A = \eta p^*(f) - \varepsilon$ for some $\varepsilon > 0$. From (3.2) we get an element $x_0 \in X$ such that $p(x_0) = \eta$ and $f(x_0) > \eta p^*(f) - \varepsilon/2$. For $\alpha \to 1$ we have $f(\alpha x_0) = \alpha f(x_0) \to f(x_0) > A + \varepsilon/2$. At the same time for any $\alpha \in (0, 1)$ we have $p(\alpha x_0) = \alpha p(x_0) < \eta$ and by assumption $f(\alpha x_0) \le A$ – a contradiction. Now if $p^*(f) = \infty$ but $A < \infty$ we get a contradiction considering once more the elements αx_0 $(0 < \alpha < 1)$ when $\alpha \to 1$.

Theorem 1.3.2 Let $\mathcal{N}$ be a convex set in the linear space X and let p be a non-symmetric seminorm defined on X. Then for every $x \in X$ we have

$$p(x, \mathcal{N}) = \sup_{f \in X', p^*(f) \le 1} [f(x) - \sup_{u \in \mathcal{N}} f(u)] \tag{3.5}$$

and in the case $p(x, \mathcal{N}) > 0$ there is a functional $f_0 \in X$, $p^*(f_0) = 1$ for which the supremum in the right-hand side of (3.5) is attained.

Proof Denote the right-hand side of (3.5) by $N(x, \mathcal{N})$. It is easy to see that

$$N(x, \mathcal{N}) \le p(x, \mathcal{N}), \tag{3.6}$$

because if $u \in \mathcal{N}$ and $p^*(f) \le 1$, then (3.3) gives

$$f(x) - \sup_{y \in \mathcal{N}} f(y) \le f(x) - f(u) = f(x - u) \le p^*(f)p(x - u) \le p(x - u)$$

and we only have to take the necessary supremum on f and infimum on u.

The inequality opposite to (3.6) will be proved under the assumption $p(x, \mathcal{N}) = \mu > 0$ (in Theorem 1.3.1 this means $x \notin \bar{\mathcal{N}}$), because, in the other case, the obvious inequality $N(x, \mathcal{N}) \ge 0$ together with (3.6) gives (3.5). We need the following assertion, known as the *separation theorem* (see [3B, p. 446], [9B, p. 130]).

Let A and B be convex sets in the linear space X, $A \cap B = \varnothing$, and at least one of both is a convex body (i.e. A contains a point x such that for every

$z \in X$ there is $\varepsilon > 0$ such that $x + tz \in A$ whenever $|t| < \varepsilon$). Then there exists a non-zero functional $f \in X'$ which separates A and B, i.e. f is such that

$$f(x) \leq f(y), \qquad \forall\, x \in A, \qquad \forall\, y \in B.$$

Consider the set

$$S(x, \mu) = \{y : y \in X,\ p(x - y) < \mu\},$$

where $\mu = p(x, \mathcal{N}) > 0$. It is easy to check (see [9B, p. 126]) that $S(x, \mu)$ is a convex body in X. $S(x, \mu) \cap \mathcal{N} = \varnothing$ because $p(x - u) \geq \mu$ for any $u \in \mathcal{N}$. From the separation theorem we get a non-zero functional $f \in X'$ such that

$$f(u) \leq f(y), \qquad \forall\, u \in \mathcal{N}, \qquad \forall\, y \in S(x, \mu). \tag{3.7}$$

We shall show that $p^*(f) < \infty$ (in Theorem 1.3.1, where $p^*(f) = \|f\|$, this is obvious).

It is easy to check the equality (in the sense of coincidence of sets)

$$\{z : p(z) < \mu\} = x - S(x, \mu), \tag{3.8}$$

which means that if $p(z) < \mu$ then $z = x - y$ where $y \in S(x, \mu)$ and conversely if $z = x - y$ where $y \in S(x, \mu)$ then $p(z) < \mu$. Fix $u' \in \mathcal{N}$ and let $p(v) < 1$; then from (3.8) we have $\mu v = x - y$ where $y \in S(x, \mu)$ and using (3.7) we get

$$f(x) - \mu f(v) = f(x - \mu v) = f(y) \geq f(u').$$

Hence for every $v \in X$ with $p(v) < 1$ we have

$$f(v) \leq (f(x) - f(u'))/\mu = c < \infty,$$

which means $p^*(f) < \infty$ because of (3.4).

For $f_0 = f/p^*(f)$ we have $f_0 \in X'$, $p^*(f_0) = 1$ and

$$f_0(u) \leq f_0(y), \qquad \forall\, u \in \mathcal{N}, \qquad \forall\, y \in S(x, \mu).$$

Now using (3.8) and (3.4) we obtain

$$\sup_{u \in \mathcal{N}} f_0(u) \leq \inf_{y \in S(x,\mu)} f_0(y) = \inf_{p(z)<\mu} f_0(x - z) = f_0(x) - \sup_{p(z)<\mu} f_0(z)$$
$$= f_0(x) - \mu p^*(f_0) = f_0(x) - p(x, \mathcal{N})$$

and hence

$$p(x, \mathcal{N}) \leq f_0(x) - \sup_{u \in \mathcal{N}} f_0(u).$$

Therefore $p(x, \mathcal{N}) \leq N(x, \mathcal{N})$ and Theorem 1.3.2 and 1.3.1 are proved. □

In the case when $\mathcal{N}$ is a subspace the right-hand sides of (3.1) and (3.5) become simpler. Set

$$\mathcal{N}^{\perp} = \{\, f : f \in X',\ f(u) = 0 \qquad \forall\, u \in \mathcal{N}\}.$$

Theorem 1.3.3 Let $\mathcal{N}$ be a subspace in the linear space X and let p be a non-symmetric seminorm defined on X. Then for every $x \in X$ we have

$$p(x, \mathcal{N}) := \inf_{u \in \mathcal{N}} p(x - u) = \sup \{f(x) : f \in \mathcal{N}^{\perp}, p^*(f) \le 1\}; \tag{3.9}$$

and in particular if X is a normed space then

$$E(x, \mathcal{N}) := \inf_{u \in \mathcal{N}} \|x - u\| = \sup \{f(x) : f \in \mathcal{N}^*, \|f\| \le 1\}. \tag{3.10}$$

If $p(x, \mathcal{N}) > 0$ $(E(x, \mathcal{N}) > 0)$ then the supremum in (3.9) and (3.10) is attained for a functional $f_0 \in \mathcal{N}^{\perp}$ such that $p^*(f_0) = 1$ $(\|f_0\| = 1)$.

Indeed, if $f \in X' \backslash \mathcal{N}^{\perp}$ then $f(u') = \alpha \neq 0$ for some $u' \in \mathcal{N}$ and hence for every $x \in X$ we have

$$f(x) - \sup_{u \in \mathcal{N}} f(u) \le f(x) - \sup_{\lambda \in \mathbb{R}} f(\lambda u') = -\infty.$$

This means that one has only to consider the functionals f from $\mathcal{N}^{\perp}$ in the supremums in (3.1) and (3.4). But for such functionals we have $\{f(u) : u \in \mathcal{N}\} = 0$ and so relations (3.5) and (3.1) are transferred into (3.9) and (3.10) respectively. The attainment of the supremums follows from the corresponding fact in Theorem 1.3.2.

Remark If X is a normed space and $p(x) \le c\|x\|$ for every $x \in X$, then the space X' can be replaced with X^* in the statements of Theorems 1.3.2 and 1.3.3 because the condition $p^*(f) \le 1$ implies

$$\|f\| \le \sup_{p(x) \le c} f(x) = c \sup_{p(x) \le 1} f(x) = cp^*(f) \le c.$$

1.3.2 The conjugate case for finite dimensional substances If $\mathcal{N} = \mathcal{N}_N$ is an N-dimensional subspace with a basis $\{x_1, x_2, \ldots, x_N\}$, then relation (3.10) can be written as

$$\inf_{\alpha_k} \left\| x - \sum_{k=1}^{N} \alpha_k x_k \right\| = \sup \{f(x) : f \in X^*, \|f\| \le 1, f(x_k) = 0, k = 1, 2, \ldots, N\}. \tag{3.11}$$

For the best approximation in the conjugate space X^* we have the following relation which is in some sense conjugate to (3.11).

Theorem 1.3.4 Let X be a normed space and let $f, f_1, f_2, \ldots, f_N$ be a fixed system of functionals in X^*. Then

$$\inf_{\alpha_k} \left\| f - \sum_{k=1}^{N} \alpha_k f_k \right\| = \sup \{ f(x) \colon x \in X, \|x\| \leq 1, f_k(x) = 0, k = 1, 2, \ldots, N \}. \tag{3.12}$$

Proof Set

$$H = \{x \colon x \in X, f_k(x) = 0, k = 1, 2, \ldots, N\}$$

and let f_H be the restriction of the functional f on H. Every functional of the type

$$\varphi = f - \sum_{k=1}^{N} \alpha_k f_k, \tag{3.13}$$

where the α_k are arbitrary, is an extension of f_H to the whole space X. We shall show that every extension of f_H to X is of the form (3.13).

Without loss of generality we may assume that functionals $f_1, f_2, \ldots, f_N$ are linearly independent. There are (see e.g. [10B p. 210]) such elements $\{x_1, x_2, \ldots, x_N\}$ that $f_k(x_i) = 0$ $(k \neq i)), f_k(x_k) = 1$. If $x \in X$ then

$$y = x - \sum_{k=1}^{N} f_k(x) x_k$$

belongs to H; hence every $x \in X$ can be represented as

$$x = y + \sum_{k=1}^{N} f_k(x) x_k, \qquad y \in H. \tag{3.14}$$

Let us fix $f \in X^*$ and denote some extension of f_H from H to X by φ. Using (3.14) we have for every $x \in X$

$$\varphi(x) = f(y) + \varphi\left(\sum_{k=1}^{N} f_k(x) x_k\right) = f\left(x - \sum_{k=1}^{N} f_k(x) x_k\right) + \sum_{k=1}^{N} f_k(x) \varphi(x_k)$$

$$= f(x) - \sum_{k=1}^{N} [f(x_k) - \varphi(x_k)] f_k(x),$$

i.e. the functional φ can be represented in the form (3.13) with $\alpha_k = f(x_k) - \varphi(x_k)$.

Then we have for every α_k

$$\left\| f - \sum_{k=1}^{N} \alpha_k f_k \right\| = \|\varphi\| \geq \|f_H\|, \tag{3.15}$$

because the norm of a functional does not decrease after its extension. On the other hand, the Hahn–Banach theorem [10B, p. 173] asserts that there is an extension of f_H from H to X which preserves the norm, i.e. there are coefficients α_k^* $(k = 1, 2, \ldots, N)$ such that

$$\left\| f - \sum_{k=1}^{N} \alpha_k^* f_k \right\| = \|f_H\| = \sup\{f(x) \colon x \in H, \|x\| = 1\}. \tag{3.16}$$

From the equality (3.16) and the validity of (3.15) for every α_k we get (3.12). □

1.3.3 General criteria for external elements Theorems 1.3.1–1.3.3 allow us to derive the necessary and sufficient conditions of a general type which characterize the extremal elements, in particular the element of best approximation in $\mathcal{N}$.

Proposition 1.3.5 Let $\mathcal{N}$ be a convex set in the linear space X and let p be a non-symmetric seminorm defined on X. If $p(x, \mathcal{N}) > 0$ then the relation

$$p(x - u_0) = \inf_{u \in \mathcal{N}} p(x - u) \tag{3.17}$$

for the element $u_0 \in X$ holds true if and only if there is a functional $f_0 \in X'$ such that:

(1) $p^*(f_0) = 1$,
(2) $p(x - u_0) = f_0(x - u_0)$,
(3) $f_0(u_0) = \sup_{u \in \mathcal{N}} f_0(u)$.

If $\mathcal{N}$ is a subspace then conditions (2) and (3) can be replaced by

(2′) $p(x - u_0) = f_0(x)$,
(3′) $f_0(u) = 0, \forall\, u \in \mathcal{N}$.

Proof Suppose (3.17) holds. Then Theorem 1.3.2 gives a functional $f_0 \in X'$ such that $p^*(f_0) = 1$ and

$$p(x - u_0) = f_0(x) - \sup_{u \in \mathcal{N}} f_0(u). \tag{3.18}$$

Functional f_0 satisfies condition (3), because if we assume on the contrary that

$$f_0(u_0) < \sup_{u \in \mathcal{N}} f_0(u),$$

then using (3.3) we get

$$p(x-u_0)=p^*(f_0)p(x-u_0)\geq f_0(x-u_0)>f_0(x)-\sup_{u\in\mathcal{N}} f_0(u),$$

which contradicts (3.18). From (3) and (3.18) we immediately get (2).

Conversely, assume that $p(x,\mathcal{N})>0$ and for $u_0\in\mathcal{N}$ there is a functional $f_0\in X'$ satisfying conditions (1)–(3). Then for every $u\in\mathcal{N}$ we have

$$p(x-u_0)=f_0(x-u_0)=f_0(x-u)+[f_0(u)-f_0(u_0)],$$

and from $f_0(u)\leq f_0(u_0)$ (see (3)) we get

$$p(x-u_0)\leq f_0(x-u_0)=p^*(f_0)p(x-u)=p(x-u),$$

i.e. (3.17) is fulfilled.

In the case when $\mathcal{N}$ is a subspace and (3.17) is fulfilled for some $u_0\in\mathcal{N}$, the functional f_0 from Theorem 1.3.3 satisfies (1), (2′) and (3′) because equality (3.18) can be written as $p(x-u_0)=f_0(x)$ in view of (3.9). Conversely, if the functional $f_0\in X'$ satisfies (1), (2′) and (3′), then for every $u\in\mathcal{N}$ we have

$$p(x-u)\geq f_0(x-u)=f_0(x)=p(x-u_0),$$

i.e. (3.17) holds true. □

Finally we state the most important case for our considerations in Proposition 1.3.5 i.e. when X is a normed space, $p(x)=\|x\|$ and $\mathcal{N}$ is a subspace of X.

Corollary 1.3.6 Let $\mathcal{N}$ be a subspace of the normed space X. Then $u_0\in\mathcal{N}$ is an element of best approximation for $x\in X\backslash\mathcal{N}$ in $\mathcal{N}$ if and only if there is a functional $f_0\in X^*$ such that

$$\|f_0\|=1,\qquad \|x-u_0\|=f_0(x),\qquad f_0(u)=0\qquad \forall\, u\in\mathcal{N}.$$

1.4 Duality in functional spaces

1.4.1 Functional spaces The general theorems proved in Section 1.3 take a more concrete form in the standard functional spaces with regard to the form of the linear functionals. Let us introduce some standard notation:

$C[a,b]$ is the space of all functions $x(t)$ continuous on $[a,b]$ with the uniform norm

$$\|x\|_{C[a,b]}=\max_{a\leq t\leq b}|x(t)|; \tag{4.1}$$

$L_p[a,b]$ is the space of all functions $x(t)$ which are pth order integrable on $[a,b]$ with the norm

$$\|x\|_{L_p[a,b]}=\left(\int_a^b|x(t)|^p\,\mathrm{d}t\right)^{1/p}; \tag{4.2}$$

$L^\infty[a,b]$ is the space of all functions $x(t)$ which are measurable and essentially bounded on $[a,b]$ with the norm

$$\|x\|_{L_\infty[a,b]} = \operatorname*{ess\ sup}_{a\le t\le b} |x(t)|; \tag{4.3}$$

$V[a,b]$ is the linear space of all functions $x(t)$ of bounded variation, defined on $[a,b]$ i.e.

$$\bigvee_a^b x < \infty. \tag{4.4}$$

By C, L_p $(1 \le p \le \infty)$ and V we shall denote the corresponding spaces of functions $x(t)$ defined on $\mathbb{R}$ and with period 2π. The norms in C and L_p and the inclusion $x \in V$ are determined by relations (4.1)–(4.4) with $[a,b] = [0,2\pi]$.

The spaces $L_p[a,b]$ and L_p with $1 < p < \infty$ are strictly normalized, i.e. the norm (4.2) for these ps is strictly convex. Indeed, let $x, y \in L_p[a,b]$,

$$\|x\|_{L_p[a,b]} = \|y\|_{L_p[a,b]} = 1 \tag{4.5}$$

and for $0 < \alpha < 1$

$$\|\alpha x + (1-\alpha)y\|_{L_p[a,b]} = 1.$$

Then the inequality

$$\|\alpha x + (1-\alpha)y\|_{L_p[a,b]} \le \alpha\|x\|_{L_p[a,b]} + (1-\alpha)\|y\|_{L_p[a,b]}, \qquad 1 < p < \infty,$$

turns out to be an equality, which is only possible (see Appendix A1) when $\alpha x(t) = c(1-\alpha)y(t)$ $(c > 0)$ almost everywhere (a.e.) on $[a,b]$. Now (4.5) implies $x(t) = y(t)$, therefore the unit ball $\|x\|_{L_p[a,b]} = 1$ does not contain segments when $1 < p < \infty$.

The spaces $C[a,b]$, $L_1[a,b]$ (and also the spaces C, L_1 with 2π-periodic functions) are not strictly normalized; the construction of the corresponding examples is not difficult.

1.4.2 Duality in $L_p[a,b]$ We shall apply the general theorems for best approximation from Section 1.3 in the case when $X = L_p[a,b]$ $(1 \le p \le \infty)$. In this section we write $\|\bullet\|_p$ instead of $\|\bullet\|_{L_p[a,b]}$.

Proposition 1.4.1 If $\mathcal{N}$ is a convex set in $L_p[a,b]$ $(1 \le p < \infty)$ then we have for every $x \in L_p[a,b]\backslash\bar{\mathcal{N}}$

$$\inf_{u\in\mathcal{N}} \|x-u\|_p = \sup_{y\in L_{p'}[a,b],\|y\|_{p'}\le 1} \left[\int_a^b x(t)y(t)\,dt - \sup_{u\in\mathcal{N}} \int_a^b u(t)y(t)\,dt\right], \qquad p' = \frac{p}{p-1}. \tag{4.6}$$

In particular, if $\mathcal{N}$ is a subspace then

$$\inf_{u\in\mathcal{N}} \|x-u\|_p = \sup\left\{\int_a^b x(t)y(t)\,\mathrm{d}t\colon \|y\|_{p'} \le 1, y \perp \mathcal{N}\right\}, \tag{4.7}$$

where $y \perp \mathcal{N}$ means

$$\int_a^b u(t)y(t)\,\mathrm{d}t = 0, \qquad \forall\, u \in \mathcal{N}.$$

The supremums in (4.6) and (4.7) are attained for some functions $y \in L_{p'}[a, b]$ with norm $\|y\|_{p'} = 1$.

Indeed every functional $f \in L_p^*[a, b]$ $(1 \le p < \infty)$ can be represented as [10B, p. 193]

$$f(x) = \int_a^b x(t)y(t)\,\mathrm{d}t, \tag{4.8}$$

where $y \in L_{p'}[a, b]$ and

$$\|f\| = \|y\|_{p'}. \tag{4.9}$$

On the other hand, every function $y \in L_{p'}[a, b]$ defines via (4.8) a functional $f \in L_p^*[a, b]$ satisfying (4.9). Therefore we obtain Proposition 1.4.1 by rewriting (3.1) and (3.10) and replacing $f(x)$ by the expression (4.8) and the norm $\|f\|$ by $\|y\|_{p'}$.

Now we turn to the duality relations in $L_\infty[a, b]$. Formula (4.8) with $x \in L_\infty[a, b]$, $y \in L_1[a, b]$ does not represent all linear functionals from $L_\infty^*[a, b]$, but the validity of (4.7) for all $1 \le p \le \infty$ for finite dimensional subspace $\mathcal{N}$ can be established using Theorem 1.3.5.

Proposition 1.4.2 If $x_1(t), x_2(t), \ldots, x_N(t)$ are linearly independent functions in $L_p[a, b]$ $(1 \le p \le \infty)$ then for every $x \in L_p[a, b]$ we have $(p^{-1} + p'^{-1} = 1)$

$$\inf_{\alpha_k} \left\| x - \sum_{k=1}^N \alpha_k x_k \right\|_p = \sup\left\{\int_a^b x(t)y(t)\,\mathrm{d}t\colon \|y\|_{p'} \le 1,\right.$$
$$\left.\int_a^b y(t)x_k(t)\,\mathrm{d}t = 0, k = 1, 2, \ldots, N\right\}. \tag{4.10}$$

Proof Define on $L_{p'}[a, b]$ the bounded linear functionals

$$f_k(y) = \int_a^b y(t)x_k(t)\,\mathrm{d}t, \qquad k = 1, 2, \ldots, N, \qquad y \in L_{p'}[a, b].$$

Let us fix the function $x \in L_p[a, b]$ and also define the functional $f \in L^*_{p'}[a, b]$ by

$$f(y) = \int_a^b y(t)x(t)\, dt, \qquad y \in L_{p'}[a, b].$$

The functional

$$\varphi_\alpha = f - \sum_{k=1}^{N} \alpha_k f_k, \tag{4.11}$$

where $\alpha = \{\alpha_1, \alpha_2, \ldots, \alpha_N\}$ is an arbitrary set of N real numbers, and also belongs to $L^*_{p'}[a, b]$ and (see Appendix A1)

$$\|\varphi\| = \sup_{\|y\|_{p'} \le 1} \left| \int_a^b \left[x(t) - \sum_{k=1}^{N} \alpha_k x_k(t) \right] y(t)\, dt \right| = \left\| x - \sum_{k=1}^{N} \alpha_k x_k \right\|_p .$$

So for every α_k we have

$$\left\| f - \sum_{k=1}^{N} \alpha_k f_k \right\| = \left\| x - \sum_{k=1}^{N} \alpha_k x_k \right\|_p$$

and by Theorem 1.3.5

$$\min_{\alpha_k} \left\| x - \sum_{k=1}^{N} \alpha_k x_k \right\|_p =$$

$$\sup \left\{ \int_a^b x(t)\, y(t)\, dt \colon \|y\|_{p'} \le 1, \quad f_k(y) = 0, k = 1, 2, \ldots, N \right\},$$

which is equivalent to (4.10), if we remember the definitions of the functionals f and f_k. □

Corollary 1.4.3 For the best approximation by constants of the function $x \in L_p[a, b]$ $(1 \le p \le \infty)$ we have $(p^{-1} + p'^{-1} = 1)$

$$\inf_{\lambda \in \mathbb{R}} \|x(t) - \lambda\|_p = \sup \left\{ \int_a^b x(t) y(t)\, dt \colon \|y\|_{p'} \le 1, \int_a^b y(t)\, dt = 0 \right\}.$$

1.4.3 Characterization of the element of best approximation in $L_p[a, b]$ Using the general Proposition 1.3.5 and the form of the linear bounded functionals in $L_p[a, b]$ $(1 \le p < \infty)$, we obtain immediately:

Proposition 1.4.4 Let $\mathcal{N}$ be a convex set in $L_p[a, b]$ $(1 \le p < \infty)$ and $x \in L_p[a, b] \backslash \bar{\mathcal{N}}$. Then the relation

$$\|x - u_0\|_p = \inf_{u \in \mathcal{N}} \|x - u\|_p \tag{4.12}$$

holds for some $u_0 \in \mathcal{N}$ if and only if there is a function $y_0 \in L_{p'}[a, b]$ $(p' = p/(p-1))$ such that

(1) $\|y_0\|_{p'} = 1$,

(2) $\|x - u_0\|_p = \int_a^b [x(t) - u_0(t)] y_0(t) \, dt$,

(3) $\int_a^b u_0(t) y_0(t) \, dt = \sup_{u \in \mathcal{N}} \int_a^b u(t) y_0(t) \, dt$.

In particular, if $\mathcal{N}$ is a subspace then conditions (2) and (3) take the form

(2′) $\|x - u_0\|_p = \int_a^b x(t) y_0(t) \, dt$,

(3′) $\int_a^b u(t) y_0(t) \, dt = 0, \ \forall \, u \in \mathcal{N}$.

This criterion relates the characterization of the element of best approximation u_0 to the existence of an implicitly defined function y_0. In the case when $\mathcal{N}$ is a subspace we can eliminate y_0 from the conditions (1), (2′) and (3′) and we obtain a more effective criterion.

Theorem 1.4.5 Let $\mathcal{N}$ be a subspace in $L_p[a, b]$ $(1 \le p < \infty)$ and $x \in L_p[a, b] \backslash \mathcal{N}$. A sufficient and (when $p = 1$ under the additional condition meas $\{t : t \in [a, b], x(t) = u_0(t)\} = 0$) necessary condition for an element $u_0 \in \mathcal{N}$ to satisfy (4.12) is

$$\int_a^b u(t) \, |x(t) - u_0(t)|^{p-1} \operatorname{sgn} [x(t) - u_0(t)] \, dt = 0, \qquad \forall \, u \in \mathcal{N}. \tag{4.13}$$

The proof of the sufficiency is very simple and we separate it out in the next statement.

Proposition 1.4.6 If the function u_0 from the subspace $\mathcal{N}$ of $L_p[a, b]$ $(1 \le p < \infty)$ satisfies (4.13) then (4.12) is fulfilled.

Indeed, using (4.13) for every $u \in \mathcal{N}$ we obtain

$$\|x - u_0\|_p^p = \int_a^b [x(t) - u_0(t)]|x(t) - u_0(t)|^{p-1} \operatorname{sgn}[x(t) - u_0(t)] \, dt$$

$$= \int_a^b [x(t) - u(t)]|x(t) - u_0(t)|^{p-1} \operatorname{sgn}[x(t) - u_0(t)] \, dt$$

$$\leq \int_a^b |x(t) - u(t)| \, |x(t) - u_0(t)|^{p-1} \, dt. \tag{4.14}$$

If $p = 1$ then we immediately obtain $\|x - u_0\|_1 \leq \|x - u\|_1$ for any $u \in \mathcal{N}$ which is (4.12). If $1 < p < \infty$ then Hölder's inequality applied to the right-hand side of (4.14) gives

$$\|x - u_0\|_p^p \leq \|x - u\|_p \, \|x - u_0\|_p^{p/p'} = \|x - u\|_p \, \|x - u_0\|_p^{p-1},$$

i.e. $\|x - u_0\|_p \leq \|x - u\|_p$ for every $u \in \mathcal{N}$ and we obtain (4.12) in this case too.

We shall prove the necessity in Theorem 1.4.5 separately for $1 < p < \infty$ and $p = 1$.

Let $1 < p < \infty$ and (4.12) be fulfilled for some $u_0 \in \mathcal{N}$. From Proposition 1.4.4 there is $y_0 \in L_{p'}[a, b]$ $(p' = p/(p-1))$ for which conditions (1), (2′) and (3′) are fulfilled. Then

$$\|x - u_0\|_p = \int_a^b x(t) y_0(t) \, dt = \int_a^b [x(t) - u_0(t)] \, y_0(t) \, dt.$$

Now applying Hölder's inequality we obtain

$$\|x - u_0\|_p \leq \int_a^b |x(t) - u_0(t)| \, |y_0(t)| \, dt \leq \|x - u_0\|_p \, \|y_0\|_{p'} \leq \|x - u_0\|_p$$

and so we have only equalities in the above chain. Hence

$$\int_a^b [x(t) - u_0(t)] \, y_0(t) \, dt = \int_a^b |x(t) - u_0(t)| \, |y_0(t)| \, dt$$
$$= \|x - u_0\|_p \, \|y_0\|_{p'}$$

Therefore

$$[x(t) - u_0(t)] \, y_0(t) \geq 0 \qquad \text{a.e. on } [a, b] \tag{4.15}$$

and (see Proposition A1.1)

$$|y_0(t)| = \alpha |x(t) - u_0(t)|^{p-1}, \qquad (\alpha > 0).$$

From the last inequality it follows that the set of points for which only one of the multipliers in (4.15) vanishes has a measure zero and hence sgn $y_0(t) = \operatorname{sgn}[x(t) - u_0(t)]$ a.e. in $[a, b]$. Therefore

$$y_0(t) = |y_0(t)| \operatorname{sgn} y_0(t) = \alpha|x(t) - u_0(t)|^{p-1} \operatorname{sgn}[x(t) - u_0(t)],$$

which together with (3′) gives (4.13).

Consider now the case $p = 1$. Let $x \in L_1[a, b]$ and let u_0 be a function from $\mathcal{N}$ for which

$$\|x - u_0\|_1 = \inf_{u \in \mathcal{N}} \|x - u\|_1$$

and let $y_0 \in L_\infty[a, b]$ satisfy $\|y_0\|_\infty = 1$ and conditions (2′) and (3′) with $p = 1$. Then

$$\|x - u_0\|_1 = \int_a^b x(t)y_0(t)\,\mathrm{d}t = \int_a^b [x(t) - u_0(t)]\,y_0(t)\,\mathrm{d}t$$

and so

$$\int_a^b [x(t) - u_0(t)]\,y_0(t)\,\mathrm{d}t = \int_a^b |x(t) - u_0(t)|\,\mathrm{d}t. \tag{4.16}$$

If the set of all points $t \in [a, b]$ for which $x(t) = u_0(t)$ is of zero measure, then equality (4.16) is possible only if a.e. on $[a, b]$ we have

$$y_0(t) = \operatorname{sgn}[x(t) - u_0(t)].$$

From here and condition (3′) we get

$$\int_a^b u(t) \operatorname{sgn}[x(t) - u_0(t)]\,\mathrm{d}t = 0, \qquad \forall\, u \in \mathcal{N},$$

which is (4.13) for $p = 1$. This completes the proof of Theorem 1.4.5. □

1.4.4 Duality in $C[a, b]$ For brevity in this section we shall write $\|\bullet\|_{C[a,b]} = \|\bullet\|_C$. The famous theorem by F. Riesz states that every linear bounded functional f on $C[a, b]$ (i.e. $f \in C[a, b]$) can be represented as a Lebesgue–Stieltjes integral:

$$f(x) = \int_a^b x(t)\,\mathrm{d}g(t), \qquad x \in C[a, b], \tag{4.17}$$

where $g \in V[a, b]$, $g(t + 0) = g(t)$ for $a < t < b$ and $\|f\| = \bigvee_a^b g$.

Hence for every subset $\mathcal{N}$ of $C[a, b]$ and every $x \in C[a, b]$ we have

$$\sup\{[f(x) - \sup_{u \in \mathcal{N}} f(u)] : f \in C^*[a, b], \|f\| \le 1\}$$

$$\le \sup\left\{\left[\int_a^b x(t)\,\mathrm{d}g(t) - \sup_{u \in \mathcal{N}} \int_a^b u(t)\,\mathrm{d}g(t)\right] : g \in V[a, b], \bigvee_a^b g \le 1\right\}. \tag{4.18}$$

On the other hand every $g \in V[a, b]$ defines (via (4.17)) a functional $f \in C^*[a, b]$ and

$$\left| \int_a^b x(t)\, dg(t) \right| \leq \|x\|_C \bigvee_a^b g. \tag{4.19}$$

Hence $\|f\| \leq \bigvee_a^b g$ and we have equality in (4.18). So when $X = C[a, b]$, Theorems 1.3.1 and 1.3.3 (see (3.10)) take the following form.

Proposition 1.4.7 If $\mathcal{N}$ is a convex set in $C[a, b]$, then for every function $x \in C[a, b]$ we have

$$\inf_{u \in \mathcal{N}} \|x - u\|_C = \sup \left\{ \left[\int_a^b x(t)\, dg(t) - \sup_{u \in \mathcal{N}} \int_a^b u(t)\, dg(t) \right] : \bigvee_a^b g \leq 1 \right\}. \tag{4.20}$$

In particular, if $\mathcal{N}$ is a subspace of $C[a, b]$ then

$$\inf_{u \in \mathcal{N}} \|x - u\|_C = \sup \left\{ \int_a^b x(t)\, dg(t) : \bigvee_a^b g \leq 1, \int_a^b u(t)\, dg(t) = 0, \forall\, u \in \mathcal{N} \right\}. \tag{4.21}$$

For $x \in C[a, b] \backslash \bar{\mathcal{N}}$ the supremums in (4.20) and (4.21) are attained for some functions $g \in V[a, b]$ with variation one on $[a, b]$.

The general criterion (Proposition 1.3.5) for $X = C[a, b]$ takes the form:

Proposition 1.4.8 Let $\mathcal{N}$ be a convex set in $C[a, b]$ and $x \in C[a, b] \backslash \bar{\mathcal{N}}$. Then the function $u_0 \in \mathcal{N}$ satisfies

$$\|x - u_0\|_C = \inf_{u \in \mathcal{N}} \|x - u\|_C \tag{4.22}$$

if and only if there exists a $g_0 \in V[a, b]$ such that

(1) $\bigvee_a^b g_0 \leq 1$,

(2) $\|x - u_0\|_C = \int_a^b [x(t) - u_0(t)]\, dg_0(t)$,

(3) $\int_a^b u_0(t)\, dg_0(t) = \sup_{u \in \mathcal{N}} \int_a^b u(t)\, dg_0(t)$.

In particular, if $\mathcal{N}$ is a subspace of $C[a, b]$ then conditions (2) and (3) can be replaced by

(2′) $\|x - u_0\|_C = \int_a^b x(t)\, dg_0(t)$,

(3′) $\int_a^b u(t)\,\mathrm{d}g_0(t) = 0, \forall\, u \in \mathcal{N}$.

The necessity follows immediately from Proposition 1.3.5 and Riesz's theorem. On the other hand, if $u_0 \in \mathcal{N}$ we can find a $g_0 \in V[a, b]$ satisfying conditions (1)–(3) such that for every $u \in \mathcal{N}$, we obtain using also (4.19):

$$\begin{aligned}\|x - u_0\|_C &= \int_a^b [x(t) - u_0(t)]\,\mathrm{d}g_0(t) \\ &= \int_a^b [x(t) - u(t)]\,\mathrm{d}g_0(t) + \left[\int_a^b u(t)\,\mathrm{d}g_0(t) - \int_a^b u_0(t)\,\mathrm{d}g_0(t)\right] \\ &\le \int_a^b [x(t) - u(t)]\,\mathrm{d}g_0(t) \le \|x - u\|_C \bigvee_a^b g_0 = \|x - u\|_C\end{aligned}$$

and hence (4.22) is true. □

In the case when $\mathcal{N}$ is a finite dimensional subspace of $C[a, b]$ one can eliminate function g_0 from conditions (2′) and (3′) (using non-trivial facts from analysis) and obtain a more effective criterion [25A, p. 157] containing the famous Chebyshev alternation theorem. We shall give the traditional elementary proof of this theorem in Chapter 2.

1.4.5 Non-symmetric problems In the space $L_p[a, b]$, $p \ge 1$, we can consider the problem of best approximation with non-symmetric distance. As in Sections 1.4.2 and 1.4.3 we set $\|\bullet\|_p = \|\bullet\|_{L_p[a,b]}$.

Let $\alpha > 0$ and $\beta > 0$. For any $x \in L_p[a, b]$ we set

$$x_\pm(t) = \max\{\pm x(t), 0\},$$
$$\|x\|_{p;\alpha,\beta} = \|\alpha x_+ + \beta x_-\|_p.$$

Obviously $\|x\|_{p;1,1} = \|x\|_p$. If $\mathcal{N}$ is a subset of $L_p[a, b]$, then the quantity

$$E(x, \mathcal{N})_{p;\alpha,\beta} = \inf_{u \in \mathcal{N}} \|x - u\|_{p;\alpha,\beta}$$

is called the *best* (α,β)*-approximation* of x by functions from $\mathcal{N}$.

Using standard arguments (see Section 1.1.2) we can easily see that, for every $x \in L_p[a, b]$, there is an element of best (α,β)-approximation in a closed locally compact set $\mathcal{N}$. If $\mathcal{N}$ is convex and $1 < p < \infty$ then this element is unique.

The quantity $p(x) = \|x\|_{p;\alpha,\beta}$ is obviously a non-symmetric norm in $L_p[a, b]$ such that $p(x) \le \max\{\alpha,\beta\}\,\|x\|$. From Theorem 1.3.2, and taking into account the remark after Theorem 1.3.3, we get

$$E(x, \mathcal{N})_{p;\alpha,\beta} = \sup \{[f(x) - \sup_{u \in \mathcal{N}} f(u)]: f \in L_p^*[a, b], \sup_{\|v\|_{p;\alpha,\beta} \le 1} f(v) \le 1\}.$$

Using an extremal relation (A1.16) we get the following statement:

Proposition 1.4.9 Let $1 \le p < \infty$, $0 < \alpha, \beta < \infty$ and let $\mathcal{N}$ be a convex set in $L_p[a, b]$. Then we have for every $x \in L_p[a, b]$ $(p' = p/(p-1))$

$$E(x, \mathcal{N})_{p;\alpha,\beta} = \sup \left\{ \left[\int_a^b x(t)g(t)\,dt - \sup_{u \in \mathcal{N}} \int_a^b u(t)g(t)\,dt \right]: \right.$$
$$\left. g \in L_{p'}[a, b], \|\alpha^{-1}g_+ + \beta^{-1}g_-\|_{p'} \le 1 \right\}. \quad (4.23)$$

In particular, when $\mathcal{N}$ is a subspace we have

$$E(x, \mathcal{N})_{p;\alpha,\beta} = \sup \left\{ \int_a^b x(t)g(t)\,dt: g \perp \mathcal{N}, \|\alpha^{-1}g_+ + \beta^{-1}g_-\|_{p'} \le 1 \right\}. \quad (4.24)$$

The supremums in (4.23) and (4.24) are attained for functions $g \in L_{p'}[a, b]$ with norm $\|g\|_{p';\alpha^{-1},\beta^{-1}} = 1$.

Let us now consider the best one-sided approximations for functions in $L_p[a, b]$, $p \ge 1$. We fix a set $\mathcal{N} \subset L_p[a, b]$ and define for $x \in L_p[a, b]$ the sets

$$\mathcal{N}_x^+ = \{u: u \in \mathcal{N}, u(t) \le x(t), a \le t \le b\},$$
$$\mathcal{N}_x^- = \{u: u \in \mathcal{N}, u(t) \ge x(t), a \le t \le b\}$$

(the inequalities $u \le x$ and $u \ge x$ are satisfied a.e.). Set

$$E^{\pm}(x, \mathcal{N})_p = \begin{cases} \inf \{\|x - u\|_p: u \in \mathcal{N}_x^{\pm}\}, \mathcal{N}_x^{\pm} \ne \varnothing \\ \infty, \mathcal{N}_x^{\pm} = \varnothing. \end{cases}$$

In the cases described in Section 1.1.2 we obtain the existence and uniqueness of the element of the best one-sided approximation (if $\mathcal{N}_x^{\pm}$ is not empty) in a standard way.

The next statement plays a basic role in our considerations.

Theorem 1.4.10 If $\mathcal{N} \subset L_p[a, b]$ $(1 \le p < \infty)$ is locally compact, then for every $x \in L_p[a, b]$ the limits

$$\lim_{\beta \to \infty} E(x, \mathcal{N})_{p;\alpha,\beta} = E^+(x, \mathcal{N})_p, \quad (4.25)$$
$$\lim_{\alpha \to \infty} E(x, \mathcal{N})_{p;\alpha,\beta} = E^-(x, \mathcal{N})_p \quad (4.26)$$

hold monotonically on α or β.

For a better explanation of the limits in (4.25) and (4.26), let us note that from the equality

$$\|x-u\|_{p;1,\beta}^p = \|(x-u)_+\|_p^p + \beta^p\|(x-u)_-\|_p^p,$$

we conclude that for large β the minimization on $u \in \mathcal{N}$ should lead to functions u for which the negative part $(x-u)_-$ is small and in the limit to functions u for which $u(t) \le x(t)$ a.e.

Proof We shall prove only (4.25); the proof of (4.26) is similar. Here and later we denote by u_β a function in $\mathcal{N}$ (assuming that it exists) for which

$$E(x,\mathcal{N})_{p;1,\beta} = \|x-u_\beta\|_{p;1,\beta}.$$

If $\beta_1 < \beta_2$ then

$$\begin{aligned}E(x,\mathcal{N})_{p;1,\beta_1} &\le \|x-u_{\beta_2}\|_{p;1,\beta_1} \le \|(x-u_{\beta_2})_+ + \beta_2(x-u_{\beta_2})_-\|_p\\ &= \|x-u_{\beta_2}\|_{p;1,\beta_2} = E(x,\mathcal{N})_{p;1,\beta_2},\end{aligned}$$

i.e. the quantities $E(x,\mathcal{N})_{p;1,\beta}$ do not decrease on β. Now because

$$E(x,\mathcal{N})_{p;1,\beta} \le \inf_{u\in\mathcal{N}_x^+} \|x-u\|_{p;1,\beta} = E^+(x,\mathcal{N})_p, \tag{4.27}$$

the limit on the left-hand side of (4.25) exists and is not greater than $E^+(x,\mathcal{N})_p$. The equality (4.25) will be proved if we show that

$$E^+(x,\mathcal{N})_p \le A =: \lim_{\beta\to\infty} E(x,\mathcal{N})_{p;1,\beta}. \tag{4.28}$$

If $A = \infty$ then (4.28) is fulfilled. If A is finite, then for large enough β we have

$$\|u_\beta\|_p \le \|x-u_\beta\|_p + \|x\|_p = E(x,\mathcal{N})_{p;1,\beta} + \|x\|_p \le 2A + \|x\|_p$$

and the local compactness of $\mathcal{N}$ means that a sequence $\{\beta_k\}$ $(\beta_k \to \infty)$ exists as well as an element $u_\infty \in \mathcal{N}$ such that $u_{\beta_k} \to u_\infty$ when $k \to \infty$.

We shall show that $u_\infty \in \mathcal{N}_x^+$, in other words

$$\|(x-u_\infty)_-\|_p = 0. \tag{4.29}$$

Using

$$E(x,\mathcal{N})_{p;1,\beta_k} = \|(x-u_{\beta_k})_+ + \beta_k(x-u_{\beta_k})_-\|_p$$

and the monotonicity of the non-symmetric approximations on β we have

$$\|(x-u_{\beta_k})_-\|_p \le \beta_k^{-1}E(x,\mathcal{N})_{p;1,\beta_k} \le \beta_k^{-1}A.$$

Now taking the limit for $k \to \infty$ we obtain (4.29). Finally for $\beta_k \ge 1$ we have

$$\begin{aligned}E^+(x,\mathcal{N})_p &\le \|x-u_\infty\|_p = \lim_{k\to\infty}\|x-u_{\beta_k}\|_p\\ &\le \lim_{k\to\infty}\|x-u_{\beta_k}\|_{p;1,\beta_k} = \lim_{k\to\infty}E(x,\mathcal{N})_{p;1,\beta_k},\end{aligned}$$

which proves (4.28) and completes the proof of the theorem. □

Using Theorem 1.4.10 we can establish the following analog of Proposition 1.4.9 for one-sided approximations.

Proposition 1.4.11 Let $1 \le p < \infty$ and let $\mathcal{N}$ be a convex set in $L_p[a, b]$. Then we have for every $x \in L_p[a, b]$ $(p' = p/(p-1))$

$$E^{\pm}(x, \mathcal{N})_p = \sup\left\{\left[\int_a^b x(t)g(t)\,dt - \sup_{u \in \mathcal{N}} \int_a^b u(t)g(t)\,dt\right] : g \in L_{p'}[a, b], \|g_\pm\|_{p'} \le 1\right\}. \quad (4.30)$$

In particular, when $\mathcal{N}$ is a subspace we have

$$E^{\pm}(x, \mathcal{N})_p = \sup\left\{\int_a^b x(t)g(t)\,dt \colon g \perp \mathcal{N}, \|g_\pm\|_{p'} \le 1\right\}. \quad (4.31)$$

Proof Denote the right-hand sides of (4.23) and (4.30) by $N_{\alpha,\beta}$ and N^+ respectively (x is a fixed function). Because of Theorem 1.4.10 it is sufficient to prove that

$$\lim_{\beta \to \infty} N_{1,\beta} = N^+, \qquad \lim_{\alpha \to \infty} N_{\alpha,1} = N^-.$$

We shall only prove the first inequality. If $\beta_2 > \beta_1 \ge 1$ then

$$\|g_+\|_{p'} \le \|g_+ + \beta_2^{-1} g_-\|_{p'} \le \|g_+ + \beta_1^{-1} g_-\|_{p'}$$

and so $N_{1,\beta}$ does not decrease on β and $N_{1,\beta} \le N^+$, hence the limit $\lim_{\beta \to \infty} N_{1,\beta}$ exists and is not greater than N^+.

Let $N^+ < \infty$. For every $\varepsilon > 0$ there is $g_\varepsilon \in L_{p'}[\mathrm{a}, \mathrm{b}]$, such that $\|(g_\varepsilon)_+\|_{p'} < 1$ and

$$\int_a^b x(t)g_\varepsilon(t)\,dt - \sup_{u \in \mathcal{N}} \int_a^b u(t)g_\varepsilon(t)\,dt > N^+ - \varepsilon.$$

For all large enough β we have $\|(g_\varepsilon)_+ + \beta^{-1}(g_\varepsilon)_-\|_{p'} \le 1$ and so for such β we have $N_{1,\beta} > N^+ - \varepsilon$. Because of the arbitrariness of $\varepsilon > 0$ we obtain in the case $N^+ < \infty$ that $N_{1,\beta} \to N^+$ as $\beta \to \infty$ (we also note that $\|g_+ + \beta^{-1} g_-\|_{p'} \to \|g_+\|_{p'}$). If $N^+ = \infty$ similar reasoning shows that $N_{1,\beta} \to \infty$.

The case when $\mathcal{N}$ is a subspace can be treated as in Proposition 1.4.9. □

Concerning the characterization of the element of best non-symmetric approximation, we can establish the following analog of Theorem 1.4.5 for (α,β)-approximations starting with the general criterion (Proposition 1.3.6) and using Proposition A1.3 instead of Proposition A1.1.

Proposition 1.4.12 Let $\mathcal{N}$ be a finite dimensional subspace of $L_p[a, b]$ $(p \geq 1)$ and $x \in L_p[a, b] \backslash \mathcal{N}$. In the case $p > 1$ the function $u_0 \in \mathcal{N}$ satisfies

$$\|x - u_0\|_{p;\alpha,\beta} = E(x, \mathcal{N})_{p;\alpha,\beta}, \qquad 0 < \alpha, \beta < \infty,$$

if and only if

$$\int_a^b u(t)\, |x(t) - u_0(t)|^{p-1} [\alpha^p \operatorname{sgn}(x(t) - u_0(t))_+ - \beta^p \operatorname{sgn}(x(t) - u_0(t))_-]\, dt = 0$$

for every $u \in \mathcal{N}$. For the case when $p = 1$ the first condition is sufficient and, under the additional assumption $x(t) \neq u_0(t)$ a.e., necessary.

1.4.6 On the duality in the spaces of periodic functions Statements 1.4.1–1.4.12 remain correct if we replace $[a, b]$ by $[0, 2\pi]$ and $L_p[a, b]$, $C[a, b]$ by L_p, C – the spaces with 2π-periodic functions. For L_p this is obvious, because $L_p[0, 2\pi]$ is the restriction of L_p on $[0, 2\pi]$ and every $x \in L_p[0, 2\pi]$ can be included in L_p if we continue it as a 2π-periodic function. Concerning $C[0, 2\pi]$ and C we can not assert the same, but we can argue in the following manner. Propositions 1.4.7 and 1.4.8 are proved (with $[a, b] = [0, 2\pi]$) for *every* function $x \in C[0, 2\pi]$ and so they are correct for every $x \in C[0, 2\pi]$ satisfying the condition $x(0) = x(2\pi)$ and allowing periodic continuous continuation on the whole axis. Of course, all this is correct for spaces of functions of every period.

1.5 Duality for best approximation of classes of functions

1.5.1 Classes of differentiable functions and general facts Some duality relations for the best approximation of single functions from $L_p[a, b]$ or $C[a, b]$ were given in Section 1.4. One is able to calculate exactly the supremums on the right-hand sides of (4.6), (4.7) and, in some exceptional cases, (4.20); the efficiency of the duality theorems becomes apparent when one starts solving the problems of the best approximation for *classes of functions*.

In particular, if the class of functions is given by imposing bounds on the norm (in different metrics) of the rth derivative, then the problem of the best approximation by finite dimensional subspaces can be reduced to finding the supremum of the norm of functions from sets determined by the orthogonality conditions of the approximating subspace.

By $C^m[a, b]$ we denote the class of all m-times continuously differentiable functions, $m = 1, 2, \ldots$; by $L_p^m[a, b]$ $(1 \leq p \leq \infty)$ or $V^m[a, b]$ we denote the class of all functions x defined on $[a, b]$ such that the $(m - 1)$ derivative $x^{(m-1)}$ is absolutely continuous on $[a, b]$ and $x^{(m)} \in L_p[a, b]$ or $x^{(m)} \in V[a, b]$

respectively. In other words $C^m[a, b]$, $L_p^m[a, b]$ and $V^m[a, b]$ are the sets of all mth integrals of functions from $C[a, b]$, $L_p[a, b]$ and $V[a, b]$ respectively.

By C^m, L_p^m, V^m we denote the corresponding sets of 2π-periodic functions, i.e. the sets of mth 2π-periodic integrals of functions $x \perp 1$ (i.e. $\int_0^{2\pi} x(t)\, dt = 0$) from C, L_p and V respectively. Let us note immediately that the functions from these sets have the integral representation

$$x(t) = \frac{1}{2\pi}\int_0^{2\pi} x(u)\, du + \frac{1}{\pi}\int_0^{2\pi} B_m(t-u)x^{(m)}(u)\, du, \qquad m = 1, 2, \ldots, \quad (5.1)$$

where $x^{(m)}$ belongs to C, L_p and V respectively, and where

$$B_m(t) = \sum_{k=1}^{\infty} \frac{\cos(kt - \pi m/2)}{k^m}, \qquad m = 1, 2, \ldots, \quad (5.2)$$

are the 2π-periodic Bernoulli functions whose properties will be given in Section 3.1. In order to obtain (5.1) we write the Fourier series for x, i.e.

$$x(t) = a_0/2 + \sum_{k=1}^{\infty} (a_k \cos kt + b_k \sin kt)$$

and use the identity

$$a_k \cos kt + b_k \sin kt = \frac{1}{\pi}\int_0^{2\pi} x(u) \cos k(t-u)\, du$$

$$= \frac{1}{\pi k^m}\int_0^{2\pi} x^{(m)}(u) \cos[k(t-u) - \pi m/2]\, du; \quad (5.3)$$

where the second equality is obtained by integration by parts and by making use of the periodicity.

Changing the order of the above arguments, one can easily see that every 2π-periodic function x of the type

$$x(t) = c + \frac{1}{\pi}\int_0^{2\pi} B_m(t-u)\varphi(u)\, du,$$

where $\varphi \in C$, L_p or V, belongs to C^m, L_p^m and V^m respectively, c being the mean value of x on the period $[0, 2\pi]$ and $x^{(m)}(t) = \varphi(t)$ a.e.

For a function $x \in L_p$ we set $\|x\|_p = \|x\|_{L_p[0,2\pi]}$ and $E(x, \mathcal{N})_p = E(x, \mathcal{N})_{L_p[0,2\pi]}$; we shall use the same notations for $\|x\|_{L_p[a,b]}$ and $E(x, \mathcal{N})_{L_p[a,b]}$ when this will not lead to any ambiguity.

Now we introduce the following classes of functions:

$$KW_p^m[a, b] = \{x\colon x \in L_p^m[a, b], \|x^{(m)}\|_p \le K\},$$
$$1 \le p \le \infty, \qquad m = 1, 2, \ldots; \quad (5.4)$$

$$KW_V^m[a,b] = \{x: x \in V^m[a,b], \bigvee_a^b x^{(m)} \le K\},$$

$$m = 1, 2, \ldots, \qquad V^0[a,b] = V[a,b]; \quad (5.5)$$

KW_p^m and KW_V^m will denote the corresponding classes of 2π-periodic functions. When $K = 1$ we shall simply write $W_p^m[a,b]$, $W_V^m[a,b]$, W_p^m and W_V^m.

Let us remark that classes $KW_\infty^m[a,b]$ and KW_∞^m can be characterized using the Lipschitz condition. Let $W^mKH^1[a,b]$ be the class of functions x from $C^m[a,b]$ ($C^0[a,b] = C[a,b]$) such that

$$|x^{(m)}(t') - x^{(m)}(t'')| \le K|t' - t''|, \qquad t', t'' \in [a,b],$$

and respectively

$$W^mKH^1 = \{x: x \in C^m, |x^{(m)}(t') - x^{(m)}(t'')| \le K|t' - t''|\}.$$

Instead of $W^0KH^1[a,b]$ and W^0KH^1 we write $KH^1[a,b]$ and KH^1.

Then we have the following equalities (in the sense of coincidence of sets)

$$KW_\infty^m[a,b] = W^{m-1}KH^1[a,b], \qquad KW_\infty^m = W^{m-1}KH^1, \qquad m = 1, 2, \ldots.$$

Obviously it is enough to check these equalities only for $m = 1$. If $x \in KW_\infty^1[a,b]$ then for every $t', t'' \in [a,b]$ we have

$$|x(t') - x(t'')| = \left| \int_{t'}^{t''} x'(\mathrm{u})\, d\mathrm{u} \right| \le K|t' - t''|,$$

i.e. $x \in KH^1[a,b]$. On the other hand, if $x \in KH^1[a,b]$, then x is absolutely continuous on $[a,b]$ and from

$$\left| \frac{x(t+h) - x(t)}{h} \right| \le K, \qquad t + h \in [a,b],$$

we get $|x'(t)| \le K$ a.e. on $[a,b]$.

From (5.4) and (5.5) we see that the corresponding classes defined by this relationship contain all algebraic polynomials of degree $(m-1)$ and we should take this into account when stating the extremal problems for best approximation.

Let P_n denote the set of all algebraic polynomials of degree not greater than n. If X is a normed function space, $\mathcal{N}$ is a subspace of X, $\mathcal{M}$ is a subset of X containing P_n, then the quantity

$$E(\mathcal{M}, \mathcal{N})_X = \sup_{x \in \mathcal{M}} E(x, \mathcal{N})_X$$

can be finite if and only if $\mathcal{N}$ contains P_n. Indeed, if the polynomial q of degree n does not belong to $\mathcal{N}$, then $E(q, \mathcal{N})_X = \alpha > 0$ and for every $\lambda > 0$

we have by Proposition 1.1.1 $E(\lambda q, \mathcal{N})_X = \lambda\alpha$. But the polynomial λq belongs to $\mathcal{M}$ together with q and hence

$$E(\mathcal{M}, \mathcal{N})_X = \sup_{\lambda>0} E(\lambda q, \mathcal{N})_X = \infty.$$

Therefore it makes sense to consider the problem for evaluating the quantities $E(KW_p^m[a, b], \mathcal{N})_X$ and $E(KW_V^m[a, b], \mathcal{N})_X$, where X is $L_p[a, b]$ or $C[a, b]$, only if $\mathcal{N}$ contains P_{m-1}. Classes KW_p^m and KW_V^m contain every constant and we shall require that the constant belongs to the approximating sets in the periodic case.

Let us note that we can assume that $K = 1$ without loss of generality when solving problems of the best approximation of classes (5.4) and (5.5) and their periodic analogs by *subspaces* $\mathcal{N}$. The reason is that from the positive homogeneity of the functional of best approximation (Proposition 1.1.1) we have

$$E(KW_p^m[a, b], \mathcal{N})_{L_q[a, b]} = KE(W_p^m[a, b], \mathcal{N})_{L_q[a, b]},$$
$$E(KW_p^m, \mathcal{N})_q = KE(W_p^m, \mathcal{N})_q.$$

Another general fact for best approximations of different classes of functions is:

Proposition 1.5.1 If $\mathcal{N}$ is a subspace in $L_q[a, b]$ or L_q $(1 \le q < \infty)$, then

$$E(W_V^{m-1}[a, b], \mathcal{N})_{L_q[a, b]} = E(W_1^m[a, b], \mathcal{N})_{L_q[a, b]}, \qquad m = 1, 2, \ldots, \tag{5.6}$$

and

$$E(W_V^{m-1}, \mathcal{N})_q = E(W_1^m, \mathcal{N})_q, \qquad m = 1, 2, \ldots, \tag{5.7}$$

respectively.

Indeed, we obtain from the obvious inclusion $W_1^m[a, b] \subset W_V^{m-1}[a, b]$ $((\bullet)_q = (\bullet)_{L_q[a, b]})$

$$E(W_1^m[a, b], \mathcal{N})_q \le E(W_V^{m-1}[a, b], \mathcal{N})_q. \tag{5.8}$$

On the other hand, if $x \in W_V^{m-1}[a, b]$ then its Stekhlov function x_h belongs to $W_1^m[a, b]$ (see Appendix A2) and we conclude from Proposition 1.1.1

$$E(x, \mathcal{N})_q \le E(x - x_h, \mathcal{N})_q + E(x_h, \mathcal{N})_q \le \|x - x_h\|_q + E(W_V^{m-1}[a, b], \mathcal{N})_q.$$

When $h \to 0$ from Proposition A2.1 we have $\|x - x_h\|_q \to 0$ and so

$$E(x, \mathcal{N})_q \le E(W_V^{m-1}[a, b], \mathcal{N})_q, \qquad \forall\, x \in W_V^{m-1}[a, b],$$

i.e. we get an inequality which is opposite to (5.8) and (5.6) is proved. The proof of (5.7) is similar. □

1.5.2 Best approximation of classes of functions on the interval When we consider the best approximation for functions which are defined on the interval $[a, b]$, we shall connect the approximating finite dimensional subspace $\mathcal{N}$ of $L_q[a, b]$ with the set

$$W^0_{q'}[a, b; \mathcal{N}] = \{h: h \in L_{q'}[a, b], \|h\|_{q'} \leq 1, h \perp \mathcal{N}\}, \tag{5.9}$$

where $h \perp \mathcal{N}$ means

$$\int_a^b h(t)u(t)\, \mathrm{d}t = 0, \qquad \forall\, u \in \mathcal{N}.$$

Now we can write (4.10) as

$$E(x, \mathcal{N})_q = \sup\left\{\int_a^b x(t)h(t)\, \mathrm{d}t: h \in W^0_{q'}[a, b; \mathcal{N}]\right\}. \tag{5.10}$$

If $x \in L^r_1[a, b]$ $(r = 1, 2, \ldots,)$ then Taylor's formula gives

$$x(t) = \sum_{k=0}^{r-1} \frac{x^{(k)}(a)}{k!} t^k + \frac{1}{(r-1)!}\int_a^b (t-u)_+^{r-1} x^{(r)}(u)\, \mathrm{d}u, \tag{5.11}$$

where

$$(t-u)_+^\nu = \begin{cases} (t-u)^\nu, & t \geq u, \\ 0, & t < u, \end{cases} \qquad \nu = 0, 1, \ldots$$

Let $m \geq r - 1$ and $P_m \subset \mathcal{N}$ which implies that for every $h \in W^0_{q'}[a, b; \mathcal{N}]$ we have $h \perp P_m$. Then the integral in (5.10) for $x \in L^r_1[a, b]$ can be represented using (5.11) as

$$\int_a^b x(t)h(t)\, \mathrm{d}t = \frac{1}{(r-1)!}\int_a^b \left[\int_a^b (t-u)_+^{r-1} h(t)\, \mathrm{d}t\right] x^{(r)}(u)\, \mathrm{d}u.$$

From

$$(t-u)_+^\nu = (t-u)^\nu - (-1)^\nu (u-t)_+^\nu$$

using once more the orthogonality $h \perp P_m$ $(m \geq r - 1)$ we obtain

$$\int_a^b x(t)h(t)\, \mathrm{d}t = (-1)^r \int_a^b \psi(u)x^{(r)}(u)\, \mathrm{d}u. \tag{5.12}$$

where

$$\psi(t) = \frac{1}{(r-1)!}\int_a^b (t-v)_+^{r-1} h(v)\, \mathrm{d}v. \tag{5.13}$$

We now need the following simple but useful lemma:

Lemma 1.5.2 Let $m = 0, 1, \ldots$, be fixed and $h \in L_p[a, b]$ $(1 \le p \le \infty)$. If

$$g(t) = \frac{1}{m!} \int_a^b (t-u)_+^m h(u)\, du, \qquad a \le t \le b, \tag{5.14}$$

then

$$g \in L_p^{m+1}[a, b],\ g^{(m+1)}(t) = h(t) \text{ a.e.}$$

The conditions

$$g^{(k)}(a) = g^{(k)}(b) = 0, \qquad k = 0, 1, \ldots, m, \tag{5.15}$$

are satisfied if and only if $h \perp P_m$.

Indeed from

$$g^{(k)}(t) = \frac{1}{(m-k)!} \int_a^b (t-u)_+^{m-k} h(u)\, du, \qquad k = 0, 1, \ldots, m, \tag{5.16}$$

we obtain $g^{(m+1)}(t) = h(t)$ a.e. and so $g \in L_p^{m+1}[a, b]$. From $u \ge a$ we have $(a-u)_+^{\nu} = 0$ and hence $g^{(k)}(a) = 0$ $(k = 0, 1, \ldots, m)$. If $h \perp P_m$ then $g^{(k)}(b) = 0$ $(k = 0, 1, \ldots, m)$ because $(b-u)_+^{m-k} = (b-u)^{m-k}$ for every $u \in [a, b]$. Conversely, if (5.15) is fulfilled then we have for every $p \in P_m$

$$\int_a^b h(t)p(t)\, dt = \int_a^b g^{(m+1)}(t)p(t)\, dt = (-1)^{m+1} \int_a^b g(t)p^{(m+1)}(t)\, dt = 0.$$

Going back to (5.12) and comparing (5.13) with (5.16) we see that $\psi(t) = g^{(m-r+1)}(t)$ where g is the function (5.14). If for $m = 0, 1, \ldots$, we set

$$\begin{aligned} W_p^{m+1}[a, b; \mathcal{N}]_0 = \{g\colon g \in W_p^{m+1}[a, b],\ g^{(m+1)} \perp P_m, \\ g^{(k)}(a) = g^{(k)}(b) = 0,\ k = 0, 1, \ldots, m\} \end{aligned} \tag{5.17}$$

then it is obvious that there is a one-to-one correspondence between the set of $(m - r + 1)$th derivatives of functions from $W_p^{m+1}[a, b; \mathcal{N}]_0$ and the set $W_p^0[a, b; \mathcal{N}]$.

Therefore, under the assumptions $P_m \subset \mathcal{N}$, $x \in L_1^r[a, b]$ and $r \le m + 1$, equality (5.10) can be rewritten in the form

$$E(x, \mathcal{N})_q = \sup \left\{ \int_a^b x^{(r)}(t) g^{(m-r+1)}(t)\, dt\colon g \in W_{q'}^{m+1}[a, b; \mathcal{N}]_0 \right\}. \tag{5.18}$$

Up to now we have simply transformed the right-hand side of (5.10) using the facts that $x \in W_p^r[a, b]$ and $P_m \subset \mathcal{N}$ $(m \ge r - 1)$. Now we can take the supremum on these with $x \in L_p^r[a, b]$ and $\|x^{(r)}\|_p \le 1$ in (5.18) and obtain an expression for $E(W_p^r[a, b], \mathcal{N})_q$. We have to use the relation (Appendix A1)

$$\sup_{\|\psi\|_p \le 1} \int_a^b \psi(t)\varphi(t)\, dt = \|\varphi\|_{p'}, \qquad \varphi \in L_{p'}[a, b],\ \frac{1}{p} + \frac{1}{p'} = 1. \tag{5.19}$$

Applying (5.19) in the integral in (5.18) we obtain:

Proposition 1.5.3 If the finite dimensional subspace $\mathcal{N}$ of $L_q[a, b]$ contains the algebraic polynomials of degree m, then we have for $r = 1, 2, \ldots, m+1$

$$E(W_p^r[a, b], \mathcal{N})_q = \sup\{\|g^{(m-r+1)}\|_{p'}: g \in W_{q'}^{m+1}[a, b; \mathcal{N}]_0\},$$
$$1 \leq p, q \leq \infty, \qquad 1/p + 1/p' = 1/q + 1/q' = 1. \quad (5.20)$$

1.5.3 The periodic case Now $\|\bullet\|_p$ and $E(x, \mathcal{N})_p$ stand for $\|\bullet\|_{L_p[0,2\pi]}$ and $E(x, \mathcal{N})_{L_p[0,2\pi]}$. If $\mathcal{N}$ is a finite dimensional subspace of L_q and $x \in L_q$ then Proposition 1.4.2 gives

$$E(x, \mathcal{N})_q = \sup\left\{\int_0^{2\pi} x(t)h(t)\,\mathrm{d}t: h \in W_{q'}^0(\mathcal{N})\right\}, \tag{5.21}$$

where

$$W_p^0(\mathcal{N}) = \{h: h \in W_p^0, h \perp \mathcal{N}\}, \qquad W_p^0 = \{h: h \in L_p, \|h\|_p \leq 1\} \tag{5.22}$$

and $h \perp \mathcal{N}$ means

$$\int_0^{2\pi} h(t)u(t)\,\mathrm{d}t = 0, \qquad \forall\, u \in \mathcal{N}.$$

Under the assumption that $\mathcal{N}$ contains the constants, we obtain from $h \perp \mathcal{N}$

$$\int_0^{2\pi} h(t)\,\mathrm{d}t = 0,$$

which allows us to consider $h(t) = g^{(m)}(t)$, $g \in L_{p'}^m$. If $x \in L_1^m$ $(m = 1, 2, \ldots)$ then, using integration by parts and the periodicity, we obtain

$$\int_0^{2\pi} x(t)h(t)\,\mathrm{d}t = (-1)^m \int_0^{2\pi} x^{(m)}(t)g(t)\,\mathrm{d}t \tag{5.23}$$

and $g \in W_{q'}^m(\mathcal{N})$, where

$$W_p^m(\mathcal{N}) = \{\mathrm{g}: \mathrm{g} \in \mathrm{L}_p^m(\mathcal{N}), \|g^{(m)}\|_p \leq 1, g^{(m)} \perp \mathcal{N}\}. \tag{5.24}$$

In (5.24) h determines g up to an additive constant, but this is not important because $x^{(m)} \perp 1$. For definiteness we may also assume that $g(0) = 0$ or $g \perp 1$. Taking into account that $W_p^m(\mathcal{N})$ contains the function $-g$ together with g, we obtain from (5.21) and (5.23)

$$E(x, \mathcal{N})_q = \sup\left\{\int_0^{2\pi} x^{(m)}(t)g(t)\,\mathrm{d}t: g \in W_{q'}^m(\mathcal{N})\right\}. \tag{5.25}$$

Taking a supremum on $x \in W_p^m$ in (5.25) we cannot use (5.19) because now we have the additional restriction $x^{(m)} \perp 1$. But we are under the conditions of Corollary 1.4.3 which, with the notation

$$E_1(g)_p = \inf_{\lambda \in \mathbb{R}} \|g - \lambda\|_p,$$

together with (5.25) gives:

Proposition 1.5.4 If the finite dimensional subspace $\mathcal{N}$ of $L_q[a, b]$ contains constants, then we have for $m = 1, 2, \ldots,$

$$E(W_p^m, \mathcal{N})_q = \sup \{E_1(g)_{p'}: g \in W_{q'}^m(\mathcal{N})\},$$
$$1 \le p, q \le \infty, \qquad 1/p + 1/p' = 1/q + 1/q' = 1, \quad (5.26)$$

where $W_{q'}^m(\mathcal{N})$ is determined by (5.24).

1.5.4 Duality for convolution classes Using duality we can treat large classes of functions defined by convolutions. Let K and φ be locally summable 2π-periodic functions, i.e. $K \in L_1$, $\varphi \in \mathrm{L}_1$. The function

$$x(t) = \int_0^{2\pi} K(t-u)\varphi(u)\, \mathrm{d}u = \int_0^{2\pi} K(u)\varphi(t-u)\, du \qquad (5.27)$$

is called a convolution of the functions K and φ and denoted by $x = K * \varphi$. Obviously x is also 2π-periodic. The dependence of x on the properties of K and φ is clarified by:

Proposition 1.5.5 If $\varphi \in L_p$ $(1 \le p \le \infty)$, then $x \in L_p$ and

$$\|x\|_p \le \|K\|_1 \|\varphi\|_p \qquad (5.28)$$

and in the case $\varphi \in \mathrm{L}_\infty$ the convolution x is continuous on $\mathbb{R}$. If $\varphi \in L_p$ $(1 < p < \infty)$, $K \in L_{p'}$ $(p' = p/(p-1))$ then x is continuous and

$$\|x\|_C \le \|K\|_{p'} \|\varphi\|_p. \qquad (5.29)$$

Proof If $\varphi \in L_p$ $(1 \le p < \infty)$ then Minkowski's general inequality (A1.19) gives

$$\|x\|_p \le \int_0^{2\pi} \left(\int_0^{2\pi} |K(u)\varphi(t-u)|^p\, \mathrm{d}t \right)^{1/p} \mathrm{d}u$$

$$= \int_0^{2\pi} |K(u)| \left(\int_0^{2\pi} |\varphi(t-u)|^p\, \mathrm{d}t \right)^{1/p} \mathrm{d}u = \|K\|_1 \|\varphi\|_p.$$

In the last inequality we used $\|\varphi\|_p = \|\varphi(\bullet - u)\|_p$ for every u because of the periodicity of φ.

Now let $\varphi \in \mathrm{L}_\infty$. For every t and h we have

$$|x(t+h) - x(t)| = \left| \int_0^{2\pi} [K(t+h-u) - K(t-u)]\varphi(u)\, \mathrm{d}u \right|$$
$$\le \|\varphi\|_\infty \|K(\bullet + h) - K(\bullet)\|_1$$

and the continuity of x follows because the translation is a continuous operator in L_p $(1 \le p < \infty)$, i.e. for $g \in L_p$ we have (see e.g. [15B, p. 407])

$$\lim_{h \to 0} \|g(\bullet + h) - g(\bullet)\|_p = 0. \tag{5.30}$$

It is also obvious that $\|x\|_C \le \|K\|_1 \|\varphi\|_\infty$.

Finally if $\varphi \in L_p$ $(1 < p < \infty)$, $K \in L_{p'}$ then Hölder's inequality (A1.1) and (5.30) give

$$|x(t+h) - x(t)| \le \|K\|_{p'} \|\varphi(\bullet + h) - \varphi(\bullet)\|_p \to 0$$

when $h \to 0$. We obtain (5.29) if we apply Hölder's inequality directly to (5.27).

In the convolution $K * \varphi$ one of the functions is usually assumed to be fixed and is named a *kernel*: if different restrictions are put on the second function different classes of functions are obtained. The properties of these classes depend both on the restrictions on the second function and on the properties of the kernel.

If we consider K to be the kernel of convolution (5.27) we introduce the following classes:

$$K * W_p^0 = \{x = K * \varphi \colon \varphi \in L_p, \|\varphi\|_p \le 1\},$$
$$K * W_p^0(\mathcal{N}) = \{x = K * \varphi \colon \varphi \in W_p^0(\mathcal{N})\},$$

where the sets W_p^0, $W_p^0(\mathcal{N})$ are defined in (5.22).

Theorem 1.5.6 Let $\mathcal{N}$ be a subspace of L_q $(1 \le q \le \infty)$, let x be the convolution (5.27) and let at least one of the following conditions hold:

(1) K is either an even or an odd function;
(2) If $\psi(t)$ belongs to $\mathcal{N}$ then $\psi(-t)$ also belongs to $\mathcal{N}$.

Then for $1 \le q \le p$ we have

$$E(K * W_p^0, \mathcal{N})_q = \sup \{\|x\|_{p'} \colon x \in K * W_{q'}^0(\mathcal{N})\},$$
$$1/p + 1/p' = 1/q + 1/q' = 1. \tag{5.31}$$

If $K \in L_q \cap L_{p'}$ (in particular when $K \in L_\infty$), then (5.31) is fulfilled for every p and q.

Proof Let $x \in K * W_p^0$. Then Proposition 1.5.5 gives $x \in L_p$ and so $x \in L_q$ if $p \ge q$; the last inclusion is also correct when $K \in L_q$. Using Proposition 1.4.1 (see (4.7)) we can write

$$E(x, \mathcal{N})_q = \sup_{h \in W_{q'}^0(\mathcal{N})} \int_0^{2\pi} x(t) h(t) \, dt, \tag{5.32}$$

and hence

$$E(K*W_p^0,\mathcal{N})_q=\sup_{\|\varphi\|_p\le 1}\ \sup_{h\in W_{q'}^0(\mathcal{N})}\int_0^{2\pi}\left(\int_0^{2\pi}K(u)\varphi(t-u)\,du\right)h(t)\,dt. \quad (5.33)$$

Now we are going to apply Fubini's theorem (see e.g. [12B, p. 336]).

Consider the integral

$$\int_0^{2\pi}|K(u)|\left(\int_0^{2\pi}|\varphi(t-u)h(t)|\,dt\right)du. \quad (5.34)$$

If $q\le p$ then $\varphi\in L_q$ which together with $h\in L_{q'}$ and Hölder's inequality gives

$$\int_0^{2\pi}|\varphi(t-u)h(t)|\,dt\le\|\varphi\|_q\|h\|_{q'};$$

so the integral (5.34) is finite. If $K\in L_q$ then

$$\int_0^{2\pi}|K(t-u)h(t)|\,dt\le\|K\|_q\|h\|_{q'}$$

and for every $\varphi\in L_1$ the integral

$$\int_0^{2\pi}|\varphi(u)|\left(\int_0^{2\pi}|K(t-u)h(t)|\,dt\right)du \quad (5.35)$$

is also finite.

The finiteness of (5.34) for $q\le p$ and (5.35) for $K\in L_q$ allows us to apply Fubini's theorem and interchange the order of integration in (5.33):

$$E(K*W_p^0,\mathcal{N})_q=\sup_{h\in W_{q'}^0(\mathcal{N})}\ \sup_{\|\varphi\|_p\le 1}\int_0^{2\pi}\varphi(u)\psi(u)\,du, \quad (5.36)$$

where

$$\psi(u)=\int_0^{2\pi}K(t-u)h(t)\,dt. \quad (5.37)$$

Using generalized Minkowski's inequality we easily see (cf. the proof of Proposition 1.5.5) that $\psi\in L_p$ whenever K or h belongs to L_p. Therefore either $K\in L_q\cap L_{p'}$ or $K\in L_1$ but $p\ge q$ (so $h\in L_{q'}\subset L_{p'}$) implies $\psi\in L_{p'}$. Then (5.19) gives

$$\sup_{\|\varphi\|_p\le 1}\int_0^{2\pi}\varphi(u)\psi(u)\,du=\|\psi\|_{p'}. \quad (5.38)$$

Let condition (1) be satisfied. Then function (5.37) is the convolution $\psi=\pm K*h$ and the replacement of (5.38) in (5.36) gives (5.31).

Let condition (2) be satisfied. Then for every function $\psi \in \mathcal{N}$

$$\int_0^{2\pi} \psi(t)h(-t)\,\mathrm{d}t = \int_0^{2\pi} \psi(-t)h(t) = 0,$$

i.e. both functions $h(t)$ and $h(-t)$ belong to $W^0_{q'}(\mathcal{N})$. But then using (5.38) and (5.37) we obtain

$$E(K * W^0_p, \mathcal{N})_q = \sup_{h \in W^0_{q'}(\mathcal{N})} \|\psi\|_{p'} = \sup_{h \in W^0_{q'}(\mathcal{N})} \|\psi(-\bullet)\|_{p'}$$

$$= \sup_{h \in W^0_{q'}(\mathcal{N})} \left\| \int_0^{2\pi} K(\bullet - t)h(t)\,\mathrm{d}t \right\|_{p'}. \qquad (5.39)$$

We have a convolution of K with $h \in W^0_p(\mathcal{N})$ inside the norm in the last expression and so the right-hand sides of (5.39) and (5.31) coincide and the theorem is proved. □

Corollary 1.5.7 If either condition (1) or (2) of Theorem 1.5.6 is satisfied then

$$E(K * W^0_p, \mathcal{N})_p = \sup\{\|x\|_{p'}: x \in K * W^0_{p'}(\mathcal{N})\},\ 1 \le p \le \infty,$$
$$1/p + 1/p' = 1. \qquad (5.40)$$

Remark If $K \perp 1$, i.e. $\int_0^{2\pi} K(t)\,\mathrm{d}t = 0$, then the convolution class $K * W^0_p$ does not contain the constants and it is natural to consider the class $K * W^0_p$ together with class $\{a\} + K * W^0_{p,0}$, where $a \in \mathbb{R}$ and $W^0_{p,0} = \{\varphi: \varphi \in W^0_p, \varphi \perp 1\}$. Classes W^r_p defined in Section 1.5.1 are exactly of this kind (see formula (5.1)). If the finite dimensional subspace $\mathcal{N}$ contains the constants then using Corollary 1.4.3 we can write instead of (5.31)

$$E(\{a\} + K * W^0_{p,0}, \mathcal{N})_q = \sup\{E_1(x)_{p'}: x = K * \varphi, \varphi \in W^0_{q',0}, \varphi \perp \mathcal{N}\}.$$

In particular if $K(t) = B_m(t)$ (see (5.2)) then we obtain (5.26).

1.5.5 Generalization for the non-symmetric case Theorem 1.5.6 can be generalized for the non-symmetric case both in the definition of the convolution class and the functional of best approximation.

For fixed $K \in L_1$ and $F \subset L_1$ we have denoted by $K * F$ the class of all convolutions $x = K * \varphi$, where $\varphi \in F$. In the previous section F was, for example, the unit ball in L_p; now it will be the non-symmetric set

$$F_{p;\gamma,\delta} = \{x: x \in L_p, \|\gamma x_+ + \delta x_-\|_p \le 1\}, \qquad \gamma > 0, \delta > 0.$$

As in Section 1.4.5 we set $\|x\|_{p;\alpha,\beta} = \|\alpha x_+ + \beta x_-\|_p$ and for $x \in L_q$ and $\mathcal{N} \subset L_q$ we define

$$E(x, \mathcal{N})_{q;\alpha,\beta} = \inf_{u \in \mathcal{N}} \|x - u\|_{q;\alpha,\beta}, \qquad \alpha > 0, \beta > 0.$$

Proposition 1.5.8 Let $K \in L_1$, $1 \le p \le \infty$, $1 \le q < \infty$, $\alpha, \beta, \gamma, \delta$ be positive numbers, let $\mathcal{N}$ be a subspace of L_q. Then for $q \le p$ we have

$$E(K * F_{p;\gamma,\delta}, \mathcal{N})_{q;\alpha,\beta} = \sup\{\|K(-\bullet) * \varphi\|_{p';\gamma^{-1},\delta^{-1}} \colon \varphi \in \mathcal{N}^{\perp}_{q';\alpha^{-1},\beta^{-1}}\}, \tag{5.41}$$

where

$$\mathcal{N}^{\perp}_{p;\alpha,\beta} = \{g \colon g \in L_p, \|g\|_{p;\alpha,\beta} \le 1, g \perp \mathcal{N}\}.$$

If $K \in L_q \cap L_{p'}$, then (5.41) holds no matter what the relation between p and q is. If $\psi(-t)$ belongs to $\mathcal{N}$ together with $\psi(t)$ then we can replace $K(-\bullet) * \varphi$ by $K * \varphi$ in the right-hand side of (5.41).

Proof From Proposition 1.4.9 for $x \in L_q$ we have

$$E(x, \mathcal{N})_{q;\alpha,\beta} = \sup\left\{\int_0^{2\pi} x(t)g(t)\,\mathrm{d}t \colon g \in \mathcal{N}^{\perp}_{q';\alpha^{-1},\beta^{-1}}\right\} \tag{5.42}$$

Then we have

$$\begin{aligned}
E(K * F_{p;\gamma,\delta}, \mathcal{N})_{q;\alpha,\beta} &=: \sup\{E(x,\mathcal{N})_{q;\alpha,\beta} \colon x \in K * F_{p;\gamma,\delta}\} \\
&= \sup_{\|\varphi\|_{p;\alpha,\beta} \le 1} \sup_{g \in \mathcal{N}^{\perp}_{q';\alpha^{-1}\beta^{-1}}} \int_0^{2\pi} g(t)\left(\int_0^{2\pi} K(t-u)\varphi(u)\,\mathrm{d}u\right)\mathrm{d}t \\
&= \sup_g \sup_\varphi \int_0^{2\pi} \varphi(t)\left(\int_0^{2\pi} K(t-u)g(u)\,\mathrm{d}u\right)\mathrm{d}t \\
&= \sup_g \|K(-\bullet) * g\|_{p';\gamma^{-1},\delta^{-1}}.
\end{aligned} \tag{5.43}$$

The last inequality follows from (A1.16). The validity of the last statement in Proposition 1.5.8 follows as in Theorem 1.5.6. □

Remark If $K \perp 1$ then as in the symmetric case it makes sense to consider $K * F$ together with the class $\{a\} + K * \hat{F}$, where $a \in \mathbb{R}$ and $\hat{F} = \{\varphi \colon \varphi \in F, \varphi \perp 1\}$. Under the assumption that the approximating space $\mathcal{N}$ contains the constants, if we repeat the arguments from (5.43) but using at the end (5.42) with $\mathcal{N} = \{a\}$, $a \in \mathbb{R}$, instead of (A1.16) we obtain

$$E(\{a\} + K * \hat{F}_{p;\gamma,\delta}, \mathcal{N})_{q;\alpha,\beta} = \sup\{E_1(K(-\bullet) * g)_{p';\gamma^{-1},\delta^{-1}} \colon g \in \mathcal{N}^{\perp}_{q';\alpha^{-1},\beta^{-1}}\}, \tag{5.44}$$

where $E_1(\bullet)_{p;\alpha,\beta}$ is the best (α,β)-approximation by constants in L_p.

The corresponding relationships for the best one-sided approximation can be obtained by taking the limit in Proposition 1.5.8 and using the following general fact.

Proposition 1.5.9 Let $\mathcal{M}, \mathcal{N} \subset L_p$ $(1 \le p < \infty)$ and let $\mathcal{N}$ be locally compact. Then

$$\lim_{\beta\to\infty} \sup_{x\in\mathcal{M}} E(x,\mathcal{N})_{p;1,\beta} = \sup_{x\in\mathcal{M}} E^+(x,\mathcal{N})_p,$$
$$\lim_{\alpha\to\infty} \sup_{x\in\mathcal{M}} E(x,\mathcal{N})_{p;\alpha,1} = \sup_{x\in\mathcal{M}} E^-(x,\mathcal{N})_p. \tag{5.45}$$

Proof We have deduced the corresponding relationships for individual functions (see Theorem 1.4.10). Let

$$E^+(\mathcal{M},\mathcal{N})_p =: sup\,\{E^+_{(x,\,\mathcal{N})p}\colon x\in\mathcal{M}\} = A < \infty.$$

Then for every $\varepsilon > 0$ there is $x_\varepsilon \in \mathcal{M}$ and a number $\beta_\varepsilon > 0$ such that for $\beta \geq \beta_\varepsilon$ we have

$$E(\mathcal{M},\mathcal{N})_{p;1,\beta} =: \sup_{x\in\mathcal{M}} E(x,\mathcal{N})_{p;1,\beta} \geq E(x_\varepsilon,\mathcal{N})_{p;1,\beta} > A - \varepsilon.$$

because of Theorem 1.4.10. On the other hand, for every $x \in \mathcal{M}$ we have

$$E(x,\mathcal{N})_{p;1,\beta} \leq E^+(x,\mathcal{N})_p \leq A,$$

i.e. $E(\mathcal{M},\mathcal{N})_{p;1,\beta} \leq A$.

If $E^+(\mathcal{M},\mathcal{N})_p = \infty$ then for every $N > 0$ for some function $x_N \in \mathcal{M}$ we have $E^+(x_N,\mathcal{N})_p \geq 2N$ and for $\beta \geq \beta_N$ we have

$$E(\mathcal{M},\mathcal{N})_{p;1,\beta} \geq E(x_N,\mathcal{N})_{p;1,\beta} > N,$$

i.e. $E(\mathcal{M},\mathcal{N})_{p;1,\beta} \to \infty$ if $\beta \to \infty$. We can prove (5.45) in the same manner. □

Therefore from Proposition 1.5.8 with $\alpha = 1$ and $\beta \to \infty$ or $\beta = 1$ and $\alpha \to \infty$ we establish:

Proposition 1.5.10 Let $K \in L_1$, $1 \leq p \leq \infty$, $q \geq 1$, let φ, δ be positive numbers, $\mathcal{N}$ be a finite subspace of L_p. Then under the additional conditions of Proposition 1.5.8 we have

$$E^\pm(K * F_{p;\gamma,\delta},\mathcal{N})_q = \sup\,\{\|K(-\bullet) * \varphi\|_{p';\gamma^{-1},\delta^{-1}}\colon \varphi \in L_{q'}, \|\varphi\|_{q'} \leq 1, \varphi \perp \mathcal{N}\}. \tag{5.46}$$

If $K(t) \perp 1$ then the remark after Proposition 1.5.8 is also correct for one-sided approximations.

Comments

Section 1.1 A more complete presentation of the problems of the existence and uniqueness of the element of best approximation in linear normed spaces can be found in the books of Tikhomirov [25A], Kiesewetter [8A] and Singer [22A]. For strictly normalized spaces see also Akhiezer [2A]. The properties of the Chebyshev system of functions are considered in the monograph by Dzyadyk [6A].

Section 1.2 The range of the extremal problems in approximation theory can be essentially extended by considering, for example, other types of N-widths. It seems that the most complete information on this subject can be found in [25A]. The N-width problem (2.8) was formulated by Kolmogorov [2], the notion for a linear N-width was introduced by Tikhomirov [2].

Section 1.3 Theorem 1.3.1 seems to be the most important case of general duality relations in convex analysis which can be found, for example, in the paper by Ioffe and Tikhomirov [1] and also in their monograph [8B] (see also the papers by Garkavi [1, 3]). The simple proof of Theorem 1.3.1, using only the separation theorem, was communicated to the author by Tikhomirov. Theorem 1.3.2 belongs to V. F. Babenko [4].

Non-symmetric analogs of Theorem 1.3.1 were also considered in the book by Korneichuk, Ligun and Doronin [11A].

The monograph by Gol'shtein (6B] is devoted to the problems for duality of extremal problems. Theorem 1.3.4 is proved by S. M. Nikol'skii [6]. For the characterization of the element of best approximation in convex sets see Garkavi [2, 3], Nikol'skii [1] and Havinson [1].

Section 1.4 Statements 1.4.2 and 1.4.3 for $p = 1$ and $p = \infty$ were first proved by S. M. Nikol'skii [6] who was the first to apply duality relationships effectively in solving extremal problems in approximation theory. Proposition 1.4.6 for $p = 1$ was obtained by Markov [1, 11B]. Duality in L_p for the best one-sided approximation of a fixed element is considered in [14B] and for (α,β)-approximations see V. F. Babenko [3, 4].

Section 1.5 Duality relationships for convolution classes were first considered by S. M. Nikol'skii [6]. The non-symmetric case for classes of functions is considered in [14B] (one-sided approximations) and in V. F. Babenko's papers [3–7].

Exercises

1. Prove that every closed set $\mathcal{N}$ in the normed space X is an existence set if and only if X is reflexive (i.e. $(X^*)^* = X$).

2. Show that the set of all functions x from $C[0,1]$ which satisfy $\int_0^1 x(t)\,\mathrm{sgn}\cos \pi t\,\mathrm{d}t = 0$ is convex and closed but not an existence set.

3. If $\mathcal{N}$ is a convex set in the normed space X then prove that the set $\mathcal{N}_x$ of the elements of best approximation of x in $\mathcal{N}$ is also convex.

4. Prove that the operator of best approximation by constants in $C[a, b]$ is not additive.

5. Show that $C[a, b]$ and $L_1[a, b]$ are not strictly normalized.

6. Prove that for X reflexive with A a convex subset of X^* and every $f \in X^*$ we have

$$\inf_{\varphi \in A} \|f - \varphi\| = \sup \{[f(x) - \sup_{\varphi \in A} \varphi(x)]: x \in X, \|x\| \leq 1\}.$$

7. Show that the statement of Exercise 6 may not be correct in a non-reflexive space. Hint: $X = L_1[-1, 1]$, $A = C[-1, 1]$.

8. Let p_{n_1} and p_{n_2} be the algebraic polynomials of best approximation in $L_p[a, b]$ $(1 < p < \infty)$ for the function $x \in L_p[a, b]$ in the space of all algebraic polynomials of degree n_1 and n_2 respectively. Prove that the difference $p_{n_1} - p_{n_2}$ is either identically zero or has at least $\min\{n_1, n_2\} + 1$ changes of the sign on $[a, b]$.

2

Polynomials and spline functions as approximating tools

In real situations the problem of approximating the function $f(t)$ consists of replacing it following a given rule by a closed (in different senses) function $\varphi(t)$ from an *a priori* fixed set $\mathcal{N}$ and estimating the error. The final result and the difficulties in obtaining it depend heavily on the choice of the approximating set $\mathcal{N}$ and the method of approximation, i.e. rules which determine how the function φ corresponds to f.

In choosing the approximating set, besides ensuring the necessary precision, one also needs to have functions φ which are simple and easy to study and calculate. Algebraic polynomials and (in the periodic case) trigonometric polynomials possess the simplest analytic structure. The main focus in approximation theory since it became an independent branch of analysis has been directed at the problem of approximation by polynomials. But, around the 1960s, spline functions started to play a more prominent role in approximation theory. They have definite advantages, in comparison with polynomials, for computer realizations and, moreover, it turns out that they are the best approximation tool in many important cases.

In this chapter we give the general properties of polynomials and polynomial splines that it will be necessary to know in the rest of the book. Let us note first that this introductory material is different in nature for polynomials and splines. The reason lies not only in their differing structures but also in the fundamental difference in the linear methods used for polynomial or spline approximation. Considering polynomials (algebraic or trigonometric), our main attention is concentrated, besides the classical Chebyshev theorem, on the linear methods based on the Fourier series and their analogs. For spline functions, we have to go deeper into non-trivial and important interpolation problems of existence and uniqueness of interpolants. Moreover, we decided that it was also appropriate to give some information about their analytic representations.

2.1 Polynomials of best approximation

2.1.1 Existence and uniqueness In the spaces $C[a, b]$ and $L_p[a, b]$ $(1 \le p \le \infty)$ the set of all algebraic polynomials

$$p(t) = \sum_{k=0}^{m} \alpha_k t^k \tag{1.1}$$

of fixed degree m is a linear manifold $\mathcal{N}_{m+1}^{A}$ of dimension $m+1$, $\dim \mathcal{N}_{m+1}^{A} = m+1$. In the spaces C and L_p of 2π-periodic functions by $\mathcal{N}_{2m+1}^{T}$ we shall denote the linear manifold of trigonometric polynomials

$$\tau(t) = \frac{\alpha_0}{2} + \sum_{k=1}^{m} (\alpha_k \cos kt + \beta_k \sin kt) \tag{1.2}$$

of degree m; its dimension is $2m + 1$ because the functions

$$1, \cos t, \sin t, \cos 2t, \sin 2t, \ldots, \cos mt, \sin mt \tag{1.3}$$

are linearly independent. Indeed, if

$$\frac{\alpha_0}{2} + \sum_{k=1}^{m} (\alpha_k \cos kt + \beta_k \sin kt) \equiv 0,$$

then multiplying both sides by the functions from system (1.3) and integrating from 0 to 2π we get that all coefficients α_k and β_k are zero.

If we mention the set of algebraic polynomials of degree m we always assume that it contains all polynomials of degree $n < m$, i.e. polynomials of the type (1.1) such that $\alpha_k = 0, n < k \le m$. The same convention is assumed for trigonometric polynomials too.

Applying Corollary 1.1.4 for subspaces $\mathcal{N}_{m+1}^{A}, \mathcal{N}_{2m+1}^{T}$ we obtain:

Proposition 2.1.1 For every $m = 0, 1, \ldots,$ and for every function $f \in C[a, b]$ or $L_p[a, b]$ $(1 \le p \le \infty)$ there is an algebraic polynomial of best approximation from $\mathcal{N}_{m+1}^{A}$ in the metric $C[a, b]$ or $L_p[a, b]$. Similarly, for $m = 0, 1, \ldots,$ and for every function $f \in C$ or L_p there is a trigonometric polynomial of best approximation from $\mathcal{N}_{2m+1}^{T}$ in C or L_p respectively.

The uniqueness of the polynomial of best approximation from $\mathcal{N}_{m+1}^{A}$ or $\mathcal{N}_{2m+1}^{T}$ in the spaces $L_p[a, b]$ or L_p for $1 < p < \infty$ follows from the strict convexity of the norm (see Proposition 1.1.5). In the spaces $C[a, b]$ and C, whose norm is not strictly convex, the uniqueness of the element of best approximation in $\mathcal{N}_{m+1}^{A}$ or $\mathcal{N}_{2m+1}^{T}$ follows from the inner properties of these subspaces. A. Haar's theorem [2A, p. 80] gives that, if $\{\varphi_1, \varphi_2, \ldots, \varphi_N\}$ is a linearly independent system of functions in $C[a, b]$, then the set of generalized polynomials

$$q(t) = \sum_{k=1}^{N} \alpha_k \varphi_k(t) \tag{1.4}$$

is a Chebyshev system in $C[a, b]$ if and only if every non-zero generalized polynomial (1.4) has no more than $N - 1$ zeros on $[a, b]$. A similar fact holds in the trigonometric case in C, but the zeros have to be counted on the period, i.e. on $[0, 2\pi)$. It is known (see e.g. [7A, p. 22]) that the algebraic polynomial (1.1), $p \neq 0$, has no more than m zeros and the trigonometric polynomial (1.2), $\tau \neq 0$, has on the period $[0, 2\pi)$ not more than $2m$ zeros. Therefore subspaces $\mathcal{N}^{\mathrm{A}}_{m+1}$ and $\mathcal{N}^{\mathrm{T}}_{2m+1}$ are Chebyshev sets in $C[a, b]$ and C respectively, i.e. in each of them the polynomial of best approximation is unique. However, we shall obtain these properties as simple corollaries from Chebyshev's theorem which is proved later.

In spaces $L_1[a, b]$ and L_1, which are also not strictly normalized, the uniqueness of the polynomials of best approximation from $\mathcal{N}^{\mathrm{A}}_{m+1}$ and $\mathcal{N}^{\mathrm{T}}_{2m+1}$ is guaranteed if the approximated function is continuous (see e.g. [2A, p. 91]).

2.1.2 Chebyshev and de la Vallée–Poussin theorems Let us turn our attention to the characterization of the polynomials of best approximation. In $L_p[a, b]$ $(1 \leq p < \infty)$ for $\mathcal{N} = \mathcal{N}^{\mathrm{A}}_{m+1}$ we cannot add anything (in the sense of making it more specific) to the general criterion (Theorem 1.4.5). We shall also use Theorem 1.4.5 for the characterization of the trigonometric polynomial of best approximation in L_p $(1 \leq p < \infty)$. Concerning the uniform metric the most powerful criterion for the polynomial of best approximation still remains the classical alternating Chebyshev theorem [2B]. A detailed proof of this theorem will be given in the trigonometric case.

Theorem 2.1.2 A necessary and sufficient condition for the trigonometric polynomial τ_* of degree $n - 1$ to be a polynomial of best approximation to $f \in C$ in $\mathcal{N}^{\mathrm{T}}_{2n-1}$ is the existence of $2n$ points t_k: $0 \leq t_1 < t_2 < \cdots < t_{2n} < 2\pi$ in which the modulus of the difference $\delta(t) = f(t) - \tau_*(t)$ attains the maximal value $\|\delta\|_C$ with alternating signs, i.e.

$$\delta(t_1) = -\delta(t_2) = \delta(t_3) = -\delta(t_4) = \cdots = -\delta(t_{2n}) = \pm\|\delta\|_C. \tag{1.5}$$

Proof The necessity follows by contradiction: if the difference δ has less than $2n$ alternations on the period $[0, 2\pi)$ then we can decrease the deviation $\|\delta\|_C$ of τ_* from f by adding to τ_* an appropriate polynomial of degree $n - 1$.

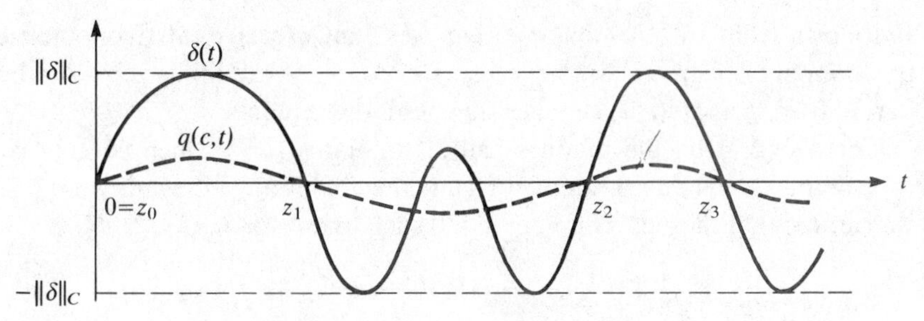

Fig. 2.1 To the proof of theorem 2.1.2.

Although this is a simple idea, its realization in strict mathematical language needs some calculations.

Let

$$\|f - \tau_*\|_C = \inf_{\tau \in \mathcal{N}^{\mathrm{T}}_{2n-1}} \|f - \tau\|_C =: d. \tag{1.6}$$

Obviously the difference $\delta = f - \tau_*$ has zeros and we may assume that $\delta(0) = \delta(2\pi) = 0$ because the conditions of the theorem are invariant under the translation of the argument. Set

$$e^+ = \{t: t \in [0, 2\pi), \delta(t) = d\},$$
$$e^- = \{t: t \in [0, 2\pi), \delta(t) = -d\}, e = e^+ \cup e^-.$$

The sets e^+ and e^- are not empty because $\mathcal{N}^{\mathrm{T}}_{2n-1}$ contains the constants.

Let us choose the points z_k: $0 = z_0 < z_1 < \cdots < z_m < z_{m+1} = 2\pi$ so that:

(1) $\delta(z_k) = 0$ (k $= 0, 1, \ldots, m$);

(2) every interval

$$\Delta_k = [z_{k-1}, z_k], \qquad k = 1, 2, \ldots, m, \tag{1.7}$$

contains at least one e-point (i.e. a point from the set e) and all points on the interval are of one and the same sign: either e^+-points or e^--points;

(3) the signs of the e-points on consecutive intervals Δ_k and Δ_{k+1} are opposite.

The necessity will be proved if we prove that the number of intervals (1.7) is not less than $2n$, i.e. that $m + 1 \geq 2n$. We shall show that $\nu \geq n$, where $\nu = [(m + 1)/2]$.

Assume that, on the contrary, $\nu \leq n - 1$ and set

$$q(c, t) = c \sin\frac{t - z_0}{2} \sin\frac{t - z_1}{2} \ldots \sin\frac{t - z_{2\nu-1}}{2}. \tag{1.8}$$

It follows from the elementary identities that $q(c, t)$ is a trigonometric polynomial of degree ν and hence $z_0, z_1, \ldots, z_{2\nu-1}$, are all its zeros on the period $[0, 2\pi)$ and $q(c, t)$ changes sign at all these points.

Let Δ^+ (Δ^-) be the union of all Δ_k intervals (1.7) which contain e^+ (e^-)-points. From the continuity of δ and the definition of the intervals (1.7) we can see that there is a number $h > 0$ such that

$$-d + h \leq \delta(t) \leq d, \qquad t \in \Delta^+, \tag{1.9}$$

$$-d \leq \delta(t) \leq d - h, \qquad t \in \Delta^-. \tag{1.10}$$

Let us choose the coefficient $c = c_*$ in (1.8) to be such that

$$\operatorname{sgn} \delta(t) = \operatorname{sgn} q(c_*, t), \qquad t \in e, \tag{1.11}$$

$$\|q(c_*, \bullet)\|_C < h/2. \tag{1.12}$$

Polynomial $\tau_*(t) + q(c_*, t)$ is of degree $n - 1$ and we shall get a contradiction with (1.6) if we prove that

$$\|f - (\tau_* + q(c_*))\|_C = \|\delta - q(c_*)\|_C < d. \tag{1.13}$$

Let $\Delta_k \in \Delta^+$. Then (1.11) gives $q(c_*, t) > 0$ in all inner points of Δ_k and, because of $\delta(z_i) = q(c_*, z_i) = 0$, (1.9) and (1.11) give

$$\max_{t \in \Delta_k} |\delta(t) - q(c_*, t)| < d. \tag{1.14}$$

The validity of the same inequality for $\Delta_k \in \Delta^-$ is obtained in a similar way using (1.10)–(1.12). Hence inequality (1.14) holds for every one of the intervals Δ_k ($k = 1, 2, \ldots, m + 1$), which is equivalent to (1.13). The necessity is proved.

The sufficiency can easily be obtained. For $f \in C$ let the polynomial $\tau_* \in \mathcal{N}^T_{2n-1}$ satisfy the condition of the theorem while (1.6) is not fulfilled, i.e. for some $\tau \in \mathcal{N}^T_{2n-1}$ we have

$$\|f - \tau\|_C < \|f - \tau_*\|_C = \|\delta\|_C, \tag{1.15}$$

where, as before, $\delta = f - \tau_*$. Then the polynomial

$$\tau_0(t) = \tau(t) - \tau_*(t) = \delta(t) - [f(t) - \tau(t)]$$

is also of degree $n - 1$ and, because of (1.15) and (1.5), takes values with alternating signs at the points $t_1, t_2, \ldots, t_{2n}$. Therefore τ_0 has at least $2n - 1$ zeros on the period $[0, 2\pi)$, which is impossible. The theorem is proved. □

For the approximation on the interval we have:

Theorem 2.1.3 The algebraic polynomial p_* of degree $n - 1$ is the polynomial of best uniform approximation for $f \in C[a, b]$ from $\mathcal{N}^A_n$ if and only if there are $n+1$ points t_k: $a \le t_1 < t_2 < \cdots < t_{n+1} \le b$ at which the difference $\delta = f - p_*$ attains the value $\pm\|\delta\|_{C[a,\,b]}$ with alternating signs.

The necessity is obtained in the same way as for Theorem 2.1.2. Consider the set e of all points on the interval $[a, b]$ in which the difference δ takes the values $\pm\|\delta\|_{C[a,\,b]}$ and choose the intervals

$$[a, z_1], [z_1, z_2], \ldots, [z_m, b]; \qquad \delta(z_k) = 0, \qquad k = 1, 2, \ldots, m,$$

containing the e-points of one and the same signs (and also the sign of e-points should change for consecutive intervals). After that, arguing by contradiction, we introduce the polynomial

$$q(c, t) = c(t - z_1)(t - z_2) \ldots (t - z_m)$$

and obtain that $m \ge n = 1$.

The sufficiency is proved as in the periodic case but using the fact that every non-zero polynomial from $\mathcal{N}_n^{\mathrm{A}}$ has no more than $n - 1$ zeros on $[a, b]$.

Remark The system of points t_k from the conditions of Theorems 2.1.2 and 2.1.3 is sometimes called the *Chebyshev alternation* for the function $f \in C$ ($f \in C[a, b]$ respectively).

Corollary 2.1.4 For every function $f \in C$ ($f \in C[a, b]$) the trigonometric (algebraic) polynomial of best uniform approximation from $\mathcal{N}_{2n-1}^{\mathrm{T}}$ ($\mathcal{N}_n^{\mathrm{A}}$) is unique.

We shall consider only the trigonometric case. Assume that we have for $f \in C$

$$E(f, \mathcal{N}_{2n-1}^{\mathrm{T}})_C = \|f - \tau_1\|_C = \|f - \tau_2\|_C, \qquad \tau_1, \tau_2 \in \mathcal{N}_{2n-1}^{\mathrm{T}}.$$

From Proposition 1.1.2 polynomial $[\tau_1(t) + \tau_2(t)]/2$ is also the polynomial of best approximation for f. By Theorem 2.1.2 there are $2n$ points t_k ($k = 1, 2, \ldots, 2n$) on the period such that

$$[f(t_k) - \tau_1(t_k)] + [f(t_k) - \tau_2(t_k)] = \pm 2E(f, \mathcal{N}_{2n-1}^{T})_C. \tag{1.16}$$

Inside each set of square brackets there is a quantity not greater in absolute value than $E(f, \mathcal{N}_{2n-1}^{\mathrm{T}})_C$. Therefore (1.16) is possible only when

$$f(t_k) - \tau_1(t_k) = f(t_k) - \tau_2(t_k) = \pm E(f, \mathcal{N}_{2n-1}^{\mathrm{T}})_C,$$

i.e. $\tau_1(t_k) = \tau_2(t_k)$ ($k = 1, 2, \ldots, 2n$). Hence the polynomial $\tau_1 - \tau_2$ of degree $n - 1$ has at least $2n$ zeros, i.e. τ_1 and τ_2 coincide. □

The following statement is known as the de la Vallée–Poussin [1] theorem. It is useful for obtaining the lower bounds for the best uniform polynomial approximation.

Proposition 2.1.5 Let $f \in C$ and suppose that for some $\tau_* \in \mathcal{N}_{2n-1}^{\mathrm{T}}$ there exist $2n$ points t_k: $0 \le t_1 < t_2 < \cdots < t_{2n} < 2\pi$ at which the difference $\delta(t) = f(t) - \tau_*(t)$ takes values with alternating signs, i.e.

$$\operatorname{sgn} \delta(t_k) = -\operatorname{sgn} \delta(t_{k+1}) \neq 0, \qquad k = 1, 2, \ldots, 2n-1;$$

then

$$E(f, \mathcal{N}_{2n-1}^{\mathrm{T}})_C \ge \min_{1 \le k \le 2n} |\delta(t_k)|.$$

Analogously, if the difference $\delta(t) = f(t) - p_*(t)$, where $f \in C[a, b]$ and $p_* \in \mathcal{N}_n^{\mathrm{A}}$, takes in n points $t_k \in [a, b]$ values with alternating signs then

$$E(f, \mathcal{N}_n^{\mathrm{T}})_{C[a,b]} \ge \min_{1 \le k \le n} |\delta(t_k)|.$$

Proof The proof is simple and is based on the same idea used in the proof of sufficiency in Chebyshev's theorem. For example in the periodic case, assuming that for some $\tau \in \mathcal{N}_{2n-1}^{\mathrm{T}}$

$$\|f - \tau\|_C < \eta =: \min_{1 \le k \le 2n} |\delta(t_k)|$$

holds, we get

$$|f(t_k) - \tau(t_k)| < \eta \le |f(t_k) - \tau_*(t_k)|, \qquad k = 1, 2, \ldots, 2n. \tag{1.17}$$

If we set

$$\tau_0(t) = \tau(t) - \tau_*(t) = \delta(t) - [f(t) - \tau(t)], \tag{1.18}$$

then (1.17) implies $\operatorname{sgn} \tau_0(t_k) = \operatorname{sgn} \delta(t_k)$ $(k = 1, 2, \ldots, 2n)$ and so the trigonometric polynomial τ_0 of degree $n - 1$ has to have at least $2n - 1$ zeros on the period, which is impossible because (1.17) and (1.18) show that $\tau_0 \neq 0$. □

2.1.2 A remark about even functions Let us note another important fact.

Proposition 2.1.6 If f is an even function from C or L_p then there exists an even trigonometric polynomial $\tau_* \in \mathcal{N}_{2n-1}^{\mathrm{T}}$ such that

$$\|f - \tau_*\|_X = E(f, \mathcal{N}_{2n-1}^{\mathrm{T}})_X, \tag{1.19}$$

where X is C or L_p respectively.

Indeed, let τ be the polynomial of best approximation for f in $\mathcal{N}_{2n-1}^{\mathrm{T}}$. Then the polynomial $\tau_*(t) = [\tau(t) + \tau(-t)]/2$ is even and also belongs to $\mathcal{N}_{2n-1}^{\mathrm{T}}$. Using the fact that f is even, we obtain

$$\|f - \tau_*\|_X \le \tfrac{1}{2}\|f(\bullet) - \tau(\bullet)\|_X + \tfrac{1}{2}\|f(\bullet) - \tau(-\bullet)\|_X$$
$$= \|f(\bullet) - \tau(\bullet)\|_X = E(f, \mathcal{N}^{\mathrm{T}}_{2n-1})_X,$$

i.e. equality (1.19) holds.

A similar property holds for the best approximation by algebraic polynomials on the interval $[-a, a]$.

2.2 Linear methods for polynomial approximation, Lebesgue constants

2.2.1 General facts Let $\mathcal{N}_N$ be an N-dimensional subspace of the normed space $X = X[a, b]$ of functions defined on $[a, b]$. Every linear operator A_N: $X \to \mathcal{N}_N$ maps a function $f \in X$ to a generalized polynomial

$$A_N(f, t) = \sum_{k=1}^{N} \mu_k(f)\varphi_k(t), \tag{2.1}$$

where μ_k $(k = 1, 2, \ldots, \mathrm{N})$ are linear functionals defined on X and $\varphi_1(t)$, $\varphi_2(t), \ldots, \varphi_N(t)$ is a basis in $\mathcal{N}_N$. The choice of the basis and the set of functionals can be influenced by the approximating function f which we are supposed to use for constructing the approximation method.

A typical example is the interpolation operator. One says that the subspace $\mathcal{N}_N$ *interpolates at the points* $t_1, t_2, \ldots, t_N$ in $[a, b]$ if for any set of prescribed numbers $y_1, y_2, \ldots, y_N$ there exists a unique function $\varphi \in \mathcal{N}_N$ such that $\varphi(t_k) = y_k$ $(k = 1, 2, \ldots, N)$. If we choose as basis functions the fundamental functions $l_k(t)$ defined by the equations

$$l_k(t_i) = \delta_{ki} = \begin{cases} 1, & k = i, \\ 0, & k \neq i, \end{cases}$$

we can construct the linear method

$$A^{\mathrm{I}}_N(f, t) = \sum_{k=1}^{N} f(t_k) l_k(t).$$

Therefore the approximating function $A^{\mathrm{I}}_N(f, t)$ is uniquely determined by the values of f at the N points t_k and we have

$$A^{\mathrm{I}}_N(f, t_k) = f(t_k), \qquad k = 1, 2, \ldots, N.$$

We have to pay special attention to the case when X $(=H)$ is a Hilbert space with the scalar product (f, g) and $\{\varphi_k(t)\}_1^N$ is an orthonormal basis in $\mathcal{N}_N$. If $f \in H$ and $c_k(f) = (f, \varphi_k)$ are the Fourier coefficients of f for the system $\{\varphi_k\}$, then we have for every $\psi \in \mathcal{N}_N$: $\psi(t) = \alpha_1\varphi_1(t) + \alpha_2\varphi_2(t) + \cdots + \alpha_N\varphi_N(t)$,

$$\|f-\psi\|_H^2 = (f-\psi, f-\psi) = (f,f) - 2(f,\psi) + (\psi,\psi)$$
$$= \|f\|_H^2 - 2\sum_{k=1}^{N} \alpha_k c_k(f) + \sum_{k=1}^{N} \alpha_k^2$$
$$= \|f\|_H^2 - \sum_{k=1}^{N} c_k(f)^2 + \sum_{k=1}^{N} (\alpha_k - c_k(f))^2.$$

The last sum has a minimal value (equals zero) only when $\alpha_k = c_k(f)$. Hence we set

$$g_N(f,t) = \sum_{k=1}^{N} c_k(f)\varphi_k(t) \tag{2.2}$$

so we have for every $f \in H$

$$\|f - g_N(f)\|_H = E(f, \mathcal{N}_N)_H = \|f\|_H^2 - \sum_{k=1}^{N} c_k(f)^2, \tag{2.3}$$

i.e. the best approximation from $\mathcal{N}_N$ to every $f \in H$ is realized by the Fourier sum (2.2).

Considering a linear method $A_N(f,t)$ defined by an arbitrary linear operator $A_N: X \to \mathcal{N}_N$ in a non-Hilbert space one can in general only write the trivial inequality $\|f - A_N(f)\|_X \geq E(f, \mathcal{N}_N)_X$. If A_N is a linear projection from X to $\mathcal{N}_N$ then we can also bound the error $\|f - A_N\|_X$ from above. Indeed, let $\varphi(f, \bullet)$ denote the function of best approximation for f in $\mathcal{N}_N$, i.e. $\|f - \varphi(f)\|_X = E(f, \mathcal{N}_N)_X$. Then we have $A_N f = A_N(f - \varphi(f)) + \varphi(f)$ for the linear projector A_N and from

$$\|A_N(f - \varphi(f))\|_X \leq \|A_N\| \, \|f - \varphi(f)\|_X = \|A_N\| E(f, \mathcal{N}_N)_X$$

we obtain

$$\|f - A_N f\|_X \leq (1 + \|A_N\|) E(f, \mathcal{N}_N)_X. \tag{2.4}$$

If X is the space of all functions f which are continuous on $[a, b]$ (or on the axis) then $A_N(f, t)$ for a fixed t is a linear functional defined on X and we obtain using similar reasoning

$$|f(t) - A_N(f,t)| \leq (1 + \sup_{f \in X, |f(t)| \leq 1} |A_N(f,t)|) E(f, \mathcal{N}_N)_X. \tag{2.5}$$

In approximation theory a lot of attention has been paid to the investigation of sequences of linear methods defined by some prescribed rule. Let $\varphi_1(t), \varphi_2(t), \ldots,$ be a sequence of linearly independent functions in the Banach space X, let $\mathcal{N}_N$ $(N = 1, 2, \ldots)$ be the subspaces generated by the first N functions of this sequence, let $A_N: X \to \mathcal{N}_N$ be linear operators, which are of type (2.1), where the linear functionals $\mu_k = \mu_k^N$ are defined on X in some uniform way. In the definition of μ_k one may incorporate a parameter which, being at our disposition, allows us to build a variety of sequences of linear operators. For example, if $\{\varphi_k\}_1^\infty$ is an orthonormal system and $c_k(f)$ are the Fourier coefficients of f on this system, then every triangular matrix of coefficients $\Lambda = \{\lambda_k^N\}$ $(k = 1, 2, \ldots, N; N = 1, 2, \ldots)$ define a sequence of linear methods

$$A_N(f, \Lambda, t) = \sum_{k=1}^{N} \lambda_k^N c_k(f) \varphi_k(t), \qquad N = 1, 2, \ldots.$$

Another way of defining families of sequences of linear methods, which is frequently used, is based on interpolation. Let $X = C[a, b]$ and each of the subspaces $\mathcal{N}_N$ $(N = 1, 2, \ldots)$, $\dim \mathcal{N}_N = N$, interpolate at every set of N points in $[a, b]$. Then every matrix

$$T = \{t_k^N\}: k = 1, 2, \ldots, N; \qquad N = 1, 2, \ldots, \qquad t_k^N \in [a, b]$$

of knots of interpolation defines a sequence of linear methods of approximation of $f \in C[a, b]$:

$$A_N(f, T, t) = \sum_{k=1}^{N} f(t_k^N) l_k^N(t), \qquad N = 1, 2, \ldots,$$

where l_k^N $(k = 1, 2, \ldots, N)$ is a basis in $\mathcal{N}_N$ of fundamental (i.e. satisfying the condition $l_k^N(t_i^N) = \delta_{ki}$) functions. Moreover

$$A_N(f, T, t_k^N) = f(t_k^N); \qquad k = 1, 2, \ldots, N; \qquad N = 1, 2, \ldots.$$

The first problem arising in the investigation of the properties of a sequence $\{A_N\}$ of linear methods of approximation in a Banach space X is the problem of convergence: is it true that $A_N f \to f$ (in the sense $\|A_N f - f\|_X \to 0$ when $N \to \infty$) for every $f \in X$? In a family of sequences of linear methods one looks for conditions for defining the sequence parameters which would ensure convergence. The most general criterion for convergence is based on the Banach–Steinhaus Theorem (see e.g. [10B, p. 151]) and it asserts that $A_N f \to f$ for every $f \in X$ iff: (1) the sequence of norms $\|A_N\|$ $(N = 1, 2, \ldots)$ is bounded; and (2) the convergence $A_N f \to f$ takes place for f from some dense subset in X. (Later on we shall refer to this assertion as the Banach–Steinhaus criterion.)

The problem of the rate of convergence of $A_N f$ for a given class $\mathcal{M}$, i.e. of the rate at which the sequence

$$\sup_{f \in \mathcal{M}} \|A_N f - f\|_X, \qquad N = 1, 2, \ldots, \tag{2.6}$$

tends to zero when $N \to \infty$ is more delicate. The supremum in (2.6) for every N is rarely identified except in the case when we are dealing with spaces of periodic functions. We often have to be satisfied by finding just order estimates for the sequence (2.6) when $N \to \infty$, but, even if the exact order is found, the problem of determining the exact constant in the leading order term of the asymptotic expansion (in other words – the problem of the exact asymptotics of (2.6)) still has to be solved.

2.2.2 Partial sums of Fourier series We shall consider linear operators with sets of algebraic or trigonometric polynomials covering a range of values. We start with Fourier expansions in the trigonometric system with which a variety of linear methods for the approximation of periodic functions is connected.

For $f \in L_1$ let

$$a_k = \frac{1}{\pi} \int_0^{2\pi} f(t) \cos kt \, dt, \qquad k = 0, 1, \ldots$$

and

$$b_k = \frac{1}{\pi} \int_0^{2\pi} f(t) \sin kt \, dt, \qquad k = 1, 2, \ldots$$

denote the Fourier coefficients of f in the trigonometric system and

$$S_n(f, t) = a_0/2 + \sum_{k=1}^{n} (a_k \cos kt + b_k \sin kt), \qquad n = 1, 2, \ldots,$$

$$S_0(f, t) = a_0/2$$

be the nth partial sum of the Fourier series of f. Using

$$a_k \cos kt + b_k \sin kt = \frac{1}{\pi} \int_0^{2\pi} f(u) \cos k(t-u) \, du, \tag{2.7}$$

we get

$$S_n(f, t) = \frac{1}{\pi} \int_0^{2\pi} D_n(t-u) f(u) \, du = \frac{1}{\pi} \int_0^{2\pi} D_n(u) f(t-u) \, du, \tag{2.8}$$

where

$$D_n(t) = 1/2 + \sum_{k=1}^{n} \cos kt = \frac{\sin (n+1/2)t}{2 \sin (t/2)} \tag{2.9}$$

is the Dirichlet kernel.

For every t we have

$$\sup_{|f(u)|\le 1} \left| \int_0^{2\pi} D_n(t-u)f(u)\,du \right| = \int_0^{2\pi} |D_n(t-u)|\,du = \int_0^{2\pi} |D_n(u)|\,du, \quad (2.10)$$

and the supremum will not decrease if we assume additionally that f is continuous. Hence the norms of the linear functionals $S_n(f,t)$ in spaces L_∞ and C do not depend on t and they are equal to

$$\mathcal{L}_n = \frac{1}{\pi}\int_0^{2\pi} |D_n(u)|\,du, \qquad n = 1, 2, \ldots.$$

These numbers are known in approximation theory as Lebesgue constants. If we denote the norm of the linear operator $A\colon X \to X$ by $\|A\|_{(X)}$, then we can write

$$\|S_n\|_{(L_\infty)} = \|S_n\|_{(C)} = \mathcal{L}_n, \qquad n = 1, 2, \ldots. \quad (2.11)$$

It is also easy to show that

$$\|S_n\|_{(L_1)} = \mathcal{L}_n, \qquad n = 1, 2, \ldots. \quad (2.12)$$

Indeed, if $f \in L_1$, $\|f\|_1 \le 1$, then

$$\int_0^{2\pi} |S_n(f,t)|\,dt \le \frac{1}{\pi}\int_0^{2\pi}\int_0^{2\pi} |D_n(t-u)|\,|f(u)|\,du\,dt$$

$$= \frac{1}{\pi}\int_0^{2\pi} |f(u)| \int_0^{2\pi} |D_n(t-u)|\,dt\,du \le \mathcal{L}_n\|f\|_1 \le \mathcal{L}_n$$

and hence $\|S_n\|_{(L_1)} \le \mathcal{L}_n$. In order to prove that we actually have an equality we consider for every $\varepsilon > 0$ the function $f_\varepsilon(u) = 1/\varepsilon$ $(0 < u < \varepsilon)$, $f_\varepsilon(u) = 0$ $(\varepsilon \le u \le 2\pi)$, $f_\varepsilon(u + 2\pi) = f_\varepsilon(u)$. We have $\|f_\varepsilon\|_1 = 1$ and for $\varepsilon \to 0$

$$|S_n(f_\varepsilon, t)| = \frac{1}{\pi\varepsilon}\left| \int_0^{\varepsilon} D_n(t-u)\,du \right| \to \frac{1}{\pi}|D_n(t)|.$$

The following theorem is of importance, especially in connection with the Banach–Steinhaus criterion (mentioned in Section 2.2.1) for the convergence of sequences of linear operators.

Theorem 2.2.1 For $n \to \infty$ we have the asymptotic relation

$$\mathcal{L}_n = \frac{4}{\pi^2}\ln n + O(1) \quad (2.13)$$

Proof Using the fact that the function $1/t - (1/2)\operatorname{ctg}(t/2)$ is bounded for $|t| \le \pi$ we have

$$\mathscr{L}_n = \frac{1}{\pi}\int_0^{2\pi} |D_n(u)|\,\mathrm{d}u = \frac{1}{\pi}\int_0^{2\pi} |\sin nt| \frac{1}{2}\left|\operatorname{ctg}\frac{t}{2}\right| \mathrm{d}t + O(1)$$

$$= \frac{2}{\pi}\int_0^{\pi} |\sin nt|/t\,\mathrm{d}t + O(1) = \frac{2}{\pi}\sum_{k=1}^{n}\int_{(k-1)\pi/n}^{k\pi/n} |\sin nt|/t\,\mathrm{d}t + O(1)$$

$$= \frac{2}{\pi}\int_0^{\pi/n}\left(\sum_{k=1}^{n-1}\frac{1}{t+k\pi/n}\right)\sin nt\,\mathrm{d}t + O(1).$$

The sum in the brackets where $t \in [0, \pi/n]$ lies between the numbers

$$\frac{n}{\pi}\left(1+\frac{1}{2}+\cdots+\frac{1}{n-1}\right) \quad \text{and} \quad \frac{n}{\pi}\left(\frac{1}{2}+\frac{1}{3}+\cdots+\frac{1}{n}\right)$$

and hence is equal to $\pi^{-1}n\,[\ln n + O(1)]$. We now obtain (2.13) by evaluating the integral of $\sin nt$ over $(0, \pi/n)$.

From (2.11)–(2.13) we obtain:

Proposition 2.2.2 For $n \to \infty$ we have the asymptotic equalities

$$\|S_n\|_{(L_\infty)} = \|S_n\|_{(C)} = \frac{4}{\pi^2}\ln n + O(1).$$

Therefore the sequence of the norms of operators S_n is unbounded in C or L_1 and, in view of the Banach–Steinhaus criterion, there are functions f from C or L_1 for which the sequence of the partial sums of Fourier series $S_n(f, t)$ does not converge to f in the corresponding metric. Moreover, from the unboundedness (for fixed t) of the norm of linear functionals $S_n(f, t)$ $(n = 1, 2, \ldots)$ it follows that for every t there exists $f \in C$ with divergence in the t Fourier series.

The situation is different in L_p when $1 < p < \infty$. Using [17B, p. 423] that we have for $f \in L_p$

$$\|S_n(f)\|_p \le M_p\|f\|_p, \qquad 1 < p < \infty, \qquad n = 1, 2, \ldots,$$

and using the density of trigonometric polynomials in L_p $(1 \le p < \infty)$ we obtain that for every $f \in L_p (1 < p < \infty)$ the sequence $S_n(f)$ converges to f in L_p $(1 < p < \infty)$. For $p = 2$ this follows immediately from (2.3).

2.2.3 Projection and interpolation operators It is well known [25A, p. 191] that if P_n is a linear projection from C or L_p $(1 \le p \le \infty)$ to $\mathcal{N}^{\mathrm{T}}_{2n+1}$ then

$$\|P_n\|_{(X)} \ge \|S_n\|_{(X)}, \qquad n = 1, 2, \ldots, \tag{2.14}$$

where X is C or L_p. From here and Proposition 2.2.2 it follows that the effect of divergence in C and L_1 holds for every sequence of projectors on $\mathcal{N}^{\mathrm{T}}_{2n+1}$

($n = 0, 1, \ldots$). In particular this is true for a sequence of trigonometric polynomials of Lagrange interpolation.

It is also well known that the subspace $\mathcal{N}^{\mathrm{T}}_{2n+1}$ interpolates on every prescribed set of $2n+1$ different points on the period $[0, 2\pi)$. The interpolation polynomial can be easily written in explicit form. For given points $0 \le t_1 < t_2 < \cdots < t_{2n+1} < 2\pi$ and a set of numbers $Y = \{y_1, y_2, \ldots, y_{2n+1}\}$ we set

$$\tau_n(Y, t) = \sum_{k=1}^{2n+1} y_k \frac{\sin \frac{t - t_1}{2} \cdots \sin \frac{t - t_{k-1}}{2} \sin \frac{t - t_{k+1}}{2} \cdots \sin \frac{t - t_{2n+1}}{2}}{\sin \frac{t_k - t_1}{2} \cdots \sin \frac{t_k - t_{k-1}}{2} \sin \frac{t_k - t_{k+1}}{2} \cdots \sin \frac{t_k - t_{2n+1}}{2}}. \tag{2.15}$$

It is easy to check that the numerator of every term of the sum (2.15) is a trigonometric polynomial of degree n and that $\tau_n(Y, t_k) = y_k$.

For given $n+1$ points t_k on the interval $[0, \pi)$: $0 = t_0 < t_1 < \cdots < t_n < \pi$ we set

$$\begin{aligned} \tau_n(Y, t) &= \sum_{k=1}^{n} y_k \frac{(\cos t - \cos t_0) \ldots (\cos t - \cos t_{k-1})}{(\cos t_k - \cos t_0) \ldots (\cos t_k - \cos t_{k-1})} \\ &\quad \times \frac{(\cos t - \cos t_{k+1}) \ldots (\cos t - \cos t_n)}{(\cos t_k - \cos t_{k+1}) \ldots (\cos t_k - \cos t_n)} \\ &= \sum_{k=0}^{n} \alpha_k \cos kt \end{aligned}$$

and obtain an even trigonometric polynomial of degree n such that $\tau_n(Y, t_k) = y_k$. For given n points t_k on the interval $(0, \pi)$: $0 < t_1 < t_2 < \cdots < t_n < \pi$ we set

$$\begin{aligned} \tau_n(Y, t) &= \sum_{k=1}^{n} y_k \frac{\sin t\,(\cos t - \cos t_0) \ldots (\cos t - \cos t_{k-1})}{\sin t_k\,(\cos t_k - \cos t_0) \ldots (\cos t_k - \cos t_{k-1})} \\ &\quad \times \frac{(\cos t - \cos t_{k+1}) \ldots (\cos t - \cos t_n)}{(\cos t_k - \cos t_{k+1}) \ldots (\cos t_k - \cos t_n)} = \sum_{k=1}^{n} \beta_k \sin kt \end{aligned}$$

and obtain an odd trigonometric polynomial of degree n such that $\tau_n(Y, t_k) = y_k$ ($k = 1, 2, \ldots, n$).

In this way a triangular matrix $\{t_k^n\colon k = 1, 2, \ldots, 2n+1;\ n = 0, 1, \ldots\}$ with the following interpolation knots given on $[0, 2\pi)$:

$$0 \le t_1^n < t_2^n < \cdots < t_{2n+1}^n < 2\pi, \qquad n = 0, 1, 2, \ldots, \tag{2.16}$$

generates a sequence of linear operators mapping C onto $\mathcal{N}^{\mathrm{T}}_{2n+1}$ of the type

$$\tau_n(f, t) = \sum_{k=1}^{2n+1} f(t_k^n) w_{n,k}(t), \qquad n = 0, 1, 2, \ldots, \tag{2.17}$$

where $w_{n,k}$ are the fundamental trigonometric Lagrange polynomials: $w_{n,k}(t_j^n) = \delta_{kj}$. These polynomials can be written in the form

$$w_{n,k}(t) = \frac{1}{2} \frac{w_n(t)}{w_n'(t_k^n) \sin[(t - t_k^n)/2]}, \qquad k = 1, 2, \ldots, 2n + 1,$$

where

$$w_n(t) = \sin\frac{t - t_1^n}{2} \sin\frac{t - t_2^n}{2} \ldots \sin\frac{t - t_{2n+1}^n}{2}.$$

Because the interpolating polynomial is unique we have $\tau_n(f) = f$ for every $f \in \mathcal{N}^{\mathrm{T}}_{2n+1}$, i.e. τ_n is a projector onto $\mathcal{N}^{\mathrm{T}}_{2n+1}$, and so the norms of the operators τ_n $(n = 1, 2, \ldots)$ are unbounded in C or L_1 (see (2.14)). Therefore for every matrix of knots (2.16) there is a function $f \in C$ for which the sequence of trigonometric polynomials (2.17) does not converge uniformly. In connection with this, we should mention that for every function $f \in C$ there exists a matrix of knots (2.16) such that the polynomials (2.17) converge uniformly to f: it is enough for t_k^n to take these $2n + 1$ points in $[0, 2\pi)$ at which the polynomial of best approximation from $\mathcal{N}^{\mathrm{T}}_{2n+1}$ coincides with the function f. Such points always exist because of the Chebyshev alternation (Theorem 2.1.2) but in general they cannot be found in an efficient way.

2.2.4 λ-methods of summation of Fourier series and interpolating polynomials To overcome the disadvantage of the fact that Fourier series of continuous functions can diverge, different methods for averaging the partial sums were introduced. Fejér [1] has proved that the arithmetic means of the partial sums of Fourier series

$$\mathcal{F}_n(f, t) = \frac{1}{n+1}[S_0(f, t) + S_1(f, t) + \cdots + S_n(f, t)], \qquad n = 0, 1, \ldots, \tag{2.18}$$

converge uniformly to f provided $f \in C$. Rogosinski [1] has established that under some restrictions on the sequence $\{\beta_n\}$ the means

$$\mathcal{R}(f, \beta_n, t) = \tfrac{1}{2}[S_n(f, t + \beta_n) + S_n(f, t - \beta_n)], \qquad n = 0, 1, \ldots, \tag{2.19}$$

possess the same property. In particular the uniform convergence of sums (2.19) for $f \in C$ is guaranteed for $\beta_n = \pi/(2n)$. Later a similar averaging was considered by Bernstein [3].

Sums (2.18) and (2.19) are representative of a wide class of linear methods of approximation defined by the operator

$$U_n(f, \lambda, t) = a_0/2 + \sum_{k=1}^{n} \lambda_k^n (a_k \cos kt + b_k \sin kt), \qquad n = 1, 2, \ldots, \tag{2.20}$$

where a_k and b_k are Fourier coefficients of f and

$$\lambda^n = \{\lambda_1^n, \lambda_2^n, \ldots, \lambda_n^n\} \tag{2.21}$$

is a set of numbers determining the particular method. It is easy to check that $\lambda_k^n = 1 - k/(n+1)$ give Fejér means (2.18) and $\lambda_k^n = \cos k\beta_n$ give Rogosinski means (2.19).

The approximation method in which function f for fixed n is related to the polynomial (2.20) by the numbers (2.21) is called the *λ-method*. One often considers sequences of polynomials (2.20) defined by a triangular matrix

$$\Lambda = \{\lambda_k^n \colon k = 1, 2, \ldots, n; \qquad n = 1, 2, \ldots\} \tag{2.22}$$

in which all the rows are given in an uniform way (e.g. the matrices for Féjer and Rogosinski means). In this case we talk about a sequence of λ-methods or the λ-process for the approximation of function f.

Using (2.7) we can write polynomials (2.20) as

$$U_n(f, \lambda, t) = \frac{1}{\pi} \int_0^{2\pi} K_n(\lambda, t-u) f(u)\, du = \frac{1}{\pi} \int_0^{2\pi} K_n(\lambda, u) f(t-u)\, du, \tag{2.23}$$

where

$$K_n(\lambda, t) = \frac{1}{2} + \sum_{k=1}^{n} \lambda_k^n \cos kt \tag{2.24}$$

is a polynomial called the *kernel of the λ-method*.

In this way λ-methods can be included in the wider class of linear methods defined by the convolution $K * f = \int_0^{2\pi} K(t-u) f(u)\, du$ of the approximating function f with a kernel K (not necessarily a polynomial).

As in the case of Fourier means (see (2.10)) the integral representation (2.23) easily implies

$$\|U_n(\lambda)\|_{(C)} = \frac{1}{\pi} \int_0^{2\pi} |K_n(\lambda, u)|\, du, \qquad n = 1, 2, \ldots.$$

Now the Banach–Steinhaus criterion implies that

$$\lim_{n \to \infty} \|U_n(f, \lambda) - f\|_C = 0$$

for every $f \in C$ if and only if

$$\int_0^{2\pi} |K_n(\lambda, u)| \le M, \qquad n = 1, 2, \ldots, \tag{2.25}$$

and

$$\lim_{n\to\infty} \lambda_k^n = 1, \qquad k = 1, 2, \ldots \tag{2.26}$$

(the limits (2.26) guarantee the uniform convergence of sequence (2.20) for the functions 1, $\cos t$, $\sin t$, $\cos 2t$, $\sin 2t$, . . .).

Interpolating trigonometric polynomials, which possess the same drawback as Fourier means with respect to uniform convergence, can be averaged in a similar way. In the case of uniform knots of interpolation using the triangle matrix (2.22) for this purpose one can find an analogy with the Fourier means basis polynomial (2.20). From (2.9) we see that

$$D_n(0) = n + 1/2, \qquad D_n\left(\frac{2\nu\pi}{2n+1}\right) = 0, \qquad \nu = 1, 2, \ldots, 2n,$$

and so the polynomial

$$\bar{\tau}_n(f, t) = \frac{2}{2n+1} \sum_{k=0}^{2n} f(\bar{t}_\nu^n) D_n(t - \bar{t}_\nu^n), \qquad \bar{t}_\nu^n = \frac{2\nu\pi}{2n+1}, \tag{2.27}$$

is the only trigonometric polynomial of degree n coinciding with f at the points $\bar{t}_\nu^n$. Using the first of the equalities (2.9) we can write $\bar{\tau}_n(f, t)$ as

$$\bar{\tau}_n(f, t) = \frac{\alpha_0^n}{2} + \sum_{k=1}^{n} (\alpha_k^n \cos kt + \beta_k^n \sin kt), \tag{2.28}$$

where

$$\alpha_k^n = \frac{2}{2n+1} \sum_{\nu=0}^{2n} f(\bar{t}_\nu^n) \cos \nu \bar{t}_k^n, \qquad k = 0, 1, \ldots, n,$$

$$\beta_k^n = \frac{2}{2n+1} \sum_{\nu=0}^{2n} f(\bar{t}_\nu^n) \sin \nu \bar{t}_k^n, \qquad k = 0, 1, \ldots, n. \tag{2.29}$$

The sequence of polynomials

$$\bar{U}_n(f, \lambda, t) = \alpha_0^n/2 + \sum_{k=1}^{n} \lambda_k^n (\alpha_k^n \cos kt + \beta_k^n \sin kt), \qquad n = 1, 2, \ldots, \tag{2.30}$$

converges uniformly to $f \in C$ under the same conditions as λ_k^n (see (2.25) and (2.26)) which ensure the convergence of polynomials (2.20). In particular, for $\lambda_k^n = 1 - k/(n+1)$ and $\lambda_k^n = \cos k\beta_n$ we obtain interpolation analogs of the Féjer and Rogosinski means.

Polynomials (2.27) and (2.30) depend on the values of f only at $2n + 1$ points which makes the norms of the functionals $\bar{\tau}_n(f, t)$ and $\bar{U}_n(f, \lambda, t)$ dependent on t (cf. the norms of the partial sums of Fourier series $S_n(f, t)$

and λ-means (2.20) which do not depend on t). The quantity

$$\bar{\mathscr{L}}_n(t) = \sup_{|f(t)|\leq 1} |\bar{\tau}_n(f,t)| = \frac{2}{2n+1}\sum_{k=0}^{2n} |D_n(t - \bar{t}_\nu^n)| \tag{2.31}$$

is called the *Lebesgue function* of operator (2.27). Obviously $\bar{\mathscr{L}}_n(\bar{t}_\nu^n) = 1$ and from (2.11) and the general inequality (2.14) for projectors onto $\mathcal{N}_{2n+1}^{\mathrm{T}}$ it follows that $\max_t \bar{\mathscr{L}}_n(t) \geq \mathscr{L}_n$. More precise information is given by the following result:

Theorem 2.2.3 The asymptotic relation

$$\bar{\mathscr{L}}_n(t) = \frac{2}{\pi}\left|\sin\frac{2n+1}{2}t\right| \ln n + O(1) \tag{2.32}$$

holds uniformly on t for $n \to \infty$.

Proof $\bar{\mathscr{L}}_n(t)$ is an even function of period $2h$ ($h = \pi/(2n+1)$) because the kernel D_n is even and 2π-periodic and the knots $\bar{t}_\nu^n$ are uniformly distributed. Therefore it is enough to prove (2.32) for $t \in (0, h]$ only. Obviously

$$\bar{\mathscr{L}}_n(t) = \frac{1}{n}\sum_{k=-n}^{n} |D_n(t - 2kh)| + O(1).$$

Because

$$|D_n(t-2kh)| = \left|\sin\frac{2n+1}{2}t\right| [2\sin\tfrac{1}{2}(t-2kh)]^{-1}$$

and the function $u^{-1} - \frac{1}{2}\operatorname{cosec}(u/2)$ is bounded for $|u| \leq \pi$, we have for $0 < t \leq h$

$$\bar{\mathscr{L}}_n(t) = \left|\sin\frac{2n+1}{2}t\right| \frac{1}{n}\left[\sum_{k=1}^{n}\frac{1}{|t-2kh|} + \sum_{k=-n}^{-1}\frac{1}{|t-2kh|}\right] + O(1)$$

$$= \left|\sin\frac{2n+1}{2}t\right| \frac{1}{nh}\sum_{k=1}^{2n}\frac{1}{k} + O(1) = \frac{2}{\pi}\left|\sin\frac{2n+1}{2}t\right| \ln n + O(1).$$

Comparing (2.13) and (2.32) we see that for $n \to \infty$ the following asymptotic relation holds uniformly on t

$$\bar{\mathscr{L}}_n(t) = \frac{\pi}{2}\left|\sin\frac{2n+1}{2}t\right| \mathscr{L}_n + O(1). \qquad \square$$

Under some restrictions on the matrix (2.22) we have a similar relationship between the norms of operators (2.20) (from C to C) and the Lebesgue functions of polynomials (2.30) (see [26A, Ch. 7]).

2.2.5 Algebraic analogs The ways described in Section 2.2.4 for constructing linear methods for approximating 2π-periodic functions have a natural analog in the approximation of functions on $[a, b]$ by algebraic polynomials. For technical reasons the change of variables

$$x = \cos t, \qquad -\infty < t < \infty, \qquad -1 \leq x \leq 1, \tag{2.33}$$

makes it more suitable to work in $[-1, 1]$ instead of $[a, b]$. If the function f is defined on $[-1, 1]$ then $\varphi(t) = f(\cos t)$ is defined on the real axis and is even and 2π-periodic. Conversely, for every even $\varphi \in C$ the function $f(x) = \varphi(\arccos x)$ is uniquely determined and belongs to $C[-1, 1]$. Relation (2.33) establishes a one-to-one correspondence between the space $\mathcal{N}^{\mathrm{A}}_{n+1}$ of all algebraic polynomials of degree n and the space of all even trigonometric polynomials of degree n, because $\cos^k t$ is an even trigonometric polynomial and $\cos(k \arccos x)$ is an algebraic polynomial (*Chebyshev polynomial*).

The algebraic analogs of partial sums of Fourier series are the Fourier–Chebyshev means

$$Q_n(f, x) = \frac{c_0}{2} + \sum_{k=1}^{n} c_k \cos(k \arccos x), \tag{2.34}$$

where

$$c_k = \frac{2}{\pi} \int_0^{\pi} f(\cos t) \cos kt \,\mathrm{d}t = \frac{2}{\pi} \int_{-1}^{1} \frac{f(x) \cos(k \arccos x)}{(1 - x^2)^{1/2}} \,\mathrm{d}x. \tag{2.35}$$

Replacing (2.35) in (2.34) we get

$$Q_n(f, x) = \frac{2}{\pi} \int_0^{\pi} f(\cos t) \left[\frac{1}{2} + \sum_{k=1}^{n} \cos kt \cos(k \arccos x)\right] \mathrm{d}t$$

$$= \frac{1}{\pi} \int_{-\pi}^{\pi} f(\cos t) \left[\frac{1}{2} + \sum_{k=1}^{n} \cos k(t - \arccos x)\right] \mathrm{d}t. \tag{2.36}$$

From the first representation in (2.36) we see that

$$\|Q_n\|_{(C[-1,1])} \geq \sup_{|f(u)| \leq 1} |Q_n(f, 1)| = \|S_n\|_{(C)} = \mathcal{L}_n,$$

and using the second one in (2.36) we get that $\|Q_n\|_{(C[-1,1])} \leq \mathcal{L}_n$ and hence taking into account Theorem 2.2.1 we have

$$\|Q_n\|_{(C[-1,1])} = \mathcal{L}_n = \frac{4}{\pi^2} \ln n + O(1).$$

This means that the sequence $Q_n(f, t)$ does not converge uniformly for any $f \in C[-1, 1]$. In this respect one considers, together with polynomials (2.34), their λ-means (analogs of (2.20))

$$V_n(f, \lambda, x) = \frac{c_0}{2} + \sum_{k=1}^{n} \lambda_k^n c_k \cos(k \arccos x), \qquad n = 1, 2, \ldots, \tag{2.37}$$

which converge uniformly for every $f \in C[-1, 1]$ under some restrictions on the triangular matrix (2.22). In particular, for $\lambda_k^n = 1 - k/(n+1)$ and $\lambda_k^n = \cos k\beta_n$ we obtain the algebraic analogs of Fejér and Rogosinski means.

Subspace $\mathcal{N}_{n+1}^{\mathrm{A}}$ interpolates on every set of $n+1$ different points. Hence every triangle matrix of interpolating knots in $[-1, 1]$: $-1 \le x_1^n < x_2^n < \cdots < x_{n+1}^n \le 1$ $(n = 0, 1, \ldots)$ determines for $f \in C[-1, 1]$ a sequence $A_n(f, x)$ $(n = 0, 1, \ldots)$ of algebraic polynomials of degree n satisfying the conditions

$$A_n(f, x_k^n) = f(x_k^n); \qquad k = 1, 2, \ldots, n+1; \qquad n = 1, 2, \ldots,.$$

Polynomial $A_n(f, x)$ can be written in the form

$$A_n(f, x) = \sum_{k=1}^{n+1} f(x_k^n) l_{n,k}(x), \tag{2.38}$$

where $l_{n,k}(x)$ are the fundamental functions which are usually represented as

$$l_{n,k}(x) = \frac{l_n(x)}{(x - x_k^n) l_n'(x_k^n)}, \qquad l_n(x) = c(x - x_1^n) \ldots (x - x_{n+1}^n).$$

If $\bar{x}_k^n = \cos[(2k-1)\pi/2(n+1)]$ $(k = 1, 2, \ldots, n+1)$ are the zeros of Chebyshev polynomial $\cos[(n+1)\arccos x]$ then

$$l_n(x) = \cos[(n+1)\arccos x], \qquad l_n'(\bar{x}_k^n) = \frac{(-1)^{k-1}(n+1)}{\sin\{[(2k-1)/2(n+1)]\pi\}},$$

and in this case the interpolation polynomial takes the form

$$\bar{A}_n(f, x) = \frac{1}{n+1} \cos[(n+1)\arccos x] \sum_{k=1}^{n+1} f(\bar{x}_k^n) \frac{(-1)^{k-1} \sin \dfrac{2k-1}{2(n+1)}\pi}{x - \cos \dfrac{2k-1}{2(n+1)}\pi}. \tag{2.39}$$

Simple calculations allows us to obtain the following representation for $\bar{A}_n(f, x)$ (see e.g. [26A, p. 187])

$$\bar{A}_n(f, x) = \gamma_0^n/2 + \sum_{\nu=1}^{n} \gamma_\nu^n \cos(\nu \arccos x),$$

where

$$\gamma_\nu^n = \frac{1}{n+1} \sum_{k=1}^{n+1} f(\bar{x}_k^n) \cos(\nu \arccos \bar{x}_k^n), \qquad \nu = 0, 1, \ldots, n.$$

This representation should first be compared with (2.34) and then with (2.28). A natural analog of λ-means (2.30) and (2.37) are the polynomials

$$\bar{V}_n(f, \lambda, x) = \frac{\gamma_0^n}{2} + \sum_{\nu=1}^{n} \lambda_\nu^n \gamma_\nu^n \cos(\nu \arccos x).$$

2.2.6 Linear positive methods A special place among the methods of approximation defined by linear operators $A_N: X \to \mathcal{N}_N$ must be given to the methods defined by linear positive operators A_N, i.e. operators for which $f(t) \geq 0$ implies $A_N(f, t) \geq 0$. In the case when the operator is given by

$$A_N(f, t) = \int_a^b K(t, u) f(u) \, du,$$

this condition is equivalent to $K(t, u) \geq 0$ and the positivity of

$$A_N(f, t) = \sum_{k=1}^{N} f(t_k) \varphi_k(t)$$

is equivalent to $\varphi_k(t) \geq 0$ $(k = 1, 2, \ldots, N)$.

The problem of convergence of the sequence of linear positive operators with values in subspaces of algebraic or trigonometric polynomials can be solved using three 'test' functions. If $A_n: C[a, b] \to \mathcal{N}_{n+1}^{\mathrm{A}}$ $(A_n: C \to \mathcal{N}_{2n+1}^{\mathrm{T}})$ $(n = 1, 2, \ldots)$ is a sequence of linear positive operators then the sequence of polynomials $A_n(f, t)$ converges uniformly to f for every $f \in C[a, b]$ $(f \in C)$ provided this happens for the functions $1, t, t^2$ $(1, \cos t, \sin t)$. This assertion is due to Korovkin [12A, p. 21–4].

For operators $U_n(f, \lambda, t): C \to \mathcal{N}_{2n+1}^{\mathrm{T}}$ (cf. (2.23)) positivity means

$$K_n(\lambda, t) = \frac{1}{2} + \sum_{k=1}^{n} \lambda_k^n \cos kt \geq 0.$$

In this case the relation $\lim_{n \to \infty} \lambda_1^n = 1$ is a sufficient condition for the uniform convergence $U_n(f, \lambda, t) \to f(t)$ (see [14A, p. 71]). In particular we can deduce the convergence of Fejér means because $\lambda_1^n = 1 - 1/(n+1)$.

2.3 Polynomial splines

2.3.1 Definitions and examples The function s which we assume to be defined and continuous on $[a, b]$ is called a *polynomial spline* (or simply *spline*) of degree $m = 1, 2, \ldots$ with *knots* t_i $(i = 1, 2, \ldots, n)$:

$$a < t_1 < t_2 < \cdots < t_n < b,$$

if on every interval $[a, t_1]$, $[t_i, t_{i+1}]$ $(i = 1, 2, \ldots, n-1)$, $[t_n, b]$ s is an algebraic polynomial of degree at most m and at every point t_i some derivative $s^{(\nu)}(t_i)$ $(1 \le \nu \le m)$ may be discontinuous. Moreover it is assumed that at least one of the polynomial components of the spline has the precise degree m.

A spline s of degree m has a *defect* k_i $(1 \le k_i \le m)$ at the point t_i if the functions $s, s', \ldots, s^{(m-k_i)}$ are continuous at t_i but the derivative $s^{(m-k_i+1)}$ is discontinuous at t_i. The number

$$k = \max_{1 \le i \le n} k_i$$

is called the *defect of the spline s*.

We shall primarily deal with linear manifolds of splines with knots at fixed points. The fixed system of points

$$\Delta_N[a, b]\colon a = t_0 < t_1 < t_2 < \cdots < t_N = b \tag{3.1}$$

will be called the *partition* of the interval $[a, b]$.[1]

By $S_m^k(\Delta_N[a, b])$ we denote the set of all splines of degree m defined on $[a, b]$ with defect $k_i \le k$ at the knots t_i $(i = 1, 2, \ldots, N-1)$ from partition (3.1). Hence $s \in S_m^k(\Delta_N[a, b])$ iff $s \in C^{m-k}[a, b]$ and on every interval (t_{i-1}, t_i) $(i = 1, 2, \ldots, N)$ $s^{(m)}$ is some constant c_i. Obviously $S_m^k(\Delta_N[a, b])$ is a linear manifold; we shall call it the set of splines of degree m with defect k with respect to partition (3.1). Instead of $S_m^1(\Delta_N[a, b])$ we shall write $S_m(\Delta_N[a, b])$.

It is convenient to consider the case when the function s itself is discontinuous at point t_i from partition (3.1) as a spline with defect $k_i = m + 1$. In particular step functions on $[a, b]$ with possible discontinuities at t_i $(i = 1, 2, \ldots, N-1)$ will be called *splines of zero degree* and the set of these functions will be denoted by $S_0(\Delta_N[a, b])$.

The value of the spline and its derivatives at the points of the partition will not be of importance for our future investigation. For definiteness, unless stated otherwise, we shall assume that

$$s^{(\nu)}(t_i) = [s^{(\nu)}(t_i + 0) + s^{(\nu)}(t_i - 0)]/2, \qquad i = 1, 2, \ldots, N-1.$$

[1] *Translator's note*: We prefer the term partition to the term mesh because $\Delta_N[a, b]$ will bear two meanings in the text: from one point of view $\Delta_N[a, b]$ is the set of points $\{t_0, t_1, \ldots, t_N\}$ and from the other – the set of intervals $\{[t_{i-1}, t_i]\colon i = 1, 2, \ldots, N\}$).

If spline s is defined only on $[a, b]$ then we set

$$s^{(\nu)}(a) = s^{(\nu)}(a+0), s^{(\nu)}(b) = s^{(\nu)}(b-0), \qquad \nu = 0, 1, \ldots, m.$$

We shall also work with periodic splines defined on the real line which can be obtained if we require that spline $s \in S_m^k(\Delta_N[a, b])$ satisfies

$$s^{(\nu)}(a) = s^{(\nu)}(b), \qquad \nu = 0, 1, \ldots, m-k \tag{3.2}$$

which will allow it to continue to be $(b-a)$-periodic without loss of smoothness. In this case points $t_0 = a$ and $t_N = b$ should be included in the knots of the spline.

When we consider periodic splines we shall usually take the period $[0, 2\pi]$ to be the basic interval and by $S_m^k(\Delta_N)$, $S_m^1(\Delta_N) = S_m(\Delta_N)$, we shall denote the linear manifold of 2π-periodic splines of degree m with defect k with respect to partition

$$\Delta_N = \Delta_N[0, 2\pi]: 0 = t_0 < t_1 < t_2 < \cdots < t_N = 2\pi. \tag{3.3}$$

So $s \in S_m^k(\Delta_N)$ means $s \in C^{m-k}$ and $s^{(m)}(t) = c_i$ $(t_{i-1} < t < t_i,\ i = 1, 2, \ldots, N)$.

The spline $s \in S_m(\Delta_N[a, b])$ or $s \in S_m(\Delta_N)$ for which

$$s^{(m)}(t) = (-1)^i \varepsilon, \qquad t_{i-1} < t < t_i, i = 1, 2, \ldots, N,$$

where $\varepsilon = 1$ or $\varepsilon = -1$, is called a *perfect* spline.

A *monospline* is a spline $s \in S_m^k(\Delta_N[a, b])$ or $s \in S_m^k(\Delta_N)$ such that $s^{(m)}(t) = c \neq 0$ at all points t which are not knots. Therefore for a monospline s we have that $s^{(m-1)}$ is a linear function on every interval (t_{i-1}, t_i) with identical slopes and discontinuities in the knots.

Here are some examples.

(1) The set of all polygons on $[a, b]$ with possible vertices at the points $t_1, \ldots, t_N$ from partition (3.1) is the set $S_1(\Delta_N[a, b])$ of splines of degree 1 and defect 1.

(2) Set

$$x_+^m = \begin{cases} x^m & x > 0, \\ 0, & x \leq 0. \end{cases} \qquad m = 0, 1, \ldots, \tag{3.4}$$

For a fixed ζ $(a < \zeta < b)$ function $s(t) = (t - \zeta)_+^m$ belonging to $C^{m-1}[a, b]$ can be treated as a spline of degree m and defect 1 in the only knot $t_1 = \zeta$. Let us mention some obvious properties of x_+^m (sometimes called the *truncated power function*):

$$(x_+^m)' = m x_+^{m-1}, m \geq 2; \qquad (x_+^1)' = x_+^0, x \neq 0;$$
$$x_+^m = x^m - (-1)^m (-x)_+^m; \qquad x^n x_+^m = x_+^{m+n}, n \geq -m.$$

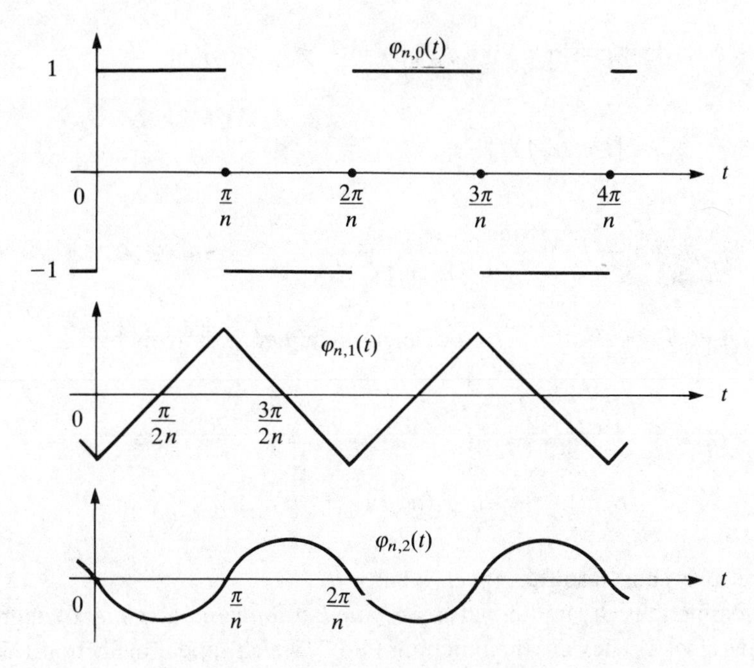

Fig. 2.2 Functions $\varphi_{n,m}(t)$ (Euler perfect splines).

(3) Let

$$t_k = a + kh_N, \qquad h_N = (b-a)/N, \qquad k = 0, 1, \ldots, N,$$

be the uniform partition of $[a, b]$. Construct on $[a, b]$ the splines $s_{N,m}(t)$ of degree n and defect 1 by setting

$$s_{N,0}(t) = (-1)^{k-1}, \qquad t_{k-1} < t < t_k, \qquad k = 1, 2, \ldots, N,$$

$$s_{N,0}(t_k) = 0, \qquad k = 0, 1, \ldots, N,$$

$$s_{N,m}(t) = \int_{\gamma_m}^{t} s_{N,m-1}(u)\,\mathrm{d}u, \qquad a \le t \le b, \qquad m = 1, 2, \ldots,$$

where $\gamma_m = a + h_N/2$ for odd m and $\gamma_m = a$ for even m.

The functions $s_{N,m}$ are called *perfect Euler splines*. If $N = 2n$ then $s_{N,m}^{(\nu)}(a) = s_{N,m}^{(\nu)}(b)$ $(\nu = 0, 1, \ldots, m-1)$ and the function $s_{2n,m}$ can continue to be $(b-a)$-periodic without loss of smoothness. For $[a, b] = [0, 2\pi]$ we obtain 2π-periodic perfect Euler splines with respect to the partition $\{k\pi/n\}$ $(k = 0, \pm 1, \pm 2, \ldots)$ which will play a very important role in our investigation of extremal problems in classes of periodic functions and for which we reserve, in this book, the notation $\varphi_{n,m}(t)$. So, by definition,

$$\varphi_{n,0}(t) = \operatorname{sgn} \sin nt, \qquad \varphi_{n,m}(t) = \int_{\gamma_m}^{t} \varphi_{n,m-1}(u)\, \mathrm{d}u,$$

$$\gamma_m = [1 - (-1)^m]\frac{\pi}{4n},$$

$$\varphi_{n,m}(t) = \frac{4}{\pi n^m} \sum_{\nu=0}^{\infty} \frac{\sin[(2\nu+1)nt - \pi m/2]}{(2\nu+1)^{m+1}}, \qquad m = 1, 2, \ldots .$$

(4) Let B_m be the 2π-periodic function defined iteratively by

$$B_1(t) = \begin{cases} (\pi - t)/2, & 0 < t < 2\pi, \\ 0, & t = 0, \end{cases}$$

$$B_m(t) = \int_{\alpha_m}^{t} B_{m-1}(u)\, \mathrm{d}u, \qquad m = 2, 3, \ldots, \tag{3.5}$$

where α_m is chosen so that $\int_0^{2\pi} B_m(t)\, \mathrm{d}t = 0$.

The functions B_m are known as *periodic Bernoulli monosplines* of degree m and defect 2. They are the functions (1.5.2) which appear in Section 1.5 in the integral representation of differentiable periodic functions. There is more detailed consideration of Euler and Bernoulli splines in Section 3.1.

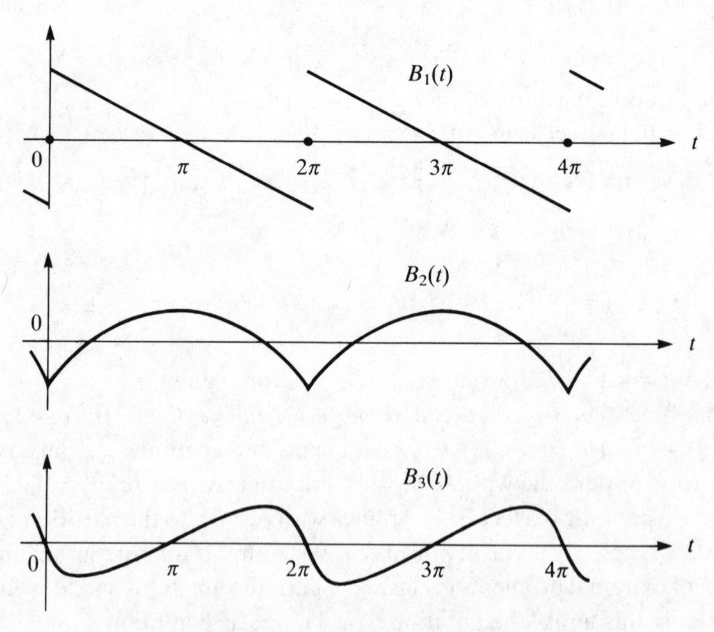

Fig. 2.3 Functions $B_m(t)$ (Bernoulli monosplines).

2.3.2 An analytic representation The most general (but not always very convenient) method for the analytic representation of splines is based on the truncated power functions (3.4).

Proposition 2.3.1 Spline s of degree m with respect to partition (3.1) with defect k_i $(0 \le k_i \le m+1)$ at point t_i $(i = 1, 2, \ldots, N-1)$ can be uniquely represented as

$$s(t) = \sum_{\nu=0}^{m} c_\nu (t-a)^\nu + \sum_{i=1}^{N-1} \sum_{j=0}^{k_i-1} \alpha_{ij}(t-t_i)_+^{m-j}, \qquad a \le t \le b; \tag{3.6}$$

moreover

$$c_\nu = s^{(\nu)}(a)/\nu!, \qquad \nu = 0, 1, \ldots, m;$$
$$\alpha_{ij} = [s^{(m-j)}(t_i+0) - s^{(m-j)}(t_i-0)]/(m-j)!,$$
$$i = 1, 2, \ldots, N-1, \qquad j = 0, 1, \ldots, k_i - 1.$$

On the other hand every function of the type (3.6), where $\alpha_{i,k_i-1} \ne 0$ $(i = 1, 2, \ldots, N-1)$ and $a < t_1 < t_2 < \cdots < t_{N-1} < b$ is a spline on $[a, b]$ of degree m with defect k_i at t_i.

Proof Set

$$p(t) = \sum_{\nu=0}^{m} \frac{1}{\nu!} s^{(\nu)}(a)(t-a)^\nu \tag{3.7}$$

and consider the function

$$g(t) = \int_a^t [s(u) - p(u)]\,\mathrm{d}u = \int_a^b [s(u) - p(u)](t-u)_+^0\,\mathrm{d}u, \qquad a \le t \le b. \tag{3.8}$$

On every interval (t_{i-1}, t_i) $(i = 1, 2, \ldots, N)$ the difference $s - p$ is a polynomial of degree m and hence integrating m-times by parts gives

$$\int_{t_{i-1}}^{t_i} [s(u) - p(u)](t-u)_+^0\,\mathrm{d}u$$
$$= -\sum_{j=0}^{m} \frac{1}{(j+1)!} [s^{(j)}(u) - p^{(j)}(u)](t-u)_+^{j+1} \Big|_{t_{i-1}+0}^{t_i-0}$$
$$= -\sum_{j=0}^{m} \frac{1}{(j+1)!} \{[s^{(j)}(t_i-0) - p^{(j)}(t_i)](t-t_i)_+^{j+1}$$
$$- [s^{(j)}(t_{i-1}+0) - p^{(j)}(t_{i-1})](t-t_{i-1})_+^{j+1}\}.$$

Taking a sum on i we obtain

$$g(t) = \sum_{i=0}^{N-1} \sum_{j=0}^{m} \frac{1}{(j+1)!} [s^{(j)}(t_i + 0) - p^{(j)}(t_i)](t - t_i)_+^{j+1}$$

$$- \sum_{i=1}^{N} \sum_{j=0}^{m} \frac{1}{(j+1)!} [s^{(j)}(t_i - 0) - p^{(j)}(t_i)](t - t_i)_+^{j+1}$$

$$= \sum_{i=1}^{N-1} \sum_{j=0}^{m} \frac{1}{(j+1)!} [s^{(j)}(t_i + 0) - s^{(j)}(t_i - 0)](t - t_i)_+^{j+1}$$

$$+ \sum_{j=0}^{m} \frac{1}{(j+1)!} [s^{(j)}(a + 0) - p^{(j)}(a)](t - a)_+^{j+1}$$

$$+ \sum_{j=0}^{m} \frac{1}{(j+1)!} [s^{(j)}(b - 0) - p^{(j)}(b)](t - b)_+^{j+1}$$

The last two sums are zero on $[a, b]$: the first one because of the definition of polynomial (3.7) and the second one because $(t - b)_+^{j+1} = 0$ for $t \le b$. Therefore

$$g(t) = \sum_{i=1}^{N-1} \sum_{j=0}^{m} \frac{1}{(j+1)!} [s^{(j)}(t_i + 0) - s^{(j)}(t_i - 0)](t - t_i)_+^{j+1}, \qquad a \le t \le b.$$

Taking a derivative of the above identity and using $g' = s - p$ (see (3.8)) we get

$$s(t) = p(t) + \sum_{i=1}^{N-1} \sum_{j=0}^{m} \frac{1}{j!} [s^{(j)}(t_i + 0) - s^{(j)}(t_i - 0)](t - t_i)_+^{j}, \qquad a \le t \le b,$$

and, in order to get (3.6), we have to observe that the terms corresponding to $j = 0, 1, \ldots, m - k_i$ vanish because of the continuity of $s^{(j)}(t)$ at t_i.

The uniqueness of representation (3.6) follows from the linear independence in $[a, b]$ of the system of functions

$$(t - a)^\nu, \ \nu = 0, 1, \ldots, m;$$
$$(t - t_i)_+^{m-j}, \qquad j = 0, 1, \ldots, k_i - 1, \qquad i = 1, 2, \ldots, N - 1, \qquad (3.9)$$

for different fixed $t_1, t_2, \ldots, t_{N-1} \in [a, b]$. The last assertion follows from the definition of the spline and the function (3.4) □

Corollary 2.3.2 A spline $s \in S_m^k(\Delta_N[a, b])$ has the representation

$$s(t) = \sum_{\nu=0}^{m} c_\nu (t-a)^\nu + \sum_{i=1}^{N-1} \sum_{j=0}^{k-1} \alpha_{ij} (t-t_i)_+^{m-j}, \tag{3.10}$$

where

$$c_\nu = s^{(\nu)}(a)/\nu!, \qquad \alpha_{ij} = [s^{(m-j)}(t_i+0) - s^{(m-j)}(t_i-0)]/(m-j)!,$$

and this is the only possible representation on these basis functions.

Corollary 2.3.3 For a fixed partition (3.1) the dimension of the linear manifold $S_m^k(\Delta_N[a, b])$ is $Nk + m - k + 1$.

The perfect spline s of degree m with respect to partition (3.1) can be written as

$$s(t) = \sum_{\nu=0}^{m} c_\nu (t-a)^\nu + \alpha \left[t^m + 2 \sum_{i=1}^{N-1} (-1)^i (t-t_i)_+^m \right], \qquad \alpha = \pm 1/m!; \tag{3.11}$$

hence

$$s^{(m)}(t) = \pm \left[1 + 2 \sum_{i=1}^{N-1} (-1)^i (t-t_i)_+^0 \right] (t \neq t_i)$$

and it is obvious that $s^{(m)}(t)$ alternates between ± 1 on the intervals of the partition.

For a monospline s of degree m with respect to partition (3.1) with defects k_i at t_i we have $s^{(m)}(t_i + 0) = s^{(m)}(t_i - 0)$ $(i = 1, 2, \ldots, N-1)$ and hence

$$s(t) = ct^m + \sum_{\nu=0}^{m-1} c_\nu (t-a)^\nu + \sum_{i=1}^{N-1} \sum_{j=1}^{k_i-1} \alpha_{ij} (t-t_i)_+^{m-j}, \tag{3.12}$$

where c is a non-zero constant.

In the periodic case representations (3.10)–(3.12) are of course valid but their coefficients should also satisfy equations (3.2). It is more convenient to represent periodic splines by Bernoulli monosplines. On $[0, 2\pi]$ we consider partition (3.3) having in mind that for a 2π-periodic spline the knots are all points $t_i \pm 2\pi n$ $(i = 0, 1, \ldots, N;\ n = 0, 1, \ldots,)$ and the defects in t_i and $t_i \pm 2\pi n$ are equal, in particular, $k_0 = k_N$.

For Bernoulli monosplines (3.5) (see also (1.5.2)) which have degree m, defect 2 and the only knots at $t_0 = 0$ on $[0, 2\pi)$ we have

$$\int_0^{2\pi} B_m(t)\,\mathrm{d}t = 0, \qquad B_m^{(\nu)}(t) = B_{m-\nu}(t), \qquad \nu = 1, 2, \ldots, m-1. \tag{3.13}$$

Lemma 2.3.4 For every partition (3.3) the function

$$g(t) = \sum_{k=1}^{N} \alpha_k B_1(t - t_k),$$

where $\alpha_1 + \alpha_2 + \cdots + \alpha_N = 0$, is a constant on every interval (t_{i-1}, t_i).

Indeed, for a fixed interval (t_{i-1}, t_i), function $B_1(t - t_k)$ has the form $c_k - t$ (see (3.5)). Then we have for $t_{i-1} < t < t_i$

$$g(t) = \sum_{k=1}^{N} \alpha_k(c_k - t) = \sum_{k=1}^{N} \alpha_k c_k = c.$$

Proposition 2.3.5 Every 2π-periodic spline s of degree m with respect to partition (3.3) with defect k_i $(0 \le k_i \le m + 1)$ at knots t_i $(i = 1, 2, \ldots, N)$ can be uniquely represented as

$$s(t) = \beta + \sum_{i=1}^{N} \sum_{j=0}^{k_i - 1} \beta_{ij} B_{m-j+1}(t - t_i) \tag{3.14}$$

where

$$\beta = \int_0^{2\pi} s(t)\, \mathrm{d}t, \tag{3.15}$$

$$\beta_{ij} = s^{(m-j)}(t_i + 0) - s^{(m-j)}(t_i - 0), \tag{3.16}$$

$$i = 1, 2, \ldots, N; \qquad j = 0, 1, \ldots, k_i - 1,$$

and coefficients β_{i0} are related by

$$\beta_{10} + \beta_{20} + \cdots + \beta_{N0} = 0. \tag{3.17}$$

Every function of the type (3.14), where $\beta_{i,k_i-1} \neq 0$ $(i = 1, 2, \ldots, N - 1)$, $0 < t_1 < t_2 < \cdots < t_N = 2\pi$ and (3.17) is satisfied, is a 2π-periodic spline of degree m with defect k_i at t_i.

Proof Set

$$g(t) = \int_0^{2\pi} [s(u) - \beta] B_1(t - u)\, \mathrm{d}u = \int_0^{2\pi} s(u) B_1(t - u)\, \mathrm{d}u,$$

where β is determined by (3.15) and remark that (see (1.5.1))

$$g'(t) = s(t) - \beta. \tag{3.18}$$

On every interval (t_{i-1}, t_i) $(i = 1, 2, \ldots, N)$ we have $s^{(m+1)}(t) = 0$; hence integrating m times by parts and (3.13) give

$$g(t) = \sum_{i=1}^{N} \int_{t_{i-1}}^{t_i} s(u)B_1(t-u)\,du = -\sum_{i=1}^{N}\sum_{\nu=0}^{m} s^{(\nu)}(u)B_{\nu+2}(t-u)\Big|_{t_{i-1}+0}^{t_i-0}.$$

Therefore using the coincidence of the values of the functions under summation we obtain

$$g(t) = \sum_{i=1}^{N}\sum_{\nu=0}^{m} [s^{(\nu)}(t_i+0) - s^{(\nu)}(t_i-0)]B_{\nu+2}(t-t_i).$$

If we take a derivative of this identity and compare the result with (3.18) we obtain

$$s(t) = \beta + \sum_{i=1}^{N}\sum_{\nu=0}^{m} [s^{(\nu)}(t_i+0) - s^{(\nu)}(t_i-0)]B_{\nu+1}(t-t_i).$$

For $\nu = 0, 1, \ldots, m-k_i$ the differences in square brackets are zero (because of the continuity of $s^{(\nu)}(t)$ at t_i), i.e. s can be written in the form (3.14) where β, β_{ij} are defined by (3.15) and (3.16) respectively. (3.17) is necessarily fulfilled because the sum of the jumps of a 2π-periodic step function on the period $[0, 2\pi)$ is zero. The uniqueness of representation (3.14) follows from the linear independence of the system of functions

$$1;\ B_{m-j+1}(t-t_i);\ j = 0, 1, \ldots, k_i - 1;\ i = 1, 2, \ldots, N, \tag{3.19}$$

for fixed different knots t_i.

Now let s be any function of the type (3.14), where points t_i are fixed and (3.17) holds. Having in mind that $B_r^{(r)}(u) = -\frac{1}{2}$ for $0 < u < 2\pi$, from (3.14) we obtain

$$s^{(m)}(t) = \sum_{k=1}^{N} \beta_{k0}B_1(t-t_k), \qquad t_{i-1} < t < t_i, \qquad i = 1, 2, \ldots, N,$$

and in view of Lemma 2.3.4 the right-hand side is a constant on every interval (t_{i-1}, t_i). By direct calculation we verify that $\beta_{i,k_i-1} \neq 0$ implies that s has defect k_i at t_i. □

Corollary 2.3.6 A spline $s \in S_m^k(\Delta_N)$ has the representation

$$s(t) = \beta + \sum_{i=1}^{N}\sum_{j=0}^{k-1} \beta_{ij}B_{m-j+1}(t-t_i), \qquad \sum_{i=1}^{N} \beta_{i0} = 0, \tag{3.20}$$

where

$$\beta = \int_0^{2\pi} s(t)\,dt,$$

$\beta_{ij} = s^{(m-j)}(t_i + 0) - s^{(m-j)}(t_i - 0), i = 1, 2, \ldots, N; j = 0, 1, \ldots, k - 1$, and this is the only possible representation on these basis functions.

Corollary 2.3.7 For a fixed partition (3.1) the dimension of the linear manifold $S_m^k(\Delta_N)$ for every $m = 0, 1, \ldots$ is Nk.

Indeed the number of basic functions in (3.19) is $Nk + 1$ but the coefficients are connected by (3.17).

A periodic function can change sign an even number of times only on the period, hence it makes sense to consider perfect periodic splines with respect to partition (3.3) for even N only, $N = 2n$. Moreover, condition $s^{(m)} \perp 1$ implies that the knots have to satisfy

$$\sum_{k=1}^{N} (-1)^k (t_k - t_{k-1}) = 0. \tag{3.21}$$

Under condition (3.21) the perfect spline $s \in S_m(\Delta_{2n})$ can be written as

$$s(t) = \beta + \varepsilon 2 \sum_{i=1}^{2n} (-1)^i B_{m+1}(t - t_i), \tag{3.22}$$

where $\varepsilon = 1$ or $\varepsilon = -1$.

A periodic monospline s of degree m with defect k with respect to partition (3.3) is determined by $s^{(m)}(t) = c$ for $t \neq t_i$. In order to ensure this condition and the periodicity of the monospline, it is necessary to set $\beta_{i1} = 0$ $(i = 1, 2, \ldots, N)$ and $\beta_{11} + \beta_{21} + \cdots + \beta_{N1} = c$ in the general representation (3.20). Therefore this spline has the form

$$s(t) = \beta + \sum_{i=1}^{N} \sum_{j=1}^{k-1} \beta_{ij} B_{m-j+1}(t - t_i), \qquad \sum_{i=1}^{N} \beta_{i1} = c,$$

where the numbers β, β_{ij} are determined as in (3.20). In particular, a monospline s of degree m with minimal defect $k = 2$ has the form

$$s(t) = \beta + c \sum_{i=1}^{N} \beta_i B_m(t - t_i), \qquad \sum_{i=1}^{N} \beta_i = 1,$$

where

$$\beta = \int_0^{2\pi} s(t)\,dt, \qquad \beta_i = s^{(m-1)}(t_i + 0) - s^{(m-1)}(t_i - 0).$$

2.4 Spline interpolation

2.4.1 General remarks It is clear from the analytic representation of a spline (see Section 2.3) that every spline of degree m is defined up to a polynomial of degree m (up to a constant in the periodic case) by the jumps in the discontinuous derivatives at the knots. It is also possible to determine the spline uniquely by prescribing its values (and the values of some derivatives) at fixed points. Such splines are called interpolating splines.

The spline s from the linear manifold $S_m^k(\Delta_N[a,b])$, i.e. of degree m defect k with respect to the partition

$$\Delta_N[a,b]\colon a = t_0 < t_1 < \cdots < t_N = b, \tag{4.1}$$

consists of N pieces of algebraic polynomials of degree m which match up to the $(m-k)$th derivatives at the $(N-1)$ knots $t_1, t_2, \ldots, t_{N-1}$. Therefore s has $(Nk + m - k + 1)$ free parameters. These parameters are in our disposal and we may require the spline to satisfy some interpolation condition of the type $s^{(\nu)}(\tau) = y_\nu$, where for $a < \tau < b$ it is natural to assume $0 \le \nu \le m - k$ because $s \in C^{m-k}[a,b]$. In the cases $\tau = a$ or $\tau = b$ these conditions are usually called *boundary* conditions and one allows $0 \le \nu \le m$. Boundary conditions can also be given as a (linear) equation connecting the values $s^{(\nu)}(a)$ and $s^{(\mu)}(b)$ $(0 \le \nu, \mu \le m)$. An important example of non-interpolating boundary conditions is the *periodicity* condition

$$s^{(\nu)}(a) = s^{(\nu)}(b), \qquad \nu = 0, 1, \ldots, m.$$

Obviously the problem of the existence of some $s \in S_m^k(\Delta_N[a,b])$ satisfying some interpolating or boundary conditions will be correctly posed if the number of these conditions is not greater than $Nk + m - k + 1$, i.e. the number of free parameters of the spline; in order to have uniqueness one usually needs the coincidence of these numbers.

While the problem of existence in algebraic and trigonometric interpolation can be solved rather easily (using, for example, the Vandermondian), the problem of existence in spline interpolation is more delicate. The difficulties arise from the fact that $S_m^k(\Delta_N[a,b])$ is not a Chebyshev set and from the strong dependence of the spline on the partition – this can be seen easily even for $m = 1$, i.e. interpolation by polygons with fixed knots.

We shall prove two general theorems (with interpolation and periodic conditions) which will guarantee the existence and unqueness in all simpler cases considered in this book.

The proof of the existence and uniqueness of interpolating splines usually follows the following scheme. First (this is the most difficult part technically) using the specific character of the spline one proves that every spline from $S_m^k(\Delta_N[a,b])$ satisfying zero interpolating conditions is identically zero. From this, taking into account the linear independence of the basic functions, one concludes that the interpolating spline is unique (if it exists) and the homogeneous system of equations determined by the zero interpolating conditions only has a zero solution with respect to the free coefficients. But then the non-homogeneous system (for given interpolating conditions) will have a unique solution, i.e. the interpolating spline will exist and will be unique.

In the first step of the scheme described above (and for other investigations) we need assertions typified in Rolle's theorem connected with the evaluation of the number of zeros or sign changes of a function or its derivatives.

Let us first introduce some notation. Let $f \in C[a, b]$ and for $n \geq 2$

$$f(x_1) = f(x_2) = \cdots = f(x_n) = 0, \qquad a \leq x_1 < x_2 < \cdots < x_n \leq b.$$

Points $x_1, x_2, \ldots, x_n$ will be called *separated zeros* of function f if f is not identically zero on every interval (t_k, t_{k+1}) $(k = 1, 2, \ldots, n-1)$. The maximal number of separated zeros of f on $[a, b]$ will be denoted by $\nu(f; [a, b])$; this number will always be finite in the cases considered. If $f \in C[a, b]$ has no separated zeros but has a zero and is not identically zero then we set $\nu(f; [a, b]) = 1$. If $f \in C[a, b]$ and $|f(t)| > 0$ for $a \leq t \leq b$ then we set $\nu(f; [a, b]) = 0$. The same definitions hold if we replace $[a, b]$ by $[a, b)$, $(a, b]$ or (a, b).

We say that the function f defined at every point of $[a, b]$ *essentially changes sign* on $[a, b]$ exactly n times if there exist $(n+1)$ points z_j on (a, b): $a < z_0 < z_1 < \cdots < z_n < b$ and a number $\varepsilon > 0$ such that for almost all $u \in (-\varepsilon, \varepsilon)$ we have

$$\operatorname{sgn} f(z_j + u) = -\operatorname{sgn} f(z_{j-1} + u) \neq 0, \qquad j = 1, 2, \ldots, n,$$

and there are no $n+2$ points on (a, b) with this property. The number n shall be called the *number of essential changes of sign* and the notation $\mu(f; [a, b])$ shall be attached to it. Equality $\mu(f; [a, b]) = n$ means that the integral $F(t) = \int_a^t f(u)\, du + c$ has on $[a, b]$ exactly $(n+1)$ monotonic intervals which alternately increase and decrease. We set $\mu(f; [a, b]) = 0$ if F is a monotonic non-constant function and $\mu(f; [a, b]) = -1$ if $F(t)$ is a constant.

For piecewise continuous functions (i.e. functions with a finite number of discontinuities on $[a, b]$) the above definition is somewhat simpler. It reads: a function f which is piecewise continuous on $[a, b]$ has n essential changes of sign on $[a, b]$ if there are $(n+1)$ points z_j: $a < z_0 < z_1 < \cdots < z_n < b$ such that f is continuous at each one of them and $\operatorname{sgn} f(z_j) = -\operatorname{sgn} f(z_{j-1}) \neq 0$ $(j = 1, 2, \ldots, n)$.

We note that the definition does not take non-essential changes of sign into account, i.e. such changes which do not affect the antiderivative; in other words $\mu(f; [a, b])$ remains the same if we change the value of f on a set of measure zero. Note also that $\mu(f; [a, b]) = \mu(f; (a, b))$.

We shall use the following assertion which can be easily verified by drawing a graph. For more details see [10A, Section 1.2].

Proposition 2.4.1 Let $f \in L_1[a, b]$ and

$$F(t) = \int_a^t f(u)\,\mathrm{d}u + c, \qquad a \le t \le b.$$

Then:

(1) if x_1 and x_2 $(a \le x_1 < x_2 \le b)$ are two separated zeros of F then f essentially changes sign on (x_1, x_2) at least once, i.e. $\mu(f; (x_1, x_2)) \ge 1$;

(2) the following inequalities hold

$$\mu(F; [a, b]) \le \nu(F; [a, b]) \le \mu(f; (a, b)) + 1,$$

moreover, if $F(a) = 0$ or $F(b) = 0$ then

$$\mu(F; [a, b]) \le \mu(f; (a, b)),$$

and if $F(a) = F(b) = 0$ then

$$\mu(F; [a, b]) \le \mu(f; (a, b)) - 1.$$

For a $2l$-periodic function it is natural to count the zeros and the essential changes of sign on a period $[a, a + 2l)$ or $(a, a + 2l]$. If f is a continuous $2l$-periodic function, then *the number of separated zeros of f on the period* is defined by

$$\nu(f; 2l) = \nu(f; (a, a + 2l)), \qquad f(a) \ne 0,$$

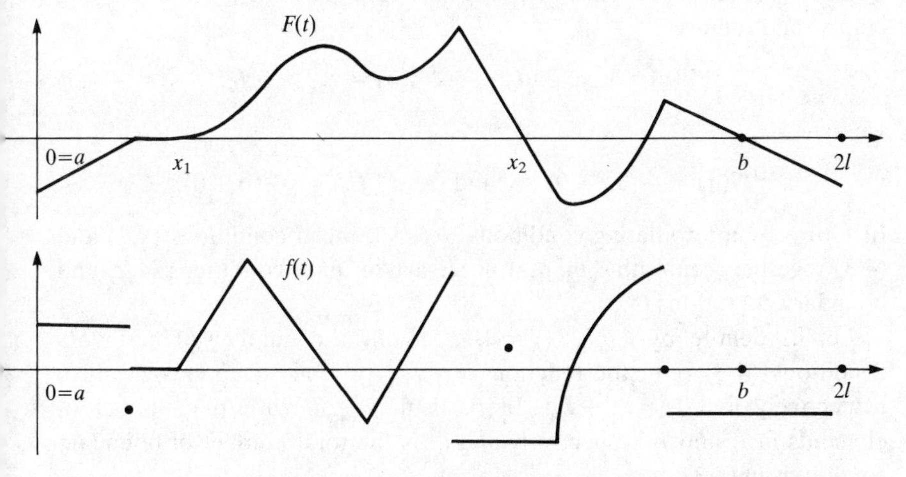

Fig. 2.4 To the statements 2.4.1 and 2.4.2. $\mu(f; (\mathrm{a}, \mathrm{b})) = 5$; $\mu(F; [a, b]) = 3$; $\nu(F; [a, b]) = 4$; $\mu(f; 2l) = 6$; $\mu(F; 2l) = \nu(F; 2l) = 4$.

which obviously does not depend on a provided $f(a) \neq 0$. For a continuous $2l$-periodic function f, the *number of essential changes of sign on the period* is given by

$$\mu(f; 2l) = \sup_a \mu(f; (a, a + 2l)).$$

In particular for a partially continuous function f we have

$$\mu(f; 2l) = \mu(f; (a, a + 2l)),$$

where a can be every point of continuity of f such that $f(a) \neq 0$. Obviously the number $\mu(f; 2l)$ is always even. For a continuous $2l$-periodic function f we get by definition $\mu(f; 2l) \leq \nu(f; 2l)$. We have [10A, p. 21]:

Proposition 2.4.2 Let f be a $2l$-periodic locally summable function defined on the real line

$$F(t) = \int_0^t f(u)\, du + c,$$

and $F(t) \not\equiv 0$, $F(0) = F(2l)$. Then $\nu(F; 2l) \leq \mu(f; 2l)$, $\mu(F; 2l) \leq \mu(f; 2l)$.

2.4.2 Interpolation on an interval Let the L points τ_n: $a < \tau_1 < \tau_2 < \cdots < \tau_L < b$ be fixed on (a, b). We want to find a spline $s \in S_m^k(\Delta_N[a, b])$ such that

$$s^{(\nu)}(\tau_n) = y_{n\nu}; \qquad \nu = 0, 1, \ldots, \gamma_n; \qquad n = 1, 2, \ldots, L, \tag{4.2}$$

where the numbers $y_{n\nu}$ and γ_n $(0 \leq \gamma_n \leq m - k)$ are given. In order to give suitable boundary conditions we consider two subsets I_a and I_b of $Q_{m+1} = \{0, 1, \ldots, m\}$, each of which may coincide with Q_{m+1} or may be empty, and require

$$s^{(\mu)}(a) = y_{a\mu}, \mu \in I_a; \qquad s^{(\kappa)}(b) = y_{b\kappa}, \kappa \in I_b, \tag{4.3}$$

where

$$s^{(\mu)}(a) = s^{(\mu)}(a + 0) \qquad \text{and} \qquad s^{(\kappa)}(b) = s^{(\kappa)}(b + 0).$$

In future by interpolating conditions we shall mean conditions (4.2) and (4.3) together, separating them if necessary to inner conditions (4.2) and boundary conditions (4.3).

Let us denote by $l_{a\nu}, l_{b\nu}$ $(\nu = 0, 1, \ldots, m)$ the number of boundary conditions (4.3) only for functions $s, s', \ldots, s^{(\nu)}$ at the points a and b respectively and $l_\nu = l_{a\nu} + l_{b\nu}$. In particular $l_{a\nu}, l_{b\nu}$ are the number of elements in I_a and I_b respectively and l_m is the total number of boundary conditions in (4.3).

In the interpolating problems considered the existence and uniqueness of the interpolating spline will depend on the distribution of the interpolating conditions both on the vertical (from s to $s^{(m)}$) and on the horizontal (from a to b). We shall characterize this distribution by:

(1) d_q $(q = 0, 1, \ldots, m)$ which is the number of interpolating conditions for $s^{(q)}$ on $[a, b]$, obviously

$$\sum_{q=0}^{m} d_q = \sum_{n=1}^{L} (\gamma_n + 1) + l_m;$$

(2) β_{ai} and β_{bi} $(i = 1, 2, \ldots, N-1)$ which are the number of all interpolating conditions for $s, s', \ldots, s^{(m)}$ on the semi-intervals $[a, t_i)$ and $(t_i, b]$ determined by the knots from partition (4.1).

Lemma 2.4.3 If $s \in S_m^k(\Delta_N[a, b])$ satisfies conditions

$$s^{(\nu)}(\tau_n) = 0, \qquad \nu = 0, 1, \ldots, \gamma_n, \qquad n = 1, 2, \ldots, L, \tag{4.4}$$

$$s^{(\mu)}(a) = s^{(\kappa)}(b) = 0, \qquad \mu \in I_a, \kappa \in I_b, \tag{4.5}$$

where

$$\sum_{n=1}^{L} (\gamma_n + 1) + l_m = Nk + m - k + 1, \tag{4.6}$$

$$d_0 + d_1 + \cdots + d_p \geq p + 1, \qquad p = 0, 1, \ldots, m, \tag{4.7}$$

$$\beta_{ai} \geq ik, \beta_{bi} \geq (N - i)k, \qquad i = 1, 2, \ldots, N-1, \tag{4.8}$$

then $s \equiv 0$ on [a, b].

Proof By contradiction assume that $s \not\equiv 0$ on $[a, b]$. Let us consider the following cases separately:

(1) $|s(t)| > 0$ a.e on $[a, b]$, i.e. s takes values 0 only in interpolating points;
(2) $|s(t)| > 0$ a.e on $[a, t_r]$ $(1 \leq r \leq N-1)$ but $s \equiv 0$ on $[t_r, t_{r+1}]$;
(3) $|s(t)| > 0$ a.e on $[t_r, b]$ $(1 \leq r \leq N-1)$ but $s \equiv 0$ on $[t_{r-1}, t_r]$;
(4) $|s(t)| > 0$ a.e on $[t_j, t_{j+r}]$ $(1 \leq j < j + r \leq N-1)$ but $s \equiv 0$ on $[t_{j-1}, t_j]$ and $[t_{j+r}, t_{j+r+1}]$.

In case (1) points $\tau_1, \tau_2, \ldots, \tau_L$ and also a (if $0 \in I_a$) and b (if $0 \in I_b$) are separated zeros of s and hence $\nu(s; [a, b]) \geq d_0$. If s' is continuous on $[a, b]$ then by Proposition 2.4.1 (point 1)) the number of the separated zeros of s'

on $[a, b]$ is bounded from below by the sum of $d_0 - 1$ and d_1 – the number of zeros of s' at $a, \tau_1, \ldots, \tau_L, b$. Therefore, using (4.7), we get

$$\nu(s'; [a, b]) \geq d_0 + d_1 - 1 \geq 1,$$

and applying the same reasoning consecutively we obtain

$$\nu(s^{(j)}; [a, b]) \geq \sum_{q=0}^{j} d_q - j \geq 1, \qquad j = 0, 1, \ldots, m - k. \tag{4.9}$$

We have to stop at $j = m - k$ because the derivative $s^{(m-k+1)}$ may be discontinuous. Noticing that

$$\sum_{q=0}^{m-k} d_q = \sum_{n=1}^{L} (\gamma_n + 1) + l_{m-k},$$

from (4.6) we get

$$\nu(s^{(m-k)}); [a, b]) \geq Nk - l_m + l_{m-k} + 1.$$

The derivative of $s^{(m-k)}$ essentially changes sign at least once on every interval between two neighboring separated zeros of $s^{(m-k)}$. Hence

$$\nu(s^{(m-k+1)}; (a, b)) \geq Nk - l_m + l_{m-k}. \tag{4.10}$$

Let us now estimate from the number $\nu(s^{(m-k+1)}; (a, b))$ starting with $s^{(m)}$. On every interval (t_{i-1}, t_i) of partition (4.1) function $s^{(m-\nu)}$ is a polynomial of degree m and cannot have more than ν sign changes. Going from $s^{(j+1)}$ to $s^{(j)}$ the number of sign changes on (t_{i-1}, t_i) may increase at most with l (see Proposition 2.4.1 (2)) and every boundary condition $s^{(j)}(a) = 0$ or $s^{(j)}(b) = 0$ decreases with l the maximal possible number of sign changes on (a, t_1) and (t_{N-1}, b) respectively. Therefore

$$\mu(s^{(m-k+1)}; (t_{i-1}, t_i)) \leq k - 1, \qquad i = 2, 3, \ldots, N - 1,$$

$$\mu(s^{(m-k+1)}; (a, t_1)) \leq k - 1 - (l_{am} - l_{a,m-k}),$$

$$\mu(s^{(m-k+1)}; (t_{N-1}, b)) \leq k - 1 - (l_{bm} - l_{b,m-k}).$$

Taking into account that derivative $s^{(m-k+1)}$ may also change sign in the knots $t_1, t_2, \ldots, t_{n-1}$, we get

$$\nu(s^{(m-k+1)}; (a, b)) \leq N(k - 1) - (l_m - l_{m-k}) + (N - 1) = Nk - l_m + l_{m-k} - 1,$$

which contradicts (4.10).

We shall only sketch the proofs for cases (2)–(4) because the arguments are similar. In case (2) let the points of interpolation on (a, t_r) be $\tau_1, \tau_2, \ldots, \tau_{L_r}$. Consider equations

$$s^{(j)}(t_r) = 0, \qquad j = 0, 1, \ldots, m - k, \tag{4.11}$$

as boundary conditions at the right end-point of the interval $[a, t_r]$, we can use estimate (4.9) interpreting d_q as the number of interpolating conditions for $s^{(q)}$ on $[a, t_r]$. With $\gamma_r = (\gamma_1 + 1) + \cdots + (\gamma_{L_r} + 1)$ we have

$$\nu(s^{(m-k)};[a, t_r]) \geq \gamma_r + l_{a,m-k} + (m - k + 1) - (m - k) = \gamma_r + l_{a,m-k} + 1.$$

It is easy to see that this estimate is also correct when there are no interpolating points in (a, t_r) and $\gamma_r = 0$. From the first condition in (4.8) we have $\gamma_r + l_{am} = \beta_{ar} \geq rk$ and hence

$$\nu(s^{(m-k)}; [a, t_r]) \geq rk - l_{am} + l_{a,m-k} + 1$$

which implies

$$\mu(s^{(m-k+1)}; (a, t_r)) \geq rk - l_{am} + l_{a,m-k}.$$

On the other hand, starting from $s^{(m)}$ and estimating from above as in case (1) the number of essential changes of sign of $s^{(m-k+1)}$ on (a, t_r), we shall come to the contradicting estimate

$$\mu(s^{(m-k+1)}; (a, t_r)) \leq r(k-1) + (r-1) - (l_{am} - l_{a,m-k})$$
$$= rk - l_{am} + l_{a,m-k-1}.$$

In case (3) we get a contradiction in a similar way using the second condition in (4.8).

Finally consider case (4). Let the number of interpolating conditions (4.4) on (t_j, t_{j+r}) be L_* (0 is also a possible value for L_*). From (see (4.8))

$$\beta_{a,j+r} \geq (j + r)k, \qquad \beta_{bj} \geq (N - j)k,$$

we see that on $[t_0, t_j]$ we have at least $(j + r)k - L_*$ interpolating conditions and on $[t_{j+r}, t_N]$ we have at least $(N - j)k - L_*$ interpolating conditions. But all interpolating conditions (4.4) and (4.5) are $Nk + m - k + 1$ because of (4.6). Therefore

$$[(j + r)k - L_*] + [(n - j)k - L_*] + L_* \leq Nk + m - k + 1$$

or

$$rk - L_* \leq m - k + 1. \tag{4.12}$$

Arguing as above, from the boundary conditions

$$s^{(\nu)}(t_j) = s^{(\nu)}(t_{j+r}) = 0, \qquad \nu = 0, 1, \ldots, m - k,$$

we get for the number of separated zeros of $s^{(m-k)}$ on $[t_j, t_{j+r}]$

$$\nu(s^{(m-k)}; [t_j, t_{j+r}]) \geq L_* + m - k + 2.$$

Using (4.12) we get

$$\nu(s^{(m-k+1)}; [t_j, t_{j+r}]) \geq L_* + m - k + 1 \geq rk.$$

Estimating the same quantity from above as in cases (1)–(3) we obtain

$$\nu(s^{(m-k+1)}; [t_j, t_{j+r}]) \leq r(k-1) + r - 1 = rk - 1$$

– a contradiction. This completes the proof. □

Now we easily obtain:

Theorem 2.4.4 Let $N \geq 2$, $1 \leq k \leq m$ and for a fixed partition (4.1) let the L points τ_n: $a < \tau_1 < \tau_2 < \cdots < \tau_L < b$, the integer numbers $\gamma_1, \gamma_2, \ldots, \gamma_L$ $(0 \leq \gamma_n \leq m-k)$ and the index sets I_a and I_b of the boundary conditions satisfy relations (4.6)–(4.8). Then for every $y_{n\nu}, y_{a\mu}, y_{b\kappa}$ there exists a unique spline $s \in S_m^k(\Delta_N[a, b])$ satisfying interpolating conditions (4.2) and (4.3). In particular, for every $f \in C^\gamma[a, b]$, where $\gamma = \max_{1 \leq n \leq L} \gamma_n, \gamma \leq m - k$, there exists a unique spline $s \in S_m^k(\Delta_N[a, b])$ satisfying the interpolating conditions

$$s^{(\nu)}(\tau_n) = f^{(\nu)}(\tau_n), \qquad \nu = 0, 1, \ldots, \gamma_n, \qquad n = 1, 2, \ldots, L,$$

and boundary conditions (4.3), where for $\mu \leq \gamma, \kappa \leq \gamma$ we can set $y_{a\mu} = f^{(\mu)}(a), y_{b\kappa} = f^{(\kappa)}(b)$.

Proof In view of Corollary 2.3.2 every spline $s \in S_m^k(\Delta_N[a, b])$ can be uniquely represented as

$$s(t) = \sum_{r=0}^{m} c_r(t-a)^r + \sum_{i=1}^{N-1} \sum_{j=0}^{k-1} \alpha_{ij}(t - t_i)_+^{m-j} \tag{4.13}$$

and so the system of equations (4.4), (4.5) can be considered as a system of linear equations with respect to the coefficients c_r, α_{ij}. Lemma 2.4.3 and the uniqueness of the above representation mean that the corresponding homogeneous system has a unique zero solution provided (4.6)–(4.8) are fulfilled. Hence the determinant of the system is not zero and for every $y_{n\nu}, y_{a\mu}, y_{b\kappa}$ the non-homogeneous system (4.2), (4.3) has a unique solution. □

Remarks

(1) Lemma 2.4.3 and Theorem 2.4.4 also hold for $N = 1$ (i.e. when s is a polynomial of degree m) provided the boundary conditions are effective, i.e. they will decrease the maximal possible number of sign changes going from $s^{(j+1)}$ to $s^{(j)}$. It is easy to check that this will happen if $l_m - l_{m-\nu} \leq \nu$ for $\nu = 1, 2, \ldots, m$.

(2) The proofs of Lemma 2.4.3 and Theorem 2.4.4 work even if there are no inner interpolating conditions. In this case equation (4.6) takes the form $l_m = Nk + m - k + 1$ and (4.7)–(4.8) are fulfilled at the expense of boundary conditions (4.5).

Let us mention two special cases of Theorem 2.4.4. First consider splines of defect 1. Let us give $2r$ boundary conditions requiring that $s \in S_m(\Delta_N[a, b])$ satisfies (for $j = 1, 2, \ldots, r, r \le m + 1$) the equations

$$s^{(q_j)}(a) = y_{aj}, \qquad s^{(q_j)}(b) = y_{bj}, \tag{4.14}$$

where $0 \le q_1 < q_2 < \cdots < q_r \le m$ and y_{aj}, y_{bj} are given numbers. Denote by l_ν the number of those from conditions (4.14) for which $q_j \le \nu$ (in particular $l_m = 2r$).

Proposition 2.4.5 Let $N \ge 2, 0 \le r \le m + 1$ and let the L points τ_n: $a < \tau_1 < \tau_2 < \cdots < \tau_L < b$, be situated so that for every $p > r$ the intervals (t_i, t_{i+p}) $(0 \le i \le N - p)$ from partition (4.1) contain at least $(p - r)$ of points τ_n. If the conditions

$$L + 2r = N + m, \tag{4.15}$$

$$L + l_\nu \ge \nu + 1, \qquad \nu = 0, 1, \ldots, m - 1, \tag{4.16}$$

are fulfilled then for every function $f \in C[a, b]$ there exists a unique spline $s \in S_m(\Delta_N[a, b])$ satisfying interpolating conditions

$$s(\tau_n) = f(\tau_n), \qquad n = 1, 2, \ldots, L,$$

and for $r \ge 1$ the boundary conditions (4.14), where we can set $y_{a1} = f(a), y_{b1} = f(b)$ provided $q_1 = 0$.

Indeed conditions (4.15) and (4.16) are equivalent in this case to conditions (4.6) and (4.7) respectively. Condition (4.8) is fulfilled because for every $p > r$ the intervals (t_i, t_{i+p}) contain at least $(p - r)$ interpolating points.

Let us note that inequalities (4.16) are necessarily fulfilled if q_j are, for example, chosen in one of the following ways:

$$q_j = j - 1, \qquad j = 1, 2, \ldots, r, \qquad r \le m,$$

$$q_j = 2(j - 1), \qquad j = 1, 2, \ldots, r, \qquad 2r \le m + 1, \tag{4.17}$$

$$q_j = 2j - 1, \qquad j = 1, 2, \ldots, r, \qquad 2r \le m. \tag{4.18}$$

Conditions (4.17) for odd m and $r = (m + 1)/2$ and conditions (4.18) for even m and $r = m/2$ are sometimes called the *Leedstone boundary conditions* (see e.g. [5A, p. 28]).

The second partial case is the *Hermite* spline interpolation. In this case the interpolating points τ_n coincide with the points from partition (4.1) and the behaviour of the spline on every interval (τ_{n-1}, τ_n) is uniquely determined by its values and the values of some of its derivatives only at the points τ_{n-1} and τ_n.

Let us first note that point a can be included between the interpolating points τ_n: $a = \tau_0$, provided the index set I_a for boundary conditions (4.3) is a segment, i.e. $I_a = \{0, 1, \ldots, \gamma_a\}$. The same is correct for point b. From Theorem 2.4.4 with $m = 2r - 1$, $k = r$, $\tau_n = t_n$, $\gamma_n = r - 1$ $(n = 0, 1, \ldots, N)$ we get:

Proposition 2.4.6 For every $f \in C^{r-1}[a, b]$ there is a unique $s \in S^r_{2r-1}(\Delta_N[a, b])$ such that

$$s^{(\nu)}(t_i) = f^{(\nu)}(t_i), \qquad \nu = 0, 1, \ldots, r-1, \qquad i = 0, 1, \ldots, N. \quad (4.19)$$

By the way, this result can be directly obtained using the fact that every polynomial of degree $2r - 1$ is uniquely determined by the conditions

$$p^{(\nu)}(t_{i-1}) = f^{(\nu)}(t_{i-1}), \qquad p^{(\nu)}(t_i) = f^{(\nu)}(t_i), \qquad \nu = 0, 1, \ldots, r-1.$$

The splines $s(t) = s(f, t)$ from $S^r_{2r-1}(\Delta_N[a, b])$ which are uniquely determined by the interpolating conditions (4.19) are called *Hermitian*.

Let us remark that splines from Proposition 2.4.6 are of odd degree. For even m a statement analogous to Proposition 2.4.6 is not valid. In this case a Hermitian spline interpolation can be done by introducing new knots.

Let points $z_i = (t_{i-1} + t_i)/2$ be added to the points from partition (4.1) and let us denote the extended partition by $\Delta'_N[a, b]$. Consider the set $S^r_{2r}(\Delta'_N[a, b])$ of all splines on $[a, b]$ of degree $2r$, with defect r at knots $t_1, t_2, \ldots, t_{N-1}$ and defect 1 at knots $z_1, z_2, \ldots, z_N$. In view of (3.6) every $s \in S^r_{2r}(\Delta'_N[a, b])$ can be given by

$$s(t) = \sum_{\nu=0}^{2r} c_\nu (t-a)^\nu + \sum_{i=1}^{N-1} \sum_{j=0}^{r-1} \alpha_{ij}(t - t_i)_+^{2r-j} + \sum_{i=1}^{N} \beta_i (t - z_i)_+^{2r}.$$

Proposition 2.4.7 For every $f \in C^r[a, b]$ there is a unique $s \in S^r_{2r}(\Delta'_N[a, b])$ such that

$$s^{(j)}(t_i) = f^{(j)}(t_i), \qquad j = 0, 1, \ldots, r, \qquad i = 0, 1, \ldots, N. \qquad (4.20)$$

This proposition may be considered as a consequence of Theorem 2.4.4 – one can easily verify that the new partition satisfies the requirements of the theorem. One can also prove the proposition by directly checking that the spline $s \in S^r_{2r}(\Delta'_N[a, b])$ is uniquely determined on every interval $[t_{i-1}, t_i]$ by conditions (4.20).

2.4.3 The periodic case It is natural to use 2π-periodic splines for interpolating periodic functions, i.e. splines $s \in S^k_m(\Delta_N)$ constructed with respect to partition (3.3) and satisfying periodic conditions $s^{(\nu)}(0) = s^{(\nu)}(2\pi)$ $(\nu = 0, 1, \ldots, m-k)$ which allow the smooth periodic continuation of the spline. Talking about Hermitian interpolation, one can solve the problems of existence and uniqueness exactly as in the non-periodic case because we construct the Hermitian spline locally, on every interval from the partition. In other situations essential modifications appear because of the periodic conditions: for example the dependence of the hypotheses for existence on the distribution of the zeros of the perfect spline constructed with respect to the same partition: these zeros may be 'forbidden' points for interpolation.

In connection with the necessarily even number of sign-changes of a periodic spline, it is natural (but not obligatory) to construct these splines with respect to partitions by dividing the period into an even number of intervals. Let $S_m(\Delta_{2n})$ be the linear manifold of all 2π-periodic splines of degree m and defect 1 with respect to partition

$$\Delta_{2n}\colon 0 = t_0 < t_1 < \cdots < t_{2n} = 2\pi. \tag{4.21}$$

If $s \in S_m(\Delta_{2n})$ then obviously

$$s^{(m)}(t) = \sum_{i=1}^{2n} c_i \psi_i(t), \qquad t \neq t_k, \tag{4.22}$$

where

$$\psi_i(t) = \begin{cases} 1, & t_{i-1} < t < t_i, \\ 0, & t \in [0, 2\pi] \backslash (t_{i-1}, t_i), \end{cases} \qquad \psi_i(t + 2\pi) = \psi_i(t),$$

$$i = 1, 2, \ldots, 2n.$$

It is clear that the step function $s^{(m)}$ can have no more than $2n$ essential changes of sign on the period and so Proposition 2.4.2 implies:

Proposition 2.4.8 For $s \in S_m(\Delta_{2n})$ we have

$$\mu(s; 2\pi) \leq \nu(s; 2\pi) \leq \mu(s'; 2\pi) \leq \cdots \leq \mu(s^{(m)}; 2\pi) \leq 2n.$$

We shall call partition (4.21) *m-normal* if there is a perfect spline $\varphi_m(t) = \varphi_m(t; \Delta_{2n})$ in $S_m(\Delta_{2n})$ with $2n$ simple zeros x_k on the period: $0 \leq x_1 < x_2 < \cdots < x_{2n} < 2\pi$. Such splines can exist only for some partitions, but if they exist then the splines are uniquely (up to the sign) determined by the partition and have no more zeros on the period because of Proposition 2.4.8. The uniform partition $\Delta_{2n} = \bar{\Delta}_{2n}$ $(t_i = i\pi/n)$ is m-normal for every $m = 0, 1, \ldots,$; the corresponding perfect spline is the Euler spline $\varphi_{n,m}$. Its zeros are the points $i\pi/n$ for even m and the points $(2i-1)\pi/(2n)$ for odd m.

Theorem 2.4.9 Let partition (4.21) be m-normal and let points τ_k satisfy

$$x_k < \tau_k < x_{k+1}, \qquad k = 1, 2, \ldots, 2n, \qquad x_{2n+1} = x_1 + 2\pi,$$

where x_k are the zeros of the corresponding perfect spline φ_m. Then for every set of numbers $y_1, y_2, \ldots, y_{2n}$ there is a unique spline $s \in S_m(\Delta_{2n})$ satisfying

$$s(\tau_k) = y_k, \qquad k = 1, 2, \ldots, 2n. \tag{4.23}$$

The proof is based on:

Lemma 2.4.10 Under the conditions of Theorem 2.4.9, if the spline $s \in S_m(\Delta_{2n})$ satisfies

$$|s(\tau_k)| \le |\varphi_m(\tau_k)|, \qquad k = 1, 2, \ldots, 2n. \tag{4.24}$$

then the coefficients c_i from representation (4.22) of $s^{(m)}$ satisfy

$$|c_i| \le 1, \qquad i = 1, 2, \ldots, 2n.$$

Proof In the case $m = 0$, i.e. when $s(t) = c_1\psi_1(t) + c_2\psi_2(t) + \cdots + c_{2n}\psi_{2n}(t)$, the conclusion is obvious. Let $m \ge 1$ and assume that for some $s \in S_m(\Delta_{2n})$ together with (4.24) we have

$$\max_{1 \le i \le 2n} |c_i| = |c_\nu| > 1. \tag{4.25}$$

If we set

$$s_*(t) = \varepsilon s(t)/|c_\nu|, \qquad \varepsilon = \pm 1, \tag{4.26}$$

then

$$s_*^{(m)}(t) = \sum_{i=1}^{2n} c_i^* \psi_i(t), \qquad t \ne t_k,$$

where $c_i^* = \varepsilon c_i / |c_{(\nu)}|$. Hence $|c_i^*| \le 1$ $(i = 1, 2, \ldots, 2n)$ and $c_\nu^* = \varepsilon \operatorname{sgn} c_{(\nu)} = \pm 1$.

Let us choose ε so that

$$s_*^{(m)}(t) = \varphi_m^{(m)}(t), \qquad t_{\nu-1} < t < t_\nu,$$

and set

$$\delta(t) = \varphi_m(t) - s_*(t).$$

Obviously $\delta \in S_m(\Delta_{2n})$ and the mth derivative $\delta^{(m)}$ being a constant on every interval (t_{i-1}, t_i) $(i = 1, 2, \ldots, 2n)$ is zero on $(t_{\nu-1}, t_\nu)$. Therefore $\delta^{(m)}$ can essentially change the sign on the period not more than $2n - 2$ times and then Proposition 2.4.8 implies $\mu(\delta, 2\pi) \le 2n - 2$. On the other hand in view of (4.24)–(4.26) we have

$$|s_*(\tau_k)| < |\varphi_m(\tau_k)|, \qquad k = 1, 2, \ldots, 2n,$$

and because of sgn $\varphi_m(\tau_k) = -\text{sgn}\, \varphi_m(\tau_{k+1})$ we obtain $\mu(\delta, 2\pi) \geq 2n$ which is a contradiction. This proves the lemma. □

Proof of Theorem 2.4.9 Now it is easy to show that every spline s from $S_m(\Delta_{2n})$ satisfying, under the conditions of the theorem, zero interpolating conditions:

$$s(\tau_k) = 0, \qquad k = 1, 2, \ldots, 2n, \tag{4.27}$$

is identically zero. Indeed, if this is not so, then at least one of the coefficients c_i in (4.22) is not zero, say $c_1 \neq 0$. But the spline $s(K, t) = Ks(t)$, $K \in \mathbb{R}$, belongs to $S_m(\Delta_{2n})$ and satisfies together with s conditions (4.27). We have

$$s^{(m)}(K, t) = K \sum_{i=1}^{2n} c_i \psi_i(t), \qquad t \neq t_k.$$

Therefore for every K

$$|s(K, \tau_k)| = 0 \leq |\varphi_m(\tau_k)|, \qquad k = 1, 2, \ldots, 2n,$$

and from Lemma 2.4.10 we get $|Kc_1| \leq 1$ which is impossible because $c_1 \neq 0$.

Therefore under the conditions of Theorem 2.4.9 (4.27) implies that $s \equiv 0$ provided $s \in S_m(\Delta_{2n})$. From here and the uniqueness of the representation (Corollary 2.3.6)

$$s(t) = \alpha_0 + \sum_{i=1}^{2n} \alpha_i B_{m+1}(t - t_i),$$

where

$$\alpha_1 + \alpha_2 + \cdots + \alpha_{2n} = 0, \tag{4.28}$$

it follows that the homogeneous system (4.27), (4.28) which is linear with respect to $\alpha_0, \alpha_1, \ldots, \alpha_{2n}$ has only the zero solution. Hence the solution of the non-homogeneous system (4.23), (4.28) exists and is unique. The theorem is proved. □

Let us formulate the important case for uniform partition $\bar{\Delta}_{2n}$: $\{i\pi/n\}_1^{2n}$, when $\varphi_m = \varphi_{n,m}$ the perfect Euler spline with zeros

$$\theta_{mk} = \frac{k\pi}{n} + [1 - (-1)^m] \frac{\pi}{4n}, \qquad k = 0, \pm 1, \pm 2, \ldots,. \tag{4.29}$$

We shall denote by $S_{2n,m}$ the linear manifold $S_m(\bar{\Delta}_{2n})$ of splines of degree m defect 1 with respect of the uniform partition $\bar{\Delta}_{2n}$.

Corollary 2.4.11 If the points τ_k satisfy inequalities $\theta_{mk} < \tau_k < \theta_{m,k+1}$, where the θ_{mk} are given by (4.29), then for every number $y_1, y_2, \ldots, y_{2n}$ there is a unique spline $s \in S_{2n,m}$ satisfying $s(\tau_k) = y_k$, $(k = 1, 2, \ldots, 2n)$.

2.4.4 Representations by fundamental splines and B-splines If a spline s from $S_m^k(\Delta_N[a,b])$ or $S_m^k(\Delta_N)$ is uniquely determined by interpolating conditions, then it can be uniquely represented by the fundamental splines – similarly to the Lagrange or Hermite formulae for interpolating polynomials.

Let us start with the periodic case. Let the space $S_m^k(\Delta_N)$ interpolate at a system of $L = Nk$ points τ_n on the period: $0 \le \tau_1 < \tau_2 < \cdots < \tau_L < 2\pi$.

Then to every point τ_n $(n = 1, 2, \ldots, L)$ there corresponds a spline $s_n \in S_m^k(\Delta_N)$, satisfying $s_n(\tau_j) = \delta_{nj}$. The system of fundamental splines $\{s_n\}_1^L$ is linearly independent and hence is a basis in $S_m^k(\Delta_N)$ (because $L = Nk = \dim S_{km}(\Delta_N)$). Therefore for every $f \in C$ the spline

$$s(f,t) = \sum_{n=1}^{L} f(\tau_n) s_n(t) \tag{4.30}$$

which obviously satisfies

$$s(f,\tau_n) = f(\tau_n), \qquad n = 1, 2, \ldots, L,$$

is this unique spline from $S_m^k(\Delta_N)$ that coincides with f at points τ_n.

In the general case – multiple interpolation and boundary interpolating conditions – one defines fundamental splines in a similar way: for any interpolating condition there is a corresponding spline taking the value 1 for this interpolating condition and value 0 for the remaining ones. Therefore, under the conditions of Theorem 2.4.4, the interpolating and boundary conditions (4.2), (4.3) determine the fundamental splines $s_{n,\nu}, s_{a,\mu}, s_{b,\kappa}$ satisfying the equations

$$\begin{array}{ll}
s_{n,\nu}^{(i)}(\tau_j) = \delta_{nj}\delta_{\nu i}, & \nu = 0, 1, \ldots, \gamma_n, i = 0, 1, \ldots, \gamma_j, n, j = 1, 2, \ldots, \mathrm{L}; \\
s_{n,\nu}^{(\mu)}(a) = 0,\ \mu \in I_a; & s_{n,\nu}^{(\kappa)}(b) = 0,\ \kappa \in I_b; \\
s_{a,\mu}^{(\nu)}(\tau_n) = s_{b,\kappa}^{(\nu)}(\tau_n) = 0, & \nu = 0, 1, \ldots, \gamma_n, n = 1, 2, \ldots, L; \\
s_{a,\mu}^{(i)}(a) = \delta_{mi},\ \mu, i \in I_a; & s_{b,\kappa}^{(j)}(b) = \delta_{\kappa,j},\ \kappa, j \in I_b.
\end{array}$$

The interpolating spline s from Theorem 2.4.4 has the following representation by the fundamental splines

$$s(t) = \sum_{n=1}^{L} \sum_{\nu=0}^{\gamma_n} f^{(\nu)}(\tau_n) s_{n,\nu}(t) + \sum_{\mu \in I_a} y_{a\mu} s_{a,\mu}(t) + \sum_{\kappa \in I_b} y_{b\kappa} s_{b,\kappa}(t),$$

where instead of $y_{a\mu}$ and $y_{b\kappa}$ one may take the values $f^{(\mu)}(a)$ and $f^{(\kappa)}(b)$ if this makes sense.

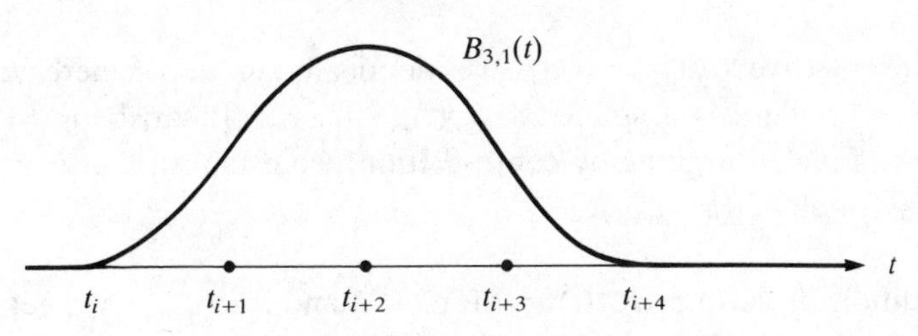

Fig. 2.5 B-spline order $m = 3$.

For $m \geq 2$ the fundamental splines have a complicated structure and from the computational point of view it is more convenient B-splines to be used as basic splines. They are different from zero only on a given interval of the partition which depends on m.

Let us extend partition (4.1) for a fixed $m = 1, 2, \ldots$, with the points

$$t_{-j} = t_0 - jh, \qquad t_{N+j} = t_N + jh, \qquad j = 1, 2, \ldots, m,$$

where $h = (b - a)/N$ (although the following constructions do not depend on the value of h). We get the system of points

$$t_{-m} < t_{-m+1} < \cdots < t_{-1} < t_0 < \cdots < t_N < t_{N+1} < \cdots < t_{N+m}. \tag{4.31}$$

Proposition 2.4.12 For every $i = -m, -m + 1, \ldots, N - 1$ there is a spline $B_{m,i}$ of degree m defect 1 with respect to the partition $t_i < t_{i+1} < \cdots < t_{i+m+1}$ satisfying the conditions

$$\begin{gathered} B^{(\nu)}_{m,i}(t_i) = \mathrm{B}^{(\nu)}_{m,i}(t_{i+m+1}) = 0, \qquad \nu = 0, 1, \ldots, m - 1; \\ B_{m,i}(t) > 0, \qquad t_i < t < t_{i+m+1}. \end{gathered} \tag{4.32}$$

These conditions determine the spline up to a constant multiplier.

The next lemma will be useful in other cases too.

Lemma 2.4.13 If $N \leq m$ and spline $s \in S_m(\Delta_N[a, b])$ satisfies conditions $s^{(\nu)}(a) = s^{(\nu)}(b) = 0$, $(\nu = 0, 1, \ldots, m - 1)$ then $s \equiv 0$ on $[a, b]$.

Indeed, assuming that $s \not\equiv 0$, we have $\nu(s; [a, b]) \geq 2$ and applying Proposition 2.4.1(1) $(m - 1)$ times we get $\nu(s^{(m-1)}; [a, b]) \geq m + 1 \geq N + 1$, which is impossible because the polygon $s^{(m-1)}$ has only N parts.

Proof of Proposition 2.4.12 The existence is proved following the familiar scheme. Take $i = 0$. First using Proposition 2.4.1 and Lemma 2.4.13 we deduce that a spline s of degree m defect 1 with respect to partition $t_0 < t_1 < \cdots < t_{m+1}$ which satisfies

$$s^{(\nu)}(t_0) = s^{(\nu)}(t_{m+1}) = 0, \qquad \nu = 0, 1, \ldots, m - 1, \tag{4.33}$$

and $s(\tau) = 0$ (fixed τ, $t_0 < \tau < t_{m+1}$) is identically zero. From here we get that for every y there is a spline $s \in S_m(\Delta_{m+1}[t_0, t_{m+1}])$ satisfying (4.33) and $s(\tau) = y$. Finally, arguing by contradiction, we establish that if $y > 0$ then $s(t) > 0$ for all $t \in (t_0, t_{m+1})$. □

In addition by setting $B_{m,i}(t) = 0$ for $t < t_i$ and $t > t_{i+m+1}$ we get a spline defined on the real line which is called a *B-spline* of degree m with respect to the set $t_i < t_{i+1} < \cdots < t_{i+m+1}$. In order to ensure the uniqueness of the B-spline one can normalize it by

$$\int_{-\infty}^{\infty} B_{m,i}(t)\, \mathrm{d}t = 1$$

or in some other way.

Proposition 2.4.14 Any system of $N + m$ B-splines $B_{m,i}$ $(i = -m, -m + 1, \ldots, N - 1)$ of degree m with respect to partition (4.31) with supports $[t_i, t_{i+m+1}]$ is a basis in $S_m(\Delta_N[a, b])$.

Proof Let

$$s(t) =: \sum_{i=-m}^{N-1} c_i B_{m,i}(t) = 0, \qquad a \le t \le b. \tag{4.34}$$

From $s(t) = 0$ for $t \le t_{-m}$ we get

$$s^{(\nu)}(t_{-m}) = s^{(\nu)}(t_0) = 0, \qquad \nu = 0, 1, \ldots, m - 1,$$

and by Lemma 2.4.13 $s(t) = 0$ for $t_{-m} \le t \le t_0$ and hence $s(t) = 0$ for all $t \le b$. Using the fact that $B_{m,-m}(t) > 0$ and $B_{m,i}(t) = 0$ $(i > -m)$ on (t_{-m}, t_{-m+1}) we conclude that $c_{-m} = 0$. Hence

$$s(t) = \sum_{i=-m+1}^{N-1} c_i B_{m,i}(t) = 0, \qquad t \le b.$$

For $t_{-m+1} < t < t_{-m+2}$ we have $B_{m,-m+1}(t) > 0$ and $B_{m,i}(t) = 0$ $(i > -m+1)$ and hence $c_{-m+1} = 0$ and so on. Therefore (4.34) implies $c_i = 0$ $(i = -m, -m + 1, \ldots, N - 1)$ which proves the proposition. □

If a spline $s \in S_m(\Delta_N[a, b])$ is uniquely determined by $N + m$ interpolating conditions, then a substitution of these conditions in the representation

$$s(t) = \sum_{i=-m}^{N-1} c_i B_{m,i}(t), \qquad a \le t \le b,$$

will give a system of linear equations with respect to c_i and, because of the finiteness of the supports of B-splines, the non-zero elements in the matrix of the system will be situated near the main diagonal.

Assuming partition (4.31) is $(b-a)$-periodic and is continued on the real line for $m \leq N-1$ we get

$$B_{m,i}^{(\nu)}(t_i) = B_{m,i}^{(\nu)}(t_i + b - a) = 0, \qquad \nu = 0, 1, \ldots, m-1,$$

and this allows us to continue $B_{m,i}$ periodically from $[t_i, t_i + b - a]$ to the real line. We get periodic B-splines $\tilde{B}_{m,i}(t)$, where $\tilde{B}_{m,i\pm N}(t) = \tilde{B}_{m,i}(t)$.

B-splines $\tilde{B}_{m,i}(t)$ $(i = 0, 1, \ldots, N-1)$ are a basis in the space of $(b-a)$-periodic splines of degree m defect 1 with respect to partition $\Delta_N[a, b]$.

2.5 On the existence of perfect splines with prescribed zeros

2.5.1 Introductory notes: Borsuk's theorem When solving some problems of spline approximation and especially in the search for lower bounds of N-widths of classes of functions one faces the problem of the existence of a perfect spline with prescribed zeros. In this section we give the necessary results in this direction. Although we consider spline interpolation here, the main idea differs totally from the one in the previous section.

In the problems of interpolation considered in Section 2.4 a spline s from $S_m(\Delta_N[a, b])$ and $S_m(\Delta_N)$ takes (under some hypothesis) given values at given points due to the choice of constants $c_i = s^{(m)}(t)$ $(t_{i-1} < t < t_i)$. But for a perfect spline $s \in S_m(\Delta_N[a, b])$ we have $s^{(m)}(t) = 0$ $(t_{i-1} < t < t_i)$ and hence the problem of the existence of a perfect spline vanishing at prescribed points will be meaningful if we start to consider a spline with a fixed number of *free* knots.

A set of splines (in particular perfect ones) with free knots is not a linear manifold; hence the methods used in Section 2.4 are not applicable and we are forced to consider arguments based on new ideas.

Our approach to solving the above problem is based on a deep, although intuitively clear, topological fact known as Borsuk's theorem (see Borsuk [1]).

Theorem 2.5.1 Let $S^N = \{\xi\colon \xi \in \mathbb{R}^{N+1}, \|\xi\| = r\}$, where $\|\bullet\|$ is some norm in $\mathbb{R}^{N+1}$, and let $\eta\colon S^N \to \mathbb{R}^N$, $\eta(\xi) = \{\eta_1(\xi), \eta_2(\xi), \ldots, \eta_N(\xi)\}$, be a continuous and odd $(\eta(-\xi) = -\eta(\xi))$ vector field on S^N. Then there is a zero in the vector field η, i.e. point $\xi^* \in S^N$ such that $\eta(\xi^*) = 0$.

The proof can be found in [25A, p. 84] for example.

2.5.2 The periodic case We start with the periodic case in which the main arguments are more transparent. The knots t_i of the perfect spline s of degree m are the points at which the derivative $s^{(m)}$ changes sign. The number of such points is necessarily even for a periodic spline. Denote by $\Gamma_{2n,m}$ the set of all periodic splines s of degree m with not more than $2n$ knots on the period $[0, 2\pi)$. From the analytic representation (3.22) and relation (3.21) for t_i we see that every $s \in \Gamma_{2n,m}$ has $2n$ free parameters and we can use them for satisfying $2n$ zero interpolating conditions. We shall prove the following general assertion.

Theorem 2.5.2 Let $0 \leq x_1 < x_2 < \cdots < x_L < 2\pi$; let n, m be positive integers and let an integer number γ_k $(0 \leq \gamma_k \leq m-1)$ be attributed to each point x_k, such that

$$\sum_{k=1}^{L} (\gamma_k + 1) = 2n. \tag{5.1}$$

Then there is a perfect spline $s \in \Gamma_{2n,m}$ such that

$$s^{(\nu)}(x_k) = 0, \qquad \nu = 0, 1, \ldots, \gamma_k, \qquad k = 1, 2, \ldots, L. \tag{5.2}$$

This spline is unique in $\Gamma_{2n,m}$ up to a sign change, it has exactly $2n$ knots on $[0, 2\pi)$ and it has no other zeros than the x_k.

Proof In the space $\mathbb{R}^{2n+1}$ of vectors $\boldsymbol{\xi} = \{\xi_1, \xi_2, \ldots, \xi_{2n+1}\}$ with norm

$$\|\boldsymbol{\xi}\| = |\xi_1| + |\xi_2| + \cdots + |\xi_{2n+1}|$$

we consider the sphere

$$S^{2n} = \{\boldsymbol{\xi}: \boldsymbol{\xi} \in \mathbb{R}^{2n+1}, \|\boldsymbol{\xi}\| = 2\pi\}$$

centered at the origin. To every $\boldsymbol{\xi} \in S^{2n}$ there correspond the following points

$$t_0 = 0, \qquad t_i = |\xi_1| + |\xi_2| + \cdots + |\xi_i|, \qquad i = 1, 2, \ldots, 2n+1 \tag{5.3}$$

in $[0, 2\pi]$. If $\xi_i = 0$ then $t_i = t_{i-1}$, but not all coordinates ξ_i are zero because $\|\xi\| = 2\pi$. Obviously $t_{2n+1} = 2\pi$.

Define on $[0, 2\pi)$ a function

$$\psi_0(\xi, t) = \begin{cases} \operatorname{sgn} \xi_i, & t_{i-1} < t < t_i, \qquad \xi_i \neq 0; \\ 0, & t = t_i, \qquad i = 0, 1, \ldots, 2n, \end{cases} \tag{5.4}$$

and note that it changes sign at most $2n$ times on $(0, 2\pi)$. For $0 \leq t \leq 2\pi$ set

$$\psi_\nu(\boldsymbol{\xi}, t) = \int_0^t \psi_{\nu-1}(\boldsymbol{\xi}, u)\, du - \frac{1}{2\pi} \int_0^{2\pi} \int_0^t \psi_{\nu-1}(\boldsymbol{\xi}, u)\, du\, dt, \tag{5.5}$$

$$\nu = 1, 2, \ldots, m-1;$$

$$s(\boldsymbol{\xi}, t) = \int_{x_L}^{t} \psi_{m-1}(\boldsymbol{\xi}, u)\, du. \tag{5.6}$$

Obviously $s(\boldsymbol{\xi}, t)$ is uniquely determined by $\boldsymbol{\xi} \in S^{2n}$,

$$s^{(j)}(\boldsymbol{\xi}, t) = \psi_{m-j}(\boldsymbol{\xi}, t), \qquad j = 1, 2, \ldots, m;\ 0 \leq t \leq 2\pi, \tag{5.7}$$

and for $j = 1, 2, \ldots, m-1$

$$\int_0^{2\pi} s^{(j)}(\boldsymbol{\xi}, t)\, dt = 0. \tag{5.8}$$

Therefore s is a perfect spline on $[0, 2\pi]$ of degree m with at most $2n$ knots, and it satisfies $s(\boldsymbol{\xi}, x_L) = 0$. We cannot yet continue it periodically while preserving the smoothness because, for $j = m$, the relation (5.8) or equivalently $s^{(m-1)}(\boldsymbol{\xi}, 0) = s^{(m-1)}(\boldsymbol{\xi}, 2\pi)$, may not be fulfilled. Borsuk's theorem 2.5.1 will give us a $\boldsymbol{\xi} \in S^{2n}$ such that both (5.8) and (5.2) hold.

Define on S^{2n} *a* $2n$-dimensional vector field $\eta(\xi)$ by

$$\eta_0(\boldsymbol{\xi}) = \int_0^{2\pi} \eta_0(\boldsymbol{\xi}, t)\, dt = 0, \tag{5.9}$$

$$\eta_{k,\nu}(\boldsymbol{\xi}) = s^{(\nu)}(\boldsymbol{\xi}, x_k), \qquad \nu = 0, 1, \ldots, \gamma_k, \qquad k = 1, 2, \ldots, L-1, \tag{5.10}$$

and if $\gamma_L \geq 1$

$$\eta_{L,\nu}(\boldsymbol{\xi}) = s^{(\nu)}(\boldsymbol{\xi}, x_L), \qquad \nu = 1, 2, \ldots, \gamma_L. \tag{5.11}$$

Because of (5.1) the equations (5.9)–(5.11) correspond to a $\boldsymbol{\xi} \in S^{2n}$ with exactly $2n$ numbers $\eta_0(\boldsymbol{\xi}), \eta_{k,\nu}(\boldsymbol{\xi})$ – the components of $\eta(\boldsymbol{\xi})$.

We shall prove that the field η is odd and continuous on S^{2n}. Oddness of the field follows immediately from the oddness of functions (5.9)–(5.11) with respect to $\boldsymbol{\xi}$: $s^{(j)}(-\boldsymbol{\xi}, t) = -s^{(j)}(\boldsymbol{\xi}, t)$ and hence $\eta_0(-\boldsymbol{\xi}) = -\eta_0(\boldsymbol{\xi}), \eta_{k,\nu}(-\boldsymbol{\xi}) = -\eta_{k,\nu}(\boldsymbol{\xi})$.

In order to prove the continuity let us remark first that for any two systems of points $\{t_i'\}, \{t_i''\}$ which are constructed with respect to (5.3) for the vectors $\boldsymbol{\xi}'$ and $\boldsymbol{\xi}''$ from S^{2n} respectively we have

$$|t_i' - t_i''| \leq \sum_{\nu=1}^{i} |\xi_\nu' - \xi_\nu''| \leq \|\boldsymbol{\xi}' - \boldsymbol{\xi}''\|, \qquad i = 1, 2, \ldots, 2n.$$

Hence, denoting the set of points $t \in [0, 2\pi]$ in which $\psi_0(\boldsymbol{\xi}', t) \neq \psi_0(\boldsymbol{\xi}'', t)$ by e, we have

$$\operatorname{meas} e \leq \sum_{i=1}^{2n} |t_i' - t_i''| \leq 2n\|\boldsymbol{\xi}' - \boldsymbol{\xi}''\|.$$

Bearing in mind that $|\psi_0(\boldsymbol{\xi}, t)| \leq 1$ we obtain

$$\int_0^{2\pi} |\psi_0(\xi', t) - \psi_0(\xi'', t)| \, dt \leq 4n\|\xi' - \xi''\|. \tag{5.12}$$

From this estimate, (5.5) and (5.6) we conclude

$$\|s^{(j)}(\xi', \bullet) - s^{(j)}(\xi'', \bullet)\|_C \leq M\|\xi' - \xi''\|, \qquad j = 0, 1, \ldots, m-1, \tag{5.13}$$

where M is a constant independent of ξ' and ξ''. Now the continuity of η_0 and $\eta_{k,\nu}$ follows from (5.12) and (5.13) respectively.

Now applying Theorem 2.5.1 we obtain $\xi^* \in S^{2n}$ such that $\eta(\xi^*) = 0$. Recalling formulae (5.6), (5.9)–(5.11) this means that

$$\int_0^{2\pi} s^{(m)}(\xi^*, t) \, dt = \int_0^{2\pi} \eta_0(\xi^*, t) \, dt = 0, \tag{5.14}$$

i.e.

$$s^{(m-1)}(\xi^*, 0) = s^{(m-1)}(\xi^*, 2\pi),$$

and

$$s^{(\nu)}(\xi^*, x_k) = 0, \qquad \nu = 0, 1, \ldots, \gamma_k, \qquad k = 1, 2, \ldots, L.$$

Equality (5.14) allows the 2π-periodic continuation of $s(\xi^*, t)$ without loss of smoothness and hence we have a perfect spline from $\Gamma_{2n,m}$ satisfying (5.2).

Before starting with the proof of unicity we shall formulate a statement whose easy proof is left to the reader (see Section 2.4.1).

Proposition 2.5.3 Let $f \in C^1, f \not\equiv 0$ and let f have n distinct zeros on $[0, 2\pi)$, m of which are not simple. Then $\nu(f'; 2\pi) \geq n + m$.

Assume that not only s but also the spline $s_1 \in \Gamma_{2n,m}$ also satisfies $s_1^{(\nu)}(x_k) = 0$ ($\nu = 0, 1, \ldots, \gamma_k$; $k = 1, 2, \ldots, L$) while for some $t^* \in [0, 2\pi)$ ($t^* \neq x_k$) we have $s_1(t^*) \neq s(t^*)$. We may assume that $s_1(t^*) > s(t^*) > 0$.

The function

$$\delta(t) = s(t) - \lambda s_1(t), \qquad \lambda = s(t^*)/s_1(t^*), \tag{5.15}$$

satisfies the $2n + 1$ equalities

$$\delta(t^*) = 0, \qquad \delta^{(\nu)}(x_k) = 0, \qquad \nu = 0, 1, \ldots, \gamma_k; \qquad k = 1, 2, \ldots, L,$$

and hence it has $L + 1$ distinct zeros on the period which are separated because the step function

$$\delta^{(m)}(t) = s^{(m)}(t) - \lambda s_1^{(m)}(t), \qquad 0 < t < 1,$$

does not vanish identically on any subinterval (α, β). Hence n $\nu(\delta; 2\pi) \geq L + 1$. Now using Proposition 2.5.3 for evaluating the separated zeros of $\delta', \delta'', \ldots, \delta^{(m-1)}$ we get (see (5.1))

$$\nu(\delta^{(m-1)}; 2\pi) \geq \sum_{k=1}^{L} (\gamma_k + 1) + 1 = 2n + 1.$$

But this is impossible because $\delta^{(m)}$ changes sign exactly at the points at which $s^{(m)}$ changes sign (which are no more than $2n$ on the period). This proves the unicity.

Using Proposition 2.5.3 to evaluate the number of separated zeros on the period consecutively of $s', s'', \ldots, s^{(m-1)}$ we get $\nu(s^{(m-1)}; 2\pi) \geq 2n$ and hence the number of sign changes of $s^{(m)}$, i.e. the number of knots of s on $[0, 2\pi)$, cannot be less than $2n$. So this number is exactly $2n$.

Finally if we assume that $s(t^*) = 0$, where $t^* \in [0, 2\pi)$, $t^* \neq x_k$ $(k = 1, 2, \ldots, L)$, we reach a contradiction using the same arguments as in the proof of unicity. This completes the proof of Theorem 2.5.2. □

Let us note that the same method can be employed in order to establish the existence under more general hypotheses (see Exercise 2.6).

2.5.3 The non-periodic case Let us denote the set of all perfect splines defined on $[a, b]$ of degree m with at most $N - 1$ knots on (a, b) by $\Gamma_{N,m}[a, b]$. Every spline $s \in \Gamma_{N,m}[a, b]$ has no more than $N + m - 1$ free parameters (see (3.11)), m of which are the coefficients of the polynomial $p \in P_{m-1}$. We are looking for a perfect spline $s \in \Gamma_{N,m}[a, b]$ satisfying $N + m - 1$ conditions of the type

$$s^{(\nu)}(\tau) = 0, \qquad a \leq \tau \leq b, \qquad 0 \leq \nu \leq m - 1, \tag{5.16}$$

determined by the pair (τ, ν). A set A of m conditions of type (5.16) is called *m-determining* if $p \in P_{m-1}$ and $p^{(\nu)}(\tau) = 0$, $(\tau, \nu) \in A$, implies $p \equiv 0$. This is equivalent to the existence, for every m numbers $y_{\tau\nu}$, $(\tau, \nu) \in A$, of a polynomial $p \in P_{m-1}$ such that $p^{(\nu)}(\tau) = y_{\tau\nu}$, $(\tau, \nu) \in A$.

For example the sets of pairs for Lagrange or Hermite interpolation are m-determining ($m - 1$ being the degree of the interpolating polynomial). In addition the set of m conditions of type (5.16) which give the corresponding Leedstone boundary conditions:

$\{(a, \nu), (b, \nu), \nu = 0, 2, 4, \ldots, m - 2\}$ for even m;

$\{(a, \nu), (b, \nu), \nu = 1, 3, 5, \ldots, m - 2, (t_0, 0), a \leq t_0 \leq b\}$ for odd $m \geq 3$

are m-determining. The proof is simple and we leave it to the reader.

In order to give the main result of this section we assume that for every $x_k \in [a, b]$ there is $I_k \subset Q_m := \{0, 1, \ldots, m - 1\}$, $I_k \neq \emptyset$, and l_k denotes the number of elements in I_k.

Theorem 2.5.4 Let $N = 2, 3, \ldots,$ and $m = 1, 2, \ldots,$ and for L fixed points x_k: $a \leq x_1 < x_2 < \cdots < x_L \leq b$ let the relation $l_1 + l_2 + \cdots + l_L = N + m - 1$ hold. Suppose that the sets I_k for the points x_k inside the interval $(a < x_k < b)$ be of the the type $I_k = \{0, 1, \ldots, l_k - 1\}$ and that there is an m-determining set A among the $N + m - 1$ pairs (x_k, ν_k) $(\nu_k \in I_k, k = 1, 2, \ldots, L)$. Then there exists up to the change in sign a unique spline $s \in \Gamma_{N,m}[a, b]$ which is unique up to a sign change such that

$$s^{(\nu_k)}(x_k) = 0, \qquad \nu_k \in I_k, k = 1, 2, \ldots, L. \tag{5.17}$$

Moreover s has precisely $N - 1$ knots in (a, b).

Proof We use the idea from the periodic case. Let $\mathbb{R}^N$ denote the space of vectors $\{\xi_1, \xi_2, \ldots, \xi_N\}$ with norm $\|\boldsymbol{\xi}\| = |\xi_1| + |\xi_2| + \cdots + |\xi_N|$ and $S^{N-1} = \{\boldsymbol{\xi}: \boldsymbol{\xi} \in \mathbb{R}^N, \|\boldsymbol{\xi}\| = b - a\}$. For $\boldsymbol{\xi} \in S^{N-1}$ we set $t_0 = a$, $t_i = a + |\xi_1| + \cdots + |\xi_i|$ $(i = 1, 2, \ldots, N)$ and define the function

$$\psi(\boldsymbol{\xi}, t) = \begin{cases} \operatorname{sgn} \xi_i, \; t_{i-1} < t < t_i \text{ if } \xi_i \neq 0, \mathrm{i} = 1, 2, \ldots, \mathrm{N}; \\ 0, \quad t = t_i, i = 0, 1, \ldots, N. \end{cases}$$

Obviously $\psi(\boldsymbol{\xi}, t)$ is defined on $[a, b]$ $(t_N = b)$ and it changes sign at most $N - 1$ times on (a, b). Set

$$g(\boldsymbol{\xi}, t) = \frac{1}{(m-1)!} \int_a^b (t-u)_+^{m-1} \psi(\boldsymbol{\xi}, u)\, du, \qquad a \leq t \leq b,$$

and choose the polynomial $p(\boldsymbol{\xi}, t) \in P_{m-1}$ so that

$$p^{(\nu_k)}(\boldsymbol{\xi}, x_k) = -g^{(\nu_k)}(\boldsymbol{\xi}, x_k), \qquad (x_k, \xi_k) \in A, \tag{5.18}$$

which is possible because A is m-determining. For

$$s(\boldsymbol{\xi}, t) = p(\boldsymbol{\xi}, t) + g(\boldsymbol{\xi}, t)$$

we have $s^{(m)}(\boldsymbol{\xi}, t) = \psi(\boldsymbol{\xi}, t)$, hence $s \in \Gamma_{N,m}[a, b]$. Moreover

$$s^{(\nu_k)}(\boldsymbol{\xi}, x_k) = 0, \qquad (x_k, \xi_k) \in A. \tag{5.19}$$

It is easy to see that functions $\psi(\boldsymbol{\xi}, t)$, $g(\boldsymbol{\xi}, t)$ and $p(\boldsymbol{\xi}, t)$ are odd with respect to $\boldsymbol{\xi}$ and hence

$$s^{(j)}(-\boldsymbol{\xi}, t) = -s^{(j)}(\boldsymbol{\xi}, t), \qquad j = 0, 1, \ldots, m-1.$$

The continuity of g and its first $m - 1$ derivatives with respect to $\boldsymbol{\xi} \in S^{N-1}$ becomes clear when we note that

$$|g^{(j)}(\boldsymbol{\xi}', t) - g^{(j)}(\boldsymbol{\xi}'', t)| \leq \frac{1}{(m-j-1)!} \times \int_a^b (t-u)_+^{m-j-1} |\psi(\boldsymbol{\xi}', u) - \psi(\boldsymbol{\xi}'', u)|\, du$$

and (as in the periodic case)

$$\int_a^b |\psi(\xi', u) - \psi(\xi'', u)| \, du \le 2N\|\xi' - \xi''\|.$$

The continuity of $p(\xi, t)$ and its derivatives with respect to ξ follows from the continuous dependence of $p(\xi, t)$ on the interpolating conditions (5.18) which determine it uniquely. Therefore the functions $s^{(j)}(\xi, t)$ $(j = 0, 1, \ldots, m-1)$ depend continuously (in the norm of $C[a, b]$) on $\xi \in S^{N-1}$.

Now for $\xi \in S^{N-1}$ we set

$$\eta_{k,\nu_k}(\xi) = s^{(\nu_k)}(\xi, x_k), \qquad \nu_k \in I_k, k = 1, 2, \ldots, L, \qquad (x_k, \nu_k) \notin A. \quad (5.20)$$

There are exactly $N-1$ equations in (5.20) which define an $(N-1)$-dimensional continuous odd vector field on S^{N-1}. Now Theorem 2.5.1 provides a $\xi^* \in S^{N-1}$ such that $\eta(\xi^*) = 0$, i.e.

$$s^{(\nu_k)}(\xi^*, x_k) = 0, \qquad \nu_k \in I_k, \qquad k = 1, 2, \ldots, L, \qquad (x_k, \nu_k) \notin A.$$

Taking relations (5.19) which hold for every $\xi \in S^{N-1}$ into account we conclude that the spline $s(t) = s(\xi^*, t)$ from $\Gamma_{N,m}[a, b]$ satisfies all conditions (5.17).

Unicity can be proved following the scheme from the periodic case. In order to establish the lower bound of the number of separated zeros of the derivatives of function (5.15) on $[a, b]$ one should use the arguments based on Proposition 2.4.1 which were applied to $s', s'', \ldots, s^{(m-1)}$ at the beginning of the proof of Lemma 2.4.3. Thus we arrive at the inequality

$$\nu(\delta^{(m-1)}; [a, b]) \ge N + 1,$$

which is impossible because $\delta^{(m)}$, just as $s^{(m)}$, has no more than $N-1$ essential sign changes on (a, b). In a similar way, arguing by contradiction, we obtain that spline $s(t) = s(\xi^*, t)$ has exactly $N-1$ zeros on (a, b). Theorem 2.5.4 is proved. □

It is easy to see that the proof of existence can also be applied to general index sets $I_k \in Q_m$.

Let us state three corollaries which will be used later.

Proposition 2.5.5 Let $a \le x_1 < x_2 < \cdots < x_{N+m-1} \le b$. There is a unique perfect spline $s \in \Gamma_{N,m}[a, b]$ satisfying conditions

$$s(x_k) = 0, \qquad k = 1, 2, \ldots, N + m - 1.$$

Moreover the knots $t_1, t_2, \ldots, t_{N-1}$ of s are situated so that

$$\beta_{ai} \ge i, \qquad \beta_{bi} \ge N - i, \qquad i = 1, 2, \ldots, N-1, \qquad (5.21)$$

where β_{ai} and β_{bi} are the number of zeros of s on $[a, t_i)$ and $(t_i, b]$ respectively.

Indeed, the existence and the uniqueness follow immediately from Theorem 2.5.4 because every m points x_k form an m-determining set. In order to prove (5.21) assume that for some i $(1 \le i \le N-1)$ the inequality $\beta_{ai} < i$ holds. Then $\nu(s; [t_i, b]) \ge N + m - 1$ and evaluating the number of separating zeros on $[t_i, b]$ of $s', s'', \ldots, s^{(m-1)}$ we obtain $\nu(s^{(m-1)}; [t_i, b]) \ge N - i + 1$. But this is impossible because the derivative $s^{(m)}$ essentially changes sign exactly $N - i - 1$ times. The second inequality (5.21) is proved similarly.

Proposition 2.5.6 Let $a < x_1 < x_2 < \cdots < x_{N-m-1} < b$. There is a unique perfect spline $s \in \Gamma_{N,m}[a, b]$ satisfying conditions

$$s(x_k) = 0, \qquad k = 1, 2, \ldots, N - m - 1;$$

$$s^{(j)}(a) = s^{(j)}(b) = 0, \qquad j = 0, 1, \ldots, m - 1.$$

Moreover the knots $t_1, t_2, \ldots, t_{N-1}$ of s are situated so that

$$\gamma_{ak} \ge k, \gamma_{bk} \ge N - m - k, \qquad k = 1, 2, \ldots, N - m - 1, \tag{5.22}$$

$$\beta_{ai} \ge i, \qquad \beta_{bi} \ge N - i, \qquad i = 1, 2, \ldots, N - 1, \tag{5.23}$$

where γ_{ak} and γ_{bk} are the number of knots of s on (a, x_k) and (x_k, b) respectively and β_{ai} and β_{bi} are the number of zeros of s (counting the multiplicity) on $[a, t_i)$ and $(t_i, b]$ respectively.

Here, the existence and uniqueness also follow from Theorem 2.5.4 because the zero interpolating conditions at point a form an m-determining set. We shall prove the first of the inequalities (5.22) (the proof of the second one is similar). The spline s has $k + 1$ separated zeros $a, x_1, \ldots, x_k$ on $[a, x_k]$. Applying Proposition 2.4.1(1) and taking into account that $s^{(j)}(a) = 0$ $(j = 0, 1, \ldots, m - 1)$ we get $\nu(s^{(m-1)}; [a, x_k]) \ge k + 1$ and hence $\mu(s^{(m)}; (a, x_k)) \ge k$, i.e. $\gamma_{ak} \ge k$.

Now (5.23) is an easy consequence of (5.22) and we may consider only $m < i < N - m$ because a and b are m-multiple zeros of s. For example if $\gamma_{b,i-m} \ge N - i$ then $x_{i-m} < t_i$ and hence $\beta_{ai} \ge i$.

In the third partial case, which is in some sense intermediate between the previous two – Leedstone boundary conditions, the number of zeros x_k coincides or differ by one from the number of knots t_i.

Proposition 2.5.7 Let the integers $m \ge 1, N \ge 2$ and the L points x_k: $a < x_1 < x_2 < \cdots < x_L < b$ be fixed, where $L = N - 1$ for even m and

$L = N$ for odd m. There is a unique perfect spline s in the set $\Gamma_{N,m}[a, b]$ satisfying the conditions

$$s(x_k) = 0, \qquad k = 1, 2, \ldots, L;$$

$$s^{(2j)}(a) = s^{(2j)}(b) = 0, \qquad j = 0, 1, \ldots, (m-2)/2, \qquad \text{for even } m.$$

$$s^{(2j-1)}(a) = s^{(2j-1)}(b) = 0, \qquad j = 1, 2, \ldots, (m-1)/2, \quad \text{for odd } m.$$

In particular, if $x_k = a + kh$ $(k = 1, 2, \ldots, N-1)$ for even m and $x_k = a + (k - 1/2)h$ $(k = 1, 2, \ldots, N)$ for odd m $(h = (b-a)/N)$ then

$$s(t) \equiv \varphi_{N,m}\left(\frac{2\pi}{b-a}(t-a)\right),$$

where $\varphi_{N,m}(t)$ is a Euler perfect spline (see Section 2.3.1).

Comments

Section 2.2 There are many investigations into the approximation properties of the partial sums of Fourier series – see e.g. Natanson [16A], Timan [26A] and Stepanets [24A]. In particular, the exact asymptotic of the Lebesgue constants has been investigated by Galkin [1]. Polynomial interpolation of functions is treated in detail in Goncharov [7A] and Davis [5A]. It seems that Favard [2,3] was the first to consider the λ-means of Fourier series in general form (i.e. in the type of polynomial (2.2.20)). Later on λ-means of Fourier series and interpolating polynomials were extensively studied for convergence and the estimates of the order of convergence in classes of functions – see the monographs of Timan [26A], Korovkin [12A], Dzyadyk [6A] and the survey paper of the author [31]. References for particular results will be given in the comments to Chapter 4.

Sections 2.3 and 2.4 Several monographs devoted to polynomial splines and especially to their approximation and extremal properties have appeared in recent times (Ahlberg, Nilson and Walsh [1A], Stechkin and Subbotin [23A], Zav'yalov, Kvasov and Miroshnichenko [28A], Korneichuk [10A], Schumaker [20A] etc.); in each of them the existence and uniqueness of interpolating splines are considered. In connection with Theorem 2.4.4 see the paper of Melkman [1] where a more general case is considered. Theorem 2.4.9 has appeared in Korneichuk [24], its version (Corollary 2.4.11) with uniform partition was proved earlier by Ahlberg, Nilson and Walsh [1A] for odd m and Subbotin [1] for even m (see also [9A, Section 10.7]). It seems that the investigation of the B-spline representation was started by Schoenberg [3]; for computational aspects of such representations see e.g. De Boor [4A] and [28A].

Section 2.5 The idea of using Borsuk's theorem in the proof of the existence of perfect splines with prescribed zeros is due to Ruban (see Velikin [4]). For some generalizations of perfect splines (ψ-splines – see Chapter 8) this problem has been solved, using different ideas, by Motornyi [1]. In the paper of Tikhomirov and Boyanov [1] similar results were implied from general theorems in convex analysis.

Exercises

1. Prove that the functions

$$f(t) = \frac{1}{2n-1} \sum_{k=0}^{2n-2} y_k \frac{\sin (n-1/2)(t-t_k)}{\sin [(t-t_k)/2]}, \qquad t_k = \frac{2k\pi}{2n-1};$$

$$g(t) = \frac{1}{2n} \sum_{k=0}^{2n-1} y_k \frac{\sin (n-1/2)(t-x_k)}{\sin [(t-x_k)/2]} + \frac{\cos nt}{2n} \sum_{k=0}^{2n-1} (-1)^k y_k, \; x_k = \frac{k\pi}{n},$$

are trigonometric polynomials of degree $n-1$ taking values y_k at points t_k, x_k respectively.

2. For Lebesgue constants of sums (2.19) with $\beta_n = \pi/(2n+1)$ obtain the representation

$$\left\| \mathcal{R}\left(\frac{\pi}{2n+1}\right) \right\|_{(C)} = \frac{2}{\pi} \int_0^{\pi} \frac{\sin t}{t} \mathrm{d}t - \frac{\gamma_n}{2n+1}, \qquad 0 < \gamma_n < 1.$$

3. Prove that the Lebesgue constants of de la Vallée–Poussin means

$$V_{n,m}(f,t) = \frac{1}{n-m} \{S_m(f,t) + S_{m+1}(f,t) + \cdots + S_{n-1}(f,t)\}, \qquad m < n,$$

satisfy

$$\|V_{n,m}\|_{(C)} = \frac{4}{\pi^2} \ln \frac{n}{n-m} + O(1).$$

4. Show that there is no spline s of order 2 defect 1 with respect to partition $\{a, (a+b)/2, b\}$ satisfying the conditions $s(a) = s((a+b)/2) = 0$, $s(b) = 1$, $s'(a) = s'(b)$.

5. Let $f \in L_2^r$ and s be a spline from $S_{2r-1}^k(\Delta_N[a,b])$ $(k \le r)$ satisfying the conditions $s^{(j)}(t_i) = f^{(j)}(t_i)$ $(i = 0, 1, \ldots, N; j = 0, 1, \ldots, k-1)$ (t_i – points of partition $\Delta_N[a,b]$) and if $k < r$ also $s^{(\nu)}(a) = s^{(\nu)}(b) = 0$ $(\nu = r, r+1, \ldots, 2r-k-1)$. Prove that

$$\|f^{(r)} - s^{(r)}\|_2^2 = \|f^{(r)}\|_2^2 - \|s^{(r)}\|_2^2$$

(here $\|\bullet\|_2 = \|\bullet\|_{L_2[a,b]}$).

6. Check that the existence in Theorem 2.5.2 also occurs when $s^{(\nu)}(x_k) = 0$ for not necessarily consecutive νs.

3

Comparison theorems and inequalities for the norms of functions and their derivatives

The main content of this chapter is related to the investigation of extremal problems in approximation theory which were formulated in general form in Section 1.2. The solution of these problems in particular cases usually needs very detailed investigations. They lead to some exact inequalities for norms or other characteristic of the functions or their derivatives which are of non-approximational character. Inequalities of this type are considered in the chapter. The theorems in Sections 3.2–3.4 are very general and are of independent interest. They will form the base for the proofs of the exact results in particular approximation problems.

The most substantial meaning is borne by assertions (we call them *comparison theorems*) in which the differential difference or the integral characteristics of the functions from a class $\mathcal{M}$ are evaluated by the corresponding characteristics of a standard function from the same class. We shall find that usually polynomial splines of a special kind and, in the first place, the Euler and Bernoulli splines introduced in Section 2.3 play the rôle of a standard function. These splines are also extremal elements in some approximation problems; their properties are given in Section 3.1.

3.1 Standard splines

1 The periodic perfect Euler splines For $\lambda > 0$ set

$$\varphi_{\lambda,0}(t) = \operatorname{sgn} \sin \lambda t = \frac{4}{\pi} \sum_{\nu=0}^{\infty} \frac{\sin (2\nu + 1)\lambda t}{2\nu + 1}, \tag{1.1}$$

$$\varphi_{\lambda,2i-1}(t) = \int_{\pi/(2\lambda)}^{t} \varphi_{\lambda,2i-2}(u)\, du = \frac{(-1)^i 4}{\pi \lambda^{2i-1}} \sum_{\nu=0}^{\infty} \frac{\cos (2\nu + 1)\lambda t}{(2\nu + 1)^{2i}}, \tag{1.2}$$

$$\varphi_{\lambda,2i}(t) = \int_0^t \varphi_{\lambda,2i-1}(u)\,du = \frac{(-1)^i 4}{\pi\lambda^{2i}} \sum_{\nu=0}^{\infty} \frac{\sin(2\nu+1)\lambda t}{(2\nu+1)^{2i+1}}, \tag{1.3}$$

$i = 1, 2, \ldots.$

Obviously

$$\varphi_{\lambda,m}^{(j)}(t) = \varphi_{\lambda,m-j}(t), \qquad j = 1, 2, \ldots, m, \tag{1.4}$$

$\varphi_{\lambda,t}$ has period $2\pi/\lambda$ and has zero mean value on the period. For $\lambda = n = 1, 2, \ldots$, we have the periodic Euler splines $\varphi_{n,m}$ previously defined in Section 2.3.1. Functions (1.1)–(1.3) dependent on a continuous parameter λ will be called *perfect Euler splines*. They can be written in the form

$$\varphi_{\lambda,m}(t) = \frac{4}{\pi\lambda^m} \sum_{\nu=0}^{\infty} \frac{\sin[(2\nu+1)\lambda t - \pi m/2]}{(2\nu+1)^{m+1}}, \qquad m = 0, 1, \ldots.$$

It is clear from (1.1)–(1.3) that functions $\varphi_{\lambda,m}$ for even m have simple zeros at points $k\pi/\lambda$ ($k = 0, \pm1, \pm2, \ldots$) and their graphs are symmetric with respect to points $(k\pi/\lambda, 0)$ or lines $t = (2k+1)\pi/(2\lambda)$. For odd m the functions $\varphi_{\lambda,m}$ have simple zeros at points $(2k+1)\pi/(2\lambda)$ and their graphs are symmetric with respect to points $((2k+1)\pi/(2\lambda), 0)$ or lines $t = k\pi/\lambda$. The functions $\varphi_{\lambda,m}$ have no other zeros and for $m \geq 1$ they are monotonic between every two consecutive extrema and preserve convexity (or concavity) between every two consecutive zeros. From (1.1)–(1.3) it follows immediately that

$$\varphi_{\lambda,m}(t) = \lambda^{-m}\varphi_{1,m}(\lambda t), \qquad m = 0, 1, \ldots. \tag{1.5}$$

Using the notation

$$\|\varphi_{\lambda,m}\|_\infty = \sup_{-\infty<t<\infty} |\varphi_{\lambda,m}(t)|, \qquad m = 0, 1, \ldots$$

from (1.1)–(1.3) and the properties mentioned above we conclude

$$\|\varphi_{\lambda,2i}\|_\infty = (-1)^{i+k}\varphi_{\lambda,2i}\left(\frac{(2k+1)\pi}{2\lambda}\right) = \frac{4}{\pi\lambda^{2i}} \sum_{\nu=0}^{\infty} \frac{(-1)^\nu}{(2\nu+1)^{2i+1}}, \tag{1.6}$$

$$\|\varphi_{\lambda,2i-1}\|_\infty = (-1)^{i+k}\varphi_{\lambda,2i-1}\left(\frac{k\pi}{\lambda}\right) = \frac{4}{\pi\lambda^{2i-1}} \sum_{\nu=0}^{\infty} \frac{1}{(2\nu+1)^{2i}}, \tag{1.7}$$

$i = 0, 1, \ldots.$

The sums of the series on the right-hand sides of (1.6) and (1.7) can be expressed by the *Favard constants* (see Favard [1, 2])

$$K_m = \frac{4}{\pi}\sum_{\nu=0}^{\infty}\frac{(-1)^{\nu(m+1)}}{(2\nu+1)^{m+1}}, \qquad m = 0, 1, \ldots. \tag{1.8}$$

It is not difficult to see that

$$K_0 = 1, \qquad K_1 = \pi/2, \qquad K_2 = \pi^2/8, \qquad K_3 = \pi^3/24, \ldots,$$

and

$$1 = K_0 < K_2 < \ldots < 4/\pi < \ldots < K_3 < K_1 = \pi/2. \tag{1.9}$$

Comparing (1.6), (1.7) and (1.8) we get

$$\|\varphi_{\lambda,m}\|_\infty = K_m\lambda^{-m}, \qquad m = 0, 1, \ldots, \tag{1.10}$$

and because of (1.4)

$$\|\varphi_{\lambda,m}^{(j)}\|_\infty = \|\varphi_{\lambda,m-j}\|_\infty = K_{m-j}\lambda^{-m+j}.$$

So we have the following proposition.

Proposition 3.1.1 For all $\lambda > 0$ we have

$$\|\varphi_{\lambda,m-j}\|_\infty = c_{mj}\|\varphi_{\lambda,m}\|_\infty^{1-j/m}, \qquad j = 1, 2, \ldots, m, \qquad m = 1, 2, \ldots, \tag{1.11}$$

where

$$c_{mj} = K_{m-j}/K_m^{1-j/m}. \tag{1.12}$$

In the case of a positive integer λ, i.e. $\lambda = n = 1, 2, \ldots$, the number of the smallest periods of $\varphi_{n,m}$ on $[0, 2\pi]$ is an integer and hence $\varphi_{n,m} \in C$. From

$$\|\varphi_{n,m}\|_1 =: \int_0^{2\pi} |\varphi_{n,m}(t)|\,\mathrm{d}t = \bigvee_0^{2\pi} \varphi_{n,m+1} = 4n\,\|\varphi_{n,m+1}\|_\infty$$

we get

$$\|\varphi_{n,m}\|_\infty = 4K_{m+1}n^{-m}. \tag{1.13}$$

Together with the perfect splines $\varphi_{n,m}$ we shall see that the extrema in some variational problems are the functions

$$g_{n,m}(t) = \frac{1}{4n}\varphi_{n,m}(t),$$

for which we have

$$\bigvee_0^{2\pi} g_{n,m}^{(m)} = \bigvee_0^{2\pi} g_{n,0} = 1.$$

From

$$\|g_{n,m-1}\|_1 = \bigvee_0^{2\pi} g_{n,m} = K_m n^{-m}, \qquad m = 1, 2, \ldots, \tag{1.14}$$

we see that $g_{n,m}$ satisfied an assertion similar to 3.1.1 in the metric of L_1.

Proposition 3.1.2 For all $n = 1, 2, \ldots$, we have

$$\|g^{(j)}_{n,m-1}\|_1 = \|g_{n,m-j-1}\|_1 = c_{mj}\|\varphi_{n,m-1}\|_1^{1-j/m}, \tag{1.15}$$

$$j = 1, 2, \ldots, m-1, \qquad m = 2, 3, \ldots,$$

where c_{mj} are defined by (1.12).

3.1.2 Bernoulli monosplines These 2π-periodic functions were introduced in Section 2.3 by the equations

$$B_1(t) = \begin{cases} (\pi - t)/2, & 0 < t < 2\pi, \\ 0, & t = 0, \end{cases}$$

$$B_m(t) = \int_{\beta m}^{t} B_{m-1}(u)\, du, \qquad m = 2, 3, \ldots, \tag{1.16}$$

where β_m is chosen by the condition $B_m \perp 1$. We have also met the Fourier series of B_m in Section 1.5 (see (1.5.2)). It is

$$B_{2\nu}(t) = (-1)^\nu \sum_{k=1}^{\infty} \frac{\cos kt}{k^{2\nu}}, \; B_{2\nu-1}(t) = (-1)^{\nu-1} \sum_{k=1}^{\infty} \frac{\sin kt}{k^{2\nu-1}}, \tag{1.16'}$$

$$\nu = 1, 2, \ldots.$$

and we see that the function B_m is even for even indices m and odd for odd indices m. Moreover

$$B_{2\nu}(\pi - t) = B_{2\nu}(\pi + t), \qquad B_{2\nu-1}(\pi - t) = -B_{2\nu-1}(\pi + t),$$

i.e. the graph of $B_{2\nu}$ is symmetric with respect to the lines $t = k\pi$ ($k = 0, \pm 1, \pm 2, \ldots,$) and the graph of $B_{2\nu-1}$ is symmetric with respect to the points $(k\pi, 0)$.

It follows from the recurrent definition (1.16) that

$$B^{(j)}_m(t) = B_{m-j}(t), \qquad j = 1, 2, \ldots, m-1, \tag{1.17}$$

and the functions $B_{2\nu}$ are strictly monotonic and have only one zero on each of the intervals $(0, \pi)$ and $(\pi, 2\pi)$ and the functions $B_{2\nu-1}$ vanish at points $k\pi$ ($k = 0, \pm 1, \pm 2, \ldots$) and have only one extremum on $(0, \pi)$ and $(\pi, 2\pi)$.

Some numerical characteristics of monosplines B_m can be expressed by Favard constants:

$$\max_t B_{2\nu}(t) - \min_t B_{2\nu}(t) = |B_{2\nu}(0) - B_{2\nu}(\pi)| = (\pi/2)\, K_{2\nu-1},$$
$$\|B_{2\nu-1}\|_1 = 2|B_{2\nu}(0) - B_{2\nu}(\pi)| = \pi\, K_{2\nu-1}. \tag{1.18}$$

In some cases it is more convenient to work with monosplines normalized differently from B_m (see (1.17)). With the continuous parameter $\lambda > 0$ we set

$$\mathcal{B}_{\lambda,m}(t) = -2\lambda^{-m} B_m(\lambda t) = -2\lambda^{-m} \sum_{k=1}^{\infty} \frac{\cos(k\lambda t - \pi m/2)}{k^m}. \tag{1.19}$$

Obviously $\mathcal{B}_{\lambda,m}$ is of period $2\pi/\lambda$, with the same symmetry as $B_m(\lambda t)$. From

$$\mathcal{B}_{\lambda,1}(t) = t - \pi/\lambda, \qquad 0 < t < 2\pi/\lambda,$$

we see (this is the new normalization)

$$\mathcal{B}_{\lambda,m}^{(m)} = 1, \qquad t \neq 2k\pi/\lambda, \qquad k = 0, \pm 1, \pm 2, \ldots.$$

Functions $\mathcal{B}_{\lambda,m}$ will also be called *Bernoulli monosplines*. Set

$$H_{\lambda m}^{+} = \sup_t \mathcal{B}_{\lambda,m}(t), \qquad H_{\lambda m}^{-} = \inf_t \mathcal{B}_{\lambda,m}(t), \tag{1.20}$$

$$H_{\lambda m} = (H_{\lambda m}^{+} - H_{\lambda m}^{-})/2. \tag{1.21}$$

From (1.18) and (1.19) we get

$$H_{\lambda,2\nu} = \tfrac{1}{2}|\mathcal{B}_{\lambda,2\nu}(0) - \mathcal{B}_{\lambda,2\nu}(\pi)| = \frac{\pi}{2\lambda^{2\nu}} K_{2\nu-1}, \qquad \nu = 1, 2, \ldots,$$

and from the symmetry of $\mathcal{B}_{\lambda,2\nu-1}$ we obtain ($\nu = 1, 2, \ldots,$)

$$\begin{aligned} H_{\lambda,2\nu+1} &= H_{\lambda,2\nu+1}^{+} = -H_{\lambda,2\nu+1}^{-} = \max_{0<t<\pi/\lambda} |\mathcal{B}_{\lambda,2\nu+1}(t)| \\ &= \lambda^{-2\nu-1} \max_{0<t<\pi} |\mathcal{B}_{1,2\nu+1}(t)| = \lambda^{-2\nu-1} |\mathcal{B}_{1,2\nu+1}(t_\nu)| \\ &= \frac{2}{\lambda^{2\nu+1}} \sum_{k=1}^{\infty} \frac{\sin kt_\nu}{k^{2\nu+1}}, \end{aligned}$$

where t_ν is the only point of extremum of $\mathcal{B}_{1,2\nu+1}$ in $(0, \pi)$. Noticing that $H_{\lambda 1} = \pi/\lambda$ we conclude that

$$H_{\lambda m} = A_m/\lambda^m, \qquad m = 1, 2, \ldots, \tag{1.22}$$

where the constant A_m depends only on m. This allows us to state an analogy to Propositions 3.1.1 and 3.1.2.

Proposition 3.1.3 We have

$$H_{\lambda,m-j} = \gamma_{mj} H_{\lambda m}^{1-j/m}, \qquad j = 1, 2, \ldots, m-1, \qquad m = 1, 2, \ldots,$$

where the constants γ_{mj} do not depend on λ.

3.1.3 Intermediate standard splines We can continuously 'connect' perfect splines $\varphi_{\lambda,m}$ and monosplines $\mathcal{B}_{\lambda,m}$ by the intermediate $2\pi/\lambda$-periodic splines $\varphi_{\lambda,m}(\alpha, \beta; t)$ $(\alpha > 0, \beta > 0)$ of degree m defect 1 with zero mean value on the period. They are defined by

$$\varphi_{\lambda,m}^{(m)}(\alpha, \beta; t) = \varphi_{\lambda,0}(\alpha, \beta; t) = \begin{cases} \alpha, & 0 < t < \gamma \\ -\beta, & \gamma < t < 2\pi/\lambda, \\ 0, & t = 0, \gamma, \end{cases}$$

where the number $\gamma = \gamma(\alpha, \beta)$ is chosen so that

$$\int_0^{2\pi/\lambda} \varphi_{\lambda,0}(\alpha, \beta; t)\, dt = 0,$$

i.e. $\gamma = 2\pi\beta/(\alpha\lambda + \beta\lambda)$. Therefore the partition which uniquely determines $\varphi_{\lambda,m}(\alpha, \beta; t)$ contains the points $2k\pi/\lambda$ and $\gamma + 2k\pi/\lambda$ $(k = 0, \pm 1, \pm 2, \ldots,)$ and on the period $[0, 2\pi/\lambda)$ function $\varphi_{\lambda,m}(\alpha, \beta; t)$ changes sign at exactly two points.

If $\alpha = \beta = 1$ we have the perfect Euler splines: $\varphi_{\lambda,m}(1, 1; t) = \varphi_{\lambda,m}(t)$.

It is easy to show that for every t

$$\lim_{\beta \to \infty} \varphi_{\lambda,m}(1, \beta; t) = \mathcal{B}_{\lambda,m}(t), \qquad m = 1, 2, \ldots, \tag{1.23}$$

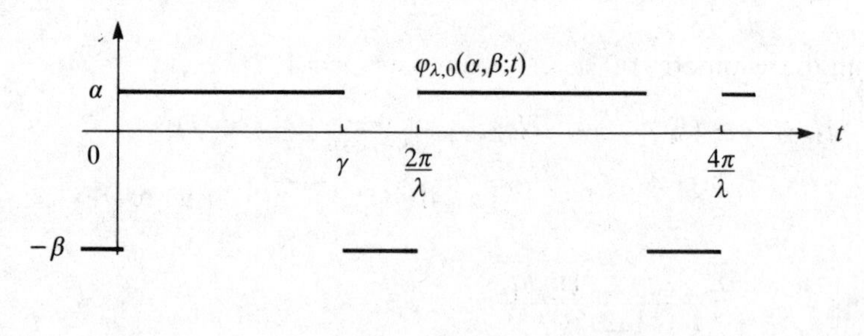

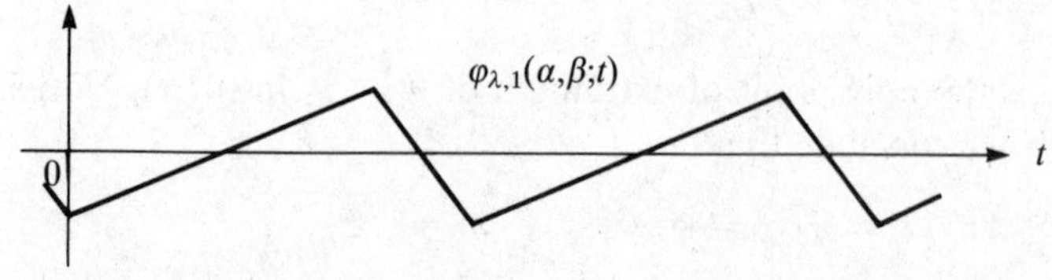

Fig. 3.1 Functions $\varphi_{\lambda,m}(\alpha, \beta; t)$.

and for $m \geq 2$ relation (1.23) holds uniformly with respect to $t \in \mathbb{R}$. Indeed (1.23) is obvious for $m = 1$, moreover $\|\varphi_{\lambda,1}(1,\beta) - \mathcal{B}_{\lambda,1}\|_1 \to 0$, $\beta \to \infty$. If $m \geq 2$ then using (1.5.1) we get

$$\pi|\varphi_{\lambda,m}(1,\beta;t) - \mathcal{B}_{\lambda,m}(t)| = \left|\int_0^{2\pi} B_{m-1}(t-u)[\varphi_{\lambda,1}(1,\beta;u) - \mathcal{B}_{\lambda,1}(u)]\,du\right|$$
$$\leq \|B_{m-1}\|_\infty \|\varphi_{\lambda,1}(1,\beta) - \mathcal{B}_{\lambda,1}\|_1 \to 0, \qquad (\beta \to \infty).$$

Let us remark that function $\varphi_{n,m}(\alpha,\beta;t)$ $(n = 1, 2, \ldots,)$ belongs to the linear manifold $S_{2n,m}(\alpha,\beta)$ of 2π-periodic splines of order m, defect 1 with respect to the partition $\{2k\pi/n, (2\pi/n)[\beta/(\alpha+\beta) + k - 1]: k = 1, 2, \ldots, n\}$.

3.2 Comparison theorems in general cases

3.2.1 Comparisons on one and the same level We start with some notation and definitions. $C(-\infty, \infty)$ denotes the set of all functions continuous on the real line and $AC(-\infty, \infty)$ denotes the set of all functions which are locally absolutely continuous on $\mathbb{R}$. In other words $f \in AC(-\infty, \infty)$ means

$$f(t) = \int_a^t \psi(u)\,du + c,$$

where ψ is a locally summable function defined on $(-\infty, \infty)$. C and AC denote the subsets of $C(-\infty, \infty)$ and $AC(-\infty, \infty)$ of 2π-periodic functions.

A function $\varphi \in C(-\infty, \infty)$ is said to be *regular* if it has a period $2l$ and if on the interval $(a, a + 2l)$ (a is a point of an absolute extremum of φ) there is a point c such that φ is strictly monotone on (a, c) and $(c, a + 2l)$. In order to emphasize the length of the period $2l$ sometimes we shall speak about φ as being *2l-regular*.

We say that a function $f \in C(-\infty, \infty)$ possesses a *μ-property* with respect to a regular function φ if for every $\alpha \in \mathbb{R}$ and on every interval of monotoni-

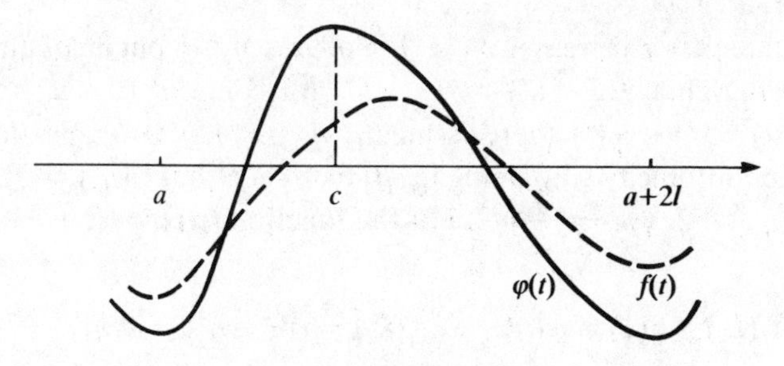

Fig. 3.2 μ-property.

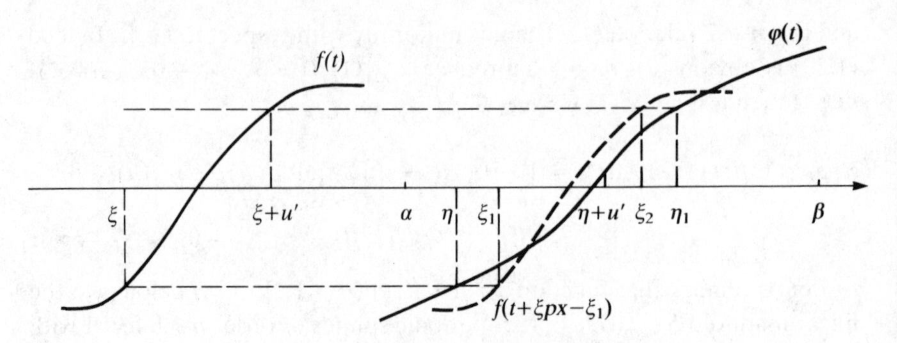

Fig. 3.3 To the proof of Lemma 3.2.1.

city of φ the difference $\varphi(t) - f(t+\alpha)$ either does not change sign or changes sign exactly once – from '+' to '−' if φ decreases or from '−' to '+' if φ increases. It is clear that f will possess the μ-property with respect to $\varphi(t+\beta)$ ($\beta \in \mathbb{R}$) if it possesses the μ-property with respect to φ.

The next lemma, which will be the base for the proofs of the main results of this section, follows immediately from the definitions.

Lemma 3.2.1 Let the function $f \in C(-\infty, \infty)$ possess the μ-property with respect to the regular function φ and

$$\min_t \varphi(t) \le f(u) \le \max_t \varphi(t), \qquad |u| < \infty. \tag{2.1}$$

If φ is monotone on $[\alpha, \beta]$ and if for some ξ and η, $\alpha \le \eta \le \beta$, we have $f(\xi) = \varphi(\eta)$ then

$$f(\xi + u) \le \varphi(\eta + u), \qquad 0 \le u \le \beta - \eta, \tag{2.2}$$

$$f(\xi - u) \ge \varphi(\eta - u), \qquad 0 \le u \le \eta - \alpha, \tag{2.3}$$

if φ increases on $[\alpha, \beta]$ and the inequalities (2.2), (2.3) are true with opposite sign if φ decreases.

Proof Assume that φ increases on $[\alpha, \beta]$ and $\alpha \le \eta \le \beta$ but inequality (2.2) does not hold. Then $f(\xi + u') > \varphi(\eta + u')$ for some u' $(0 < u' < \beta - \eta)$. Because $f(u) \le \max_t \varphi(t)$ there is an η_1, $\eta_1 = \eta + u' + \delta$, $\delta > 0$, in the interval of monotonicity containing $[\alpha, \beta]$ of φ such that $\varphi(\eta_1) = f(\xi + u')$.

Set $\xi_1 = \eta + \delta/2$, $\xi_2 = \eta_1 - \delta/2$. For the function $f_1(t) = f(t + \xi - \xi_1)$ we have

$$f_1(\xi_1) = f(\xi) = \varphi(\eta), \qquad f_1(\xi_2) = f(\xi + u') = \varphi(\eta_1),$$

and hence $f_1(\xi_1) < \varphi(\xi_1)$, $f_1(\xi_2) > \varphi(\xi_2)$ because $\eta < \xi_1 < \xi_2 < \eta_1$ and φ is strictly increasing in $[\alpha, \beta]$. Therefore the difference $[\varphi(t) - f(t + \xi - \xi_1)]$

changes sign from '+' to '−', which contradicts the μ-property. Inequality (2.2) is proved; we obtain the other cases of the lemma in a similar way. □

From Lemma 3.2.1 we derive results for the behaviour of the function f and its first derivative.

Proposition 3.2.2 Let function $f \in AC(-\infty, \infty)$ possess the μ-property with respect to the regular function $\varphi \in AC(-\infty, \infty)$ and let inequalities (2.1) hold. Then:

(1) if $f(\xi) = \varphi(\eta)$ and $\varphi'(\eta) f'(\xi) \geq 0$ (whenever the derivatives exist) then $|f'(\xi)| \leq |\varphi'(\eta)|$;
(2) we have a.e.

$$\operatorname*{ess\,inf}_t \varphi'(t) \leq f'(u) \leq \operatorname*{ess\,sup}_t \varphi'(t). \tag{2.4}$$

Proof Assume that $\varphi'(\eta) \geq 0$ and $f'(\xi) \geq 0$ in (1). From (2.2) and (2.3) for all h sufficiently small in modulus we have

$$\frac{f(\xi + h) - f(\xi)}{h} \leq \frac{\varphi(\eta + h) - \varphi(\eta)}{h},$$

and hence $f'(\xi) \leq \varphi'(\eta)$. The case $\varphi'(\eta) \leq 0, f'(\xi) \leq 0$ is similar. From (1) and (2.1) we get that (2.4) is fulfilled a.e. □

For $f \in C(-\infty, \infty)$ we define the quantities

$$\omega_+(f, \delta) = \sup\{f(t'') - f(t') \colon 0 \leq t'' - t' \leq \delta\}, \tag{2.5}$$

$$\omega_-(f, \delta) = \sup\{f(t') - f(t'') \colon 0 \leq t'' - t' \leq \delta\}, \tag{2.6}$$

which characterize the maximal increment in absolute value on intervals of length δ where the function f increases or decreases

Proposition 3.2.3 Let f satisfy the conditions of Lemma 3.2.1. Then:

(1) if $f(\xi) = \varphi(\eta)$ and $f(\xi_1) = \varphi(\eta_1)$, where $(\xi_1 - \xi)(\eta_1 - \eta) > 0$ and the points η and η_1 belong to one and the same interval of monotonicity of φ, then $|\xi_1 - \xi| \geq |\eta_1 - \eta|$;
(2) for $\delta \geq 0$ we have

$$\omega_+(f, \delta) \leq \omega_+(\varphi, \delta), \qquad \omega_-(f, \delta) \leq \omega_-(\varphi, \delta). \tag{2.7}$$

In the proof of (1) we assume that $\xi_1 > \xi$, $\eta_1 > \eta$ and $f(\xi_1) > f(\xi)$ (the other cases can be treated similarly). If we suppose that $\xi_1 - \xi < \eta_1 - \eta$ then we have for $u = \xi_1 - \xi$

$$\varphi(\eta + u) < \varphi(\eta_1) = f(\xi + u),$$

which contradicts (2.2).

Let $0 < t'' - t' \leq \delta$ and $f(t'') > f(t')$. Because of (2.1) we can find points η' and η'' ($\eta' < \eta''$) on an interval of increase of φ such that $\varphi(\eta') = f(t')$, $\varphi(\eta'') = f(t'')$. Now (1) implies $0 \leq \eta'' - \eta' \leq t'' - t' \leq \delta$ and hence

$$f(t'') - f(t') = \varphi(\eta'') - \varphi(\eta') \leq \omega_+(\varphi, \delta).$$

The second inequality in (2.7) has a similar proof. □

3.2.2 Rearrangement of a function More delicate comparison theorems are connected with the non-increasing rearrangements of functions. The rearrangements will also play an important role in the investigation of extremal problems in the following chapters. Here we give the definition and some properties.

Let f be summable on the interval (a, b) (finite or infinite) and hence measurable and finite a.e. on (a, b). For $y \geq 0$ by

$$m(f, y) = \text{meas}\,\{t: t \in (a, b), |f(y)| > y\} \tag{2.8}$$

we define a right continuous function $m(f, y)$ which is non-increasing on $[0, \infty)$ and is called the *distribution function* for $|f(t)|$. Whenever $m(f, t)$ is continuous and strictly decreasing it has an inverse function $y = r(f, t) = r(f; a, b; t)$ which is defined and strictly decreasing on $(0, b - a)$ and is called the *non-increasing rearrangement* of f.

In the general case the non-increasing rearrangement of f is defined by

$$r(f, t) = \inf\,\{y: m(f, y) \leq t\} \tag{2.9}$$

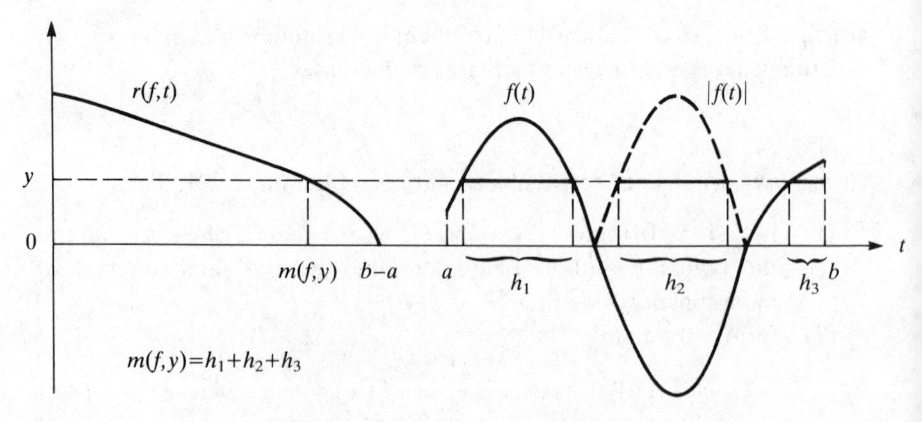

Fig. 3.4 Rearrangement $r(f, t)$.

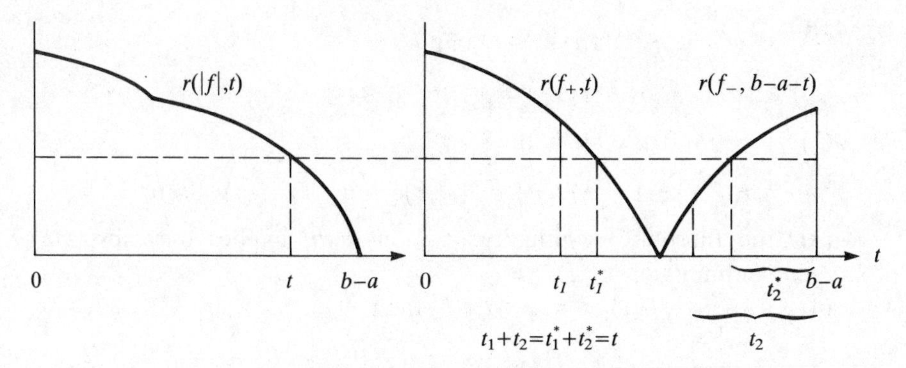

Fig. 3.5 To the proof of (3.2.17).

and is right continuous as $m(f, y)$. It follows from the definitions that

$$\operatorname{meas}\{t: t \in (0, b-a), r(f, t) > y\} = m(f, y) \tag{2.10}$$

for every $y \geq 0$, i.e. the functions $|f(t)|$ and $r(f, t)$ have one and the same distribution function and hence they are *equimeasurable*. Let us note that $r(f, t) = r(|f|, t)$. If f is defined on the real axis and is $(b-a)$-periodic then its rearrangement on (a, b) is invariant with respect to the translation, i.e. $r(f(\bullet + \alpha), t) = r(f, t)$ for every $\alpha \in \mathbb{R}$.

From (2.8) and (2.10) we get

$$\int_a^b |f(t)|\,dt = \int_0^{b-a} r(f, t)\,dt = \int_0^\infty m(f, y)\,dy, \tag{2.11}$$

and

$$\int_a^b |f(t)|^q dt = \int_0^{b-a} r^q(f, t)\,dt, \qquad 0 < q < \infty, \tag{2.12}$$

provided that $|f|^q$ is summable. We obtain the last equality from

$$\begin{aligned}\operatorname{meas}\{t: t \in (0, b-a), r(|f|^q, t) > y\} &= \operatorname{meas}\{t: t \in (a, b), |f(t)|^q > y\}\\ &= \operatorname{meas}\{t: t \in (a, b), |f(t)| > y^{1/q}\}\\ &= \operatorname{meas}\{t: t \in (0, b-a), r(f, t) > y^{1/q}\}\\ &= \operatorname{meas}\{t: t \in (0, b-a), r^q(f, t) > y\}.\end{aligned}$$

For a function f bounded on $[a, b]$ the rearrangement $r(f; a, b; t)$ is defined on the interval $[0, b-a]$ and

$$\operatorname*{ess\,sup}_{a<t<b} |f(t)| = r(f, 0).$$

The following properties are a direct implication of the definitions.

(1) $|f_1(t)| \le |f_2(t)|$ $(a \le t \le b)$ implies

$$r(f_1, t) \le r(f_2, t), \qquad 0 \le t \le b - a;$$

(2) if $f(t) \ge 0$, $a \le t \le b$, then

$$r(f + c, t) = r(f, t) + c,\ r(cf, t) = cr(f, t), \qquad \forall\, c > 0;$$

(3) the (absolute) continuity of f on $[a, b]$ implies the (absolute) continuity of $r(f, t)$;

(4) if $|f_1(t) - f_2(t)| < \epsilon$, $a \le t \le b$, then

$$|r(f_1, t) - r(f_2, t)| < \epsilon, \qquad 0 \le t \le b - a.$$

Proposition 3.2.4 For any measurable subset e of $[a, b]$ we have

$$\int_{b-a-\text{mes}\,e}^{b-a} r(f, t)\,\mathrm{d}t \le \int_e |f(t)|\,\mathrm{d}t \le \int_0^{\text{mes}\,e} r(f, t)\,\mathrm{d}t. \tag{2.13}$$

Indeed if

$$f_1(t) = \begin{cases} f(t), & t \in e, \\ 0, & t \in [a, b] \setminus e, \end{cases}$$

then $m(f_1, y) \le m(f, y)$ $(y \ge 0)$ and hence

$$\int_e |f(t)|\,\mathrm{d}t = \int_a^b |f_1(t)|\,\mathrm{d}t$$
$$= \int_0^{b-a} r(f_1, t)\,\mathrm{d}t = \int_0^{\text{mes}\,e} r(f_1, t)\,\mathrm{d}t \le \int_0^{\text{mes}\,e} r(f, t)\,\mathrm{d}t,$$

which proves the second inequality in (2.13). We get the first inequality by subtracting the inequality

$$\int_{e_1} |f(t)|\,\mathrm{d}t \le \int_0^{\text{mes}\,e_1} r(f, t)\,\mathrm{d}t,$$

from the first equality in (2.11) where $e_1 = [a, b] \setminus e$.

In view of (2.12) inequality $r(f, t) \le r(g, t)$ $(0 \le t \le b - a)$ implies $\|f\|_{L_q[a,b]} \le \|g\|_{L_q[a,b]}$ $(1 \le q \le \infty)$. The next proposition is less obvious.

Proposition 3.2.5 Let $f, g \in L_q[a, b]$ $(1 \le q < \infty)$ and

$$\int_0^t r(f, u)\,\mathrm{d}u \le \int_0^t r(g, u)\,\mathrm{d}u, \qquad 0 \le t \le b - a. \tag{2.14}$$

Then

$$\int_a^b |f(t)|^q\,\mathrm{d}t \le \int_a^b |g(t)|^q\,\mathrm{d}t. \tag{2.15}$$

If f and g are essentially bounded on $[a, b]$ then

$$\operatorname*{ess\,sup}_{a \le t \le b} |f(t)| \le \operatorname*{ess\,sup}_{a \le t \le b} |g(t)|. \tag{2.16}$$

The proof of (2.15) will be based on the next simple lemma.

Lemma 3.2.6 Let $\psi, \mu \in L_1[a, b]$, μ be non-negative and non-increasing. If

$$\int_a^t \psi(u)\,du \ge 0, \qquad a \le t \le b,$$

and the product $\psi \cdot \mu$ is integrable on $[a, b]$ then

$$\int_a^b \psi(t)\mu(t)\,dt \ge 0.$$

Indeed if μ is bounded then in view of the second mean value theorem [15B, p. 388] we have

$$\int_a^b \psi(t)\mu(t)\,dt = \mu(a+0)\int_a^\xi \psi(t)\,dt \ge 0, \qquad a < \xi < b.$$

In the general case we apply the above inequality with the functions $\mu_N(t) := \min\{\mu(t), N\}$ $(N = 1, 2, \ldots)$ instead of μ and use the Lebesgue theorem for the limit under the sign of the integral.

Proof of Proposition 3.2.5 Relation (2.12) shows that the rearrangements $r(f, t)$ and $r(g, t)$ belong to $L_q[0, b-a]$. Applying Lemma 3.2.6 with $\psi(t) = r(g, t) - r(f, t)$ and $\mu(t) = r^{q-1}(f, t)$ from (2.14) yields

$$\int_0^{b-a} r^{q-1}(f, t)[r(g, t) - r(f, t)]\,dt \ge 0,$$

i.e.

$$\int_0^{b-a} r^q(f, t)\,dt \le \int_0^{b-a} r^{q-1}(f, t) r(g, t)\,dt.$$

Applying Hölder's inequality to the last integral we obtain

$$\int_0^{b-a} r^q(f, t)\,dt \le \left[\int_0^{b-a} r^q(f, t)\,dt\right]^{1-1/q} \left[\int_0^{b-a} r^q(g, t)\,dt\right]^{1/q},$$

which implies

$$\int_0^{b-a} r^q(f, t)\,dt \le \int_0^{b-a} r^q(g, t)\,dt.$$

This inequality is equivalent to (2.15) in view of (2.12). From the definition of rearrangement (2.9) we immediately see that (2.16) is obtained by letting $t \to 0$ in (2.14). □

Let us consider the positive and negative parts of a function

$$f_+(t) = \max\{0, f(t)\}, \qquad f_-(t) = \max\{0, -f(t)\}.$$

For f defined on $[a, b]$ we have

$$r(f, t) = r(r(f_+, \bullet) + r(f_-, b - a - \bullet), t)$$

(drawing the graphs is very helpful here). Considering the graphs of monotone functions $r(f, t), r(f_+, t)$ and $r(f_-, b - a - t)$ it is easy to see that

$$\int_0^t r(f, u)\,du \geq \int_0^{t_1} r(f_+, u)\,du + \int_0^{t_2} r(f_-, u)\,du, \tag{2.17}$$

where $0 < t < b - a$, $t = t_1 + t_2$, t_1, $t_2 > 0$. Moreover for every t there are points $t_1, t_2, t = t_1 + t_2$, for which we have an equality sign in (2.17); for these points $r(f, t) = r(f_+, t_1) = r(f_-, t_2)$ provided that $f \in C[a, b]$. Using these observations we shall prove:

Proposition 3.2.7 If the functions f and g take both positive and negative values in $[a, b]$ and the inequalities

$$\int_0^t r(f_+, u)\,du \leq \int_0^t r(g_+, u)\,du, \qquad \int_0^t r(f_-, u)\,du \leq \int_0^t r(g_-, u)\,du,$$

$$0 \leq t \leq b - a.$$

hold, then

$$\int_0^t r(f, u)\,du \leq \int_0^t r(g, u)\,du, \qquad 0 \leq t \leq b - a.$$

Indeed for a fixed $t \in (0, b - a)$ and points $t_1, t_2, t_1 + t_2 = t$, chosen so that an equality holds in (2.17) we obtain

$$\begin{aligned}\int_0^t r(f, u)\,du &= \int_0^{t_1} r(f_+, u)\,du + \int_0^{t_2} r(f_-, u)\,du \\ &\leq \int_0^{t_1} r(g_+, u)\,du + \int_0^{t_2} r(g_-, u)\,du \leq \int_0^t r(g, u)\,du.\end{aligned}$$

3.2.3 Comparison theorems for rearrangements We use the definitions introduced in the beginning of the section.

Theorem 3.2.8 Let $f \in C$ possess the μ-property with respect to the $2\pi/n$-regular ($n = 1, 2, \ldots,$) function φ and

$$\int_0^{2\pi/n} \varphi(t)\,\mathrm{d}t = 0. \tag{2.18}$$

If

$$\min_u \varphi(u) \le f(t) \le \max_u \varphi(u), \qquad |t| < \infty, \tag{2.19}$$

$$\max_{a,b} \left| \int_a^b f(t)\,\mathrm{d}t \right| \le \frac{1}{2} \int_0^{2\pi/n} |\varphi(t)|\,\mathrm{d}t, \tag{2.20}$$

then

$$\int_0^t r(f_+, u)\,\mathrm{d}u \le \int_0^t r(\varphi_\pm, u)\,\mathrm{d}u, \int_0^t r(f, u)\,\mathrm{d}u \le \int_0^t r(\varphi, u)\,\mathrm{d}u, \quad 0 \le t \le 2\pi, \tag{2.21}$$

where $r(g, t) = r(g; 0, 2\pi; t)$.

Proof Relation (2.20) with $a = 0$, $b = 2\pi m$ and m large enough, implies

$$\int_0^{2\pi} f(t)\,\mathrm{d}t = 0 \tag{2.22}$$

and hence f also takes zero values. Because for every 2π-periodic function g and every a we have $r(g(a + \bullet), t) = r(g, t)$, in the proof of (2.21) we can assume that $f(0) = \varphi(0) = f(2\pi) = \varphi(2\pi) = 0$ where we are replacing f and φ by their translates if necessary.

Let $[0, \eta_+)$ and $[0, \eta_-)$ be the largest intervals on which $r(f_+, t) > 0$ and $r(f_-, t) > 0$ respectively; let $[0, \overset{*}{\eta}_+)$ and $[0, \overset{*}{\eta}_-)$ be the corresponding intervals for φ_+ and φ_-. Because $\eta_+ + \eta_- \le \overset{*}{\eta}_+ + \overset{*}{\eta}_- = 2\pi$ at least one of the inequalities $\eta_+ \le \overset{*}{\eta}_+$ and $\eta_- \le \overset{*}{\eta}_-$ holds.

Assuming that $\eta_+ \le \overset{*}{\eta}_+$ we shall prove

$$\int_0^t r(f_+, u)\,\mathrm{d}u \le \int_0^t r(\varphi_+, u)\,\mathrm{d}u, \qquad 0 \le t \le 2\pi, \tag{2.23}$$

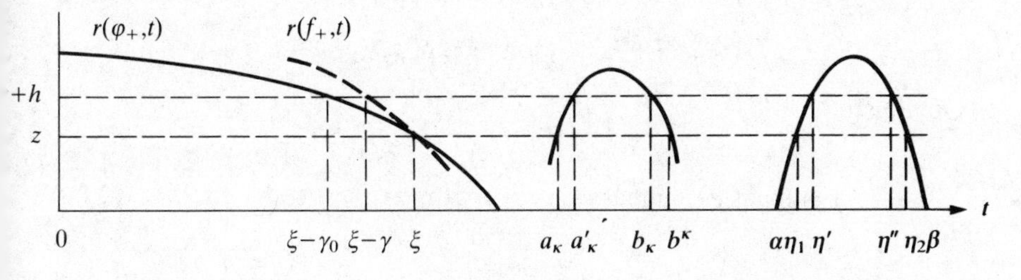

Fig. 3.6 To the proof of Theorem 3.2.8.

Set

$$\delta(t) = r(\varphi_+, t) - r(f_+, t), 0 \le t \le 2\pi,$$

and remark that it is enough to establish the validity of $\int_0^t \delta(u)\, du \ge 0$ only for these $t \in (0, 2\pi)$ for which $\delta(t) = 0$ and $\delta(u) < 0$ if $u \in (t - \varepsilon, t)$ for some $\varepsilon > 0$.

Let $t = \xi$ be such a point and

$$z = r(f_+, \xi) = r(\varphi_+, \xi). \tag{2.24}$$

If for $y \ge 0$ we set

$$e_y = \{t : t \in [0, 2\pi], f_+(t) > y\},$$

then meas $e_z = \xi$ and

$$\int_0^\xi r(f_+, t)\, dt = \int_{e_z} f_+(t)\, dt,$$

The set e_z consists of non-intersecting intervals $(a_k, b_k) \subset (0, 2\pi)$ in which f satisfies

$$f(a_k) = f(b_k) = z, \tag{2.25}$$

$$f(t) > z, \qquad a_k < t < b_k. \tag{2.26}$$

We shall show that there are no more than n such intervals. On the contrary assume that there are $n + 1$ intervals (a_k, b_k) $(k = 1, 2, \ldots, n + 1)$ from e_z in which relations (2.25) and (2.26) hold. Let us choose $h > 0$ so small that: (1) the set e_{z+h} has a non-empty intersection with each of intervals (a_k, b_k); and (2) if $r(f_+, \xi - \gamma) = z + h$ then $\delta(t) < 0$ for $\xi - \gamma < t < \xi$. It follows from (2) and the strict monotonicity of $r(\varphi_+, t)$ that

$$r(\varphi_+, \xi - \gamma_0) = z + h \tag{2.27}$$

implies $\gamma_0 > \gamma$.

On the other hand, let a_k', b_k' be points in (a_k, b_k) determined by the relations

$$f(a_k') = f(b_k') = z + h, \tag{2.28}$$

$$z < f_+(t) \le z + h, \qquad a_k < t \le a_k', \qquad b_k > t \ge b_k'.$$

Obviously

$$\gamma = \text{meas } e_z - \text{meas } e_{z+h} \ge \sum_{k=1}^{n+1} [(a_k' - a_k) + (b_k - b_k')]. \tag{2.29}$$

Function φ_+ (and φ_- too) consists of n identical 'hats' on $[0, 2\pi]$. Consider one of them with support $[\alpha, \beta]$, i.e. $\varphi(\alpha) = \varphi(\beta) = 0$, φ_+ is strictly increasing on (α, c), where $\alpha < c < \beta$, and strictly decreasing on (c, β). Because $f_+(t) \le \varphi(c)$ there are points $\eta_1, \eta_2, \eta', \eta''$ such that

$$\alpha < \eta_1 < \eta' < c < \eta'' < \eta_2 < \beta, \tag{2.30}$$

$$\varphi(\eta_1) = \varphi(\eta_2) = z, \varphi(\eta') = \varphi(\eta'') = z + h. \tag{2.31}$$

Now considering rearrangement $r(\varphi_+, t)$ and taking into account (2.24), (2.27) and (2.31) we obtain

$$\gamma_0 = n[(\eta' - \eta_1) + (\eta_2 - \eta'')]. \tag{2.32}$$

Let us compare the right-hand sides of (2.29) and (2.32). Because of (2.25), (2.28) and (2.31), the conditions of Proposition 3.2.3.(1) hold and hence

$$a_k' - a_k \ge \eta' - \eta_1, \qquad b_k - b_k' \ge \eta_2 - \eta''.$$

Therefore

$$\gamma \ge (n+1)[(\eta' - \eta_1) + (\eta_2 - \eta'')] > \gamma_0$$

which is a contradiction

We have established that there are finitely many non-intersecting intervals, say m, $m \le n$, on $[0, 2\pi]$

$$(a_k, b_k), \qquad k = 1, 2, \dots, m, \tag{2.33}$$

satisfying conditions (2.25) and (2.26). If $[c_i, d_i]$ $(i = 1, 2, \dots, p)$ are intervals in $[0, 2\pi]$ each of which contains at least one of intervals (2.33) and

$$f(c_i) = f(d_i) = 0, \qquad f(t) > 0, \qquad c_i < t < d_i,$$

then $p \le m \le n$. Let us fix $[c_i, d_i]$ and let a_ν and b_μ be the first of the left end-points and the last of the right end-points of the intervals (2.33) on $[c_i, d_i]$ respectively. If we consider once more the 'hat' of φ_+ on (α, β) and the points (2.30) then we have

$$f(c_i) = \varphi(\alpha) = 0, \qquad f(a_\nu) = \varphi(\eta_1) = z,$$
$$f(d_i) = \varphi(\beta) = 0, \qquad f(b_\mu) = \varphi(\eta_2) = z,$$

and φ increases on (α, η_1) and decreases on (η_2, β).

Because of Lemma 3.2.1

$$f(a_\nu - u) \ge \varphi(\eta_1 - u), \qquad 0 \le u \le \eta_1 - \alpha,$$

$$f(b_\mu + u) \ge \varphi(\eta_{12} + u), \qquad 0 \le u \le \beta - \eta_2,$$

and hence

$$\int_{(c_i, d_i)\setminus e_z} f(t)\,dt \ge \int_{c_i}^{a_\nu} f(t)\,dt + \int_{b_\mu}^{d_i} f(t)\,dt \ge \int_\alpha^{\eta_1} \varphi(t)\,dt + \int_{\eta_2}^{\beta} \varphi(t)\,dt. \tag{2.34}$$

From (2.20) and (2.18) we have

$$\int_{c_i}^{d_i} f(t)\,dt \le \frac{1}{2}\int_0^{2\pi/n} |\varphi(t)|\,dt = \int_\alpha^\beta \varphi(t)\,dt. \tag{2.35}$$

Using (2.34) and (2.35) we obtain

$$\begin{aligned}
\int_0^\xi r(f_+,t)\,dt &= \int_{e_z} f(t)\,dt = \sum_{k=1}^m \int_{a_k}^{b_k} f(t)\,dt \\
&= \sum_{i=1}^p \left[\int_{c_i}^{d_i} f(t)\,dt - \int_{(c_i,d_i)\setminus e_z} f(t)\,dt\right] \\
&\le p\left(\int_\alpha^\beta \varphi(t)\,dt - \int_\alpha^{\eta_1} \varphi(t)\,dt - \int_{\eta_2}^\beta \varphi(t)\,dt\right) \\
&= p\int_{\eta_1}^{\eta_2} \varphi(t)\,dt \le n\int_{\eta_1}^{\eta_2} \varphi(t)\,dt = \int_0^\xi r(\varphi_+,t)\,dt.
\end{aligned}$$

Inequality (2.23) is proved. The validity of

$$\int_0^t r(f_-,u)\,du \le \int_0^t r(\varphi_-,u)\,du, \tag{2.36}$$

for $t \in (0,\eta_-^*)$ can be established in the same manner. If $t \in (\eta_-^*,\eta_-)$ then we obtain from (2.18) and (2.22) and from $\|f_+\|_1 \le \|\varphi_+\|_1$, which follows from (2.23),

$$\int_0^t r(f_-,u)\,du \le \|f_-\|_1 = \|f_+\|_1 \le \|\varphi_+\|_1 = \|\varphi_-\|_1$$

$$\int_0^{\eta_-^*} r(\varphi_-,u)\,du = \int_0^t r(\varphi_-,u)\,du.$$

The last of the inequalities (2.21) follows from (2.23), (2.36) and Proposition 3.2.7. □

Theorem 3.2.8 together with Proposition 3.2.4 implies:

Corollary 3.2.9 Under the conditions of Theorem 3.2.8 we have

$$\|f\|_q \le \|\varphi\|_q, \qquad 1 \le q \le \infty,$$

where $\|\bullet\|_q = \|\bullet\|_{L_q[0,2\pi]}$.

3.3 Comparison theorems and exact inequalities for differentiable functions

The main assertions in the previous section were essentially based on the μ-property of the function f. There are many cases covering wide classes of functions when a μ-property with respect to one or another standard function is provided and the general results of Section 3.2 give the according theorems.

In this section we prove comparison theorems under restrictions on the rth derivative of function f which, in particular, imply exact inequalities for the norms of derivatives of f of the Kolmogorov type.

3.3.1 Comparison theorems in the symmetric case Let $L^r(-\infty, \infty)$ $(r = 1, 2, \ldots,)$ be the set of all functions f defined and $(r-1)$ times continuously differentiable on $\mathbb{R}$, such that $f^{(r-1)} \in AC(-\infty, \infty)$ and

$$\|f^{(r)}\|_\infty := \operatorname*{ess\,sup}_t |f(t)| < \infty.$$

Set

$$W^r(-\infty, \infty) = \{f: f \in L^r(-\infty, \infty), \|f^{(r)}\|_\infty \le 1\},$$

and by W^r_∞ denote as before the subset of 2π-periodic functions from $W^r(-\infty, \infty)$. The perfect splines $\varphi_{\lambda,r}$ (see Section 3.1) will play the rôle of standard functions for comparison. They belong to $W^r(-\infty, \infty)$ and to W^r_∞, provided that $\lambda = n = 1, 2, \ldots$. The main results of this section are based on the following lemma.

Lemma 3.3.1 Let $f \in W^r(-\infty, \infty)$ $(r = 1, 2, \ldots)$ and let it be such that for some $\lambda > 0$

$$\|f\|_C \le \|\varphi_{\lambda,r}\|_C,$$

where $\|\bullet\|_C = \|\bullet\|_{C(-\infty,\infty)}$. Then the function f possesses a μ-property with respect to $\varphi_{\lambda,r}$.

Proof For $r = 1$ the assertion is obvious because then $|f'(t)| \le 1$ a.e., so let us assume $r \ge 2$.

If f is a $2\pi n/\lambda$-periodic function (n being a positive integer) then the lemma admits a simple and elegant proof based on the calculation of sign changes. If f does not possess a μ-property then, for some α, the difference $\varphi_{\lambda,r}(t) - f(t + \alpha)$ changes sign on the interval (τ_1, τ_2) of monotonicity of $\varphi_{\lambda,r}$, $\tau_2 = \tau_1 + \pi/\lambda$, against the μ-property, i.e. from '+' to '−' if $\varphi_{\lambda,r}$

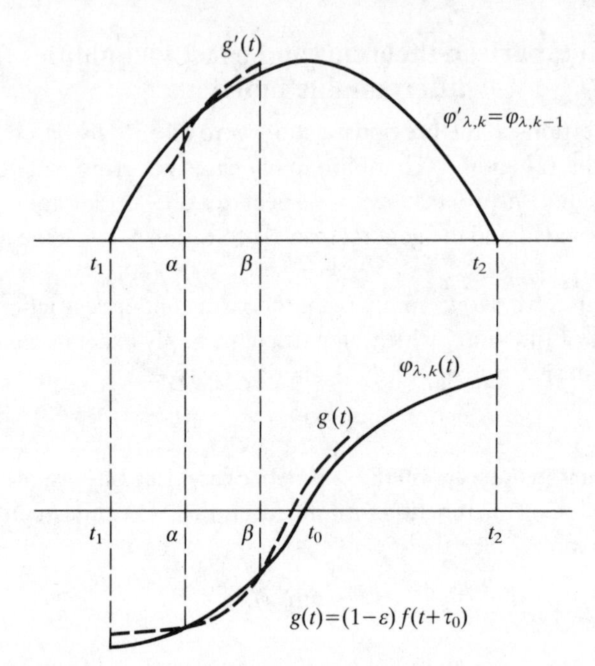

Fig. 3.7 To the proof of Lemma 3.3.1 with $r = k$.

increases and from '−' to '+' if $\varphi_{\lambda,r}$ decreases. By continuity arguments this fact will also hold for the difference

$$\delta(t) = \varphi_{\lambda,r} - (1-\varepsilon)f(t+\alpha)$$

for small $\varepsilon > 0$.

Because $(1-\varepsilon)\|f\|_C < \|\varphi_{\lambda,r}\|_C$, at points τ_1 and τ_2 the sign of δ coincides with the sign of $\varphi_{\lambda,r}$. Hence δ changes sign at least three times on $[\tau_1, \tau_2]$ which implies at least $2n+2$ sign changes of δ on the period $[\tau_1, \tau_1 + 2n\pi/\lambda)$, i.e. $\mu(\delta; 2\pi n/\lambda) \geq 2n+2$. The number of essential sign changes of a periodic function can only increase (see Proposition 2.4.2) and we obtain $\mu(\delta^{(r)}; 2\pi n/\lambda) \geq 2n+2$. on the other hand

$$\delta^{(r)}(t) = \operatorname{sgn}\sin\lambda t - (1-\varepsilon)f^{(r)}(t+\alpha)$$

and because $(1-\varepsilon)|f^{(r)}(t)| < 1$ we have $\mu(\delta^{(r)}; 2\pi n/\lambda)$ which is a contradiction.

In the general case (without the assumption of $2\pi n/\lambda$ periodicity of f) the proof is technically more complicated; we shall go by induction on r. Assume the lemma is proved for $r = k-1$ $(k = 2, 3, \ldots)$ and the inequality

$$\|f\|_C \leq \|\varphi_{\lambda,r}\|_C \tag{3.1}$$

holds for $f \in W^r(-\infty, \infty)$. First we shall show that

$$\|f'\|_C \le \|\varphi'_{\lambda,r}\|_C = \|\varphi_{\lambda,r-1}\|_C. \tag{3.2}$$

Assume on the contrary that

$$\|f'\|_C = |f'(t_*)| > \|\varphi_{\lambda,r-1}\|_C. \tag{3.3}$$

But instead of $f(t)$ we can take any $\pm f(t+\tau)$, τ being arbitrary, and hence without loss of generality we may assume that

$$f'(t_*) > \varphi_{\lambda,k-1}(t_*) = \|\varphi_{\lambda,r-1}\|_C. \tag{3.4}$$

Let t_1 and t_2 be the nearest to the left- and right-hand zeros of $\varphi_{\lambda,k-1}$ respectively. Hence $\varphi_{\lambda,k-1}$ increases on (t_1, t_*) and decreases on (t_*, t_2). In view of (3.1) we have

$$\int_{t_1}^{t_2} f'(t)\,dt \le \int_{t_1}^{t_2} \varphi_{\lambda,k-1}(t)\,dt.$$

From here and (3.4) we conclude that there is $\alpha \in (t_1, t_2)$ such that $0 < f'(\alpha) < \varphi_{\lambda,k-1}(\alpha)$; without loss of generality we may assume that $\alpha \in (t_1, t_*)$.

Set $g(t) = [\varphi_{\lambda,k-1}(t_*)/f'(t_*)]f'(t)$. Then

$$\|g\|_C = g(t_*) = \varphi_{\lambda,k-1}(t_*) = \|\varphi_{\lambda,k-1}\|_C$$

and of course $0 < g(\alpha) < \varphi_{\lambda,k-1}(\alpha)$. Now it is clear that for a small enough $\varepsilon > 0$ the difference $\varphi_{\lambda,k-1}(t) - g(t+\varepsilon)$ changes sign from + to − on the interval (t_1, t_*) of increasing $\varphi_{\lambda,k-1}$ which is impossible because $g \in W^r(-\infty, \infty)$, $\|g\|_C = \|\varphi_{\lambda,k-1}\|_C$, and because we have assumed that the lemma is proved for $r = k-1$. So we have proved (3.2) under the assumptions that the lemma be true for $r = k-1$ and that (3.1) holds for $f \in W^r(-\infty, \infty)$.

Now we shall prove the lemma for $r = k$. Assume on the contrary that there is $f \in W^k(-\infty, \infty)$ which does not possess a μ-property with respect to $\varphi_{\lambda,k}$ and hence on the interval (t_1, t_2), t_1 and t_2 being two consecutive zeros of $\varphi'_{\lambda,k}$, the difference $\varphi_{\lambda,k}(t) - f(t+\tau_0)$, τ_0 is fixed and real and does not change sign appropriately: if for definiteness $\varphi_{\lambda,k}$ increases on (t_1, t_2) then the difference $\varphi_{\lambda,k}(t) - f(t+\tau_0)$ changes sign from + to − on (t_1, t_2). By continuity arguments this fact will also hold for the difference

$$\delta(t) = \varphi_{\lambda,k}(t) - (1-\varepsilon)f(t+\tau_0)$$

for small $\varepsilon > 0$.

Because $\delta(t_1) < 0$ and $\delta(t_2) > 0$ we conclude that δ changes sign at least three times on (t_1, t_2). Then in (t_1, t_2) there is an interval (α, β) belonging totally to one of both intervals of monotonicity (t_1, t_0) or (t_0, t_1) $(t_0 = (t_1 + t_2)/2)$ of $\varphi_{\lambda,k-1}$ such that $\delta(\alpha) = \delta(\beta) = 0$ and $\operatorname{sgn} \delta(t) = \operatorname{sgn} \varphi''_{\lambda,k}(t)$ for all $t \in (\alpha, \beta)$. But this means that if, for example,

$(\alpha, \beta) \subset (t_1, t_0)$ and hence $\varphi'_{\lambda,k} = \varphi_{\lambda,k-1}$ increases on (α, β), then the difference

$$\delta'(t) = \varphi_{\lambda,k-1}(t) - (1-\varepsilon) f'(t+\tau_0)$$

changes sign on (α, β) from $+$ to $-$. But this is impossible because function $g(t) = (1-\varepsilon) f'(t-\tau_0)$ $(0 < \varepsilon < 1)$ belongs to $W^{k-1}(-\infty, \infty)$ and satisfies $\|g\|_C \leq \|\varphi_{\lambda,k-1}\|_C$ because of (3.2) and the lemma is assumed to be true for $r = k-1$. This proves the lemma. □

From Lemma 3.3.1 and Proposition 3.2.2 taking into account the symmetries of the graph of $\varphi_{\lambda,m}$, we obtain:

Theorem 3.3.2 Let $f \in W^r(-\infty, \infty)$ $(r = 1, 2, \ldots)$ and $\|f\|_C = \|\varphi_{\lambda,r}\|_C$ for some λ. Then

(1) if $f(\xi) = \varphi_{\lambda,r}(\eta)$ then $|f'(\xi)| \leq |\varphi'_{\lambda,r}(\eta)|$ (provided that $f'(\xi)$ exists when $r = 1$);
(2) the following exact (under the conditions of the theorem) inequalities hold:

$$\|f^{(k)}\|_C \leq \|\varphi^{(k)}_{\lambda,r}\|_C = \|\varphi_{\lambda,r-k}\|_C, \qquad k = 1, 2, \ldots, r-1. \tag{3.5}$$

Let us remark that statement (2) was established during the proof of Lemma 3.3.1 (see (3.2)). It also easily follows from statement (1): if we assume that $|f'(\xi)| > \|\varphi_{\lambda,r-1}\|_C$, then $|f'(\xi)| > |\varphi'_{\lambda,r}(\eta)|$, where η is chosen so that $\varphi_{\lambda,r}(\eta) = f(\xi)$ – a contradiction. The exactness of (3.5) follows from the fact that $\varphi_{\lambda,r} \in W^r(-\infty, \infty)$ for every $\lambda > 0$.

The geometric sense of statement (1) of Theorem 3.3.2 is that a function $f \in W^r(-\infty, \infty)$, $\|f\|_C = \|\varphi_{\lambda,r}\|_C$, on a fixed level cannot change faster than the perfect spline $\varphi_{\lambda,r}$.

Let f be continuous on the real line and let

$$\omega(f, \delta) = \sup\{|f(t') - f(t'')|: |t' - t''| \leq \delta\}, \qquad \delta \geq 0, \tag{3.6}$$

be its modulus of continuity (we postpone the detailed study of this modulus to Section 6.1). From Lemma 3.3.1, Proposition 3.2.3, Theorem 3.3.2(2) and the symmetry of the graph of $\varphi_{\lambda,r}$ we obtain:

Proposition 3.3.3 Let $f \in W^r(-\infty, \infty)$ $(r = 1, 2, \ldots)$ and $\|f\|_C \leq \|\varphi_{\lambda,r}\|_C$ for some λ. Then

(1) if $f(\xi) = \varphi_{\lambda,r}(\eta)$, $f(\xi_1) = \varphi_{\lambda,r}(\eta_1)$ and if the points η, η_1 belong to one and the same interval of monotonicity on $\varphi_{\lambda,r}$, then $|\xi - \xi'| \geq |\eta - \eta'|$;
(2) for all $\delta \geq 0$ we have $\omega(f^{(k)}, \delta) \leq \omega(\varphi_{\lambda,r-k}, \delta)$ $(k = 0, 1, \ldots, r-1)$.

Of course Theorem 3.3.2 and Proposition 3.3.3 hold for 2π-periodic functions $f \in W^r(-\infty, \infty)$, i.e. $f \in W^r_\infty$. The periodicity may affect only the problem of exactness. For example, inequalities (3.5) cannot be improved for $f \in W^r(-\infty, \infty)$ satisfying

$$\|f\|_C \leq \|\varphi_{\lambda,r}\|_C \tag{3.7}$$

because the function $\varphi_{\lambda,r}$ itself belongs to $W^r(-\infty, \infty)$. Inequalities (3.5) are always strict for functions $f \in W^r_\infty$ satisfying (3.7) only when $\lambda = n$ (n is a positive integer) because then $\varphi_{\lambda,r} \in W^r_\infty$.

In other words we have, for $k = 1, 2, \ldots, r-1$

$$\sup\{\|f^{(k)}\|_C : f \in W^r(-\infty, \infty), \|f\|_C \leq \|\varphi_{\lambda,r}\|_C\} = \|\varphi^{(k)}_{\lambda,r}\|_C, \qquad \lambda > 0,$$

$$\sup\{\|f^{(k)}\|_C : f \in W^r_\infty, \|f\|_C \leq \|\varphi_{n,r}\|_C\} = \|\varphi^{(k)}_{n,r}\|_C, \qquad n = 1, 2, \ldots,.$$

Turning our attention to comparing rearrangements we shall consider 2π-periodic functions and we set $r(f, t) = r(f; 0, 2\pi; t)$.

Theorem 3.3.4 If $f \in W^r_\infty$ and $\|f\|_C \leq \|\varphi_{n,r}\|_C$ ($n = 1, 2, \ldots$) then for $k = 1, 2, \ldots, r$ the exact inequalities

$$\int_0^t r((f^{(k)})_\pm, u)\, du \leq \int_0^t r((\varphi_{n,r-k})_\pm, u)\, du, \qquad 0 \leq t \leq 2\pi, \tag{3.8}$$

$$\int_0^t r(f^{(k)}, u)\, du \leq \int_0^t r(\varphi_{n,r-k}, u)\, du, \qquad 0 \leq t \leq 2\pi, \tag{3.9}$$

$$\|f^{(k)}\|_q \leq \|\varphi_{n,r-k}\|_q, \qquad 1 \leq q \leq \infty \tag{3.10}$$

hold. Under the additional condition

$$\max_{a,b} \left| \int_a^b f(t)\, dt \right| \leq 2\|\varphi_{n,r+1}\|_C = \int_0^{\pi/n} |\varphi_{n,r}(t)|\, dt \tag{3.11}$$

inequalities (3.8)–(3.10) hold and are also strict for $k = 0$, i.e. for the function f itself.

The proof consists of applying Theorem 3.2.8, Lemma 3.3.1 and Theorem 3.3.2 (2). From Lemma 3.3.1 f possesses the μ-property with respect to the $2\pi/n$-regular function $\varphi = \varphi_{n,r}$ for which (2.18) holds; moreover conditions (3.11) and (2.20) are equivalent in this case because of the symmetry of the graph of $\varphi_{n,r}$ and $\|f\|_C \leq \|\varphi_{n,r}\|_C$ implies (2.19). Hence applying Theorem 3.2.8 we get (3.8) and (3.9) when $k = 0$.

Now from $f' \in W^{r-1}_\infty$ and $\|f'\|_C \leq \|\varphi_{n,r-1}\|_C$ (see Theorem 3.3.2 (2)) we conclude from Lemma 3.3.1 that f' possesses a μ-property with respect to $\varphi_{n,r-1}$. By noticing that

$$\max_{a,b}\left|\int_a^b f'(t)\,\mathrm{d}t\right| \le 2\|f\|_C \le 2\|\varphi_{n,r}\|_C$$

we see that Theorem 3.2.8 is applicable for functions f' and $\varphi_{n,r-1}$ and we also get (3.8) and (3.9) for $k = 1$. The same reasoning consecutively establishes the validity of (3.8) and (3.9) for $k = 2, 3, \ldots, r-1$. For $k = r$ these inequalities are obvious because then $|f^{(r)}(t)| \le 1$ a.e. and $f^{(r)} \perp 1$.

Inequality (3.10) follows from (3.9) and Proposition 3.2.5. It remains only to notice that $\varphi_{n,r} \in W^r_\infty$ and (3.11) is fulfilled as an equality for $f = \varphi_{n,r}$.

3.3.2 Kolmogorov's inequality and its integral analogs In 1939 A. N. Kolmogorov [3] stated and solved the following problem: for given positive numbers A_0 and A_r find the supremum of $\|f^{(k)}\|_C$ $(1 \le k \le r-1)$ on all functions $f \in L^r(-\infty, \infty)$ satisfying $\|f\|_C \le A_0$, $\|f^{(r)}\|_\infty \le A_r$. Here $\|\bullet\|_C = \|\bullet\|_{C(-\infty,\infty)}$, $\|\bullet\|_\infty = \|\bullet\|_{L_\infty(-\infty,\infty)}$. The answer is given by:

Theorem 3.3.5 For every function $f \in L^r(-\infty, \infty)$ $(r = 2, 3, \ldots,)$ with finite uniform norm and for every $k = 1, 2, \ldots, r-1$ we have

$$\|f^{(k)}\|_C \le c_{rk}\|f\|_C^{1-k/r}\,\|f^{(r)}\|_\infty^{k/r}, \tag{3.12}$$

where

$$c_{rk} = K_{r-k}/K_r^{1-k/r}, \tag{3.13}$$

K_m being the Favard constants (1.8). Inequality (3.12) becomes an equality for $f(t) = \gamma\varphi_{\lambda,r}(t+\alpha)$, where $\gamma, \alpha \in \mathbb{R}$, λ being positive.

Proof Without loss of generality we can assume that

$$\|f^{(r)}\|_\infty = 1 \tag{3.14}$$

because the validity of (3.12) will not be affected if we multiply f by any non-zero constant. Under this assumption we have to prove

$$\|f^{(k)}\|_C \le c_{rk}\|f\|_C^{1-k/r}, \qquad k = 1, 2, \ldots, r-1.$$

Fix $f \in L^r(-\infty, \infty)$ to satisfy (3.14), i.e. $f \in W^r(-\infty, \infty)$, and choose λ so that $\|\varphi_{\lambda,r}\|_C = \|f\|_C$. This is always possible because $\|\varphi_{\lambda,r}\|_C$ is a continuous function of λ (see (1.10)) and because

$$\lim_{\lambda\to 0}\|\varphi_{\lambda,r}\|_C = \infty, \qquad \lim_{\lambda\to\infty}\|\varphi_{\lambda,r}\|_C = 0.$$

Now Theorem 3.3.2 gives $\|f^{(k)}\|_C \le \|\varphi_{\lambda,r-k}\|_C$ which together with Proposition 3.1.1 gives

$$\|f^{(k)}\|_C \leq c_{rk} \|\varphi_{\lambda,r}\|_C^{1-k/r} = c_{rk} \|f\|_C^{1-k/r}$$

and (3.12) is proved. The fact that (3.12) is an equality for the splines $\gamma\varphi_{\lambda,r}(t+\alpha)$ follows from Proposition 3.1.1 and the invariance of (3.12) with respect of multiplication of f with a non-zero constant. □

Going back to the Kolmogorov problem we can write Theorem 3.3.5 in the form

$$\sup \{\|f^{(k)}\|_C : f \in L^r(-\infty, \infty), \|f\|_C \leq A_0, \|f^{(r)}\|_\infty \leq A_r\} = \|f_*^{(k)}\|_C,$$

$$k = 1, 2, \ldots, r-1,$$

where $f_*(t) = A_r \varphi_{\lambda,r}(t)$ and the parameter λ is chosen so that $\|f_*\|_C = A_0$.

Let us also remark that the statement of Theorem 3.3.5 is equivalent to the validity of the relations (for every $\lambda > 0$):

$$\sup_{f \in W^r(-\infty,\infty)} \frac{\|f^{(k)}\|_C^r}{\|f\|_C^{r-k}} = \frac{\|\varphi_{\lambda,r-k}\|_C^r}{\|\varphi_{\lambda,r}\|_C^{r-k}} = \frac{K_{r-k}^r}{K_r^{r-k}}, \qquad k = 1, 2, \ldots, r-1,$$

where as before the K_m are the Favard constants.

It was discovered by Stein [1] in 1957 that inequality (3.12) remains valid if $f \in L_p^r(-\infty, \infty)$ and all norms are calculated in $L_p(-\infty, \infty)$; moreover the inequality is exact with the same constant c_{rk} if $p = 1$.

We shall prove here for $p = 1$ the periodic case of this fact using Stein's scheme involving the sets V^m $(m = 1, 2, \ldots)$ of 2π-periodic functions f such that $f^{(m-1)} \in AC$ and $f^{(m)}$ is of bounded variation on $[0, 2\pi]$, i.e. $f^{(m)} \in V := V^0$.

Proposition 3.3.6 If $f \in L_1^r$ $(r = 2, 3, \ldots)$ then

$$\|f^{(k)}\|_1 \leq c_{rk} \|f\|_1^{1-k/r} \|f^{(r)}\|_1^{k/r}, \qquad k = 1, 2, \ldots, r-1, \tag{3.15}$$

where $\|\bullet\|_1 = \|\bullet\|_{L_1[0,2\pi]}$. If $f \in V^{r-1}$ then

$$\|f^{(k)}\|_1 \leq c_{rk} \|f\|_1^{1-k/r} \left[\bigvee_0^{2\pi} f^{(r-1)} \right]^{k/r}, \qquad k = 1, 2, \ldots, r-1. \tag{3.16}$$

Inequalities (3.15), (3.16) with constants (3.13) are exact and (3.16) becomes an equality for $f(t) = \gamma\varphi_{n,r-1}(t+\alpha)$.

Proof Fix $k = 1, 2, \ldots, r-1$ and set $\psi(t) = \operatorname{sgn} f^{(k)}(t)$ and

$$F(t) = \int_0^{2\pi} f(t+u)\psi(u)\,du. \tag{3.17}$$

Because of

$$\left|\frac{F(t+h)-F(t)}{h}-\int_0^{2\pi} f'(t+u)\psi(u)\,du\right|$$
$$\le \int_0^{2\pi}\left|\frac{f(t+u+h)-f(t+u)}{h}-f'(t+u)\right|du<\varepsilon(h),$$

where $\varepsilon(h)\to 0$ when $h\to 0$, the derivative F' exists and

$$F'(t)=\int_0^{2\pi} f'(t+u)\psi(u)\,du.$$

By repeating differentiation the same arguments give

$$F^{(j)}(t)=\int_0^{2\pi} f^{(j)}(t+u)\psi(u)\,du, \qquad j=1,2,\ldots,r. \tag{3.18}$$

Moreover $f^{(r)}$ is continuous since

$$|F^{(r)}(t+\mathrm{h})-F^{(r)}(t)|=\int_0^{2\pi}|f^{(r)}(t+u+h)-f^{(r)}(t+u)|\,du$$

and since the right-hand side tends to zero when $h\to 0$ according to the Lebesgue theorem (see e.g. [1B, p. 52]).

From (3.12) we obtain

$$|F^{(k)}(0)|\le c_{rk}\|F\|_C^{1-k/r}\|F^{(r)}\|_C^{k/r}. \tag{3.18$'$}$$

But from (3.17), (3.18) and the definition of ψ we have

$$F^{(k)}(0)=\|f^{(k)}\|_1, \qquad \|F\|_C\le\|f\|_1, \qquad \|F^{(r)}\|_C\le\|f^{(r)}\|_1,$$

and the replacement of these quantities in (3.18′) gives (3.15).

For $f\in L_1^r$ we have

$$\|f^{(r)}\|_1=\bigvee_0^{2\pi} f^{(r-1)}$$

and for such functions (3.16) also holds. Now we shall show that (3.16) holds on V^{r-1} too. If $f\in V^{r-1}$ and f_h is its Stekhlov function, then $f_h\in L_1^r$ and (3.15) holds for f_h. But (cf. Section A2)

$$\|f_h\|_1\le\|f\|_1, \qquad \|f_h^{(r)}\|_1\le\bigvee_0^{2\pi} f^{(r-1)},$$

and hence

$$\|f_h^{(k)}\|_1 \le c_{rk}\|f\|_1^{1-k/r}\left[\bigvee_0^{2\pi} f^{(r-1)}\right]^{k/r},$$

which proves (3.16) because $\|f_h^{(k)}\|_1 \to \|f^{(k)}\|_1$ for $h \to 0$.

Equality in (3.16) is attained for $f(t) = g_{n,r-1}(t) = \varphi_{n,r-1}(t)/(4n)$ whose $(r-1)$th derivative has variation 1 on $[0, 2\pi]$ (see Proposition 3.1.2) and hence for every function of the type $\gamma\varphi_{n,r-1}(t)$. □

In the terms of classes W_1^m, W_V^m (introduced in Section 1.5), Proposition 3.3.6 can be written as

$$\sup_{f\in W_1^r}\frac{\|f^{(k)}\|_1^r}{\|f\|_1^{r-k}} = \sup_{f\in W_V^{r-1}}\frac{\left[\bigvee_0^{2\pi} f^{(k-1)}\right]^r}{\|f\|_1^{r-k}} = \frac{K_{r-k}^r}{K_r^{r-k}},$$

$$k = 1, 2, \ldots, r-1, \qquad r = 2, 3, \ldots,.$$

It was mentioned above that the inequality

$$\|f^{(k)}\|_q \le c\|f\|_q^{1-k/r}\|f^{(r)}\|_q^{k/r}, \qquad k = 1, 2, \ldots, r-1, \tag{3.19}$$

with constant $c = c_{rk}$ holds for any $f \in L_q^r$ for all $q \ge 1$. However, although this constant is exact for $q = 1, \infty$, it does not seem to be the best if $1 < q < \infty$. At least for $q = 2$ the exact constant in (3.19) is $c = 1$.

Proposition 3.3.7 If $f \in L_2^r$ $(r = 2, 3, \ldots,)$ then

$$\|f^{(k)}\|_2 \le \|f\|_2^{1-k/r}\,\|f^{(r)}\|_2^{k/r}, \qquad k = 1, 2, \ldots, r-1, \tag{3.20}$$

becomes an equality for $f(t) = \gamma\cos m(t+\alpha)$.

Proof Relation (3.20) follows from Parseval's equation and Hölder's inequality for a series which has the form:

$$\sum_{\nu=1}^{\infty}\alpha_\nu\beta_\nu \le \left(\sum_{\nu=1}^{\infty}\alpha_\nu^p\right)^{1/p}\left(\sum_{\nu=1}^{\infty}\beta_\nu^{p'}\right)^{1/p'}, \qquad p \ge 1, \qquad \frac{1}{p}+\frac{1}{p'} = 1. \tag{3.21}$$

Indeed, if $f \in L_2^r$ and

$$f(t) = a_0/2 + \sum_{\nu=1}^{\infty}(a_\nu\cos\nu t + b_\nu\sin\nu t),$$

then

$$\|f\|_2^2 = \pi\left[a_0^2/2 + \sum_{\nu=1}^{\infty}(a_\nu^2 + b_\nu^2)\right], \qquad \|f^{(k)}\|_2^2 = \pi\sum_{\nu=1}^{\infty}\nu^{2k}(a_\nu^2 + b_\nu^2).$$

We set for $k = 1, 2, \ldots, r-1$

$$\alpha_\nu = (a_\nu^2 + b_\nu^2)^{1-k/r}, \qquad \beta_\nu = [\nu^{2r}(a_\nu^2 + b_\nu^2)]^{k/r},$$

and we obtain from (3.21) with $p = r/(r-k)$

$$\|f^{(k)}\|_2^2 = \pi \sum_{\nu=1}^{\infty} \alpha_\nu \beta_\nu \le \pi \left(\sum_{\nu=1}^{\infty} \alpha_\nu^{r/(r-k)} \right)^{1-k/r} \left(\sum_{\nu=1}^{\infty} \beta_\nu^{r/k} \right)^{k/r}$$

$$= \left[\pi \sum_{\nu=1}^{\infty} (a_\nu^2 + b_\nu^2) \right]^{1-k/r} \left[\pi \sum_{\nu=1}^{\infty} \nu^{2r}(a_\nu^2 + b_\nu^2) \right]^{k/r}$$

$$\le \|f\|_2^{2-2k/r} \|f^{(r)}\|_2^{2k/r}.$$

For $f(t) = \gamma \cos m(t+\alpha)$ it follows that (3.20) turns out to be an equality. □

3.3.3 The non-symmetric case In Lemma 3.3.1 the μ-property with respect to the symmetric standard function $\varphi_{\lambda,r}$ was guaranteed by the conditions $\|f^{(r)}\|_\infty \le \|\varphi_{\lambda,r}^{(r)}\|_\infty = 1$ and $\|f\|_C \le \|\varphi^{(r)}\|_C$. If we take for the $2\pi/\lambda$-regular function introduced in Section 3.1.3 the function $\varphi_{\lambda,r}(\alpha, \beta; t)$ which is non-symmetric in general, then it is not difficult to find conditions on $f \in L^r(-\infty, \infty)$ which will ensure its μ-property with respect to $\varphi_{\lambda,r}(\alpha, \beta; t)$.

Let us introduce the class $W^r_{\alpha,\beta}(-\infty, \infty)$ $(\alpha > 0, \beta > 0)$ of all functions $f \in L^r(-\infty, \infty)$ such that

$$-\beta \le f^{(r)}(t) \le \alpha \text{ a.e.} \tag{3.22}$$

or equivalently

$$\|\alpha^{-1}(f^{(r)})_+ + \beta^{-1}(f^{(r)})_-\|_\infty \le 1,$$

i.e.

$$\max\{\alpha^{-1}\|(f^{(r)})_+\|_\infty, \beta^{-1}\|(f^{(r)})_-\|_\infty\} \le 1.$$

For $\alpha = \beta = 1$ this class coincides with $W^r(-\infty, \infty)$.

Lemma 3.3.8 If $f \in W^r_{\alpha,\beta}(-\infty, \infty)$ $(r = 1, 2, \ldots,)$ and if

$$\min_t \varphi_{\lambda,r}(\alpha, \beta; t) \le f(t) \le \max_t \varphi_{\lambda,r}(\alpha, \beta; t), \tag{3.23}$$

then function f possesses the μ-property with respect to $\varphi_{\lambda,r}(\alpha, \beta; t)$.

The proof is similar to the proof of Lemma 3.3.1. We shall not repeat it but we shall note only those places where the non-symmetry is essential.

For $r = 1$ the result is trivial because we have $-\beta \le f'(t) \le \alpha$ a.e. for $f \in W^1_{\alpha,\beta}(-\infty, \infty)$.

When evaluating in the periodic case the number of sign changes in the difference $\delta(t) = \varphi_{\lambda,r}(\alpha, \beta; t) - (1-\varepsilon)f(t+\gamma)$ and its derivatives on the period we should take into consideration that

$$\delta^{(r)}(t) = \varphi_{\lambda,0}(\alpha, \beta; t) - (1-\varepsilon)f^{(r)}(t+\gamma)$$

and hence $\delta^{(r)}$ essentially changes sign exactly $2n$ times on a period of length $2n\pi/\lambda$ because $0 < \varepsilon < 1$ and (3.22).

In the general case we have to use induction and repeat the arguments from the proof of Lemma 3.3.1 with the obvious changes required by the non-symmetry. In particular, with the assumption that (3.23) holds for $r = k$, we have to replace (3.2) with

$$\min_t \varphi'_{\lambda,k}(\alpha, \beta; t) \le f'(t) \le \max_t \varphi'_{\lambda,k}(\alpha, \beta; t), \tag{3.24}$$

and arguing by contradiction we have to assume that at least one of the two inequalities in (3.24) is not fulfilled.

Lemma 3.3.8 allows us to restate the comparison theorems from Section 3.1 for functions from $W^r_{\alpha,\beta}(-\infty, \infty)$ and in view of Proposition 3.2.2 (2) we can also extend them for the continuous derivatives of $f \in W^r_{\alpha,\beta}(-\infty, \infty)$.

Theorem 3.3.9 Let $f \in W^r_{\alpha,\beta}(-\infty, \infty)$ $(r = 1, 2, \ldots)$, $\psi_{\lambda,r}(t) = \varphi_{\lambda,r}(\alpha, \beta; t)$ and

$$\min_t \psi_{\lambda,r}(t) \le f(t) \le \max_t \psi_{\lambda,r}(t)$$

for some $\lambda > 0$. Then

(1) if $f(\xi) = \psi_{\lambda,r}(\eta)$ and $f'(\xi)\psi'_{\lambda,r}(\eta) \ge 0$ then

$$|f'(\xi)| \le |\psi'_{\lambda,r}(\eta)|;$$

(2) if

$$f(\xi) = \psi_{\lambda,r}(\eta), \qquad f(\xi_1) = \psi_{\lambda,r}(\eta_1), \qquad (\xi_1 - \xi)(\eta_1 - \eta) > 0$$

and if the points η and η_1 belong to the same interval of monotonicity of $\psi_{\lambda,r}$, then

$$|\xi_1 - \xi| \ge |\eta_1 - \eta|;$$

(3) the exact inequalities

$$\min_t \psi_{\lambda,r-k}(t) \le f^{(k)}(t) \le \max_t \psi_{\lambda,r-k}(t), \qquad k = 1, 2, \ldots, r-1;$$

$$\omega_+(f^{(k)}, \delta) \le \omega_+(\psi_{\lambda,r-k}, \delta), \quad \omega_-(f^{(k)}, \delta) \le \omega_-(\psi_{\lambda,r-k}, \delta), \tag{3.25}$$

$$\delta \ge 0, \qquad k = 0, 1, \ldots, r-1,$$

hold, where $\omega_\pm(g, \delta)$ were defined in (2.5) and (2.6).

Let us denote by $W^r_{\alpha,\beta}$ the class of 2π-periodic functions from $W^r_{\alpha,\beta}(-\infty, \infty)$. Functions $\varphi_{n,m}(\alpha, \beta; t)$ $(n = 1, 2, \ldots,)$ are $2\pi/n$-regular with zero mean value on the period. Therefore Theorem 3.2.8, Lemma 3.3.8 and Theorem 3.3.9 (3) imply:

Theorem 3.3.10 Let $f \in W^r_{\alpha,\beta}(-\infty, \infty)$ $(r = 1, 2, \ldots,)$, let

$$\min_t \varphi_{n,r}(\alpha, \beta; t) \le f(t) \le \max_t \varphi_{n,r}(\alpha, \beta; t) \tag{3.26}$$

for some $n = 1, 2, \ldots$, and let $k = 0, 1, \ldots, r$. For $k = 0$ we additionally assume that

$$\max_{a,b} \left| \int_a^b f(t)\,dt \right| \le \frac{1}{2} \int_0^{2\pi/n} |\varphi_{n,r}(\alpha, \beta; t)|\,dt.$$

Then the following strict inequalities hold:

$$\int_0^t r((f^{(k)})_\pm, u)\,du \le \int_0^t r(\varphi_{n,r-k}(\alpha, \beta))_\pm, u)\,du, \qquad 0 \le t \le 2\pi,$$

$$\int_0^t r(f^{(k)}, u)\,du \le \int_0^t r(\varphi_{n,r-k}(\alpha, \beta), u)\,du, \qquad 0 \le t \le 2\pi, \tag{3.27}$$

$$\|f^{(k)}\|_q \le \|\varphi_{n,r-k}(\alpha, \beta)\|_q, \qquad 1 \le q \le \infty. \tag{3.28}$$

Indeed, Lemma 3.3.8 and Theorem 3.2.8 show the statements of the theorem are valid for $k = 0$. Now f' possesses a μ-property with respect to $\varphi_{n,r-1}(\alpha, \beta; t)$ because of $f' \in W^{r-1}_{\alpha,\beta}$ and (3.25) with $k = 1$ and $\lambda = n$. Moreover, in view of (3.26) we have

$$\int_a^b f'(t)\,dt \le \max_t f(t) - \min_t f(t) \le \max_t \varphi_{n,r}(\alpha, \beta; t) - \min_t \varphi_{n,r}(\alpha, \beta; t)$$

$$= \frac{1}{2} \int_0^{2\pi/n} |\varphi_{n,r-1}(\alpha, \beta; t)|\,dt$$

and hence Theorem 3.2.8 is applicable to f' too. Inequalities (3.28) follow from (3.27) and Proposition 3.2.5. Finally we have to keep in mind that $\varphi_{n,r-k}(\alpha, \beta; t) \in W^{r-k}_{\alpha,\beta}$.

Now let us turn our attention to the Kolmogorov inequality (3.12) and let us try to find a non-symmetric analog. Inequality (3.12) says that if we know the norms $\|f\|_C$ and $\|f^{(r)}\|_C$ for $f \in L^r(-\infty, \infty)$ then we know the exact bounds for the norms $\|f^{(k)}\|_C$ of the intermediate derivatives. These bounds are determined by the symmetric standard function $c\varphi_{\lambda,r}$ whose parameters c and λ are chosen to satisfy the conditions

$$\|c\varphi_{\lambda,r}\|_C = \|f\|_C, \qquad \|c\varphi_{\lambda,r}^{(r)}\|_C = \|f^{(r)}\|_C.$$

If we would like to take into account the non-symmetric character of function f and its derivatives then it is natural to work with the spline $c\varphi_{\lambda,r}(\alpha,\beta;t)$, where parameters α and β correspond to the bounds of $f^{(r)}$. But then the norms of f and $f^{(k)}$ have to be replaced by the best approximations of these functions with constants. For this purpose we use the notation introduced earlier:

$$E_1(g)_C = \inf_{\gamma\in\mathbb{R}} \|g-\gamma\|_C, \qquad g\in C(-\infty,\infty).$$

Some other notations that are needed are:

$$M_{\lambda,r}^{+}(\alpha,\beta) = \max_t \varphi_{\lambda,r}(\alpha,\beta;t); \qquad M_{\lambda,r}^{-}(\alpha,\beta) = \min_t \varphi_{\lambda,r}(\alpha,\beta;t);$$

$$M_{\lambda,r}(\alpha,\beta) = [M_{\lambda,r}^{+}(\alpha,\beta) - M_{\lambda,r}^{-}(\alpha,\beta)]/2; \qquad M_{\lambda,0}(\alpha,\beta) = (\alpha+\beta)/2;$$

$$M_{1,r}(\alpha,\beta) = M_r(\alpha,\beta).$$

Obviously

$$E_1(\varphi_{\lambda,r}(\alpha,\beta))_C = M_{\lambda,r}(\alpha,\beta), \qquad r=1,2,\ldots.$$

Theorem 3.3.11 Let $\alpha>0$, $\beta>0$, f be a function bounded on $\mathbb{R}$ from $L^r(-\infty,\infty)$, $r=1,2,\ldots$. Then the following exact inequality

$$E_1(f^{(k)})_C \le c_{rk}(\alpha,\beta)\,[E_1(f)_C]^{1-k/r}\,\|\alpha^{-1}f_+^{(r)} + \beta^{-1}f_-^{(r)}\|_\infty^{k/r}, \quad k=1,2,\ldots,r-1, \tag{3.29}$$

holds, where

$$c_{rk}(\alpha,\beta) = \frac{M_{r-k}(\alpha,\beta)}{[M_r(\alpha,\beta)]^{1-k/r}}. \tag{3.30}$$

Equality in (3.29) is attained by every function of the type $c\varphi_{\lambda,r}(\alpha,\beta;t+\gamma)$ $(c>0,\ \gamma\in\mathbb{R})$.

Proof It suffices to prove (3.29) under the assumption that $\|\alpha^{-1}f_+^{(r)} + \beta^{-1}f_-^{(r)}\|_\infty \le 1$, i.e. $f\in W_{\alpha,\beta}^r(-\infty,\infty)$, because all functionals in (3.29) are positively homogeneous. It is easy to see from the definition of $\varphi_{\lambda,r}(\alpha,\beta;t)$ (see Section 3.1.3) that λ can be chosen to satisfy

$$M_{\lambda,r}(\alpha,\beta) = E_1(f)_C. \tag{3.31}$$

Then Theorem 3.3.9 (formula (3.25)) gives

$$M_{\lambda,r-k}^{-}(\alpha,\beta) \le f^{(k)}(t) \le M_{\lambda,r-k}^{+}(\alpha,\beta), \qquad k=1,2,\ldots,r-1.$$

These inequalities mean

$$E_1(f^{(k)})_C \le M_{\lambda,r-k}(\alpha,\beta),$$

which in view of (3.31) is equivalent to

$$E_1(f^{(k)})_C \le \frac{M_{\lambda,r-k}(\alpha,\beta)}{[M_{\lambda,r}(\alpha,\beta)]^{1-k/r}}\,[E_1(f)_C]^{1-k/r}.$$

It is easy to check that the ratio $M_{\lambda,r-k}(\alpha,\beta)/[M_{\lambda,r}(\alpha,\beta)]^{1-k/r}$ does not depend on λ and hence setting $\lambda = 1$ in the above inequality and using (3.30) we get (3.29). The last statement of the theorem is obvious. □

Setting $\alpha = \beta = 1$ in (3.29) we get the inequality (which is also exact)

$$E_1(f^{(k)})_C \le c_{rk}\,[E_1(f)_C]^{1-k/r}\,\|f^{(r)}\|_\infty^{k/r} \tag{3.32}$$

with the same constant c_{rk} as in (3.12).

The virtue of the parameters α and β in Theorems 3.3.11 can be seen in the possibility of choosing them so as to fit the bounds for the rth derivative of a function f (or for a class of functions). Thus one may obtain a better estimate for the functional $E_1(f^{(k)})_C$ than (3.32).

3.3.4 The one-sided case It was shown in Section 3.1.3 that spline $\varphi_{\lambda,m}(\alpha,\beta;t)$ $(m = 1, 2, \ldots,)$ when $\alpha = 1$, $\beta \to \infty$ is transferred to the monospline $\mathcal{B}_{\lambda,m}$ of degree m and defect 2 with knots at $2k\pi/\lambda$ (hence $\mathcal{B}_{\lambda,m}^{(m)}(t) = 1$ for $t \ne 2k\pi/\lambda$). It has to be remarked that in going to the limit one loses one degree of smoothness – $\mathcal{B}_{\lambda,m}^{(m-1)}(t)$ has discontinuities at the knots while $\varphi_{\lambda,m}^{(m-1)}(\alpha,\beta;t)$ is a continuous polygon.

Therefore the function $\mathcal{B}_{\lambda,m}$ may be $2\pi/\lambda$-regular only for $m = 2, 3, \ldots,$. The loss of smoothness also causes another problem. Splines $\varphi_{\lambda,m}(t)$ and $\varphi_{\lambda,m}(\alpha,\beta;t)$ are standard ($2\pi/\lambda$-regular) representatives of classes $W^m(-\infty,\infty)$, $W^m_{\alpha,\beta}(-\infty,\infty)$ respectively and belong to these classes. It is natural to correspond the class

$$W^m_+(-\infty,\infty) = \{f : f \in L^m(-\infty,\infty),\ \operatorname*{ess\,sup}_t f^{(m)}(t) \le 1\},$$

with the monospline $\mathcal{B}_{\lambda,m}$, but it is clear that $\mathcal{B}_{\lambda,m}$ does not belong to this class. This problem can be solved in this case by the use of the Steklov functions of $\mathcal{B}_{\lambda,m}$ (Section A2) which belong to $W^m_+(-\infty,\infty)$ for every h. Considering $\mathcal{B}_{\lambda,m}$ as a comparison function for $W^m_+(-\infty,\infty)$ we are able to get exact statements analogous to the one previously proved. Let us note that $W^m_{1,\beta}(-\infty,\infty) \subset W^m_+(-\infty,\infty)$ for every β.

Lemma 3.3.12 If $f \in W^r_+(-\infty,\infty)$ $(r = 2, 3, \ldots,)$ and

$$\min_t \mathcal{B}_{\lambda,r}(t) \le f(t) \le \max_t \mathcal{B}_{\lambda,r}(t), \tag{3.33}$$

then function f possesses a μ-property with respect to $\mathcal{B}_{\lambda,m}$.

This can be proved (as in Lemma 3.3.8) following the scheme of the proof of Lemma 3.3.1, but it is easier to obtain Lemma 3.3.12 as a consequence of Lemma 3.3.8 and the limit relation (1.23).

If a function f from $W_+^m(-\infty, \infty)$ does not possess a μ-property with respect to $\mathcal{B}_{\lambda,r}$ then the same is true for the function $(1-\varepsilon)f(t)$ for sufficiently small $\varepsilon > 0$. Then $(1-\varepsilon)f$ does not possess a μ-property with respect to $\varphi_{\lambda,r}(1,\beta;t)$ with a sufficiently large β because $\varphi_{\lambda,r}(1,\beta;t)$ tends uniformly to $\mathcal{B}_{\lambda,r}$ for $\beta \to \infty$. But for large enough β we have $(1-\varepsilon)f \in W_{1,\beta}^r(-\infty, \infty)$ and

$$\min_t \varphi_{\lambda,r}(1,\beta;t) \le (1-\varepsilon)f(t) \le \max_t \varphi_{\lambda,r}(1,\beta;t).$$

This is a contradiction to the assertion of Lemma 3.3.8.

Now from Lemma 3.3.12 and the general statements in Section 3.3.2 we immediately obtain analogs of Theorems 3.3.9 and 3.3.10 for one-sided restrictions on $f^{(r)}$. In the statements we use the notation from (1.20).

Theorem 3.3.13 Let $f \in W_+^r(-\infty, \infty)$ $(r = 2, 3, \ldots,)$ and $H_{\lambda r}^- \le f(t) \le H_{\lambda r}^+$ for some $\lambda > 0$. Then:

(1) if $f(\xi) = \mathcal{B}_{\lambda,r}(\eta)$ and $f'(\xi)\mathcal{B}_{\lambda,r}'(\eta) \ge 0$ then

$$|f'(\xi)| \le |\mathcal{B}_{\lambda,r}'(\eta)|;$$

(2) the following exact inequalities hold

$$H_{\lambda,r-k}^-(t) \le f^{(k)}(t) \le H_{\lambda,r-k}^+(t), \qquad k = 1, 2, \ldots, r-2;$$

$$\omega_\pm(f^{(k)}, \delta) \le \omega_\pm(\mathcal{B}_{\lambda,r-k}, \delta), \qquad \delta \ge 0, \qquad k = 0, 1, \ldots, r-2.$$

Denote by W_+^r the subspace of 2π-periodic functions in $W_+^r(-\infty, \infty)$.

Theorem 3.3.14 Let $f \in W_+^r$ $(r = 2, 3, \ldots,)$, let $H_{nr}^- \le f(t) \le H_{nr}^+$ for some $n = 1, 2, \ldots,$ and let $k = 0, 1, \ldots, r-1$. For $k = 0$ we additionally assume that

$$\max_{a,b} \left| \int_a^b f(t)\, dt \right| \le \frac{1}{2} \int_0^{2\pi/n} |\mathcal{B}_{n,r}(t)|\, dt.$$

Then the following exact inequalities hold:

$$\int_0^t r((f^{(k)})_\pm, u)\, du \le \int_0^t r((\mathcal{B}_{n,r-k})_\pm, u)\, du, \qquad 0 \le t \le 2\pi,$$

$$\int_0^t r(f^{(k)}, u)\, du \le \int_0^t r(\mathcal{B}_{n,r-k}, u)\, du, \qquad 0 \le t \le 2\pi,$$

$$\|f^{(k)}\|_q \le \|\mathcal{B}_{n,r-k}\|_q, \qquad 1 \le q \le \infty.$$

Finally we get a one-sided analog of the Kolmogorov inequality (3.12) which can be considered as a limit (for $\alpha = 1$, $\beta \to \infty$) case of the non-symmetric inequality (3.29).

Theorem 3.3.15 For the function $f \in L^r(-\infty, \infty)$ $(r = 2, 3, \ldots,)$ bounded on the real axis the following exact inequality holds

$$E_1(f^{(k)})_C \leq \gamma_{rk}[E_1(f)_C]^{1-k/r} \|(f^{(r)})_+\|_\infty^{k/r}, \qquad k = 1, 2, \ldots, r-1, \quad (3.34)$$

where $\gamma_{rk} = H_{r-k}/H_r^{1-k/r}$, $H_m = H_{1m}$ and $H_{\lambda m}$ are given by (3.20) and (3.21).

For the proof of (3.34) it is necessary to follow the simple arguments for establishing (3.29) in Theorem 3.3.11, but now with the help of Proposition 3.1.3 and Theorem 3.3.13. Concerning the exactness of inequality (3.34), we note that the function $\mathcal{B}_{\lambda,r}$ gives an inequality sign in (3.34) but does not belong to $L^r(-\infty, \infty)$ because $\mathcal{B}_{\lambda,r}^{(r-1)}$ is discontinuous. But the Steklov function of the monospline $\mathcal{B}_{\lambda,r}$ for every h $(0 < h < \pi/\lambda)$ belongs to $L^r(-\infty, \infty)$ and letting $h \to 0$ we see that constant γ_{rk} in (3.34) cannot be decreased.

3.4 Internal extremal properties of splines

3.4.1 Periodic splines with respect to the uniform partition The meaning of the basic results in Section 3.3 is that we can compare norms, the best approximation by constants or any other characteristics of a function $f \in L^r(-\infty, \infty)$ or its derivatives with the corresponding characteristics of regular functions which, in these cases, were standard splines. Now we shall find that Euler and Bernoulli splines not only have special places with respect to their external properties in the classes W_∞^r and W_+^r but they happen to be standard functions for a wide class of polynomial splines. The special structure of polynomial splines determines some specific extremal properties which characterize the standard splines from 'inside', i.e. distinguishes them from other splines. Many of these properties are based on the general comparison theorems from Section 3.2.

Let us start with the set $S_{2n,r}$ of 2π-periodic splines of degree r and with defect 1 at the knots $\{i\pi/n\}$ $(i = 0, 1, \ldots, 2n)$. Obviously $c\varphi_{n,r} \in S_{2n,r}$ for every $c \in \mathbb{R}$.

Proposition 3.4.1 For every spline $s \in S_{2n,r}$ the inequalities

$$\|s^{(r)}\|_\infty \leq K_r^{-1} n^r \|s\|_C, \qquad n, r = 1, 2, \ldots, \qquad (4.1)$$

$$\bigvee_0^{2\pi} s^{(r)} \le K_{r+1}^{-1} n^{r+1} \|s\|_1, \qquad r = 0, 1, \ldots, \qquad n = 1, 2, \ldots, \tag{4.2}$$

hold, where the K_m are the Favard constants (1.8). We have equalities in (4.1) and (4.2) for splines of the type $c\varphi_{n,r}$, where c can be any constant.

Proof It is obviously enough to prove (4.1) for $\|s^{(r)}\|_\infty = 1$, and (4.2) for

$$\bigvee_0^{2\pi} s^{(r)} = 1. \tag{4.3}$$

Equality $\|s^{(r)}\|_\infty = 1$ means that $s^{(r)}(t) = 1$ or $s^{(r)}(t) = -1$ for at least one interval (t_{i-1}, t_i) $(i = 1, 2, \ldots, 2n, t_j = j\pi/n)$. But then

$$\|s^{(r-1)}\|_C \ge \frac{\pi}{2n} = \frac{K_1}{n} \tag{4.4}$$

and (4.1) is proved for $r = 1$. If $r \ge 2$ then we have from (3.12) with $k = r - 1$

$$\|s^{(r-1)}\|_C \le K_1 K_r^{-1/r} \|s\|_C^{1/r}.$$

Replacing (4.4) in this inequality we get (4.1) because $\|s^{(r)}\|_\infty = 1$.

Now let (4.3) hold. With the notations $s^{(r)}(t) = c_i$ $(t_{i-1} < t < t_i)$ we have $c_{2n+1} = c_1$ and also

$$1 = \bigvee_0^{2\pi} s^{(r)} = \sum_{i=1}^{2\pi} |c_{i+1} - c_i| \le 2 \sum_{i=1}^{2n} |c_i|,$$

i.e. $|c_1| + |c_2| + \ldots + |c_{2n}| \ge \frac{1}{2}$. Therefore

$$\int_0^{2\pi} |s^{(r)}(t)| \mathrm{d}t = \frac{\pi}{n} \sum_{i=1}^{2n} |c_i| \ge \frac{\pi}{2n} = \frac{K_1}{n} \tag{4.5}$$

and (4.2) is proved for $r = 0$. For $r \ge 1$ inequality (3.16) gives

$$\|s^{(r)}\|_1 \le K_1 K_{r+1}^{-1/(r+1)} \|s\|_1^{1/(r+1)}. \tag{4.6}$$

From (4.5) and (4.6) we get

$$\|s\|_1 \ge K_{r+1} n^{-r-1},$$

i.e. inequality (4.2) holds because of (4.3). From (1.10) and (1.13) we see that (4.1) and (4.2) become equalities when $s = c\varphi_{n,r}$. □

From Proposition 3.4.1 we immediately get

$$\inf \{\|s\|_C : s \in S_{2n,r}, \|s^{(r)}\|_\infty \ge 1\} = \|\varphi_{n,r}\|_C = K_r n^{-r}, \tag{4.7}$$

$$n, r = 1, 2, \ldots,$$

$$\inf\{\|s\|_1 : s \in S_{2n,r}, \bigvee_0^{2\pi} s^{(r)} \geq 1\} = \|g_{n,r}\|_1 = K_{r+1} n^{-r-1}, \tag{4.8}$$

$$n = 1, 2, \ldots, \qquad r = 0, 1, \ldots,$$

where $g_{n,r}(t) = \varphi_{n,r}(t)/(4n)$.

Relation (4.7) means that all splines $s \in S_{2n,r}$ with $\|s^{(r)}\|_\infty \geq 1$ are outside the open ball $\{f: \|f\|_C < \|\varphi_{n,r}\|_C\}$ in C. Similarly (4.8) can be interpreted as the open ball $\{f: \|f\|_1 < \|g_{n,r}\|_1\}$ in L_1 does not contain splines from $S_{2n,r}$ whose variation on $[0, 2\pi]$ is not less than 1.

If we set $f = s$ ($s \in S_{2n,r}$) in the general inequality (3.12) and if we apply (4.1) we obtain

$$\|s^{(k)}\|_C \leq K_{r-k} K_r^{-1} n^k \|s\|_C, \qquad k = 1, 2, \ldots, r-1. \tag{4.9}$$

Using (4.2) and (3.16) in the same manner we have for every $s \in S_{2n,r}$

$$\|s^{(k)}\|_1 \leq K_{r-k+1} K_{r+1}^{-1} n^k \|s\|_1.$$

These results combined with (1.10) and (1.13) can be summarized in the following generalization of Proposition 3.4.1.

Theorem 3.4.2 We have

$$\sup_{s \in S_{2n,r}} \frac{\|s^{(k)}\|_\infty}{\|s\|_C} = \frac{\|\varphi_{n,r}^{(k)}\|_\infty}{\|\varphi_{n,r}\|_C} = \frac{K_{r-k}}{K_r} n^k, \qquad k = 1, 2, \ldots, r, \tag{4.10}$$

$$\sup_{s \in S_{2n,r}} \frac{\|s^{(k)}\|_1}{\|s\|_1} = \frac{\|\varphi_{n,r}^{(k)}\|_1}{\|\varphi_{n,r}\|_1} = \frac{K_{r-k+1}}{K_{r+1}} n^k, \qquad k = 1, 2, \ldots, r. \tag{4.11}$$

Comparing (4.10) and (3.12) we notice the following fact. For splines $s \in S_{2n,r}$ the norm $\|s^{(k)}\|_C$ can be estimated exactly by using only $\|s\|_C$, while in general we can give an exact estimate for $\|f^{(k)}\|_C$ if we know not only $\|f\|_C$ but also $\|f^{(r)}\|_\infty$. A similar observation can be made by comparing (4.11) and (3.15).

The general comparison theorems from Section 3.2 can also be transferred for splines from $S_{2n,r}$.

Lemma 3.4.3 For every spline $s \in S_{2n,r}$ and every $\alpha \in \mathbb{R}$ the function

$$\eta(t) = \varphi_{n,r}(t) - s(t + \alpha) \tag{4.12}$$

has no more than $2n$ sign changes on the period.

Proof In view of Proposition 2.4.2 the lemma will follow if we prove that $\mu(\eta^{(r)}; 2\pi) \leq 2n$, i.e. the rth derivative of the difference (4.12) has no more than $2n$ sign changes on the period.

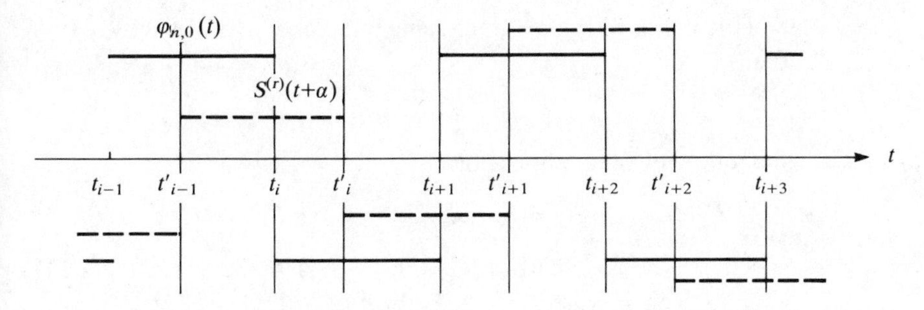

Fig. 3.8 To the proof of Lemma 3.4.3.

For $\alpha = i\pi/n$ $(i = 0, \pm 1, \pm 2, \ldots,)$ the last assertion is obvious because $\eta \in S_{2n,r}$ in this case. Let $\alpha \neq i\pi/n$, $t'_j = j\pi/n - \alpha$ $(j = 0, \pm 1, \pm 2, \ldots,)$ and hence the function $s^{(r)}(t + \alpha)$ is constant on every interval (t'_i, t'_{i+1}) of length π/n. In every such interval the function $\eta^{(r)}(t) = \operatorname{sgn} \sin nt - s^{(r)}(t + \alpha)$ can change sign not more than once, moreover if $\eta^{(r)}$ changes sign on (t'_{i-1}, t'_i) and (t'_i, t'_{i+1}) then there is no essential change of sign in t'_i (this becomes obvious if one draws the graph). Hence $\mu(\eta^{(r)}; 2\pi) \leq 2n$ in this case too.

An easy consequence of Lemma 3.4.3 is:

Proposition 3.4.4 If $s \in S_{2n,r}$ and $\|s\|_C \leq \|\varphi_{n,r}\|_C$ then the spline s possesses a μ-property with respect to $\varphi_{n,r}$.

Indeed, assuming the opposite we can conclude that for an appropriately small $\varepsilon > 0$ the difference

$$\delta(t) = \varphi_{n,r}(t) - (1 - \varepsilon)s(t + \alpha)$$

for some α has at least $2n + 2$ sign changes on the period, which contradicts Lemma 3.4.3.

Let us remark that the μ-property of a spline $s \in S_{2n,r}$ is ensured only by the condition $\|s\|_C \leq \|\varphi_{n,r}\|_C$ while in the general case for $f \in L^r(-\infty, \infty)$ (see Lemma 3.3.1) the additional requirement is $\|f^{(r)}\|_\infty \leq 1$. The reason is that the inequality $\|s\|_C \leq \|\varphi_{n,r}\|_C$ immediately implies $\|s^{(r)}\|_\infty \leq 1$ in view of (4.1). Here we meet an effect caused by the special structure of splines with uniform knots.

Starting with Proposition 3.4.4 we can reformulate the comparison theorems from Section 3.2 for splines. We have in particular:

Theorem 3.4.5 If $s \in S_{2n,r}$ and $\|s\|_C \leq \|\varphi_{n,r}\|_C = K_r n^{-r}$ then

(1) $s(\xi) = \varphi_{n,r}(\eta)$ implies $|s'(\xi)| = |\varphi'_{n,r}(\eta)|$ and if also $s(\xi_1) = \varphi_{n,r}(\eta_1)$

and the points η, η_1 are on the same interval of monotonicity of $\varphi_{n,r}$ then

$$|\xi - \xi_1| \geq |\eta - \eta_1|;$$

(2) the following inequalities hold

$$\|s^{(k)}\|_C \leq \|\varphi_{n,r-k}\|_C = K_{r-k} n^{k-r}, \qquad k = 1, 2, \ldots, r, \tag{4.13}$$

$$\omega(s^{(k)}, \delta) \leq \omega(\varphi_{n,r-k}, \delta), \qquad \delta \geq 0, \qquad k = 0, 1, \ldots, r-1. \tag{4.14}$$

Let us explain how the above inequalities for the kth ($k = 2, 3, \ldots, r-1$) derivative are obtained (cf. Propositions 3.2.2 and 3.2.3). The validity of (4.13) for $k = 1$ allows us to apply statement 1) for s' from where we get (4.13) and (4.14) for $k = 2$ and so on.

Proposition 3.4.4 and Lemma 3.2.1 enable finite differences to be used to obtain a more general and in some sense a more precise estimate for $\|s^{(k)}\|_C$, $s \in S_{2n,r}$, than (4.9).

Let us introduce the notation when $h > 0$:

$$\Delta_h^1(f, t) = \Delta_h(f, t) = f(t+h) - f(t), \tag{4.15}$$

$$\Delta_h^k(f, t) = \Delta_h^1(\Delta_h^{k-1}(f, \bullet), t) = \sum_{i=1}^{k} (-1)^{k-i} \binom{k}{i} f(t + ih), \qquad k = 2, 3, \ldots.$$

Lemma 3.4.6 For every spline $s \in S_{2n,r}$ ($r = 1, 2, \ldots$) and every h, $0 < h < \pi/n$, the following exact inequalities hold

$$\|s'\|_C \leq \frac{K_{r-1}}{n^{r-1}} \frac{\|\Delta_h(s)\|_C}{\omega(\varphi_{n,r}, h)} \leq \frac{K_{r-1}}{n^{r-1}} \frac{\omega(s, h)}{\omega(\varphi_{n,r}, h)}, \tag{4.16}$$

where $\omega(f, \delta)$ is the modulus of continuity (3.6) of function $f \in C$.

Proof Fix $s \in S_{2n,r}$ and set

$$\|s'\|_C = |s'(t_*)| = M, \tag{4.17}$$

$$q(t) = \frac{\|\varphi_{n,r}\|_C}{M} s'(t) = \frac{K_{r-1}}{M n^{r-1}} s'(t). \tag{4.18}$$

Obviously $s \in S_{2n,r-1}$ and

$$|q(t_*)| = \|q\|_C = \|\varphi_{n,r-1}\|_C. \tag{4.19}$$

In view of Proposition 3.4.4. the spline q possesses a μ-property with respect to $\varphi_{n,r-1}$ and hence with respect to $\varphi_\alpha(t) = \varphi_{n,r-1}(t+\alpha)$ for every α. We choose α so that $\varphi_\alpha(t_*) = q(t_*)$ which is always possible because of the symmetry of $\varphi_{n,r-1}$ and equalities (4.19).

Now Lemma 3.2.1 gives $|q(t_* + u)| \geq |\varphi_\alpha(t_* + u)|$ $(|u| \leq \pi/(2n))$ and hence

$$\left|\int_{t_*-h/2}^{t_*+h/2} q(t)\,\mathrm{d}t\right| \geq \left|\int_{t_*-h/2}^{t_*+h/2} \varphi_\alpha(t)\,\mathrm{d}t\right| = \omega(\varphi_{n,r}, h), \qquad 0 < h \leq \pi/n,$$

which in view of (4.17) and (4.18) gives

$$\|\varphi_{n,r-1}\|_C\,|s(t_* + h/2) - s(t_* - h/2)| \geq \|s'\|_C\,\omega(\varphi_{n,r}, h).$$

This proves the first inequality (4.16). The second one is obvious. (4.16) is an equality for $s(t) = c\varphi_{n,r}(t)$, $c =$ constant. □

From Lemma 3.4.6 we easily obtain:

Proposition 3.4.7 For every spline $s \in S_{2n,r}$ $(r = 1, 2, \ldots)$ the following exact inequalities hold

$$\|s^{(k)}\|_C \leq \frac{K_{r-k}}{2^k K_r}\, n^k\, \|\Delta^k_{\pi/n}(s)\|_C, \qquad k = 1, 2, \ldots, r. \tag{4.20}$$

Indeed for $k = 1$ inequality (4.20) coincides with (4.16) because $\omega(\varphi_{n,r}, \pi/n) = 2\|\varphi_{n,r}\|_C = 2K_r n^{-r}$. Now by induction assume (4.20) holds for $k = 1, 2, \ldots, i$. Applying (4.20) with $k = 1$ for $s^{(i)} \in S_{2n,r-i}$ we get

$$\|s^{(i+1)}\|_C \leq \frac{K_{r-i-1}}{2\,K_{r-i}}\, n\, \|\Delta_{\pi/n}(s^{(i)})\|_C. \tag{4.21}$$

The difference $s(t + \pi/n) - s(t)$ is a spline from $S_{2n,r}$ and applying inequality (4.20) for $k = i$ to it we obtain

$$\|\Delta_{\pi/n}(s^{(i)})\|_C \leq \frac{K_{r-i}}{2^i K_r}\, n^i\, \|\Delta^{i+1}_{\pi/n}(s)\|_C. \tag{4.22}$$

Replacing (4.22) in (4.21) we obtain (4.20) for $k = i + 1$. The equality in (4.20) is also realized by the splines $c\varphi_{n,r}$.

At the end of the Section we give a theorem for comparing rearrangements of splines with uniform knots.

Theorem 3.4.8 If $s \in S_{2n,r}$ and $\|s\|_C \leq \|\varphi_{n,r}\|_C$ then for $k = 1, 2, \ldots, r$ the following exact inequalities hold

$$\int_0^t r((s^{(k)})_\pm, u)\,\mathrm{d}u \leq \int_0^t r((\varphi_{n,r-k})_\pm, u)\,\mathrm{d}u, \qquad 0 \leq t \leq 2\pi, \tag{4.23}$$

$$\int_0^t r(s^{(k)}, u)\,\mathrm{d}u \leq \int_0^t r(\varphi_{n,r-k}, u)\,\mathrm{d}u, \qquad 0 \leq t \leq 2\pi, \tag{4.24}$$

$$\|s^{(k)}\|_q \le \|\varphi_{n,r-k}\|_q, \qquad 1 \le q \le \infty. \tag{4.25}$$

Under the additional condition

$$\max_{a,b} \left| \int_a^b s(t)\,dt \right| \le 2\,\|\varphi_{n,r+1}\|_C = 2\,K_{r+1}\,n^{-r-1}$$

inequalities (4.23)–(4.25) hold and are also exact for $k = 0$.

The statements in Theorem 3.4.8 follow from Propositions 3.2.5, 3.4.4 and Theorem 3.2.8 if we take into account the validity of (4.13). The exactness of all inequalities is checked on spline $\varphi_{n,r}$.

Let us note that inequality (4.25) for splines $s \in S_{2n,r}$ with norms $\|s\|_C \le \|\varphi_{n,r}\|_C$ means

$$\sup_{s \in S_{2n,r}} \frac{\|s^{(k)}\|_q}{\|s\|_C} = \frac{\|\varphi^{(k)}_{n,r}\|_q}{\|\varphi_{n,r}\|_C}, \qquad 1 \le q \le \infty, \qquad k = 1, 2, \ldots, r.$$

3.4.2 Splines with free knots In the problems considered in Section 3.4.1 the perfect spline $\varphi_{n,r}$ was an extremal element in the set $S_{2n,r}$. There are sets with a totally different structure from $S_{2n,r}$ for which $\varphi_{n,r}$ is also an extremal element. Let $\Gamma_{2n,m}$ (see Section 2.5) denote the set of 2π-periodic perfect splines of degree $m = 0, 1, \ldots,$ with not more than $2n$ free knots on the period. Therefore $s \in \Gamma_{2n,m}$ $(m = 1, 2, \ldots)$ means that $s \in C^{m-1}$, $|s^{(m)}(t)| = 1$ a.e. and $s^{(m)}$ essentially changes sign not more than $2n$ times. Obviously $\varphi_{n,m} \in \Gamma_{2n,m}$.

With minor complications we can also consider the more general situation where $\Gamma_{2n,m}$ is replaced by $\mathcal{M}_{2n,m}(\alpha, \beta)$ – the set of all 2π-periodic splines of degree m defect 1 such that $s^{(m)}$ is a step function taking only values (except at the points of discontinuity) α and $-\beta$ $(\alpha > 0, \beta > 0)$ and with not more than $2n$ essential changes of sign on the period. The standard spline from $\mathcal{M}_{2n,r}(\alpha, \beta)$ will be the function $\varphi_{n,r}(\alpha, \beta; t)$ (see Section 3.1).

Proposition 3.4.9 For every $n, r = 1, 2, \ldots,$ we have

$$\inf\{E_1(s)_C : s \in \mathcal{M}_{2n,r}(\alpha, \beta)\} = E_1(\varphi_{n,r}(\alpha, \beta))_C,$$

where $E_1(g)_X = \inf\{\|g - \lambda\|_X : \lambda \in \mathbb{R}\}$.

Proof Let $s \in \mathcal{M}_{2n,r}(\alpha, \beta)$ and $s \ne \varphi_{n,r}(\alpha, \beta)$. Then there is an interval (a, a') on which $s^{(r)}(t) = \alpha$ and $a' - a > \gamma$, where γ is the point from $(0, 2\pi/n)$ in which $\varphi_{n,0}(\alpha, \beta)$ changes sign: $\varphi_{n,0}(\alpha,\beta; t) = \alpha$ for $t \in (0, \gamma)$ and $\varphi_{n,0}(\alpha, \beta; t) = -\beta$ for $t \in (\gamma, 2\pi/n)$. If we set

$$\eta(t) = s^{(r)}(t + a) - \varphi_{n,0}(\alpha, \beta; t),$$

then $\eta(t) \geq 0$ for $0 < t < 2\pi/n$ and having in mind the structure of functions $s^{(r)}(t+a), \varphi_{n,0}(\alpha, \beta; t)$ we easily see that the difference η can essentially change sign not more than $2n-2$ times on the period. But then the function $s(t+a) + \lambda - \varphi_{n,r}(\alpha, \beta; t)$ also has for every $\lambda \in \mathbb{R}$ no more than $2n-2$ essential changes of sign on the period (see Proposition 2.4.2).

If we assume that $E_1(s)_C < E_1(\varphi_{n,r}(\alpha, \beta))_C$ then for some $\lambda = \lambda_*$, we shall have

$$\min_t \varphi_{n,r}(\alpha, \beta; t) < s(t+a) + \lambda_* < \max_t \varphi_{n,r}(\alpha, \beta; t)$$

and hence the function $s(t+a) + \lambda_* - \varphi_{n,r}(\alpha, \beta; t)$ will change sign exactly $2n$ times on the period – a contradiction. Finally we have to take into account that $\varphi_{n,r}(\alpha, \beta) \in \mathcal{M}_{2n,r}(\alpha, \beta)$.

For $\alpha = \beta = 1$ we obtain:

Corollary 3.4.10 We have

$$\inf_{s \in \Gamma_{2n,r}} E_1(s)_C = \inf_{s \in \Gamma_{2n,r}} \|s\|_C = \|\varphi_{n,r}\|_C = K_r n^{-r}. \tag{4.26}$$

Let us consider a similar situation for monosplines with free knots. Let $\mathcal{M}_{N,r}$ be the set of all monosplines s of degree r defect 2 with no more than $2n$ free knots t_i on the period and with $s^{(r)}(t) = 1$ for $t \neq t_i$. $\mathcal{B}_{N,r}$ is a monospline from $\mathcal{M}_{N,r}$ defined by (see (1.19))

$$\mathcal{B}_{N,r}(t) = -2N^{-r}B_r(Nt), \qquad N = 2, 3, \ldots, \qquad r = 1, 2, \ldots,$$

where B_r is given by (1.16) and (1.16′).

Proposition 3.4.11 For all $N = 2, 3, \ldots,$ and $r = 1, 2, \ldots,$ we have

$$\inf\{E_1(s)_\infty : s \in \mathcal{M}_{N,r}\} = E_1(\mathcal{B}_{N,r})_\infty. \tag{4.27}$$

Proof The spline $s \in \mathcal{M}_{N,1}$ has the representation $t + c$, $c =$ constant, between every two consecutive knots and hence (4.27) is obvious for $r = 1$.

Let $r > 2$ and $s \in \mathcal{M}_{N,r}$ and let us assume that $E_1(s)_C < E_1(\mathcal{B}_{N,r})_C$. Then for some $\lambda \in \mathbb{R}$ we have

$$\min_t \mathcal{B}_{N,r}(t) < s(t) + \lambda < \max_t \mathcal{B}_{N,r}(t)$$

and hence for every $\alpha \in \mathbb{R}$ the difference $\mathcal{B}_{N,r}(t) - [s(t+\alpha) + \lambda]$ changes sign at least $2N$ times on the period. Choose $\alpha = \alpha_*$ so that one of the knots of the monospline $s_*(t) = s(t + \alpha_*) + \lambda$ will coincide with a knot of $\mathcal{B}_{N,r}$, say $2\pi j/N$, and consider

$$\delta(t) = \mathcal{B}_{N,r}(t) - s_*(t).$$

At the points t which are not knots of either $\mathcal{B}_{N,r}$ or s_* we have $\mathcal{B}_{N,r}^{(r)}(t) = s_*^{(r)}(t) = 1$ and hence $\delta^{(r-1)}$ is a step function with no other possible discontinuities than the knots of $\mathcal{B}_{N,r}$ and s_*. But we have a coincidence of one of the knots of s_* with one of the knots of $\mathcal{B}_{N,r}$ and hence there are no more than $2N - 1$ points of discontinuity of $\delta^{(r-1)}$. Therefore $\delta^{(r-1)}$ has at most $2N - 1$ essential changes of sign and so δ does – a contradiction. □

In connection with (4.27) let us remark that (see Section 3.1.2)

$$E_1(\mathcal{B}_{N,2\nu})_C = \frac{1}{2}\left|\mathcal{B}_{N,2\nu}(0) - \mathcal{B}_{N,2\nu}\left(\frac{\pi}{N}\right)\right| = \frac{\pi}{2N^{2\nu}} K_{2\nu-1};$$

$$E_1(\mathcal{B}_{N,2\nu+1})_C = \|\mathcal{B}_{N,2\nu+1}\|_C, \qquad \nu = 1, 2, \ldots; \qquad E_1(\mathcal{B}_{N,1})_C = \pi/N.$$

Let us go back to the symmetric case. In the metric of L_1 we have the following analog of (4.26).

Proposition 3.4.12 For all $n, r = 1, 2, \ldots$, we have

$$\inf_{s\in\Gamma_{2n,r}} \|s\|_1 = \|\varphi_{n,r}\|_1 = 4\, K_{r+1}\, n^{-r}. \tag{4.28}$$

Proof For $s \in \Gamma_{2n,r}$ we have

$$\bigvee_0^{2\pi} s^{(r)} \le 4n, \qquad \|s^{(r)}\|_1 = 2\pi. \tag{4.29}$$

If we replace f in (3.16) with s, r with $r + 1$ and after that set $k = r$, then using (4.29) we obtain

$$(2\pi)^{r+1} \le K_1^{r+1} K_{r+1}^{-1} \|s\|_1 (4n)^r$$

or

$$\|s\|_1 \ge 4K_{r+1} n^{-r} = \|\varphi_{n,r}\|_1.$$ □

In chapter 8 we shall show, using Σ-rearrangements, that the first relation in (4.28) remains true if norm $\|\bullet\|_1$ is replaced by $\|\bullet\|_q$ $(q \ge 1)$.

3.4.3 On splines on a finite interval In this section up until now we have discussed periodic splines whose extremal properties, which are invariant under translation of the arguments, are expressed via the standard regular functions $\varphi_{\lambda,r}$, $\varphi_{\lambda,r}(\alpha, \beta)$ and $\mathcal{B}_{\lambda,r}$. Extremal problems about exact inequalities for norms of the derivatives can also be considered for splines defined on a finite interval. But in this case we cannot, in general, expect the same symmetry and regularity which takes place in the relationships for periodic splines, because the presence and type of boundary condition essentially affects the extremal properties. Some results for these problems will be given in Section 8.1; here we shall consider only a special case which can be reduced to the periodic case due to a particular choice of the boundary conditions.

Consider the linear manifold $\hat{S}_m(\Delta_N[0, l])$ of the splines of degree m defect 1 defined on $[0, l]$ with respect to partition

$$\Delta_N[0, l]: 0 = t_0 < t_1 < \ldots < t_N = l, \tag{4.30}$$

satisfying for even m the boundary conditions

$$s(0) = s(l), \qquad s^{(2\nu)}(0) = s^{(2\nu)}(l) = 0, \qquad \nu = 1, 2, \ldots, m/2 - 1, \tag{4.31}$$

and for odd m the boundary conditions

$$s^{(2\nu-1)}(0) = s^{(2\nu-1)}(l) = 0, \qquad \nu = 1, 2, \ldots, (m-1)/2. \tag{4.32}$$

Fix $s \in \hat{S}_m(\Delta_N[0, l])$ and construct on $[-l, 0]$ a spline q by setting $q(t) = s(-t)$ for even m or $q(t) = -s(-t) + 2s(0)$ for odd m. From this definition and boundary conditions (4.31) and (4.32) we easily see that

$$q^{(j)}(0) = s^{(j)}(0), \qquad j = 0, 1, \ldots, m-1,$$

$$q^{(j)}(-l) = s^{(j)}(l), \qquad j = 0, 1, \ldots, m-1.$$

Therefore if we extend s on $[-l, l]$ by setting $s(t) = q(t)$ for $-l \le t < 0$ we shall get a spline s_* of degree m defect 1 with $2N - 1$ knots on the interval $(-l, l)$ satisfying the conditions $s_*^{(j)}(-l) = s_*^{(j)}(l)$ $(j = 0, 1, \ldots, m-1)$. Hence we can extend s_* 2π-periodically on the real line without loss of smoothness. The extended spline s_* will have $2N$ knots on the period.

Setting $l = \pi$ and denoting the set $\hat{S}_m(\Delta_N[0, l])$ by $\hat{S}_{N,m}[0, \pi]$ in the case when partition (4.30) is uniform, i.e. $t_i = i\pi/N$ $(i = 0, 1, \ldots, N)$ we can, using a reduction to the trigonometric case, transfer the results from Sections 3.4.1 to 3.4.2 about $S_{2n,m}$ for the set $\hat{S}_{N,m}[0, \pi]$.

For example, for $k = 1, 2, \ldots, r$ we have

$$\sup_{s \in \hat{S}_{N,r}[0,\pi]} \frac{\|s^{(k)}\|_{L_q}}{\|s\|_C} = \frac{\|\varphi_{N,r}^{(k)}\|_{L_q}}{\|\varphi_{N,r}\|_C}, \qquad 1 \le q \le \infty,$$

$$\sup_{s \in \hat{S}_{N,r}[0,\pi]} \frac{\|s^{(k)}\|_{L_1}}{\|s\|_{L_1}} = \frac{\|\varphi_{N,r}^{(k)}\|_{L_1}}{\|\varphi_{N,r}\|_{L_1}},$$

where $\|\bullet\|_X = \|\bullet\|_{X[0,\pi]}$.

Let $\hat{\Gamma}_{N,m}[0, l]$ be the set of all perfect splines s of degree m defined on $[0, l]$ with no more than $N - 1$ knots on $(0, l)$ satisfying conditions (4.31) (even m) or (4.32) (odd m). Then the following analogs of (4.26) and (4.28) hold:

$$\inf_{s \in \hat{\Gamma}_{N,r}[0,\pi]} \|s\|_{L_p[0,\pi]} = \|\varphi_{N,r}\|_{L_p[0,\pi]}, \qquad p = 1, \infty.$$

3.5 Inequalities for polynomials

The extremal properties of both algebraic and trigonometric polynomials have been thoroughly investigated. In the following we give some exact inequalities for the derivatives of polynomials, showing in simple cases how they follow from the general comparison theorems in Section 3.2.

3.5.1 Bernstein's inequality for trigonometric polynomials We consider the inequality (see [3A, vol. 1, p. 26])

$$\|\tau_n'\|_C \leq n\|\tau_n\|_C, \tag{5.1}$$

where τ_n can be any trigonometric polynomial of degree n. There are several known proofs of (5.1). We shall prove it with the help of Proposition 3.2.2 to underline once more the generality of the approach connected with this proposition. First we prove:

Lemma 3.5.1 Every trigonometric polynomial τ_n of degree n possesses a μ-property with respect to the $2\pi/n$-regular polynomial $A \cos nt$, provided $\|\tau_n\|_C \leq |A|$.

Indeed, assuming the opposite we have to conclude that for some $\beta \in \mathbb{R}$ and for some small enough $\varepsilon > 0$ the difference

$$A \cos nt - (1 - \varepsilon)\, \tau_n(t + \beta) \tag{5.2}$$

(if $\|\tau_n\|_C < |A|$ we may take $\varepsilon = 0$) changes sign at least $2n + 2$ times on the period. This is impossible because (5.2) is a trigonometric polynomial of degree n.

Theorem 3.5.2 For every trigonometric polynomial τ_n of degree n the exact inequalities

$$\|\tau_n^{(k)}\|_C \leq n^k \|\tau_n\|_C, \qquad k = 1, 2, \ldots, \tag{5.3}$$

hold. The equality sign is realized, in particular, by the polynomial $\tau_n(t) = \cos nt$.

Obviously we have to prove (5.3) only for $k = 1$, i.e. we have to establish (5.1).

In view of Lemma 3.5.1 we apply Proposition 3.2.2 (2) to polynomial τ_n and $2\pi/n$-regular function $q_n(t) = \|\tau_n\|_C \cos nt$ and we get

$$\|\tau_n'\|_C \leq \|q_n'\|_C \leq n\|\tau_n\|_C.$$

Obviously (5.3) becomes an equality if $\tau_n(t)$ is any polynomial of the type $a \cos n(t + b)$.

3.5.2 Some generalizations and analogs If we apply Lemma 3.2.1 instead of Proposition 3.2.2 we can, using finite differences, derive a more general and precise estimate for the norm of the kth derivative of a trigonometric polynomial than (5.3).

Theorem 3.5.3 For the trigonometric polynomial τ_n of degree n and $0 < h < 2\pi/n$ the exact inequalities

$$\|\tau_n^{(k)}\|_C \leq \left(\frac{n}{2\sin(nh/2)}\right)^k \|\Delta_h^k(\tau_n)\|_C, \qquad k = 1, 2, \ldots, \tag{5.4}$$

hold, where $\Delta_h^k(f, t)$ is the kth finite difference of function f with step h (see (4.15)). The equality sign in (5.4) is attained, in particular, by the polynomial $\tau_n(t) = \cos nt$.

Proof Consider the case $k = 1$ first. We set $\|\tau_n'\|_C = M$ and let $M = |\tau_n'(t_0)|$. Without loss of generality we may assume $M = \tau_n'(t_0)$. In view of Lemma 3.5.1 the polynomial τ_n' possesses a μ-property with respect to $M \cos nt$ and hence Lemma 3.2.1 gives

$$\tau_n'(t_0 + u) \geq M \cos nu, \qquad |u| \leq \pi/n.$$

Integrating this inequality with respect to u from $-h/2$ to $h/2$ gives

$$\tau_n\left(t_0 + \frac{h}{2}\right) - \tau_n\left(t_0 - \frac{h}{2}\right) \geq \frac{2M}{n} \sin \frac{nh}{2}$$

and hence

$$M = \|\tau_n'\|_C \leq n\left(2 \sin \frac{nh}{2}\right)^{-1} \|\Delta_h^1(\tau_n)\|_C.$$

Obviously all inequalities become equalities for $\tau_n(t) = \cos nt$.

Now by induction we assume that (5.4) holds for $k = 1, 2, \ldots, j$. Then its application with $k = j$ to the trigonometric polynomial $\Delta_h(\tau_n, t)$ gives

$$\|\Delta_h(\tau_n^{(j)})\|_C \leq n^j \left(2 \sin \frac{nh}{2}\right)^{-j} \|\Delta_h^{j+1}(\tau_n)\|_C. \tag{5.5}$$

Here, we have used the fact that

$$\frac{d^j}{dt^j} \Delta_h(\tau_n, t) = \Delta_h(\tau_n^{(j)}, t), \qquad \Delta_h^j(\Delta_h(\tau_n, \bullet), t) = \Delta_h^{j+1}(\tau_n, t).$$

From the validity of Theorem 3.5.3 with $k = 1$ we obtain

$$\|\tau_n^{(j+1)}\|_C \leq n\left(2 \sin \frac{nh}{2}\right)^{-1} \|\Delta_h(\tau_n^{(j)})\|_C, \tag{5.6}$$

and now the use of inequalities (5.5) and (5.6) proves (5.4) for $k = j + 1$. The only point left is to remark that the equality sign in (5.4) for $k = 1, 2, \ldots, j$ implies the equality sign for the same polynomial for $k = j + 1$. □

Setting $h = \pi/n$ in (5.4) we get:

Corollary 3.5.4 For the trigonometric polynomial τ_n of degree n the following exact inequalities hold:

$$\|\tau_n^{(k)}\|_C \le \left(\frac{n}{2}\right)^k \|\Delta^k_{\pi/n}(\tau_n)\|_C, \qquad k = 1, 2, \ldots. \tag{5.7}$$

We note that inequality (5.7) implies (5.3) because of $\|\Delta_h^k(f)\|_C \le 2^k \|f\|_C$.

Integral analogs of inequalities (5.3) can be obtained by starting with the Riesz formula (see e.g. [17A, p. 93]):

$$\tau_n'(t) = \frac{1}{4n} \sum_{k=1}^{2n} (-1)^{k+1} \left(\sin \frac{x_k}{2}\right)^{-2} \tau_n(t + x_k), \qquad x_k = \frac{2k-1}{2n}\pi. \tag{5.8}$$

Setting $\tau_n(t) = \sin nt$ in (5.8) and after that $t = 0$ we obtain

$$n = \frac{1}{4n} \sum_{k=1}^{2n} \left(\sin \frac{x_k}{2}\right)^{-2},$$

and hence estimating the $L_q = L_q[0, 2\pi]$ norm of the right-hand side of (5.8) we obtain

$$\|\tau_n'\|_q \le n \|\tau_n\|_q, \qquad 1 \le q < \infty.$$

Applying this inequality to the derivatives of τ_n we obtain

$$\|\tau_n^{(k)}\|_q \le n^k \|\tau_n\|_q, \qquad k = 1, 2, \ldots. \tag{5.9}$$

Equality in (5.9) also holds for $\tau_n(t) = \cos nt$.

By appropriately choosing the standard function Lemma 3.5.1 allows us to transfer Comparison Theorem 3.2.8 and its Corollary 3.2.9 to trigonometric polynomials and their derivatives.

Theorem 3.5.5 If τ_n is a trigonometric polynomial of degree n and

$$q_n(t) = \|\tau_n\|_C \cos nt$$

then for $k = 1, 2, \ldots,$ the following exact inequalities hold

$$\int_0^t r((\tau_n^{(k)})_\pm, u)\, du \le \int_0^t r((q_n^{(k)})_\pm, u)\, du = 2\, n^k \|\tau_n\|_C \sin \frac{t}{2}, \tag{5.10}$$

$$\int_0^t r(\tau_n^{(k)}, u)\,du \le \int_0^t r(q_n^{(k)}, u)\,du = 4\,n^k\,\|\tau_n\|_C \sin\frac{t}{4}. \tag{5.11}$$

Under the additional condition

$$\max_{a,b}\left|\int_a^b \tau_n(t)\,dt\right| \le 2\,\|\tau_n\|_C\,n^{-1} \tag{5.12}$$

inequalities (5.10) and (5.11) hold for $k = 0$, i.e. for the polynomial τ_n itself.

Proof From

$$q_n^{(k)}(t) = n^k\,\|\tau_n\|_C \cos(nt + k\pi/2), \qquad k = 1, 2, \ldots,$$

(5.3) and Lemma 3.5.1 we see that polynomial $\tau_n^{(k)}$ possesses a μ-property with respect to the $2\pi/n$-regular function $q_n^{(k)}$. Moreover $\|\tau_n\|_C = \|q_n\|_C$ and

$$\max_{a,b}\left|\int_a^b \tau_n^{(k)}(t)\,dt\right| \le 2\,\|\tau_n^{(k-1)}\|_C \le 2\,n^{k-1}\,\|\tau_n\|_C$$

$$= 2\,\|q_n^{(k-1)}\|_C = \frac{1}{2}\int_0^{2\pi/n} |q_n^{(k)}(t)|\,dt.$$

Hence we can apply Theorem 3.2.8 to the polynomial $\tau_n^{(k)}$ and we get the inequalities in (5.10) and (5.11). The last relations in (5.10) and (5.11) are obtained by direct calculations using the definition of the rearrangement $r(f, t)$ of the periodic function f. □

If inequality (5.12) holds then Theorem 3.2.8 is applicable to the polynomial τ_n. The first inequalities in (5.10) and (5.11) became equalities for example for $\tau_n(t) = \cos nt$.

Corollary 3.5.6 For the derivatives $\tau_n^{(k)}$ $(k = 1, 2, \ldots)$ of the trigonometric polynomial τ_n of degree n the exact inequalities

$$\|\tau_n^{(k)}\|_q \le n^k\,\|\tau_n\|_C\,\|\cos(\bullet)\|_q, \qquad 1 \le q < \infty \tag{5.13}$$

hold. If τ_n satisfies (5.12) then (5.13) also holds for $k = 0$, i.e.

$$\|\tau_n\|_q \le \|\tau_n\|_C\,\|\cos(\bullet)\|_q.$$

3.5.3 Chebyshev polynomials In the comparison theorems for splines or trigonometric polynomials the standard functions were the $2\pi/n$-periodic functions $\varphi_{n,r}(t)$ and $\cos nt$. Besides the properties characterizing the regularity (in the definition in Section 3.2.1), a distinguishing feature of these functions is that their best approximation in C or L_q among the trigonometric polynomials of degree $n-1$ is realized by the zero function, i.e.

$$E(\varphi_{n,r}, \mathcal{N}^T_{2n-1})_X = \|\varphi_{n,r}\|_X, \qquad E(\cos n(\bullet), \mathcal{N}^T_{2n-1})_X = \|\cos n(\bullet)\|_X, \quad (5.14)$$

where X is C or L_q $(q \geq 1)$.

For $X = C$ equalities (5.14) are connected with the fact that $\varphi_{n,r}(t)$ and $\cos nt$ take on the period 2π alternatively $2n$ times the values $\pm\|\varphi_{n,r}\|_C$ and ± 1 respectively and we only have to remember the Chebyshev Theorem 2.1.2. For $X = L_q$ $(1 \leq q < \infty)$ equalities (5.14) follow from Proposition 1.4.6 because in view of Proposition 4.1.2 (see Chapter 4), the functions $|\varphi_{n,r}(t)|^{q-1} \operatorname{sgn} \varphi_{n,r}(t)$ and $|\cos nt|^{q-1} \operatorname{sgn} \cos nt$ are orthogonal to the subspace $\mathcal{N}^T_{2n-1}$.

The algebraic polynomials possessing the above mentioned properties play an important role in analysis, but, unlike the periodic case, their form essentially depends on the metric in X.

In the spaces $C[-1, 1]$ and $L_1[-1, 1]$ Chebyshev polynomials possess the minimal norm property, and on the interval $[-1, 1]$ they can be written as

$$T_n(t) = \cos(n \arccos t), \quad (5.15)$$

$$Q_n(t) = \frac{\sin[(n+1)\arccos t]}{(1-t^2)^{1/2}}. \quad (5.16)$$

We can verify by induction that

$$T_n(t) = 2^{n-1}t^n + a_{n-1}t^{n-1} + \ldots + a_1 t + a_0, \quad (5.17)$$

$$Q_n(t) = 2^n t^n + b_{n-1}t^{n-1} + \ldots + b_1 t + b_0. \quad (5.18)$$

Indeed for $n = 1$ we have $T_n(t) = t$ and $Q_n(t) = 2t$. Assuming that representations (5.17) and (5.18) hold for $n = 1, 2, \ldots, k-1$ we obtain

$$\begin{aligned} T_k(t) &= \cos[\arccos t + (k-1)\arccos t] = tT_{k-1}(t) - (1-t^2)Q_{k-2}(t) \\ &= t(2^{k-2}t^{k-1} + \ldots) - (1-t^2)(2^{k-2}t^{k-2} + \ldots) \\ &= 2^{k-1}t^k + a_{k-1}t^{k-1} + \ldots + a_0, \end{aligned}$$

and similarly

$$\begin{aligned} Q_k(t) &= (1-t^2)^{-1/2} \sin[\arccos t + k \arccos t] = T_k(t) + tQ_{k-1}(t) \\ &= 2^k t^k + b_{k-1}t^{k-1} + \ldots + b_0. \end{aligned}$$

Obviously $\|T_n\|_{C[-1,1]} = 1$ and

$$T_n(x^n_k) = (-1)^k, \qquad x^n_k = \cos(k\pi/n), \qquad k = 0, 1, \ldots, n,$$

i.e. polynomial T_n takes its maximal in modulus value in $n+1$ points in $[-1, 1]$ alternating the sign. But in this case Theorem 2.1.3 asserts that the polynomial of best approximation to T_n from $\mathcal{N}_n^A$ is identically zero:

$$E(T_n, \mathcal{N}_n^A)_{C[-1,1]} = \|T_n\|_{C[-1,1]} = 1. \tag{5.19}$$

From (5.19) we immediately get that the polynomial

$$\hat{T}_n(t) = 2^{1-n}\, T_n(t) = 2^{1-n} \cos(n \arccos t)$$

has the least deviation from zero in $C[-1,1]$ among all polynomials of degree n with leading coefficient 1. In other words

$$\inf\{\|p_n\|_{C[-1,1]}: p_n(t) = t^n + \alpha_{n-1}t^{n-1} + \ldots + \alpha_0\}$$
$$= \|\hat{T}_n\|_{C[-1,1]} = 2^{1-n}.$$

Let us now remark that, unlike the periodic standard functions, the zeros and extremal points of the polynomial T_n are not uniformly distributed – they condense to the end–points of the interval $[-1,1]$. Therefore the derivatives of T_n at the points t should increase when t approaches the ends of $[-1,1]$, provided that $T_n(t) = a$ for any fixed a, $-1 < a < 1$. From

$$T_n'(t) = \frac{n \sin(n \arccos t)}{(1-t^2)^{1/2}} = \frac{n \sin n\theta}{\sin\theta}, \qquad \theta = \arccos t,$$

for every $t \in [-1,1]$ and from the inequality $|\sin n\theta| \le n\,|\sin\theta|$ we obtain

$$|T_n'(t)| \le T_n'(1) = n^2, \qquad -1 \le t \le 1. \tag{5.20}$$

Let us go back to polynomial (5.16) (which is called a *Chebyshev polynomial of the second kind*). Clearly $Q_n(t) = T_{n+1}'(t)/(n+1)$ and in view of (5.20) we have

$$|Q_n'(t)| \le Q_n'(1) = n+1, \qquad -1 \le t \le 1.$$

We shall show that

$$E(Q_n, \mathcal{N}_n^A)_{L_1[-1,1]} = \int_{-1}^{1} |Q_n(t)|\, \mathrm{d}t = \int_0^\pi |\sin(n+1)\theta|\, \mathrm{d}\theta = 2, \tag{5.21}$$

i.e. the best approximant of Q_n in $L_1[-1,1]$ among all polynomials of degree $n-1$ is identically zero.

In view of Proposition 1.4.6 we only have to show that function $\operatorname{sgn} Q_n(t)$ is orthogonal on $[-1,1]$ to all functions t^k $(k = 0, 1, \ldots, n-1)$ in order to prove this assertion. Letting $t = \cos\theta$ we have

$$\int_{-1}^{1} t^k \operatorname{sgn} Q_n(t)\, \mathrm{d}t = \int_0^\pi \cos^k\theta \operatorname{sgn}\sin(n+1)\theta \, \mathrm{d}\cos\theta$$

$$= -\frac{1}{2}\int_{-\pi}^{\pi} \cos^k\theta \sin\theta \operatorname{sgn}\sin(n+1)\theta\, \mathrm{d}\theta. \tag{5.22}$$

The function $\cos^k \theta \sin \theta$ is an odd trigonometric polynomial of degree $k + 1$ and the function sgn sin $(n + 1)\theta$ has period $2\pi/(n + 1)$. Hence in view of Proposition 4.1.2 (Section 4.1) the last integral in (5.22) is zero for $k = 0, 1, \ldots, n - 1$, i.e. $\operatorname{sgn} Q_n(t) \perp \mathcal{N}_n^A$.

From (5.21) we obtain that the polynomial

$$\hat{Q}_n(t) = 2^{-n} Q_n(t) = t^n + \beta_{n-1} t^{n-1} + \ldots + \beta_n$$

has the least deviation from zero in $L_1[-1, 1]$ among all polynomials of degree n with leading coefficient 1.

3.5.4 Inequalities for the derivatives of algebraic polynomials The methods for obtaining exact inequalities for the derivatives of periodic functions were based on the μ-property in which the invariance under translation of the argument was used. The functions defined on a finite interval and the algebraic polynomials do not possess such invariance. This reduces our tools for investigating their extremal properties substantially. We shall now obtain some inequalities for polynomials using, in particular, reduction to the periodic case.

Proposition 3.5.7 If p_n is an algebraic polynomial of degree n and $\|p_n\|_{C[-1,1]} = M$ then

$$|p_n'(t)| \leq n\, M\, (1 - t^2)^{-1/2}, \qquad -1 < t < 1. \tag{5.23}$$

We have equality in the points $t = \cos\,[(2k - 1)\pi/(2n)]$ for $p_n = T_n$.

Indeed the function $\tau_n(\theta) = p_n(\cos \theta)$ is a trigonometric polynomial of degree n and $\|\tau_n\|_C = M$. From Theorem 3.5.2 we have

$$|\tau_n'(\theta)| = |p_n'(\cos \theta) \sin \theta| \leq nM. \tag{5.24}$$

The change of variables $\cos \theta = t$ gives inequality (5.23).

Relation (5.23) (known as the Bernstein inequality for algebraic polynomials), which gives an estimate for the absolute value of the derivative at every point, becomes inaccurate near the ends of the interval $[-1, 1]$. A uniform exact estimate for the set of all algebraic polynomials of degree n is given by the Markov inequality [11B].

Theorem 3.5.8 If p_n is an algebraic polynomial of degree n and $\|p_n\|_{C[-1,1]} = M$ then

$$\|p_n'\|_{C[-1,1]} \leq n^2 M. \tag{5.25}$$

We have an equality for $p_n = T_n$.

This theorem follows from:

Lemma 3.5.9 If q is an algebraic polynomial of degree $n-1$ and

$$|q(t)| \le (1-t^2)^{-1/2}, \qquad -1<t<1 \tag{5.26}$$

then for every t, $-1 \le t \le 1$, we have

$$|q(t)| \le n. \tag{5.27}$$

Proof For $|t| \le \cos[\pi/(2n)]$ inequality (5.27) immediately follows from (5.26) because in this case $(1-t^2)^{1/2} \ge \sin[\pi/(2n)] \ge 1/n$.

Let $\cos[\pi/(2n)] < |t| \le 1$. Using the Lagrange interpolating formula on Chebyshev knots $\tilde{x}_k = \tilde{x}_k^n = \cos[(2k-1)\pi/(2n)]$ $(k=1, 2, \ldots, n)$ (see (2.2.38) and (2.2.39)) we can write

$$q(t) = \sum_{k=1}^{n} q(\tilde{x}_k) \frac{T_n(t)}{T_n'(\tilde{x}_k)(t-\tilde{x}_k)} = \frac{1}{n} \sum_{k=1}^{n} (-1)^{k-1} q(\tilde{x}_k)(1-\tilde{x}_k^2)^{1/2} \frac{T_n(t)}{t-\tilde{x}_k}.$$

Using (5.26) and (5.20) and the fact that all differences $t-\tilde{x}_k$ have the same sign we obtain

$$|q(t)| \le \frac{|T_n(t)|}{n} \sum_{k=1}^{n} \left| \frac{1}{t-\tilde{x}_k} \right| = \frac{|T_n(t)|}{n} \left| \sum_{k=1}^{n} \frac{1}{t-\tilde{x}_k} \right|$$

$$= \frac{|T_n(t)|}{n} \left| \frac{T_n'(t)}{T_n(t)} \right| = \frac{1}{n} |T_n'(t)| \le \frac{1}{n} |T_n'(1)| = n.$$

The lemma is proved. □

If p_n is an arbitrary algebraic polynomial of degree n then it satisfies (5.23) and setting $q(t) = p_n'(t)/(nM)$ we see that the lemma implies Markov inequality (5.25).

A reduction to the periodic case with the help of Corollary 3.5.4 gives an improvement in inequality (5.23).

Estimating $|\tau_n'(\theta)|$ in (5.24) with the help of (5.7) setting $k=1$ and with the help of the inequality $|\cos(t+\pi/n) - \cos t| \le 2\sin[\pi/(2n)]$ we obtain

$$|p_n'(\cos\theta)\sin\theta| \le \frac{n}{2} \|p_n(\cos(\bullet + \pi/n)) - p_n(\cos(\bullet))\|_C$$

$$\le \frac{n}{2} \omega\left(p_n, 2\sin\frac{\pi}{2n}\right)_{C[-1,1]}, \tag{5.28}$$

where $\omega(f, \delta)_{C[-1,1]}$ is the modulus of continuity for $f \in C[-1,1]$. Setting $\cos\theta = t$ in (5.28) we obtain

$$|p_n'(t)| \le \frac{n}{2(1-t^2)^{1/2}} \omega\left(p_n, 2\sin\frac{\pi}{2n}\right)_{C[-1,1]}, \qquad -1<t<1. \tag{5.29}$$

Comments

Section 3.1 As can be seen from their definitions perfect Euler splines and Bernoulli monosplines are generated from the well known Euler and Bernoulli polynomials (see e.g. Danilov and others [4B]).

Section 3.2 The rearrangement of functions was introduced by Hardy and Littlewood (see [7B] and [17B, vol II]). Proposition 3.2.5 is contained in Kong-Ming Shong [1]. Theorem 3.2.8 in the symmetric case was proved by Korneikchuk [11, 9A]; after that it became clear that the same ideas also work in the non-symmetric case (see [11A, Theorem 5.5.2], V. F. Babenko [5, 8]).

Section 3.3 The proof of Lemma 3.3.1 follows a scheme described by Kolmogorov [3] (see also [9A]). Theorem 3.3.4 belongs to me [10, 11, 9A], inequality (3.20) can be found in [7B]. There are a lot of works devoted to obtaining results analogous to the Kolmogorov inequality (3.12) in spaces L_p, with mixed metrics, on the semi-axis or on a finite interval. Besides the Stein [1] result (3.15) different cases were considered in particular by Stechkin [3], Gabushin [1, 2], Kuptsov [1] and Arestov [1]. The history of the problem can be found in the appendix of [7B] written by Levin and Stechkin, later surveys are given in monographs [6A, 25A].

There is a close relationship between the problem of exact constants in the inequalities for the norms of functions and their derivatives from one side and other extremal problems, in particular, the problem of the best approximation by linear constrained operators from another (see Stechkin [5], Taikov [3] and Arestov [2]).

It seems that the first to consider non-symmetric problems of this type was L. Hörmander [1] (see also [2A]) who is the author of Theorem 3.3.15. In connection with (α, β)-restrictions see the papers of V. F. Babenko [3–7].

A series of results devoted to extensions of Kolmogorov-type inequalities to more general than $\mathrm{d}^r/\mathrm{d}x^r$ differential operators can be found in the articles of Sharma and Tzimbalario [1], Nguen Thi Hoa [1, 2], Sun Yong-sheng [3], V. F. Babenko [8].

Section 3.4 Inequalities (4.1) and (4.2) were obtained (using other arguments) by Tikhomirov [5] and Subbotin [3] respectively. Proposition 3.4.11 is due to Ligun [2].

Section 3.5 Inequalities (5.4), (5.7) and (5.29) were proved by Stechkin [1]. An analog of inequality (5.7) for entire functions of the exponential type was obtained by Nikol'skii [12], an analog of (5.4) for $r = 1$ by S. N. Bernstein [3A, vol. 2]. Inequality (5.9) is due to Zygmund [1, 17B, vol. II] who obtained it in another way. Relation (5.13) was obtained (also by a different argument) by Taikov [1].

Other inequalities for polynomials (including with exact constants) can be found in the books [2A, 3A, 6A, 7A, 17A].

Exact inequalities for algebraic and trigonometric polynomials in the L_p $(0 < p < 1)$ metric are contained in Arestov's paper [3]. Generalizations of Bernstein-type inequalities for trigonometric polynomials and splines in situations connected with linear differential operators can be found, in particular, in the articles of Shevaldin [1], V. F. Babenko [12] and Nguen Thi Hoa [2].

Other results and exercises

1. For $s \in S_{2n,r}$ we have the exact inequality

$$\bigvee_0^{2\pi} s^{(r)} \le 4n^{r+1} \|s\|_p \|\varphi_{1,r}\|_p^{-1}, \qquad 1 \le p \le \infty$$

(Subbotin [4] ($p = 1$) and Ligun [5]).

2. Let $f \in W_\infty^{r+1}$ ($r = 0, 1, \ldots,$) and $\|f\|_C = b^{r+1} \|\varphi_{1,r+1}\|_C$. Then

$$\int_0^x r(f', t)\, dt \le b^r \int_0^x r(\varphi_{1,r}, t)\, dt, \qquad 0 \le x \le 2\pi;$$

the inequality is exact (Korneichuk and Ligun [2]).

3. For $f \in V^r$ ($r = 1, 2, \ldots,$) and $k = 1, 2, \ldots, r-1$ we have the exact inequality

$$\frac{\bigvee_0^{2\pi} (f^{(k)})}{\|\varphi_{1,r-k}\|_C} \le \left(\frac{\mu(f')}{2}\right)^{\alpha(p-1)/p} \left(\frac{4\|f\|_p}{\|\varphi_{1,r}\|_p}\right)^\alpha \left[\bigvee_0^{2\pi} f^{(r)}\right]^{1-\alpha},$$

where $\mu(f')$ is the number of sign changes of f' on the period and a$\alpha = (r-k)/(r + 1/p)$ (Ligun [10]).

4. If τ_n is a trigonometric polynomial of degree n then

$$\frac{\|\tau_n^{(k)}\|_1}{\|\cos(\bullet)\|_1} \le n^k \frac{\|\tau_n\|_p}{\|\cos(\bullet)\|_p}; \qquad k = 1, 2, \ldots, n; \qquad 1 \le p \le \infty;$$

the inequality is exact (Ligun [10]).

5. For $f \in L_\infty^r$ and $k = 1, 2, \ldots, r-1$ we have the exact inequalities

$$\left(\frac{\|f^{(k)}\|_p}{2\|B_{r-k}\|_p \|(f^{(r)})_+\|_\infty}\right)^{(r-k)^{-1}} \le \left(\frac{\|f\|_C}{2E_1(B_r)_C \|(f^{(r)})_+\|_\infty}\right)^{r^{-1}},$$

(V. G. Doronin and Ligun, see [11A]). In the case of (α, β)-non-symmetry an analogous result is obtained by V. F. Babenko [5].

6. For every trigonometric polynomial τ_n of degree n we have

$$\int_0^x r(\tau_n', t)\, dt \le n \int_0^x r(\tau_n, t)\, dt, \qquad 0 \le x \le 2\pi;$$

the inequality is exact (Lorentz [1]).

7. Let $f \in L_\infty^r$ and let $\tilde{f}$ denote the function trigonometrically conjugated to f. Then for $r = 2, 3, \ldots,$; $k = 1, 2, \ldots, r-1$, we have

$$\|\tilde{f}^{(k)}\|_1 \le \|\tilde{\varphi}_{1,r-k}\|_1 (\|f\|_\infty / \|\varphi_{1,r}\|_\infty)^{1-k/r} \|f^{(r)}\|^{k/r};$$

the inequality is exact (V. F. Babenko [9]).

8. Let $f \in C(\mathbb{R}^2)$ be bounded on $\mathbb{R}^2$ together with the partial derivatives $f^{(3,0)}$ and $f^{(0,3)}$ and have locally absolutely continuous partial derivatives $f^{(2,0)}(x, y)$ and $f^{(0,2)}(x, y)$ for all fixed y and x respectively. Then

$$\|f^{(1,1)}\|_\infty \le (3)^{1/3} (\|f\|_\infty \|f^{(3,0)}\|_\infty \|f^{(0,3)}\|_\infty)^{1/3},$$

where $\|\bullet\|_\infty = \|\bullet\|_{L_\infty(\mathbb{R}^2)}$; the inequality is exact (Konovalov [1]).

9. For $s \in S_{2n,r}$ obtain using Exercise 3 the exact inequality

$$\bigvee_0^{2\pi} s^{(k)} \leq n^{k+1} \|s\|_p \left(\|\varphi_{1,r}\|_p \bigvee_0^{2\pi} \varphi_{1,r-k} \right)^{-1}, \qquad k = 0, 1, \ldots, r, \qquad 1 \leq p \leq \infty.$$

10. Prove that for every algebraic polynomial p of degree n such that $\|p\|_{C[-1,1]} \leq 1$ we have $|p(t)| \leq |T_n(t)|$ for $|t| \geq 1$.

11. Obtain inequality (4.1) as a consequence of Lemma 2.4.10.

12. Prove that inequality (3.15) remains true (but not with the exact constant) if we replace $\|\bullet\|_1$ by $\|\bullet\|_p$ $(1 < p < \infty)$.

4

Polynomial approximation of classes of functions with bounded *r*th derivative in L_p

When trying to find the supremums of the errors of different approximation methods for a given functional class $\mathfrak{M}$, upper bounds of these errors obtained using certain characteristics defining $\mathfrak{M}$ are very important. The difficulties encountered depend on the type of these characteristics. The best results in solving such problems are obtained when the set $\mathfrak{M}$ is defined by a restriction (in some metric) on the norm of some derivative.

The purpose of this chapter is obtaining exact or asymptotically exact results for the approximation of classes W_p^r and $W_p^r[a, b]$ (see Section 1.5) by means of trigonometric or algebraic polynomials respectively. In the trigonometric case the focus is on approximations by means of λ-methods (Section 2.2) constructed on the basis of Fourier series. The approximating properties of these methods in different metrics have been extensively studied for many years and it seems that the exact (or asymptotically exact) estimates which can be found (in the univariable case) have been found by now.

4.1 Minimizing the error in the class of λ-methods

4.1.1 General relations and an optimization problem We have seen in Section 2.2 that for every n and for every function $f \in L_1$ there is a corresponding trigonometric polynomial

$$U_n(f, \lambda, t) = a_0/2 + \sum_{k=1}^{n} \lambda_k^n(a_k \cos kt + b_k \sin kt)$$

$$= \frac{1}{\pi}\int_0^{2\pi} K_n(\lambda, t-u) f(u)\mathrm{d}u = \frac{1}{\pi}\int_0^{2\pi} K_n(\lambda, u) f(t-u)\mathrm{d}u, \qquad (1.1)$$

where a_k and b_k are the Fourier coefficients of f with respect to the trigonometric system where $\lambda = \{\lambda_1^n, \lambda_2^n, \ldots \lambda_n^n\}$ are prescribed and where

$$K_n(\lambda, t) = \frac{1}{2} + \sum_{k=1}^{n} \lambda_k^n \cos kt. \tag{1.2}$$

So we may consider every set λ as defining a method (λ-method) for approximating functions f by means of trigonometric polynomials of degree n.

Here we shall be concerned with estimating the approximating errors of λ-methods on periodical functionals. Under fairly general assumptions the problem of finding the supremum of the error on class $\mathcal{M}$ can be reduced to the problem of the supremum of a linear functional. In particular we have:

Proposition 4.1.1 Let $\mathcal{M}$ be a class of functions from C containing every function of the type $c + f(t+\alpha)$ $(c, \alpha \in \mathbb{R})$ provided that $f \in \mathcal{M}$. Then for every $t \in \mathbb{R}$ we have

$$\begin{aligned} \sup_{f\in\mathcal{M}} |f(t) - U_n(f, \lambda, t)| &= \sup_{f\in\mathcal{M}_0} |U_n(f, \lambda, 0)| \\ &= \sup_{f\in\mathcal{M}_0} \frac{1}{\pi} \left| \int_0^{2\pi} K_n(\lambda, u) f(u)\, du \right|, \end{aligned} \tag{1.3}$$

where $\mathcal{M}_0 = \{f : f \in \mathcal{M}, f(0) = 0\}$. If, moreover, $\mathcal{M}$ contains $[f(t) + f(-t)]/2$ together with f, then

$$\sup_{f\in\mathcal{M}} |f(t) - U_n(f, \lambda, t)| = \sup_{f\in\mathcal{M}_0} \frac{2}{\pi} \left| \int_0^{\pi} K_n(\lambda, u) f(u)\, du \right|.$$

Proof Using

$$\frac{1}{\pi} \int_0^{2\pi} K_n(\lambda, u)\, du = 1 \tag{1.4}$$

for a fixed $t = t_0$ we obtain

$$f(t_0) - U_n(f, \lambda, t_0) = \frac{1}{\pi} \int_0^{2\pi} K_n(\lambda, u)\, [f(t_0) - f(t_0 - u)]\, du.$$

If $f \in \mathcal{M}$ then the function $f_0(t) = f(t+t_0) - f(t_0)$ belongs to $\mathcal{M}_0$ and $U_n(f, \lambda, t_0) - f(t_0) = U_n(f_0, \lambda, 0)$. Therefore we have for every t

$$\sup_{f\in\mathcal{M}} |f(t) - U_n(f, \lambda, t)| \le \sup_{f\in\mathcal{M}_0} |U_n(f, \lambda, 0)| \tag{1.5}$$

and, setting $t = 0$ on the left-hand side, we see that a strict inequality in (1.5) is impossible because $\mathcal{M}_0 \subset \mathcal{M}$. This proves (1.4). The last statement of Proposition 4.1.1 follows from the evenness of $K_n(\lambda, t)$ which implies

$$\int_0^{2\pi} K_n(\lambda, u) f(u)\, du = 2 \int_0^{\pi} K_n(\lambda, u) \frac{f(u) + f(-u)}{2}\, du. \qquad \square$$

Now let $f \in L_1^r$ ($r = 1, 2, \ldots,$). Then (see relation (1.5.1))

$$f(t) = \frac{a_0}{2} + \frac{1}{\pi} \int_0^{2\pi} B_r(t-u) f^{(r)}(u)\, du, \qquad r = 1, 2, \ldots, \tag{1.6}$$

where B_r are Bernoulli functions (see Section 3.1)

$$B_r(t) = \sum_{k=1}^{\infty} \frac{\cos(kt - \pi r/2)}{k^r}. \tag{1.7}$$

Using the identity (see (1.5.3))

$$a_k \cos kt + b_k \sin kt = \frac{1}{\pi k^r} \int_0^{2\pi} f^{(r)}(u) \cos\left(k(t-u) - \frac{\pi r}{2}\right) du$$

polynomial (1.1) for $f \in L_1^r$ can be written as

$$U_n(f, \lambda, t) = \frac{a_0}{2} + \frac{1}{\pi} \int_0^{2\pi} \left[\sum_{k=1}^{n} \frac{\lambda_k^n}{k^r} \cos\left(k(t-u) - \frac{\pi r}{2}\right)\right] f^{(r)}(u)\, du. \tag{1.8}$$

Because $f^{(r)} \perp 1$, i.e.

$$\int_0^{2\pi} f^{(r)}(u)\, du = 0,$$

we can introduce an additional parameter γ and, setting

$$K_{n,r}(\gamma, \lambda; t) = \gamma + \sum_{k=1}^{n} \frac{\lambda_k^n}{k^r} \cos\left(kt - \frac{\pi r}{2}\right), \tag{1.9}$$

we have

$$U_n(f, \lambda, t) = \frac{a_0}{2} + \frac{1}{\pi} \int_0^{2\pi} K_{n,r}(\gamma, \lambda; t-u) f^{(r)}(u)\, du, \qquad n = 1, 2, \ldots. \tag{1.10}$$

For $n = 0$ we set

$$U_0(f, \lambda, t) = a_0/2, \qquad K_{0,r}(\gamma, \lambda; t) = \gamma.$$

Having in mind (1.6) and (1.10) we set

$$\Psi_{n,r}(\varphi, \lambda; t) = B_r(t) - K_{n,r}(\gamma, \lambda; t)$$

$$= \sum_{k=1}^{n} \frac{1-\lambda_k^n}{k^r} \cos\left(kt - \frac{\pi r}{2}\right) + \sum_{k=n+1}^{\infty} \frac{\cos(kt - \pi r/2)}{k^r} - \gamma \quad (1.11)$$

and represent the error at the point t as the convolution

$$f(t) - U_n(f, \lambda, t) = \frac{1}{\pi} \int_0^{2\pi} \Psi_{n,r}(\gamma, \lambda; t-u)\, f^{(r)}(u)\, du. \quad (1.12)$$

Now it is possible to estimate the error in some metric applying different approaches. If $f \in L_p^r$ $(1 \le p \le \infty)$ then taking into account the freedom in the choice of parameter γ we obtain from Proposition 1.5.5

$$\|f - U_n(f, \lambda)\|_p \le \frac{1}{\pi} \inf_{\gamma} \|\Psi_{n,r}(\gamma, \lambda)\|_1 \, \|f^{(r)}\|_p. \quad (1.13)$$

Setting

$$e_n(\mathcal{M}, \lambda)_X = \sup_{f \in \mathcal{M}} \|f - U_n(f, \lambda)\|_X, \quad (1.14)$$

where X is either L_p $(1 \le p \le \infty)$ or C, we obtain immediately from (1.13) the following estimate for the class W_p^r in the L_p metric

$$e_n(W_p^r, \lambda)_p \le \frac{1}{\pi} \inf_{\gamma} \|\Psi_{n,r}(\gamma, \lambda)\|_1, \qquad (1 \le p \le \infty). \quad (1.15)$$

Let us remark that function $\Psi_{n,r}(\gamma, \lambda; t) + \gamma$ is odd for odd r (see (1.11)) and in this case the infimum in (1.13) and (1.15) is attained for $\gamma = 0$.

In another case we can get an exact upper bound for the quantity (1.14) choosing γ appropriately. Starting once more with (1.12) and taking into account that W_p^r contains functions $\pm f(\alpha \pm t)$ together with f we deduce from Corollary 1.4.3

$$e_n(W_p^r, \lambda)_C = \sup_{f \in W_p^r} [f(0) - U_n(f, \lambda, 0)]$$

$$= \sup_{\|\varphi\|_p \le 1, \varphi \perp 1} \frac{1}{\pi} \int_0^{2\pi} \Psi_{n,r}(\gamma, \lambda; u)\, \varphi(u)\, du = \frac{1}{\pi} E_1(\Psi_{n,r}(\gamma, \lambda))_{p'}.$$

Therefore for every $n, r = 1, 2, \ldots,$ we have

$$e_n(W_p^r, \lambda)_C = \inf_{\gamma} \frac{1}{\pi} \|\Psi_{n,r}(\gamma, \lambda)\|_{p'}, \qquad 1 \le p \le \infty, \qquad \frac{1}{p} + \frac{1}{p'} = 1, \quad (1.16)$$

and here the infimum is also attained for $\gamma = 0$ if r is odd.

Relations (1.13), (1.15) and (1.16) hold for every λ-method, i.e. for every set of coefficients λ_k^n $(k = 1, 2, \ldots, n)$ of the polynomial (1.1). If these coefficients are at our disposition then the problem of finding the λ which minimizes the right-hand side of (1.15) or (1.16) naturally arises: that is we are looking for the best λ-method for a given class of functions. This problem will be considered later. Let us note that the problem of minimizing $\|\Psi_{n,r}(\gamma,\lambda)\|_q$ on γ and λ_k^n is (see (1.11)) actually the problem of the best approximation of the function B_r in metric L_q by trigonometric polynomials. For $q = 1$ this problem has a beautiful solution with powerful applications.

4.1.2 Best approximation in L_1 of Bernoulli kernels First we prove the following useful result:

Proposition 4.1.2 If $f \in L_1$ is $2\pi/n$-periodic ($n \geq 2$) then

$$\int_0^{2\pi} f(t)\cos kt\,\mathrm{d}t = \int_0^{2\pi} f(t)\sin kt\,\mathrm{d}t = 0, \qquad k = 1, 2, \ldots, n-1. \tag{1.17}$$

If moreover $\int_0^{2\pi} f(t)\,\mathrm{d}t = 0$ then f is orthogonal to all trigonometric polynomials of degree $n-1$ ($f \perp \mathcal{N}_{2n-1}^T$).

Proof Setting $t = u + 2\pi/n$ we get

$$\int_0^{2\pi} \exp(\mathrm{i}kt) f(t)\,\mathrm{d}t = \exp(2\mathrm{i}\pi k/n)\int_0^{2\pi} \exp(\mathrm{i}kt) f(t)\,\mathrm{d}t,$$

i.e.

$$(1 - \exp(2\mathrm{i}\pi k/n))\int_0^{2\pi} \exp(\mathrm{i}kt) f(t)\,\mathrm{d}t = 0.$$

But for $k = 1, 2, \ldots, n-1$ we have

$$1 - \exp(2\mathrm{i}\pi k/n) = 1 - \cos(2k\pi/n) - \mathrm{i}\sin(2k\pi/n) \neq 0,$$

and hence

$$\int_0^{2\pi} \exp(\mathrm{i}kt) f(t)\,\mathrm{d}t = 0, \qquad k = 1, 2, \ldots, n-1,$$

which is equivalent to (1.17). The second assertion in Proposition 4.1.2 follows immediately. □

The main result in this section is:

Theorem 4.1.3 Among all trigonometric polynomials of degree $n-1$ the polynomial of best L_1 approximation of function (1.7) is the polynomial $\tau_{n-1,r}$ which interpolates B_r at the zeros of $\cos nt$ (even r) or $\sin nt$ (odd r). Moreover

$$E(B_r, \mathcal{N}_{2n-1}^T)_1 = \|B_r - \tau_{n-1,r}\|_1 = \pi K_r n^{-r}, \qquad n, r = 1, 2, \ldots, \tag{1.18}$$

where the K_r are the Favard constants (3.1.8).

Proof Let r be even and for a fixed $n = 1, 2, \ldots$, let $\tau_{n-1,r}$ be the even trigonometric polynomial interpolating the even function B_r at the points

$$t_j = \frac{2j-1}{2n}\pi, \qquad j = 1, 2, \ldots, 2n \tag{1.19}$$

(see Section 2.2.3). Hence $\tau_{n-1,r} \in \mathcal{N}_{2n-1}^T$ and $\delta(t_j) = 0$ $(j = 1, 2, \ldots, 2n)$ where

$$\delta = B_r - \tau_{n-1,r}. \tag{1.20}$$

The points (1.19) are the zeros of cos nt and we are going to show that either

$$\operatorname{sgn} \delta(t) = \operatorname{sgn} \cos nt \qquad \text{or} \qquad \operatorname{sgn} \delta(t) = -\operatorname{sgn} \cos nt, \tag{1.21}$$

i.e. the only zeros of δ are points (1.19) at which δ changes sign.

Assume that this is not so. Then the even function δ either has a zero different from t_j $(j = 1, 2, \ldots, n)$ or at least one from these t_j is a multiple zero. Then Rolle's theorem implies that δ' has at least n zeros on $(0, \pi)$. Then δ' has at least $n + 2$ zeros on $[0, \pi]$ because δ' is odd and hence $\delta'(0) = \delta'(\pi) = 0$. Applying Rolle's theorem once more we see that the even function δ'' has at least $n + 1$ zeros on $[0, \pi]$. Repeating these arguments we see that the odd function $\delta^{(r-1)}$ has at least n zeros on $(0, \pi)$. From

$$\delta^{(r-1)}(t) = B_1(t) - \tau_{n-1,r}^{(r-1)}(t) = \frac{\pi - t}{2} - \sum_{k=1}^{n-1} \beta_k \sin kt, \qquad 0 < t < 2\pi,$$

we see that $\delta^{(r-1)}(\pi) = 0$. Hence the even trigonometric polynomial

$$\delta^{(r)}(t) = -\frac{1}{2} - \sum_{k=1}^{n-1} k\,\beta_k \cos kt$$

of degree $n - 1$ has at least $2n$ zeros on the period which is impossible. So (1.21) is proved.

From (1.21) and Proposition 4.1.2 it follows that for every $\tau \in \mathcal{N}_{2n-1}^T$ we have

$$\int_0^{2\pi} \tau(t) \operatorname{sgn} \delta(t)\, \mathrm{d}t = \pm \int_0^{2\pi} \tau(t) \operatorname{sgn} \cos nt\, \mathrm{d}t = 0,$$

i.e. $\operatorname{sgn} \delta \perp \mathcal{N}_{2n-1}^T$.

Now let r be odd and let $\tau_{n-1,r}$ be the odd trigonometric polynomial of degree $n-1$ $(n \geq 2)$ coinciding with B_r at the points $j\pi/n$ $(j = 1, 2, \ldots, n-1)$ or $\tau_{n-1,r} \equiv 0$ for $n = 1$. From the oddness of B_r and $\tau_{n-1,r}$ we see that the difference (1.20) satisfies

$$\delta(j\pi/n) = 0, \qquad j = 0, 1, \ldots, n.$$

As in the case of even r we obtain by contradiction that $j\pi/n$ are the only zeros of δ and they are also simple. Therefore for odd r we obtain for all t that $\operatorname{sgn} \delta(t) = \pm \operatorname{sgn} \sin nt$ and now Proposition 4.1.2 implies $\operatorname{sgn} \delta \perp \mathcal{N}^T_{2n-1}$.

Therefore for all $n = 0, 1, \ldots,$ and $r = 1, 2, \ldots,$ we have

$$\int_0^{2\pi} \tau(t) \operatorname{sgn}[B_r(t) - \tau_{n-1,r}(t)]\, dt = 0, \qquad \forall \tau \in \mathcal{N}^T_{2n-1},$$

and applying Proposition 1.4.6 we obtain

$$E(B_r, \mathcal{N}^T_{2n-1})_1 = \int_0^{2\pi} |B_r(t) - \tau_{n-1,r}(t)|\, dt, \qquad n = 0, 1, \ldots, r = 1, 2, \ldots. \tag{1.22}$$

The right-hand side of (1.22) allows a representation via the series involved in the definition of Favard constants (see (3.1.8)). For even r, $r = 2\nu$, from

$$\operatorname{sgn}[B_{2\nu}(t) - \tau_{n-1,2\nu}(t)] = \pm \operatorname{sgn} \cos nt$$

and Proposition 4.1.2 we obtain

$$\begin{aligned} E(B_{2\nu}, \mathcal{N}^T_{2n-1})_1 &= \pm \int_0^{2\pi} [B_{2\nu}(t) - \tau_{n-1,2\nu}(t)] \operatorname{sgn} \cos nt\, dt \\ &= \left| \int_0^{2\pi} B_{2\nu}(t) \operatorname{sgn} \cos nt\, dt \right|. \end{aligned} \tag{1.23}$$

It is known (see e.g. [15B, p. 432]) that for the integrable functions f and g one of which being of bounded variation we have

$$\int_0^{2\pi} f(t)g(t)\, dt = \frac{\pi}{2} a_0(f)a_0(g) + \pi \sum_{k=1}^{\infty} [a_k(f)a_k(g) + b_k(f)b_k(g)], \tag{1.24}$$

where $a_k(f), b_k(f)$ and $a_k(g), b_k(g)$ are the Fourier coefficients of f and g respectively with respect to the trigonometric system. Because

$$\operatorname{sgn} \cos nt = \frac{4}{\pi}\left(\cos nt - \frac{1}{3} \cos 3nt + \frac{1}{5} \cos 5nt - \ldots\right) \tag{1.25}$$

and because (see (1.7))

$$B_{2\nu}(t) = (-1)^\nu \sum_{k=1}^{\infty} \frac{\cos kt}{k^{2\nu}}, \qquad \nu = 1, 2, \ldots, \tag{1.26}$$

an application of (1.24) gives

$$\int_0^{2\pi} B_{2\nu}(t) \operatorname{sgn} \cos nt \, \mathrm{d}t = \pi (-1)^\nu \frac{4}{\pi} \sum_{k=0}^{\infty} \frac{(-1)^k}{[(2k+1)n]^{2\nu}(2k+1)}. \tag{1.27}$$

From (1.23) and (1.27) we obtain

$$E(B_{2\nu}, \mathcal{N}_{2n-1}^T)_1 = \frac{4}{n^{2\nu}} \sum_{k=0}^{\infty} \frac{(-1)^k}{(2k+1)^{2\nu+1}} = \frac{\pi K_{2\nu}}{n^{2\nu}}, \qquad \nu = 1, 2, \ldots. \tag{1.28}$$

If r is odd ($r = 2\nu + 1$) then $\operatorname{sgn}[B_{2\nu+1}(t) - \tau_{n-1,2\nu+1}(t)] = \pm \operatorname{sgn} \sin nt$ and from

$$B_{2\nu+1}(t) = (-1)^\nu \sum_{k=1}^{\infty} \frac{\sin kt}{k^{2\nu+1}}, \qquad \nu = 0, 1, \ldots,$$

$$\operatorname{sgn} \sin nt = \frac{4}{\pi}\left(\sin nt + \frac{1}{3} \sin 3nt + \frac{1}{5} \sin 5nt + \ldots \right),$$

from the general equality (1.24), and from Proposition 4.1.2, we have

$$\begin{aligned} E(B_{2\nu+1}, \mathcal{N}_{2n-1}^T)_1 &= \int_0^{2\pi} |B_{2\nu+1}(t) - \tau_{n-1,2\nu+1}(t)| \, \mathrm{d}t \\ &= \left| \int_0^{2\pi} B_{2\nu+1}(t) \operatorname{sgn} \sin nt \, \mathrm{d}t \right| \\ &= \frac{4}{n^{2\nu+1}} \sum_{k=0}^{\infty} \frac{1}{(2k+1)^{2\nu+2}} = \frac{\pi K_{2\nu+1}}{n^{2\nu+1}}, \qquad \nu = 0, 1, \ldots,. \end{aligned}$$

This completes the proof. □

It is interesting to know the form of the polynomial $\tau_{n-1,r} \in \mathcal{N}_{2n-1}^T$ which realizes the best L_1 approximation to B_r, i.e. to have an explicit formula for its coefficients. We shall use this for constructing the best λ-method.

First let r be even. For a fixed n we have

$$\tau_{n-1,r}(t) = \frac{\mu_{r,0}}{2} + \sum_{m=1}^{n-1} \mu_{r,m} \cos mt,$$

where the coefficients $\mu_{r,m}$ ($m = 0, 1, \ldots, n-1$) are uniquely determined by the interpolating conditions

$$\tau_{n-1,r}\left(\frac{2j-1}{2n}\pi\right) = B_r\left(\frac{2j-1}{2n}\pi\right), \qquad j = 1, 2, \ldots, n,$$

i.e.

$$\frac{\mu_{r,0}}{2} + \sum_{m=1}^{n-1} \mu_{r,m} \cos(2j-1)\frac{m\pi}{2n} = (-1)^{r/2} \sum_{k=1}^{\infty} k^{-r} \cos(2j-1)\frac{k\pi}{2n},$$

$$j = 1, 2, \ldots, n. \qquad (1.29)$$

For the calculation of $\mu_{n,r}$ we shall use the identity

$$\sum_{j=1}^{n} \cos(2j-1)t = \frac{\sin 2nt}{2 \sin t}. \qquad (1.30)$$

Later on for the case of odd r we shall need the identity

$$\sum_{j=1}^{n-1} \cos jt = \frac{\sin(n-1/2)t}{2 \sin(t/2)} - \frac{1}{2}. \qquad (1.31)$$

We obtain (1.30) by multiplying and dividing each term on the left-hand side by $\sin t$ and replacing $\cos(2j-1)t \sin t$ by a differences of sines. (1.31) is obtained in a similar manner by multiplication and division with $\sin(t/2)$.

Setting $t = m\pi/(2n)$ in (1.30) and $t = m\pi/n$ in (1.31) ($m = 0, \pm 1, \pm 2, \ldots,$) we obtain

$$\sum_{j=1}^{n} \cos(2j-1)\frac{m\pi}{2n} = \begin{cases} (-1)^s n, & m = 2sn, \\ 0, & m \neq 2sn, \qquad s = 0, \pm 1, \pm 2, \ldots, \end{cases} \qquad (1.32)$$

$$\sum_{j=1}^{n-1} \cos j\frac{m\pi}{n} = \begin{cases} n-1, & m = 2sn, \\ [(-1)^{m-1} - 1]/2, & m \neq 2sn, \qquad s = 0, \pm 1, \pm 2, \ldots. \end{cases} \qquad (1.33)$$

Let us go back to equations (1.29). Summing them for j from 1 to n and using (1.32) we obtain

$$\mu_{r,0} = (-1)^{r/2}\, 2^{1-r}\, n^{-r} \sum_{s=1}^{\infty} \frac{(-1)^s}{s^r}. \qquad (1.34)$$

For the computation of $\mu_{r,l}$ ($l = 1, 2, \ldots, n-1$) we multiply both sides of (1.29) with $\cos[(2j-1)l\pi/(2n)]$, replace the product of cosines with sums of cosines and sum for j from 1 to n. Applying (1.32) we see that on the left-hand side the only non-zero term is the one containing $\mu_{r,l}$. We obtain

$$\mu_{r,l} = (-1)^{r/2}\left[\frac{1}{l^r} + \sum_{s=1}^{\infty}(-1)^s\left(\frac{1}{(2sn+l)^r} + \frac{1}{(2sn-l)^r}\right)\right],$$
$$l = 1, 2, \ldots, n-1. \tag{1.35}$$

For odd r polynomial $\tau_{n-1,r}$ has the form

$$\tau_{n-1,r}(t) = \sum_{k=1}^{n-1} \nu_{r,k} \sin kt,$$

and its coefficients can be found from the conditions for interpolating B_r at the zeros of $\sin nt$:

$$\sum_{k=1}^{n-1} \nu_{r,k} \sin j\frac{k\pi}{n} = (-1)^{(r-1)/2}\sum_{k=1}^{\infty} k^{-r} \sin j\frac{k\pi}{n}, \qquad j = 1, 2, \ldots, n-1.$$

Multiplying both sides of these equations by $\sin(jl\pi/n)$, summing on j and using (1.33) we obtain

$$\nu_{r,l} = (-1)^{(r-1)/2}\left[\frac{1}{l^r} + \sum_{s=1}^{\infty}(-1)^s\left(\frac{1}{(2sn+l)^r} - \frac{1}{(2sn-l)^r}\right)\right],$$
$$l = 1, 2, \ldots, n-1. \tag{1.36}$$

In particular for $r = 1$ we have

$$\nu_{1,l} = \frac{\pi}{2n}\operatorname{ctg}\frac{l\pi}{2n}, \qquad l = 1, 2, \ldots, n-1. \tag{1.37}$$

4.1.3 Favard means Let us go back to the minimization of

$$\|\Psi_{n,r}(\varphi, \lambda)\|_1 = \|B_r(\bullet) - K_{n,r}(\gamma, \lambda; \bullet)\|_1$$

on all parameters $\gamma, \lambda_1^n, \ldots, \lambda_n^n$, but for sake of simplicity we shall use $n-1$ instead of n. Knowing the explicit form of the polynomials $\tau_{n-1,r}$ with the best approximation to B_r in L_1 we shall require the coincidence of $K_{n-1,r}(\gamma, \lambda)$ with them. This is possible because both polynomials $\tau_{n-1,r}$ and $K_{n-1,r}(\gamma, \lambda)$ have one and the same degree $n-1$ and both are even or odd functions together with r being an even or an odd number.

It is necessary to ensure that the identities (we omit the upper index in λ_k^{n-1})

$$\gamma + (-1)^{(r-1)/2}\sum_{k=1}^{n-1}\frac{\lambda_k}{k^r}\sin kt = \sum_{k=1}^{n-1}\nu_{r,k}\sin kt, \qquad r = 1, 3, \ldots;$$

$$\gamma + (-1)^{r/2} \sum_{k=1}^{n-1} \frac{\lambda_k}{k^r} \cos kt = \frac{\mu_{r,0}}{2} + \sum_{k=1}^{n-1} \mu_{r,k} \cos kt, \qquad r = 2, 4, \ldots,$$

hold. Obviously, for $r = 1, 3, 5, \ldots$, we have to set

$$\gamma = \gamma_r^* = 0, \qquad \lambda_k = \lambda_{r,k}^* = (-1)^{(r-1)/2} k^r \nu_{r,k}, \qquad k = 1, 2, \ldots, n-1; \quad (1.38)$$

and for $r = 2, 4, 6, \ldots$

$$\gamma = \gamma_r^* = \mu_{r,0}/2, \qquad \lambda_k = \lambda_{r,k}^* = (-1)^{r/2} k^r \mu_{r,k}, \qquad k = 1, 2, \ldots, n-1. \quad (1.39)$$

Taking (1.34)–(1.36) into account the best coefficients for the polynomial $K_{n-1,r}(\gamma, \lambda; t)$ can be explicitly given as sums of numerical series:

$$\gamma_r^* = \gamma_r^{n-1,*} = \begin{cases} (-1)^{r/2} 2^{-r} n^{-r} \sum_{s=1}^{\infty} (-1)^s s^{-r}, & r = 2, 4, \ldots, \\ 0, & r = 1, 3, \ldots; \end{cases}$$

$$\lambda_{r,k}^* = \lambda_{r,k}^{n-1,*} \quad (1.40)$$

$$= \begin{cases} 1 - k^r \sum_{s=1}^{\infty} [(2sn - k)^{-r} - (2sn + k)^{-r}], & r = 1, 3, \ldots, \\ 1 - k^r \sum_{s=1}^{\infty} (-1)^{s+1} [(2sn - k)^{-r} + (2sn + k)^{-r}], & r = 2, 4, \ldots. \end{cases} \quad (1.41)$$

In particular for $r = 1$ (see (1.37))

$$\lambda_{1,k}^* = \lambda_{1,k}^{n-1,*} = \frac{k\pi}{2n} \operatorname{ctg} \frac{k\pi}{2n}, \qquad k = 1, 2, \ldots, n-1. \quad (1.42)$$

The set of coefficients $\lambda^* = \{\lambda_{r,1}^{n-1,*}, \lambda_{r,2}^{n-1,*}, \ldots, \lambda_{r,n-1}^{n-1,*}\}$ dependent on n and r together with the parameter $\gamma_r^* = \gamma_r^{n-1,*}$ determine the *Favard λ-means* (see Favard [2, 3]) $U_{n-1}(f, \lambda^*, t) = U_{n-1,r}(f, \lambda^*, t)$ by the kernel

$$K_{n-1,r}(\gamma^*, \lambda^*, t) = \gamma_r^{n-1,*} + \sum_{k=1}^{n-1} \frac{\lambda_{r,k}^{n-1,*}}{k^r} \cos\left(kt - \frac{\pi r}{2}\right). \quad (1.43)$$

Let us note that parameter γ appears only in representation (1.10) because of $f^{(r)} \perp 1$; in the first λ-means (see Section 1.1) this parameter is missing. When talking about Favard λ^*-means let us always assume that the parameter $\gamma = \gamma_r^{n-1,*}$ is always given by (1.40).

In view of the choice (see (1.38) and (1.39)) of polynomial (1.43) we have

$$K_{n-1,r}(\gamma^*, \lambda^*, t) = \tau_{n-1,r}(t), \qquad n, r = 1, 2, \ldots, \quad (1.44)$$

and hence (see (1.10)) for $f \in L_1^r$ we have

$$U_{n-1,r}(f,\lambda^*,t) = \frac{a_0}{2} + \frac{1}{\pi}\int_0^{2\pi} \tau_{n-1,r}(u) f^{(r)}(t-u)\, du. \tag{1.45}$$

From Theorem 4.1.3

$$\|B_r(\bullet) - K_{n,r}(\gamma^*, \lambda^*; \bullet)\|_1 = E(B_r, \mathcal{N}^T_{2n-1})_1 = \pi\, K_r\, n^{-r}, \tag{1.46}$$

and now from (1.13) and (1.46) we obtain

$$\|f - U_{n-1,r}(f,\lambda^*)\|_p \le K_r\, n^{-r}\, \|f^{(r)}\|_p, \qquad 1 \le p \le \infty. \tag{1.47}$$

Therefore for every $f \in L^r_p$ $(1 \le p \le \infty)$ we have

$$E_n(f)_p =: E_n(f, \mathcal{N}^T_{2n-1})_p \le K_r\, n^{-r}\, \|f^{(r)}\|_p, \qquad n = 1, 2, \ldots. \tag{1.48}$$

Using Favard means we can obtain a better result.

Theorem 4.1.4 For every $f \in L^r_p$ $(r = 1, 2, \ldots,; 1 \le p \le \infty)$ we have

$$E_n(f)_p \le K_r\, n^{-r}\, E_n(f^{(r)})_p, \qquad n = 1, 2, \ldots. \tag{1.49}$$

For $p = 1, \infty$ inequality (1.49) cannot be improved.

Proof For $q \in \mathcal{N}^T_{2n-1}$ the function

$$G(t) = \frac{1}{\pi}\int_0^{2\pi} [B_r(u) - \tau_{n-1,r}(u)]\, q(t-u)\, du,$$

where $\tau_{n-1,r}$ is the polynomial constructed in Section 4.1.2, also belongs to $\mathcal{N}^T_{2n-1}$. Because of (1.6) and (1.45) we have

$$\begin{aligned} &f(t) - U_{n-1,r}(f,\lambda^*,t) - G(t) \\ &\quad = \frac{1}{\pi}\int_0^{2\pi} [B_r(t-u) - \tau_{n-1,r}(t-u)][f^{(r)}(u) - q(u)]\, du. \end{aligned}$$

Now we can take q to be the polynomial of best approximation for $f^{(r)}$ in L_p. Then using Proposition 1.5.5 we obtain

$$\begin{aligned} E_n(t)_p &\le \|f - U_{n-1,r}(f,\lambda^*) - G\|_p \\ &\le \frac{1}{\pi}\|B_r - \tau_{n-1,r}\|_1\, \|f^{(r)} - q\|_p = \frac{1}{\pi}\, E_n(B_r)_1\, E_n(f^{(r)})_p \\ &= K_r n^{-r} E_n(f^{(r)})_p. \end{aligned}$$

The exactness on L^r_p of the estimate (1.48) (and hence of estimate (1.49)) for $p = 1, \infty$ follows from the following property of the perfect spline $\varphi_{n,m}$ defined in Section 3.1.

Proposition 4.1.5 The best approximation to $\varphi_{n,m}$ in the subspace $\mathcal{N}^T_{2n-1}$ of trigonometric polynomials of degree $n-1$ in L_p is the constant zero, i.e.

$$E_n(\varphi_{n,m})_p = \|\varphi_{n,m}\|_p, \qquad 1 \le p \le \infty, \qquad m = 0, 1, \ldots.$$

Proof If $1 \le p < \infty$ then function $|\varphi_{n,m}(t)|^{p-1} \operatorname{sgn} \varphi_{n,m}(t)$ is orthogonal to subspace $\mathcal{N}_{2n-1}^T$ because of Proposition 4.1.6 and it remains to apply Proposition 1.4.6. In the case $p = \infty$ and $m \ge 1$ we can replace metric L_∞ with C and the necessary result follows from Chebyshev's Theorem 2.1.2 because $\varphi_{n,m}$ alternately takes the values $\pm\|\varphi_{n,m}\|_\infty$ at $2n$ points on the period $[0, 2\pi)$. The same arguments also work for $m = 0$ if we first smooth $\varphi_{n,0}$, for example with the Stekhlov function.

Obviously $\varphi_{n,r} \in L_\infty^r \subset L_p^r$ and because of $E_n(\varphi_{n,r})_\infty = \|\varphi_{n,r}\|_\infty = K_r n^{-r}\|\varphi_{n,r}^{(r)}\|_\infty$ (see Section 3.1) we cannot improve estimate (1.48) for $p = \infty$. The similar fact for $p = 1$ can be easily established with the help of Stekhlov functions (see Section A2). If

$$\Psi_h(t) = \frac{1}{2h}\int_{t-h}^{t+h} \varphi_{n,r-1}(u)\, du, \qquad h > 0,$$

then $\Psi_h \in L_1^r$, $E_n(\Psi_h)_1 = \|\Psi_h\|_1$ and

$$\lim_{h\to 0} \|\Psi_h\|_1 = \|\varphi_{n,r-1}\|_1 = 4\, K_r\, n^{-r+1},$$

and for $0 < h < 2\pi/n$ we have

$$\|\Psi_h^{(r)}\|_1 = \bigvee_0^{2\pi} \varphi_{n,r-1} = 4n.$$

4.2 The supremums of the best approximations of classes W_p^r by trigonometric polynomials

4.2.1 Best approximation of classes W_∞^r in C and classes W_1^r and W_V^r in L_1 Linear λ-methods for approximations of periodic functions were considered in Section 4.1. The Favard sum estimates obtained also give, in some cases, the exact solutions of the problem of the best approximation of classes W_p^r by trigonometric polynomials. Using the previous notation

$$E_n(f)_q = E_n(f, \mathcal{N}_{2n-1}^T)_q, \qquad E_n(\mathcal{M})_q = \sup_{f\in\mathcal{M}} E_n(f)_q,$$

the problem is as follows: find the exact value of

$$E_n(W_p^r)_q = \sup_{f\in W_p^r}\ \inf_{\tau\in\mathcal{N}_{2n-1}^T} \|f-\tau\|_q, \qquad 1 \le p, q \le \infty. \tag{2.1}$$

In three cases ($p = q = 1$, $q = 2$ and $p = q = \infty$) the solution is provided either by the partial sums of Fourier series ($q = 2$) or by Favard λ^*-means ($p = q = 1, \infty$).

Together with the quantity (see Section 1.2)

$$\mathscr{E}_n(W_p^r)_q = \mathscr{E}_n(W_p^r, \mathscr{N}_{2n-1}^T)_q = \inf_{A \in \mathscr{L}(L_q, \mathscr{N}_{2n-1}^T)} \sup_{f \in W_p^r} \|f - Af\|_q$$

($\mathscr{L}(X, \mathscr{N})$ being the set of all linear bounded operators from X to $\mathscr{N}$) we shall also consider

$$\mathscr{E}_n^\lambda(W_p^r)_q = \inf_\lambda \sup_{f \in W_p^r} \|f - U_{n-1}(f, \lambda)\|_q,$$

where $U_n(f, \lambda, t)$ is the polynomial (1.1) and the infimum is taken over all possible sets of coefficients $\lambda = \{\lambda_1^{n-1}, \lambda_2^{n-1}, \ldots, \lambda_{n-1}^{n-1}\}$. The set λ for which the infimum is attained (for fixed n) is called the best λ-method for the class W_p^r in the metric L_q. Obviously we have $E_n(W_p^r)_q \le \mathscr{E}_n(W_p^r)_q \le \mathscr{E}_n^\lambda(W_p^r)_q$. From inequality (1.47) which is valid for every $f \in L_p^r$ it follows that

$$E_n(W_p^r)_p \le \mathscr{E}_n^\lambda(W_p^r)_p \le \sup_{f \in W_p^r} \|f - U_{n-1,r}(f, \lambda^*)\|_p = K_r n^{-r},$$

$$1 \le p \le \infty, \qquad n, r, = 1, 2, \ldots. \tag{2.2}$$

On the other hand, the $2\pi/n$-periodic function $\varphi_{n,r}$ (see Section 3.1) belongs to W_∞^r and hence by Proposition 4.1.5 we have

$$E_n(W_\infty^r)_C \ge E_n(\varphi_{n,r})_C = \|\varphi_{n,r}\|_C = K_r n^{-r}. \tag{2.3}$$

The function $g_{n,m}(t) = \varphi_{n,m}(t)/(4n)$ introduced in Section 3.1 clearly belongs to W_V^r (see Section 1.5.1). From Proposition 1.1.1, Proposition 4.1.5 and (3.1.13) we have

$$E_n(g_{n,m})_1 = \frac{1}{4n} E_n(\varphi_{n,m})_1 = \frac{1}{4n} \|\varphi_{n,m}\|_1 = \frac{K_{m+1}}{n^{m+1}}$$

and hence applying Proposition 1.5.1 we obtain

$$E_n(W_1^r)_1 = E_n(W_V^{r-1})_1 \ge E_n(g_{n,r-1})_1 = K_r n^{-r}. \tag{2.4}$$

Inequalities (2.2), (2.3) and (2.4) imply:

Theorem 4.2.1 For all $n, r = 1, 2, \ldots$, we have

$$E_n(W_\infty^r)_C = \mathscr{E}_n^\lambda(W_\infty^r)_C = \sup_{f \in W_\infty^r} \|f - U_{n-1,r}(f, \lambda^*)\|_C = K_r n^{-r}, \tag{2.5}$$

$$E_n(W_1^r)_1 = \mathscr{E}_n^\lambda(W_1^r)_1 = \sup_{f \in W_1^r} \|f - U_{n-1,r}(f, \lambda^*)\|_1 = K_r n^{-r}, \tag{2.6}$$

$$E_n(W_V^{r-1})_1 = \mathscr{E}_n^\lambda(W_V^{r-1})_1 = \sup_{f \in W_V^{r-1}} \|f - U_{n-1,r}(f, \lambda^*)\|_1 = K_r n^{-r}, \tag{2.7}$$

where K_r are Favard constants.

Therefore Favard λ^*-means are the best linear methods for the classes W_∞^r in C and the classes W_1^r and W_V^r in L_1, realizing at the same time the supremum of the best approximations.

Remark We have seen that

$$E_n(W_\infty^r)_C = E_n(\varphi_{n,r})_C \qquad \text{and} \qquad E_n(W_V^{r-1})_1 = E_n(g_{n,r-1})_1, \quad (2.8)$$

i.e. the functions $\varphi_{n,r}$ and $g_{n,r-1}$ are extremal elements for problem (2.1) in the classes W_∞^r and W_V^{r-1} for the metrics C and L_1, respectively. Concerning class W_1^r, there is no function realizing the supremum $E_n(W_1^r)_1$, but it is easy to construct a sequence $\{f_m\} \subset W_1^r$ for which

$$\lim_{m\to\infty} E_n(f_m)_1 = E_n(W_1^r)_1.$$

For example we can take

$$f_m(t) = \frac{m}{2}\int_{-1/m}^{1/m} g_{n,r-1}(t+u)\,du$$

and we notice that $E_n(f_m)_1 = \|f_m\|_1$ and $\|f_m\|_1 \to \|g_{n,r-1}\|_1$ for $m \to \infty$ (see Section A2).

Let us also remark that the functions $\varphi_{n,r}$ for which the supremums in $E_n(W_\infty^r)_C$ $(n = 1, 2, \ldots,)$ are attained (see (2.8)) essentially depend on n. It follows from results due to Bernstein [1, 3A] that for a fixed m we have

$$\lim_{n\to\infty} nE_n(\varphi_{m,1})_C = 0.283\ldots,$$

while $nE_n(W_\infty^1)_C = \pi/2$. The following question naturally arises: is there a function f from W_∞^r such that $\lim_{n\to\infty} [E_n(f)_C/E_n(W_\infty^r)_C] = 1$? In 1946 Nikol'skii [8] built an $f_r \in W_\infty^r$ such that

$$\overline{\lim_{n\to\infty}}\, n^r E_n(f_r)_C = K_r = n^r E_n(W_\infty^r)_C.$$

In Chapter 7 we shall consider some refinements and generalizations of this result.

4.2.2 The case of the Hilbert metric The simplest and most complete solution to problem (2.1) is in the case $p = q = 2$ – due to specific properties of the metric L_2. The general relation (2.2.3) is applicable to this case and for every function $f \in L_2$ we can write

$$E_n(f)_2 = \|f - S_{n-1}(f)\|_2, \quad (2.9)$$

where

$$S_{n-1}(f,t) = \frac{a_0}{2} + \sum_{k=1}^{n-1} (a_k \cos kt + b_k \sin kt)$$

are the partial sums of Fourier series of f.

For $f \in L_2^r$ $(r = 1, 2, \ldots,)$ after integrating by parts r-times we obtain

$$|a_k| = \frac{1}{\pi}\left|\int_0^{2\pi} f(t)\cos kt\,\mathrm{d}t\right| = \begin{cases} k^{-r}|b_{r,k}|, & r = 1, 3, \ldots, \\ k^{-r}|a_{r,k}|, & r = 2, 4, \ldots, \end{cases}$$

where $a_{r,k}$ and $b_{r,k}$ are the Fourier coefficients of $f^{(r)}$. Similarly

$$|b_k| = \begin{cases} k^{-r}|a_{r,k}|, & r = 1, 3, \ldots, \\ k^{-r}|b_{r,k}|, & r = 2, 4, \ldots. \end{cases}$$

Therefore

$$a_k^2 + b_k^2 = k^{-2r}(a_{r,k}^2 + b_{r,k}^2)$$

and using Parseval's equality and (2.9) we obtain

$$E_n^2(f)_2 = \left\|\sum_{k=n}^{\infty} (a_k \cos kt + b_k \sin kt)\right\|_2^2 = \pi \sum_{k=n}^{\infty} (a_k^2 + b_k^2)$$

$$= \pi \sum_{k=n}^{\infty} k^{-2r}(a_{r,k}^2 + b_{r,k}^2) \leq \frac{\pi}{n^{2r}} \sum_{k=n}^{\infty} (a_{r,k}^2 + b_{r,k}^2) \leq n^{-2r}\|f^{(r)}\|_2^2.$$

If $f \in W_2^r$ then $\|f^{(r)}\|_2 \leq 1$ and hence for every $f \in W_2^r$ we have

$$E_n(f)_2 \leq n^{-r}, \qquad n = 1, 2, \ldots. \tag{2.10}$$

The function

$$f_0(t) = \pi^{-1/2} n^{-r} \sin nt$$

belongs to W_2^r and

$$E_n(f_0)_2 = \|f_0\|_2 = n^{-r}$$

because $S_{n-1}(f_0, t) \equiv 0$, i.e. inequality (2.10) cannot be improved on W_2^r.

Taking the linearity of $S_{n-1}(f, t)$ with respect to f into account we obtain:

Theorem 4.2.2 For all $n, r = 1, 2, \ldots,$ we have

$$E_n(W_2^r)_2 = \mathscr{E}_n(W_2^r)_2 = \sup_{f \in W_2^r} \|f - S_{n-1}(f)\|_2 = n^{-r}.$$

Let us remark that the Favard λ^*-means realize the best approximations in C or L_1 of the whole classes W_∞^r or W_1^r, while partial sums of Fourier series provide the best approximations in L_2 for any particular function as (2.9) shows.

4.2.3 Duality relations In Chapter 1 we gave some duality theorems for more or less general situations. We shall now obtain here some similar facts for the best approximation by trigonometric subspaces on W_p^r. The results of Section 4.1 are essential for these considerations. We set

$$W_{p,n}^r = \{f: f \in W_p^r, f \perp \mathcal{N}_{2n-1}^T\}.$$

Theorem 4.2.3 For all $n, r = 1, 2, \ldots$, we have

$$\sup_{f \in W_{\infty,n}^r} \|f\|_C = E_n(W_\infty^r)_C = K_r n^{-r}, \tag{2.11}$$

$$\sup_{f \in W_{1,n}^r} \|f\|_1 = E_n(W_1^r)_1 = K_r n^{-r}. \tag{2.12}$$

Proof Let $f \in W_{p,n}^r$. Then its Fourier series is of the type

$$f(t) = \sum_{k=n}^{\infty} (a_k \cos kt + b_k \sin kt)$$

and hence $f^{(r)}$ is also orthogonal to the subspace $\mathcal{N}_{2n-1}^T$: $f^{(r)} \perp \mathcal{N}_{2n-1}^T$. But then

$$f(t) = \frac{1}{\pi} \int_0^{2\pi} B_r(t-u) f^{(r)}(u) \, \mathrm{d}u$$

$$= \frac{1}{\pi} \int_0^{2\pi} [B_r(t-u) - \tau_{n-1,r}(t-u)] f^{(r)}(u) \, \mathrm{d}u, \tag{2.13}$$

where $\tau_{n-1,r}$ is the polynomial of degree $n-1$ of best approximation to B_r in L_1 (see Section 4.1.9). Obviously $\tau_{n-1,r}(t+\alpha)$ belongs to $\mathcal{N}_{2n-1}^T$ for every $\alpha \in \mathbb{R}$. For every $f \in W_{p,n}^r$ we obtain from (2.13), (1.18) and (1.5.28)

$$\|f\|_p \le \frac{1}{\pi} \|B_r - \tau_{n-1,r}\|_1 \|f^{(r)}\|_p = K_r n^{-r} \|f^{(r)}\|_p.$$

Therefore

$$\sup_{f \in W_{p,n}^r} \|f\|_p \le K_r n^{-r}; \qquad 1 \le p \le \infty; n, r = 1, 2, \ldots.$$

For $p = \infty$ and $p = 1$ we actually have an equality. In order to see this it is enough to set $f = \varphi_{n,r}$ in the first case and to take into account that $\varphi_{n,r} \in W_{\infty,n}^r$ and to take the Stekhlov function ψ_h of $\psi_h = g_{n,r-1}$ in the second case and take into account that $\psi_h \in W_{1,n}^r$ and $\|\psi_h\|_1 \to \|g_{n,r-1}\|_1 = K_r n^{-r}$ for $h \to 0$. The second equalities in (2.11) and (2.12) are proved in Theorem 4.2.1. □

Let us also remark the following. Starting with Proposition 1.5.4 with $\mathcal{N} = \mathcal{N}_{2n-1}^T$ we obtain the equality

$$E_n(W_p^r)_q = \sup_{g \in W_{q',n}^r} E_1(g)_{p'}, \qquad \frac{1}{p} + \frac{1}{p'} = \frac{1}{q} + \frac{1}{q'} = 1. \tag{2.14}$$

Setting $p = q = 1$ and $p = q = \infty$ and recalling (2.5) and (2.6) we come to the relations

$$\sup_{g \in W^r_{\infty,n}} E_1(g)_\infty = E_n(W^r_1)_1 = K_r n^{-r},$$

$$\sup_{g \in W^r_{1,n}} E_1(g)_1 = E_n(W^r_\infty)_C = K_r n^{-r},$$

which complement Theorem 4.2.3.

4.2.4 Best approximation of classes W^r_p in L_1 In the cases considered in Section 4.2.2 the best approximation on the class was obtained by a linear method and problem (2.1) was solved by virtue of an exact upper estimate for this method. Other arguments based on duality and comparison theorems allow us to obtain the exact solution in another case too. We have:

Theorem 4.2.4 For all $n, r = 1, 2, \ldots$, we have

$$E_n(W^r_p)_1 = \sup_{f \in W^r_{\infty,n}} \|f\|_{p'} = \|\varphi_{n,r}\|_{p'}, \qquad 1 \le p \le \infty, \qquad 1/p + 1/p' = 1, \tag{2.15}$$

where $\varphi_{n,r}$ is the $2\pi/n$-periodic perfect Euler spline of degree r.

Proof Let $f \in W^r_{\infty,n}$ and

$$f_1(t) = \int_0^t f(u)\, du + c,$$

where the constant c is chosen so that $f_1 \perp 1$. Then $f_1 \in W^{r+1}_{\infty,n}$ and by (2.11) it satisfies

$$\|f_1\|_C \le K_{r+1} n^{-r-1} = \|\varphi_{n,r+1}\|_C.$$

Applying Theorem 3.3.4 to f_1 we obtain

$$\|f_1'\|_q = \|f\|_q \le \|\varphi_{n,r}\|_q, \qquad 1 \le q \le \infty,$$

and because of $\varphi_{n,r} \in W^r_{\infty,n}$ the second relation in (2.15) is proved.

Now from (2.14) we obtain

$$E_n(W^r_p)_1 = \sup_{f \in W^r_{\infty,n}} E_1(f)_{p'} \le \sup_{f \in W^r_{\infty,n}} \|f\|_{p'},$$

which in view of the penultimate displayed expression implies

$$E_n(W^r_p)_1 \le \|\varphi_{n,r}\|_{p'}.$$

Finally we show that here we actually have an equality. For $p = 1$ this follows from (2.12). If $p > 1$ then $1 \le p' < \infty$ and using (2.14) and Proposition 4.1.5 we obtain

$$E_n(W_p^r)_1 = \sup_{f \in W^r_{\infty,n}} E_1(f)_{p'} \geq E_1(\varphi_{n,r})_{p'} = \|\varphi_{n,r}\|_{p'}.$$

This completes the proof. □

In connection with the equality $E_n(W_p^r)_1 = \|\varphi_{n,r}\|_{p'}$ we have to remark that if $p = 1$ then the supremum $E_n(W_p^r)_1$ is realized by a linear method (see (2.6)), namely by the Favard λ^*-means, while in the case $1 < p \leq \infty$ we do not know anything about the possibility of realizing the supremum by a linear method. The quantities $\mathscr{E}_n(W_p^r)_1$ and $\mathscr{E}_n^\lambda(W_p^r)_1$ are also not known.

4.2.5 The trigonometrically conjugate case Together with classes W_p^r sometimes the classes $\tilde{W}_p^r$ $(r = 1, 2, \ldots,)$ of functions trigonometrically conjugated with the functions from W_p^r are also considered. If $f \in L_1^r$ $(r = 1, 2, \ldots,)$.

$$f(t) = \frac{a_0}{2} + \sum_{k=1}^{\infty} (a_k \cos kt + b_k \sin kt),$$

then the function conjugate with f

$$\tilde{f}(t) = \sum_{k=1}^{\infty} (-b_k \cos kt + a_k \sin kt)$$

allows the representation

$$\tilde{f}(t) = \frac{1}{\pi} \int_0^{2\pi} \tilde{B}_r(u) f^{(r)}(t-u)\, du, \tag{2.16}$$

where

$$\tilde{B}_r(t) = \sum_{k=1}^{\infty} \frac{\sin(kt - \pi r/2)}{k^r}$$

is the function conjugate with Bernoulli kernel (1.7). It is useful to keep in mind that

$$\tilde{B}_1(t) = \tilde{B}_r^{(r-1)}(t) = \ln\left|2 \sin\left(\frac{t}{2}\right)\right|, \qquad t \neq 2k\pi.$$

Hence $\tilde{W}_p^r$ may be defined as the class of functions with the representation (2.16) with $f \in W_p^r$.

The arguments used in the proof of Theorem 4.1.3 allow us to prove the following statement (see e.g. [2A]).

Theorem 4.2.5 We have

$$E_n(\tilde{B}_r)_1 = \|\tilde{B}_r - q_{n-1,r}\|_1 = \pi\,\tilde{K}_r\,n^{-r}, \qquad n, r = 1, 2, \ldots, \tag{2.17}$$

where

$$\tilde{K}_r = \frac{4}{\pi}\sum_{\nu=0}^{\infty}\frac{(-1)^{\nu r}}{(2\nu+1)^{r+1}},$$

$$\tilde{K}_1 < \tilde{K}_3 < \tilde{K}_5 < \ldots < 4/\pi < \ldots < \tilde{K}_6 < \tilde{K}_4 < \tilde{K}_2 < \pi/2,$$

and $q_{n-1,r}$ is the trigonometric polynomial of degree $n-1$ interpolating $\tilde{B}_r$ at the zeros of $\sin(nt - \pi r/2)$.

Given a function $f \in L_1^r$, let us consider the trigonometric polynomial of degree $n-1$

$$Q_{n-1,r}(f, t) = \frac{1}{\pi}\int_0^{2\pi} q_{n-1,r}(t-u)\,f^{(r)}(u)\,\mathrm{d}u,$$

which obviously depends linearly on f. Then

$$\tilde{f}(t) - Q_{n-1,r}(f, t) = \frac{1}{\pi}\int_0^{2\pi}[\tilde{B}_r(u) - q_{n-1,r}(u)]\,f^{(r)}(t-u)\,\mathrm{d}u$$

and the general inequality (1.5.28) for convolutions gives together with (2.17)

$$\|\tilde{f} - Q_{n-1,r}(f)\|_p \le \frac{1}{\pi}\|\tilde{B}_r - q_{n-1,r}\|_1\,\|f^{(r)}\|_p = \frac{\tilde{K}_r}{n^r}\|f^{(r)}\|_p.$$

For $f \in W_p^r$ and hence $\tilde{f} \in \tilde{W}_p^r$ we have

$$E_n(\tilde{f})_p \le \|\tilde{f} - Q_{n-1,r}(f)\|_p \le \tilde{K}_r\,n^{-r}; \qquad 1 \le p \le \infty; \qquad n, r = 1, 2, \ldots.$$

It is easy to check that the trigonometric conjugate of $\varphi_{n,r}$ is the function

$$\tilde{\varphi}_{n,r}(t) = -\frac{4}{\pi n^r}\sum_{\nu=0}^{\infty}\frac{\cos[(2\nu+1)nt - \pi r/2]}{(2\nu+1)^{r+1}},$$

which is $2\pi/n$-periodic. Moreover, following the proof of Proposition 4.1.5 we can obtain

$$E_n(\tilde{\varphi}_{n,r})_p = \|\tilde{\varphi}_{n,r}\|_p, \qquad 1 \le p \le \infty.$$

For $p = 1$ and $p = \infty$ it is easy to calculate that

$$\|\tilde{\varphi}_{n,r}\|_\infty = \tilde{K}_r n^{-r}, \|\tilde{\varphi}_{n,r}\|_1 = \tilde{K}_{r+1} n^{-r}.$$

Because $\varphi_{n,r} \in W_\infty^r$ and $g_{n,r-1}(t) = \varphi_{n,r-1}(t)/(4n)$ belongs to W_V^{r-1} we see that $\tilde{\varphi}_{n,r}$ and $\tilde{g}_{n,r-1}$ belong to $\tilde{W}_\infty^r$ and $\tilde{W}_V^{r-1}$ respectively, and hence the following theorem is valid.

Theorem 4.2.6 We have

$$E_n(\tilde{W}_\infty^r)_C = E_n(\tilde{W}_V^{r-1})_1 = E_n(\tilde{W}_1^r)_1 = \tilde{K}_r n^{-r}, \qquad n, r = 1, 2, \ldots.$$

4.2.6 Classes with fractional *r* We say [cf. 17B, vol. 2, p. 201] that the function $\varphi \in L_1$ is the rth ($r > 0$) *derivative in the sense of Weyl* of the function f if (cf. (1.6)) $\varphi \perp 1$ and if

$$f(t) = \frac{1}{\pi} \int_0^{2\pi} B_r(t-u)\, \varphi(u)\, du, \tag{2.18}$$

where the function B_r ($r > 0$) is determined by the same equality (1.7), the series of which (containing both sines and cosines for non-integral r) converges for $r > 0$ at every point t with a possible exception at $t = 2k\pi$.

Let us denote the class of functions which have the representation (2.18), where $\varphi \in L_p$ and $\|\varphi\|_p \le 1$, by W_p^r ($r > 0, 1 \le p \le \infty$). The determination of the supremums $E_n(W_\infty^r)_C$, $E_n(W_1^r)_1$ is here also reduced to the best approximation of the kernel B_r in the L_1 metric.

But for non-integral r the arguments connected with Rolle's theorem cannot be applied and so for solving the problem other complicated ways have been found, which require in particular the investigation of integrals of absolutely monotonic functions. Here we restrict ourselves to formulating the results (references are given in the comments).

Theorem 4.2.7 For $0 < r < 1$ we have

$$E_n(W_\infty^r)_C = E_n(W_1^r)_1 = \frac{4 \sin(r\pi/2)}{\pi n^r} \sum_{\nu=0}^{\infty} \frac{1}{(2\nu+1)^{r+1}}.$$

For $r \ge 1$ we have

$$E_n(W_\infty^r)_C = E_n(W_1^r)_1 = \frac{1}{\pi} \|B_r - \tau_{n-1,r}\|_1$$

$$= \frac{4}{\pi n^r} \left| \sum_{\nu=0}^{\infty} \frac{\sin[(2\nu+1)\gamma - \pi r/2]}{(2\nu+1)^{r+1}} \right|,$$

where $\tau_{n-1,r}$ is the trigonometric polynomial of degree $n-1$ interpolating B_r at the point $(\gamma + k\pi)/n$ and γ is a number satisfying

$$\sum_{\nu=0}^{\infty} \frac{\cos\left[(2\nu+1)\gamma - \pi r/2\right]}{(2\nu+1)^r} = 0.$$

Similar results are also obtained for the convolution classes $W_\infty^{r,\beta}$, $W_1^{r,\beta}$ of functions of the type (2.18) with kernel

$$B_r^\beta(t) = \sum_{k=1}^{\infty} \frac{\cos\left(kt - \pi\beta/2\right)}{k^r}, \qquad r > 0, \qquad \beta \in \mathbb{R}.$$

These classes coincide with $\tilde{W}_\infty^r$, $\tilde{W}_1^r$ for $\beta = r + 1$.

4.3 Approximation by partial sums of Fourier series, their means and analogs

4.3.1 Kolmogorov's results Fourier means as approximating tools have attracted the attention of mathematicians for a long time. Although one cannot even guarantee their convergence for continuous functions, for smooth functions Fourier means behave much better, and of course the problem of evaluating their approximation error has always been a main topic for investigation.

In 1935 Kolmogorov [1] obtained the first asymptotically exact result for this problem which initiated a new branch in approximation theory. He proved:

Theorem 4.3.1 For the partial sums of Fourier series

$$S_n(f,t) = \frac{a_0}{2} + \sum_{k=1}^{n} (a_k \cos kt + b_k \sin kt), \qquad n = 1, 2, \ldots,$$

we have for large n

$$\sup_{f \in W_\infty^r} \|f - S_n(f)\|_C = \frac{4}{\pi^2}\frac{\ln n}{n^r} + O(n^{-r}), \qquad r = 1, 2, \ldots. \tag{3.1}$$

Proof Set

$$D_{n,r}(t) = \sum_{k=n+1}^{\infty} \frac{\cos\left(kt - \pi r/2\right)}{k^r}. \tag{3.2}$$

For $f \in L_1^r$ we can write, taking (1.5.3) into account

$$f(t) - S_n(f,t) = \sum_{k=n+1}^{\infty} (a_k \cos kt + b_k \sin kt)$$

$$= \frac{1}{\pi} \int_0^{2\pi} D_{n,r}(t-u) f^{(r)}(u) \, du \tag{3.3}$$

and, in view of the invariance of the class W_∞^r with respect to translation of the argument,

$$S_n(W_\infty^r)_C =: \sup_{f \in W_\infty^r} \|f - S_n(f)\|_C$$

$$= \sup_{\|\varphi\|_\infty \le 1} \frac{1}{\pi} \left| \int_0^{2\pi} D_{n,r}(u) \, \varphi(u) \, du \right| = \frac{1}{\pi} \int_0^{2\pi} |D_{n,r}(u)| \, du. \tag{3.4}$$

We now have to extract the dominant term from the last integral for large n and fixed r. We notice first that, in view of $(n+1)^{-r} + (n+2)^{-r} + \ldots = O(n^{-r+1})$, we have for every $c > 0$ (see (3.2))

$$\int_{-c/n}^{c/n} |D_{n,r}(t)| \, dt = O(n^{-r}). \tag{3.5}$$

Hence

$$S_n(W_\infty^r)_C = \frac{1}{\pi} \int_{\pi/n}^{2\pi - \pi/n} |D_{n,r}(u)| \, du + O(n^{-r}). \tag{3.6}$$

We shall transform (3.2) with the help of *Abel summation*: for every numerical sequence $\{u_k\}_0^\infty$, $\{v_k\}_0^\infty$ we have

$$\sum_{k=n+1}^{m} u_k v_k = \sum_{k=n+1}^{m} U_k (v_k - v_{k+1}) - U_n v_{n+1} + U_m v_{m+1}, \tag{3.7}$$

where $U_k = u_0 + u_1 + \ldots + u_k$, $n \ge 0$, $m \ge n+1$. The validity of (3.7) can be directly verified.

We set $u_k = \cos(kt - \pi r/2)$ (fixed t) in (3.7) and we also set $v_k = k^{-r}$. Elementary identities for the sums of sines and cosines (see e.g. [13B, vol. 2, p. 90]) give

$$U_k = \sum_{j=0}^{k} \cos\left(jt - \frac{\pi r}{2}\right) = A_{r,k}(t) + a_r(t),$$

where

$$A_{r,k}(t) = \frac{\sin[(k + \frac{1}{2})t - \pi r/2]}{2 \sin(t/2)}, \tag{3.8}$$

$a_r(t) = \pm\frac{1}{2}$ for even r and $a_r(t) = \pm(\frac{1}{2}) \operatorname{ctg}(t/2)$ for odd r. Setting additionally $\Delta(k^{-r}) = k^{-r} - (k+1)^{-r}$ we obtain after substituting in (3.7)

$$\sum_{k=n+1}^{m} \frac{\cos(kt - \pi r/2)}{k^r} = \sum_{k=n+1}^{m} \Delta(k^{-r})[A_{r,k}(t) + a_r(t)]$$
$$- [A_{r,n}(t) + a_r(t)](n+1)^{-r}$$
$$+ [A_{r,m}(t) + a_r(t)](m+1)^{-r}. \tag{3.9}$$

For fixed $t \neq 2k\pi$ the quantities $A_{r,m}(t) + a_r(t)$ are uniformly bounded on m and hence after taking the limit for $m \to \infty$ in (3.9) and making obvious simplifications we obtain

$$D_{n,r}(t) = -\frac{A_{r,n}(t)}{(n+1)^r} + \sum_{k=n+1}^{m} \Delta(k^{-r}) A_{r,k}(t), \qquad t \neq 2k\pi. \tag{3.10}$$

Now we are going to show that the dominant term of the asymptotics of the integral in (3.6) is determined by the first term on the right-hand side of (3.10).

First let us find the asymptotics of the integral

$$J_{n,r} =: \frac{1}{\pi} \int_{\pi/n}^{2\pi - \pi/n} |A_{n,r}(u)| \, du = \frac{2}{\pi} \int_{\pi/n}^{\pi} |A_{n,r}(u)| \, du.$$

The last equality is correct in view of $A_{r,n}(-t) = \pm A_{r,n}(2\pi - t)$. But

$$A_{r,k}(t) = \frac{\sin(kt - \pi r/2)}{t} + O(1), \qquad 0 < t \leq \pi,$$

because of the boundedness of the function $\frac{1}{2}\operatorname{ctg}(t/2) - t^{-1}$ on $(0, \pi]$ and by repeating the arguments from the proof of Theorem 2.2.1 we obtain

$$J_{n,r} = \frac{2}{\pi} \int_{\pi/n}^{\pi} \frac{|\sin(kt - \pi r/2)|}{t} \, du + O(1)$$
$$= \frac{2}{\pi} \int_{\pi/n}^{\pi} \left| \sin\left(nt - \frac{\pi r}{2}\right) \right| \left\{ \sum_{k=1}^{n-1} \frac{1}{t + k\pi/n} \right\} dt + O(1) = \frac{4}{\pi^2} \ln n + O(1). \tag{3.11}$$

We also have to prove that the integral from π/n to $2\pi - \pi/n$ from the sum in the right-hand side of (3.10) is of the order $O(n^{-r})$. In order to do this we shall apply Abel summation for the second time. In view of (3.8) it is sufficient to prove that

$$\int_{\pi/n}^{\pi} \frac{1}{t} \left| \sum_{k=n+1}^{m} \Delta(k^{-r}) \sin\left(nt - \frac{\pi r}{2}\right) \right| dt = O(n^{-r}). \tag{3.12}$$

Setting $u_k = \sin(kt - \pi r/2)$, $v_k = \Delta(k^{-r})$, $v_k - v_{k+1} = \Delta^2(k^{-r})$ in (3.7) and taking the limit for $m \to \infty$ we obtain

$$\sum_{k=n+1}^{\infty} \Delta(k^{-r}) \sin\left(nt - \frac{\pi r}{2}\right)$$

$$= -b_{r,n}(t)\,\Delta((n+1)^{-r}) + \sum_{k=n+1}^{\infty} \Delta^2(k^{-r})\, b_{r,k}(t), \qquad t \neq 2\nu\pi,$$

where

$$b_{r,k}(t) = \frac{\cos\left[(k+\frac{1}{2})t - \pi r/2\right]}{2\sin(t/2)}.$$

Because $\Delta(n^{-r}) = O(n^{-r-1})$, $\Delta^2(n^{-r}) = O(n^{-r-2})$ for $n \to \infty$, we shall prove (3.12) follows if we show that

$$I_{k,r} =: \int_{1/k}^{\pi} |b_{r,k}(t)|\, t^{-1}\, dt = O(k).$$

But for $1/k \leq t \leq \pi$

$$b_{k,r}(t) = \frac{\cos(kt - \pi r/2)}{t} + O(1)$$

and hence, for example for even r, we have

$$I_{k,r} = \int_{1/k}^{\pi} |\cos kt|t^{-2}\, dt + O(k) = k \int_{1}^{k\pi} |\cos t|t^{-2}\, dt + O(k) = O(k).$$

Therefore

$$\int_{\pi/n}^{2\pi-\pi/n} \left| \sum_{k=n+1}^{m} \Delta(k^{-r}) A_{r,k}(t) \right| dt = O(n^{-r}) \tag{3.13}$$

and now from (3.5), (3.10), (3.11) and (3.13) we obtain

$$\frac{1}{\pi}\int_0^{2\pi} |D_{n,r}(u)|\, du = \frac{4}{\pi^2}\frac{\ln n}{n^r} + O(n^{-r}). \tag{3.14}$$

In connection with the estimate (3.4) it now remains to construct a sequence of functions f_n in W_∞^r for which the asymptotically exact supremum in (3.4) is attained.

By having the above it is sufficient to construct functions $\psi_n \in L_\infty$ $(n = 1, 2, \ldots,)$ such that $\|\psi_n\|_\infty \leq 1$, $\psi_n \perp 1$ and such that

$$\frac{1}{\pi}\int_{\pi/n}^{2\pi-\pi/n} A_{n,r}(u)\,\psi_n(u)\,\mathrm{d}u = \frac{4}{\pi^2}\frac{\ln n}{n^r} + O(n-r).$$

In view of equalities (3.8), (3.11) we can set

$$\psi_n(t) = \operatorname{sgn}\sin\,(nt - \pi r/2)$$

and then f_n can be taken to be the rth periodical integral of ψ_n. □

4.3.2 On estimates for linear projectors In connection with (3.1) let us note the following. The partial sum $S_n(f, t)$ of a Fourier series is a linear projector on the subspace $\mathcal{N}^T_{2n+1}$ of trigonometric polynomials of degree n and for an estimate of the error one can use the general inequality (2.2.4). Of course an application of this inequality in particular cases will be effective only if we have exact or at least good enough estimates both for the norm of the operator and for the best approximation. In our case

$$\|f - S_n(f)\|_C \le (1 + \mathcal{L}_n)\, E(f, \mathcal{N}^T_{2n+1})_C,$$

where $\mathcal{L}_n = \|S\|_{(C)}$ and, in view of Theorems 2.2.1 and 4.2.1, for every function $f \in W^r_\infty$ $(r = 1, 2, \ldots,)$ we have

$$\|f - S_n(f)\|_C \le \frac{4K_r}{\pi^2}\frac{\ln n}{n^r} + O(n^{-r}). \tag{3.15}$$

Because (see (3.1.9)) $1 < \pi^2/8 < K_r \le \pi/2$ for $r \ge 1$ the estimate (3.15) is close enough to (3.1) but is not asymptotically exact on W^r_∞. This fact is not surprising in view of the generality of the original inequality (2.2.4). Nevertheless we shall now indicate a similar situation in which the application of (2.2.4) is asymptotically exact for the same class estimate.

The trigonometric polynomial $\bar{\tau}_n(f, t)$ (see (2.2.27)) interpolating the values of f at the uniformly distributed knots $2\nu\pi/(2n + 1)$ $(\nu = 1, 2, \ldots, 2n)$, is also a linear projector (because of the unicity) on $\mathcal{N}^T_{2n+1}$. These polynomials may be considered as analogs of Fourier means constructed on the base of discrete information:

$$\bar{\tau}_n(f, t) = \frac{\alpha_0^n}{2} + \sum_{k=1}^{n} (\alpha_k^n \cos kt + \beta_k^n \sin kt),$$

where the coefficients α_k^n, β_k^n are determined by the values of f at the interpolating points (see (2.2.29)). In view of the general inequality (2.2.5) we have for every $f \in C$

$$|f(t) - \bar{\tau}_n(f, t)| \le [1 + \bar{\mathcal{L}}_n(t)]\, E(f, \mathcal{N}^T_{2n+1})_C, \tag{3.16}$$

where the function $\bar{\mathcal{L}}_n(t)$ is determined in (2.2.31). From Theorems 2.2.3, 4.2.1 and from (3.16) for every $f \in W^r_\infty$ we obtain

$$|f(t) - \bar{\tau}_n(f, t)| \leq \frac{2K_r}{\pi} \frac{\ln n}{n^r} |\sin (n + \tfrac{1}{2})t| + O(n^{-r}). \tag{3.17}$$

It turns out that this estimate is asymptotically exact on W^r_∞ for every t. Nikol'skii [8] has constructed for every $t \neq 2\nu\pi/(2n+1)$ a sequence of functions f_n $(n = 1, 2, \ldots,)$ (depending on r and t) such that for $n \to \infty$ the asymptopic of the sequence $|f_n(t) - \bar{\tau}_n(f_n, t)|$ coincides with the right-hand side of (3.17).

4.3.3 On the λ-methods of best order: Rogosinski means One has to pay special attention to the methods among all λ-methods (containing in particular the partial sums of Fourier series) which asymptotically provide the best order of approximation for some classes of functions. However, except for the three cases investigated in Section 4.2 for which the supremums $E_n(W^r_p)$ $(p = 1, 2, \infty)$ for every $n = 1, 2, \ldots$, are attained by λ-methods and except for the forthcoming case of λ-methods with positive kernels, there are only a few asymptotically exact results with the best order being realized by a linear method constructed by Fourier means. And the reason for that is not any lack of attention to these problems but in the difficulties which arise in every particular case.

Studying the proof of Theorem 4.3.1 it is easy to see that the method which was applied there in order to determine the dominant term of the asymptote of $\|f - S_n(f)\|_C$ on W^r_∞ will not work if the whole error is of order $O(n^{-r})$. On the other hand, attempting to use the general relations (3.15) or (3.16) for obtaining exact estimates leads to the necessity of finding the zeros of functions $\Psi_{n,r}(\gamma, \lambda, t)$. This can be done only in some exceptional cases – for Favard sums when the sign of $\Psi_{n,r}(\gamma^*, \lambda^*, t)$ coincides with either the sign of $\sin (n+1)t$ or $\cos (n+1)t$. For example for $r = 1$ (3.16) implies

$$\begin{aligned} e_n(W^1_\infty, \lambda)_C &= \frac{2}{\pi} \int_0^\pi |\Psi_{n,1}(0, \lambda, t)| \, \mathrm{d}t \\ &= \frac{2}{\pi} \int_0^\pi \left| \sum_{k=1}^{n} \frac{1 - \lambda^n_k}{k} \sin kt + \sum_{k=n+1}^{\infty} \frac{\sin kt}{k} \right| \mathrm{d}t. \end{aligned} \tag{3.18}$$

If $\lambda^n_k = \dfrac{k\pi}{2(n+1)} \operatorname{ctg} \dfrac{k\pi}{2(n+1)}$ then

$$\operatorname{sgn} \Psi_{n,1}(0, \lambda, t) = \operatorname{sgn} \sin (n+1)t,$$

but for other values of the coefficients λ^n_k the points of sign changing of

$\Psi_{n,1}(0, \lambda, t)$ (if there are any) are not uniformly distributed and their determination is very difficult.

The investigation of the asymptotic behaviour of $e_n(\mathcal{M}, \lambda)_q$ can be sometimes facilitated if one goes to the infinite interval in the integral representation of polynomial $U_n(f, \lambda, t)$ (see (1.1) and (1.10)). The first who applied this idea to Fourier means was Szokefalvy-Nagy [2].

We shall illustrate this method on the Rogosinski means (2.2.19). Besides the simplicity of these sums they have good approximating properties for functions which are not very smooth when β_n is near to $\pi/(2n)$. In particular for the class W^1_∞ their uniform error is not only of the order of the best approximation but differs little from $E_{n+1}(W^1_\infty)_C$ with respect to the exact constant.

Thus, for $f \in L_1$ we consider the sequence of trigonometric polynomials

$$\begin{aligned} \mathcal{R}(f, \beta_n, t) &= \tfrac{1}{2}[S_n(f, t - \beta_n) + S_n(f, t + \beta_n)] \\ &= \frac{1}{\pi}\int_{-\pi}^{\pi} K(\beta_n, t) f(t - u)\, du, \end{aligned} \tag{3.19}$$

where

$$K(\beta_n, t) = \frac{1}{2}[D_n(t - \beta_n) + D_n(t + \beta_n)] = \frac{1}{2} + \sum_{k=1}^{n} \cos k\beta_n \cos kt,$$

and D_n is the Dirichlet kernel (2.2.9). For $\beta_n = \pi/(2n)$ we shall write $\mathcal{R}(f, \pi/(2n), t) = \mathcal{R}_n(f, t)$.

In order to pass to the infinite interval we shall use the known expansion (see e.g. [5B, p. 472])

$$\operatorname{ctg} t = \frac{1}{t} + \sum_{k=1}^{\infty} \frac{2t}{t^2 - k^2\pi^2}, \qquad t \neq m\pi, \qquad m = 0, \pm 1, \pm 2, \ldots, \tag{3.20}$$

in which the series converges uniformly on every closed interval not containing the points $m\pi$. Using (3.20) we transform the Dirichlet kernel:

$$\begin{aligned} D_n(t) &= \frac{\sin(n + \frac{1}{2})t}{2\sin(t/2)} = \frac{1}{2}\left(\sin nt \operatorname{ctg}\frac{t}{2} - \cos nt\right) \\ &= \frac{\sin nt}{t} + \sum_{k=1}^{\infty} \frac{2t \sin nt}{t^2 - 4k^2\pi^2} - \tfrac{1}{2}\cos nt, \qquad t \neq m\pi. \end{aligned} \tag{3.21}$$

For every $f \in L_1$ using (3.21) we obtain

$$\int_{-\pi}^{\pi} f(t-u)\,[D_n(u) - \tfrac{1}{2}\cos nu]\,du = \int_{-\pi}^{\pi} \frac{\sin nu}{u} f(t-u)\,du$$

$$+ \sum_{k=1}^{\infty} \int_{-\pi}^{\pi} \left(\frac{\sin nu}{u-2k\pi} + \frac{\sin nu}{u+2k\pi}\right) f(t-u)\,du$$

$$= \int_{-\infty}^{\infty} \frac{\sin nu}{u} f(t-u)\,du.$$

The last equality is obtained by a change of the variables in every term of the sum using the periodicity of f.

Therefore

$$S_n(f,t) = \frac{1}{\pi}\int_{-\pi}^{\pi} D_n(u)\,f(t-u)\,du$$

$$= \frac{1}{\pi}\int_{-\infty}^{\infty} \frac{\sin nu}{u} f(t-u)\,du + \frac{1}{2\pi}\int_{-\pi}^{\pi} \cos n(t-u)\,f(u)\,du. \tag{3.22}$$

Replacing this expression in sum (3.19) with $\beta_n = \pi/(2n)$ and noticing that this semisum is zero for the last integral in (3.22) we obtain

$$\mathcal{R}_n(f,t) = \frac{1}{2\pi}\left[\int_{-\infty}^{\infty} \frac{\sin nu}{u} f(t-u-\pi/(2n))\,du\right.$$

$$\left. + \int_{-\infty}^{\infty} \frac{\sin nu}{u} f(t-u+\pi/(2n))\,du\right]$$

$$= \frac{1}{2\pi}\int_{-\infty}^{\infty} \left(\frac{\cos nu}{u+\pi/(2n)} - \frac{\cos nu}{u-\pi/(2n)}\right) f(t-u)\,du$$

$$= \frac{1}{2\pi}\int_{-\infty}^{\infty} \left(\frac{\cos u}{u+\pi/2} - \frac{\cos u}{u-\pi/2}\right) f(t-u/n)\,du.$$

If a class of functions $\mathcal{M}$ satisfies the conditions of Proposition 4.1.1 then taking the evenness of the function

$$\psi(t) = \frac{\cos t}{t+\pi/2} - \frac{\cos t}{t-\pi/2} = -\frac{\pi\cos t}{t^2 - \pi^2/4} \tag{3.23}$$

into account, we obtain

$$e(\mathcal{M}, \mathcal{R}_n)_C =: \sup_{f\in\mathcal{M}} \|f - \mathcal{R}_n(f)\|_C = \sup_{f\in\mathcal{M}_0} |\mathcal{R}_n(f,0)| = \frac{1}{\pi}\sup_{f\in\mathcal{M}_0}\left|\int_0^{\infty} f\left(\frac{t}{n}\right)\psi(t)\,dt\right|,$$

where $\mathcal{M}_0 = \{f : f \in \mathcal{M}, f(0) = 0\}$. Under the assumption $\mathcal{M} \subset L_1^1$ setting

$$\psi_1(t) = -\int_t^\infty \psi(u)\,du \tag{3.24}$$

an integration by parts gives

$$e(\mathcal{M}, \mathcal{R}_n)_C = \frac{1}{n\pi} \sup_{f \in \mathcal{M}_0} \left| \int_0^\infty f'\left(\frac{t}{n}\right) \psi_1(t)\,dt \right|. \tag{3.25}$$

In particular for $\mathcal{M} = W^1_\infty$, i.e. when $|f'(u)| \le 1$, we obtain the inequality

$$e(W^1_\infty, \mathcal{R}_n)_C \le \frac{1}{n\pi} \int_0^\infty |\psi_1(t)|\,dt. \tag{3.26}$$

We shall prove that the estimate (3.26) is asymptotically exact. Let us first note that

$$\psi_1(t) = \mathrm{Si}\,(t + \pi/2) + \mathrm{Si}\,(t - \pi/2) = \mathrm{si}\,(t + \pi/2) + \mathrm{si}\,(t - \pi/2),$$

where

$$\mathrm{Si}\,(t) = \int_0^t \frac{\sin u}{u}\,du, \qquad \mathrm{si}\,(t) = -\int_t^\infty \frac{\sin u}{u}\,du = \mathrm{Si}\,(t) - \pi/2 \tag{3.27}$$

are functions known as *intregral sines*. The oddness of ψ_1 follows from the oddness of Si. Hence the $2\pi n$-function g defined by

$$g(t) = \frac{1}{n}\,\mathrm{sgn}\,\psi_1(t), \qquad |t| \le \pi n,$$

is zero in the mean on the period. Setting

$$g_1(t) = \int_0^t g(u)\,du, \qquad f_1(t) = g_1(nt), \qquad -\infty < t < \infty,$$

we get an even 2π-periodic function from W^1_∞ such that $f_1'(t/n) = \mathrm{sgn}\,\psi_1(t)$ for $|t| \le n\pi$. For this function

$$\mathcal{R}_n(f_1, 0) = \frac{1}{\pi} \int_0^\infty f_1\left(\frac{t}{n}\right) \psi(t)\,dt$$

$$= \frac{1}{n\pi} \int_0^\infty f_1'\left(\frac{t}{n}\right) \psi_1(t)\,dt = \frac{1}{n\pi} \int_0^\infty |\psi_1(t)|\,dt + \frac{1}{n\pi} I_n,$$

where

$$I_n = \int_{n\pi}^\infty \left[f_1'\left(\frac{t}{n}\right) - \mathrm{sgn}\,\psi_1(t) \right] \psi_1(t)\,dt,$$

and hence

$$I_n < 2\int_{n\pi}^{\infty} |\psi_1(t)|\, dt. \tag{3.28}$$

We shall show that $I_n = O(n^{-1})$. In order to do this it is sufficient to prove that $\psi_1(t) = O(t^{-2})$ for $t \to \infty$ (see (3.28)). If we set $t_k = (k+\frac{1}{2})\pi$, $t_{n-1} < t \le t_n$, then we obtain from (3.23), (3.24)

$$|\psi_1(t)| = \left| \int_t^{t_n} \psi(u)\, du + \sum_{k=n}^{\infty} \int_{t_k}^{t_{k+1}} \psi(u)\, du \right| \le \int_t^{t_{n+1}} |\psi(u)|\, du \le ct^{-2}.$$

Here, we have used that the terms in the above series are alternating in sign and decreasing in modulus. Therefore

$$\mathcal{R}_n(f_1, 0) = \frac{1}{n\pi}\int_0^{\infty} |\psi_1(t)|\, dt + 0(n^{-2}) \tag{3.29}$$

and because of $f_1 \in W^1_\infty$ from (3.26), (3.29) we obtain:

Theorem 4.3.2 For the approximations by the sums

$$\mathcal{R}_n(f, t) = [S_n(f, t - \pi/(2n)) + S_n(f, t + \pi/(2n))]/2$$

we have for $n \to \infty$ the asymptotically exact equality

$$\sup_{f \in W^1_\infty} \|f - \mathcal{R}_n(f)\|_C = \frac{1}{n\pi}\int_0^{\infty} |\mathrm{Si}\,(t + \pi/2) + \mathrm{Si}\,(t - \pi/2)|\, dt + O(n^{-2})$$

$$= \frac{1}{n}\int_0^{\infty} \left| \int_t^{\infty} \frac{\cos u}{u^2 - \pi^2/4}\, du \right| dt + O(n^{-2}). \tag{3.30}$$

We can estimate the magnitude of the constant in the leading order term in the right-hand side of (3.30) by the following result for the sums (3.19):

$$\sup_{f \in W^1_\infty} \left\| f - \mathcal{R}_n\left(f, \frac{\pi}{2n+1}\right) \right\|_C = \frac{\pi}{2n+1} + \frac{\varepsilon_n}{n+1} + \frac{c_n}{(n+1)^2}, \qquad n = 2, 3, \ldots, \tag{3.31}$$

where $0 < \varepsilon_n < 0.006$, $|c_n| \le 1$; if we replace in the left-hand side of (3.31) $\mathcal{R}_n(f, \pi/(2n+1), t)$ with $\mathcal{R}_n(f, \beta_n, t)$, where $\beta_n = (\pi/(2n+1) + o((n \ln n)^{-1})$, the right-hand side will change by a term of order $o(n^{-1})$. Relation (3.31) should be compared with the equality

$$E_{n+1}(W^1_\infty)_C = \pi/(2n+2),$$

which was obtained earlier (Theorem 4.2.1).

Integrating (3.25) by parts once more we can obtain the following asymptotic equality:

$$\sup_{f\in W^2_\infty} \|f - \mathcal{R}_n(f)\|_C = \frac{1}{\pi n^2}\int_0^\infty |(t+\pi/2)\,\mathrm{si}\,(t+\pi/2)$$

$$+ (t-\pi/2)\,\mathrm{si}\,(t-\pi/2)|\,\mathrm{d}t + O(n^{-3}).$$

But Rogosinski means are not sensitive to further increase in the smoothness of the approximating function (recalling the order of the leading term). We shall return to this question in Section 7.3.

4.3.4 Positive λ-methods A special place among the λ-means of Fourier sums is held by the operators

$$U_n(f,\lambda^+,t) = \frac{1}{\pi}\int_0^{2\pi} K_n(\lambda^+,u)\,f(t-u)\,du, \qquad n = 1,2,\ldots, \tag{3.32}$$

with non-negative kernel

$$K_n(\lambda^+,t) = \frac{1}{2} + \sum_{k=1}^{n} \lambda_k^n \cos kt \geq 0. \tag{3.33}$$

Hence $U_n(f,\lambda^+,t) \geq 0$ for $f \geq 0$. Such methods are called *positive*.

Condition (3.33) imposes very strong restrictions on the coefficients λ_k^n. It is known [25A, p. 213] – and it is far from trivial – that (3.33) implies

$$|\lambda_k^n| \leq \cos\frac{\pi}{[n/k]+2}, \qquad k = 1,2,\ldots,n, \tag{3.34}$$

and in particular

$$|\lambda_1^n| \leq \cos\frac{\pi}{n+2}, \; |\lambda_2^n| \leq \cos\frac{2\pi}{n+4} \tag{3.35}$$

(the first of these estimates can be found in [13B] too).

But it is clear that the positivity of the kernel $K_n(\lambda^+,t)$ simplifies the investigation of the approximation properties of the positive λ-methods (λ^+-methods) substantially. But for functions from L^r_∞ the positive operators do not react (in the sense of the order of the error) on the increasing smoothness of the approximating functions if $r > 2$. Let us consider the problem for finding the exact asymptotic of the error of λ-methods on the classes W^r_∞ ($r = 2,3,\ldots$). Of course only the case $r = 2$ is essential, but with a little more effort the problem can be solved for any integer $r \geq 2$. At the same time we shall get some new extremal properties of Euler splines.

The approximation possibilities of λ^+-methods on classes W^r_∞ can be clarified using the function

$$\psi_r(t) = \varphi_{1,r}(t + \alpha_r) + K_r, \tag{3.36}$$

where K_r is the Favard constant, $\varphi_{1,r}$ is the Euler spline of degree r ($\varphi_{1,r}^{(r)}(t) = \operatorname{sgn} \sin t$) and the number α_r is chosen so that $\psi_r(0) = 0$. Because of $\|\varphi_{1,r}\|_C = K_r$ we have $\psi_r(t) \geq 0$. Also it is clear that ψ_r decreases on $(-\pi, 0)$ and increases on $(0, \pi)$ and

$$\max_t \psi_r(t) = \psi_r(-\pi) = \psi_r(\pi) = 2K_r.$$

Because of $\psi_r^{(k)}(t) = \varphi_{1,r}^{(k)}(t + \alpha_r)$ $(1 \leq k \leq r - 1)$ we see, under the assumption that the corresponding derivatives exist, that

$$\psi_r'(0) = \psi_r^{(3)}(0) = 0, \qquad \psi_r''(0) = K_{r-2}, \qquad \psi_r^{(4)}(0) = -K_{r-4}.$$

Lemma 4.3.3 For every sequence of λ^+-methods we have for $n \to \infty$

$$U_n(\psi_r, \lambda^+, 0) - \psi_r(0) = U_n(\psi_r, \lambda^+, 0) \geq \frac{\pi^2 K_{r-2}}{2n^2} + o(n^{-2}). \tag{3.37}$$

Proof For $r = 2$ and $r = 4, 5, 6, \ldots$, we can prove (using relatively simple arguments) that the stronger inequalities

$$U_n(\psi_r, \lambda^+, 0) - \psi_r(0) \geq \begin{cases} (1 - \lambda_1^n)\, K_{r-2}, & r = 2, 4, \ldots, \\ (1 - \lambda_2^n)\, K_{r-2}, & r = 5, 7, \ldots, \end{cases} \tag{3.38}$$

are valid for every n which in view of (3.35) implies (3.37) for $r \neq 3$.

Considering the even case first we shall show that

$$\psi_r(t) \geq (1 - \cos t)\, K_{r-2}, \qquad |t| \leq \pi, \qquad r = 2, 4, \ldots. \tag{3.39}$$

Set $\delta(t) = \psi_r(t) - (1 - \cos t)\, K_{r-2}$. It is clear that $\delta(0) = \delta'(0) = 0$. For $r = 2$ we have $\delta'(t) = \varphi_{1,1}(t + \pi/2) - \sin t$ and hence δ' changes sign once on $[-\pi, \pi]$, (at $t = 0$) and in this case (3.39) is proved.

Let r be even and $r \geq 4$. It is easy to check that

$$\delta^{(\nu)}(0) = 0, \qquad \nu = 0, 1, 2, 3, \qquad \delta^{(4)}(0) = -K_{r-4} + K_{r-2} > 0,$$

from where we get, in particular, $\delta(t) > 0$ $(t \neq 0)$ in a neighborhood of the origin. Because of $\delta(\pi) = 2(K_r - K_{r-2}) > 0$, the assumption that δ takes a negative value somewhere on $(-\pi, \pi)$ implies that δ has at least eight zeros on this interval (counting with multiplicity). By applying Rolle's theorem consecutively (or using Proposition 2.4.1 in the appropriate form) we conclude from the periodicity of δ that the function $\delta^{(r-1)}$ has to change sign at least eight times on the period. But this is impossible because $\delta^{(r-1)}(t) = \varphi_{1,1}(t + \alpha_r) \pm K_{r-2} \sin t$ and hence $\delta^{(r-1)}$ can change sign no more than six times (draw the graph!).

Therefore we have for $r = 2, 4, 6, \ldots,$

$$U_n(\psi_r, \lambda^+, 0) - \psi_r(0) = \frac{1}{\pi}\int_{-\pi}^{\pi} K_n(\lambda^+, t)\, \psi_r(t)\, dt$$

$$\geq \frac{K_{r-2}}{\pi}\int_{-\pi}^{\pi}\left(\frac{1}{2} + \sum_{k=1}^{n} \lambda_k^n \cos kt\right)(1 - \cos t)\, dt = (1 - \lambda_1^n)\, K_{r-2},$$

and (3.38) is proved for even r.

For $r = 5, 7, \ldots,$ (3.38) is obtained in a similar manner from the inequality

$$\psi_r(t) \geq (1 - \cos 2t)K_{r-2}/4, \qquad |t| \leq \pi, \tag{3.40}$$

which in turn can be proved by contradiction, in the same way as (3.39).

In the case $r = 3$ inequality (3.40) is not valid but using better arguments connected with the application of (3.34) one can show that (3.37) also holds for $r = 3$ too. □

Because of $\psi_r \in W_\infty^r$ Lemma 4.3.3 implies that we have for every sequence of λ^+-methods

$$e_n(W_\infty^r, \lambda^+)_C =: \sup_{f \in W_\infty^r} \|f - U_n(f, \lambda^+)\|_C \geq \frac{\pi^2 K_{r-2}}{2n^2} + o(n^{-2}). \tag{3.41}$$

There are sequences of linear positive operators $U_n(f, \lambda^+, t)$ for which relation (3.41) holds with an equality sign. In order to see this let us first note that we have for $\delta \in (0, \pi)$

$$\frac{1}{\pi}\int_{\delta \leq |t| \leq \pi} K_n(\lambda^+, t)\, dt \leq \frac{1 - \lambda_1^n}{2(1 - \cos\delta)^2}\left(4 - \frac{1 - \lambda_2^n}{1 - \lambda_1^n}\right). \tag{3.42}$$

Indeed, in view of the positivity of the kernel, i.e. $K_n(\lambda^+, t) \geq 0$, the left-hand side of (3.42) is not larger than

$$\frac{1}{\pi}\int_{-\pi}^{\pi}\left(\frac{1 - \cos t}{1 - \cos\delta}\right)^2 K_n(\lambda^+, t)\, dt = \frac{1}{(1 - \cos\delta)^2}\left(\frac{3}{2} - 2\lambda_1^n + \frac{1}{2}\lambda_2^n\right),$$

and the last expression is equal to the right-hand side of (3.42).

The following theorem has an important place in the theory of the linear positive operators.

Theorem 4.3.4 If in the triangular matrix $\{\lambda_k^n\}$ $(k = 1, 2, \ldots n; n = 1, 2, \ldots)$ which defines a sequence of λ^+-methods the coefficients λ_1^n and λ_2^n satisfy

$$\lim_{n \to \infty} \frac{1 - \lambda_2^n}{1 - \lambda_1^n} = 4, \tag{3.43}$$

then we have for every $f \in C^2$ and for $n \to \infty$

$$f(x) - U_n(f, \lambda^+, x) = f''(x)\,(1 - \lambda_1^n\,(1 + o(1)).$$

Proof We may assume that $x = 0$. Let us fix $f \in C^2$. It is easy to see (e.g. by using l'Hospital's rule) that we have in a neighborhood of the origin

$$f(u) = f(0) + f'(0)\sin u + \frac{f''(0)}{2}\sin^2 u + \alpha(u)\sin t^2 u, \qquad (3.44)$$

where $\alpha(t) \to 0$ for $t \to 0$ and $\alpha \in C$. Replacing (3.44) in (3.32) and recalling (1.4) and the evenness of the kernel $K_n(\lambda, t)$ we obtain

$$f(0) - U_n(f, \lambda^+, 0) = \frac{f''(0)}{2\pi}\int_{-\pi}^{\pi} F_n(t)\,dt + \frac{1}{\pi}\int_{-\pi}^{\pi} \alpha(t)\,F_n(t)\,dt =: A_n + \frac{1}{\pi}B_n,$$

where we have set $F_n(t) = \sin^2 t\,K_n(\lambda^+, t)$. Now (3.43) implies

$$A_n = \frac{1}{4}f''(0)\,(1 - \lambda_2^n) = f''(0)\,(1 - \lambda_1^n)\,(1 + o(1))$$

and it remains to prove $B_n = o(1 - \lambda_1^n)$.

$$\xi_n = 4 - \frac{1 - \lambda_2^n}{1 - \lambda_1^n}, \qquad \delta_n = \xi_n^{1/5},$$

then (3.43) implies $\lim \delta_n = \lim \xi_n = 0$. We have

$$B_n = \int_{|t| \le \delta_n} \alpha(t)\,F_n(t)\,dt + \int_{\delta_n \le |t| \le \pi} \alpha(t)\,F_n(t)\,dt =: B'_n + B''_n.$$

If $\max_{|t| \le \delta_n} |\alpha(t)| = \gamma_n$ then $\gamma_n \to 0$ and using (3.43) we obtain

$$|B'_n| \le \gamma_n \int_{-\pi}^{\pi} F_n(t)\,dt = \frac{\pi}{2}\gamma_n\,(1 - \lambda_2^n) = o(1 - \lambda_1^n).$$

We have also $\xi_n(1 - \cos\delta_n)^{-2} \to 0$ and hence using (3.42) we obtain

$$|B''_n| \le \frac{\|\alpha\|_C}{2}\,|1 - \lambda_1^n|\,(1 - \cos\delta_n)^{-2}\,\zeta_n = o(1 - \lambda_1^n).$$

This proves the theorem. □

We shall also need an extremal property of the function (3.36) with respect to class W_∞^r.

Proposition 4.3.5 For every $t_0 \in \mathbb{R}$ and $r = 1, 2, \ldots$, we have

$$\sup_{f\in W^r_\infty} [f(t_0+t)+f(t_0-t)-2f(t_0)] = 2\,\psi_r(t), \qquad |t|\le\pi,$$

where function ψ_r is defined in (3.36).

Proof The class W^r_∞ contains every function $c+f(t+\alpha)$ together with f and hence it is sufficient to prove that

$$\sup_{f\in W^r_{\infty,0}} [f(t)+f(-t)] = 2\,\psi_r(t), \qquad |t|\le\pi, \tag{3.45}$$

where $W^r_{\infty,0}=\{f : f\in W^r_\infty, f(0)=0\}$. But (1.6) gives

$$f(t)+f(-t)=\frac{1}{\pi}\int_{-\pi}^{\pi}\Delta_r(t,u)f^{(r)}(u)\,\mathrm{d}u\le\frac{1}{\pi}\int_{-\pi}^{\pi}|\Delta_r(t,u)|\,\mathrm{d}u, \tag{3.46}$$

for $f\in W^r_{\infty,0}$, where

$$\Delta_r(t,u)=(-1)^r[B_r(u+t)+B_r(u-t)-2B_r(u)]+\eta_r(t),$$

with the number $\eta_r(t)$ for every $t\in[-\pi,\pi]$ being chosen from the condition

$$\inf_\gamma\int_{-\pi}^{\pi}|\Delta_r(t,u)-\gamma|\,\mathrm{d}u=\int_{-\pi}^{\pi}|\Delta_r(t,u)|\,\mathrm{d}u. \tag{3.47}$$

It is clear that the function $B_r(u+t)+B_r(u-t)-2B_r(u)$ in u has the same parity as r and hence $\eta_r(t)=0$ for odd r.

We shall show that for $|t|\le\pi$ and $f=\psi_r$ we have an equality in (3.46). Because of $B'_m=B_{m-1}$ and $B_1(t)=(\pi-t)/2$ $(0<t<2\pi)$ we easily see that the $(r-1)$th derivative of $\Delta_r(t,u)$ with respect to u is a step function (see also Lemma 2.3.4) and changes sign on the period exactly twice. Then by Proposition 2.4.2 and (3.47) we conclude that $\Delta_r(t,u)$ (as a function of u) changes sign exactly twice – namely at the points $k\pi$ for odd $r\ge3$ and at the points $(k+\frac{1}{2})\pi$ for even $r\ge4$. But these are exactly the points of sign change for the step function $\psi_r^{(r)}$ and for $|t|\le\pi$ we have $\psi_r^{(r)}(u)=\operatorname{sgn}\Delta_r(t,u)$. Therefore we have for $r\ge3$

$$2\psi_r(t)=\psi_r(t)+\psi_r(-t)=\frac{1}{\pi}\int_{-\pi}^{\pi}|\Delta_r(t,u)|\,\mathrm{d}u, \qquad |t|\le\pi;$$

the validity of this equality for $r=1,2$ can be directly verified. It remains to notice that $\psi_r\in W^r_{\infty,0}$. Relation (3.45) and the proposition are proved. □

From Proposition 4.1.1, from (3.45) and from the positivity of the kernel $K_n(\lambda^+,t)$ for every λ^+-method we obtain

$$e_n(W^r_\infty,\lambda^+)_C=|\psi_r(0)-U_n(\psi_r,\lambda^+,0)|=\frac{2}{\pi}\int_0^{\pi}K_n(\lambda^+,t)\psi_r(t)\,\mathrm{d}t. \tag{3.48}$$

Theorem 4.3.6 For every $r = 2, 3, \ldots$, we have the asymptotic (for $n \to \infty$) equality

$$\inf_{\lambda^+} e_n(W^r_\infty, \lambda^+)_C = \frac{\pi^2 K_{r-2}}{2n^2} + o(n^{-2}), \tag{3.49}$$

where the infimum for every n is taken over all positive λ-methods. There is a sequence of λ^+-methods given for every n with the set of coefficients $\tilde{\lambda} = \{\tilde{\lambda}^n_1, \tilde{\lambda}^n_2, \ldots, \tilde{\lambda}^n_n\}$ such that

$$e_n(W^r_\infty, \tilde{\lambda})_C = \frac{\pi^2 K_{r-2}}{2n^2} + o(n^{-2}). \tag{3.50}$$

Proof Inequality (3.41), which is valid for every sequence of λ^+-methods, gives

$$\inf_{\lambda^+} e_n(W^r_\infty, \lambda^+)_C \geq \frac{\pi^2 K_{r-2}}{2n^2} + o(n^{-2})$$

and therefore only the second assertion of the theorem remains to be proved.

Set

$$\tilde{\lambda}^n_k = \frac{2}{n+2} \sum_{\nu=1}^{n-k+1} \sin\frac{\nu\pi}{n+2} \sin\frac{(k+\nu)\pi}{n+2}$$

$$= \frac{1}{2(n+2)\sin[\pi/(n+2)]} \times$$

$$\left[(n-k+3)\sin\frac{(k+1)\pi}{n+2} - (n-k+1)\sin\frac{(k-1)\pi}{n+2}\right]. \tag{3.51}$$

Because of

$$K_n(\tilde{\lambda}, t) =: \frac{1}{2} + \sum_{k=1}^{n} \tilde{\lambda}^n_k \cos kt = \frac{1}{n+2}\left|\sum_{k=1}^{n+2} a_k z^{k-1}\right|^2$$

(see e.g. [13B, vol. 2, p. 296]), where $a_k = \sin[k\pi/(n+2)]$, $z = \exp(\mathrm{i}t) = \cos t + \mathrm{i}\sin t$, the $\tilde{\lambda}$-method is positive. Condition (3.43) is fulfilled for the first two coefficients of this method

$$\tilde{\lambda}^n_1 = \cos\frac{\pi}{n+2}, \qquad \tilde{\lambda}^n_1 = \frac{n+1}{n+2}\cos\frac{2\pi}{n+2} + \frac{1}{n+2}$$

and Theorem 4.3.4 and (3.48) give for $r \geq 3$

$$e_n(W^r_\infty, \tilde{\lambda})_C = |\psi_r(0) - U_n(\psi_r, \tilde{\lambda}, 0)|$$

$$= \psi''_r(0)(1 - \tilde{\lambda}^n_1)(1 + o(1)) = \frac{\pi^2 K_{r-2}}{2n^2} + o(n^{-2}).$$

The function ψ_2 is not from C^2 but we can apply Theorem 4.3.4 with its Steklov function $\psi_{2,h}$ and we get that (3.50) also holds for $r = 2$ because $\psi''_{2,h}(0) \to K_0$ for $h \to 0$. The theorem is proved. □

The sequence of $\tilde{\lambda}$-methods is asymptotically best for the class W^r_∞ in the space C among all sequences of positive methods given by the operators $U_n(f, \lambda^+, t)$. On the other hand, inequality (3.41) shows that the saturation order of every sequence of positive λ-methods cannot be better than $O(n^{-2})$: the increase in the smoothness of the approximating function does not imply an order of convergence which is better than $O(n^{-2})$.

4.3.5 Fejér means We shall consider the most popular positive λ-method, namely Fejér operators (see Section 2.2) separately. They are

$$\mathscr{F}_n(f, t) = \frac{1}{n+1}[S_0(f, t) + S_1(f, t) + \ldots + S_n(f, t)]$$

$$= \frac{1}{\pi}\int_{-\pi}^{\pi} F_n(t-u) f(u)\, du, \tag{3.52}$$

where

$$F_n(t) = \frac{1}{2} = \sum_{k=1}^{n}\left(1 - \frac{k}{n+1}\right)\cos kt = \frac{1}{2(n+1)}\left(\frac{\sin[(n+1)t/2]}{\sin(t/2)}\right)^2.$$

For $f(t) = 1 - \cos t$ we have

$$\mathscr{F}_n(f 0) - f(0) = \frac{1}{\pi}\int_{-\pi}^{\pi} F_n(u)(1 - \cos u)\, du = \frac{1}{n+1};$$

hence the saturation order of Fejér operators is $O(n^{-1})$ and only the problem of estimating the approximation error on functional classes with little smoothness is important.

Together with the class KH^1 of functions from C introduced in Section 1.5 which satisfy a Lipschitz condition with Lipschitz constant K ($KH^1 = KW^1_\infty$) we shall consider the more general class KH^α ($1H^\alpha = H^\alpha$) of functions $f \in C$ which satisfy Hölder's condition

$$|f(t') - f(t'')| \le K|t' - t''|^\alpha, \qquad 0 < \alpha \le 1.$$

In the case $\alpha = 1$ we can use the general formula (3.18) for estimating the error of positive λ-methods. The function

$$\Phi_n(\lambda, t) = \int_t^{\pi} K_n(\lambda, u)\, du = \frac{\pi - t}{2} - \sum_{k=1}^{n}\frac{\lambda_k^n}{k}\sin kt, \qquad 0 \le t \le \pi,$$

coincides on $(0, \pi)$ with $\Psi_{n,1}(0, \lambda; t)$ and if $\Phi_n(\lambda, t) \geq 0$ then

$$e_n(W_\infty^1, \lambda)^C = \frac{2}{\pi}\int_0^\pi \Phi_n(\lambda, t)\,dt$$
$$= \frac{4}{\pi}\left\{\sum_{\nu=1}^{N} \frac{1-\lambda_{2\nu-1}^n}{(2\nu-1)^2} + \sum_{\nu=N+1}^{\infty} \frac{1}{(2\nu-1)^2}\right\}, \qquad N = \left[\frac{n+1}{2}\right]. \tag{3.53}$$

This relation, which holds for all positive λ-methods, is also valid for kernels $K_n(\lambda, t)$ with alternating signs provided $\Phi_n(\lambda, t) \geq 0$.

In the case of the Fejér kernel, using the notation

$$e(\mathcal{M}, \mathcal{F}_n)_X = \sup_{f\in\mathcal{M}} \|f - \mathcal{F}_n(f)\|_X,$$

we obtain after some calculations in (3.53)

$$e_n(W_\infty^1, \mathcal{F}_n)_C = \frac{4}{\pi}\left\{\frac{1}{n+1}\sum_{\nu=1}^{N} \frac{1}{2\nu-1} + \sum_{\nu=N+1}^{\infty} \frac{1}{(2\nu-1)^2}\right\}$$
$$= \frac{4}{\pi(n+1)}[\ln(n+1) + \ln 2 + C + 1] + o(n^{-1}), \tag{3.54}$$

where C is the Euler constant, i.e. $C = 0.577\ldots$.

Now let $0 < \alpha < 1$. We obtain from Proposition 4.1.1

$$e(H^\alpha, \mathcal{F}_{n-1})_C = \sup_{f\in H_0^\alpha} \frac{2}{\pi}\int_0^\pi F_{n-1}(t)f(t)\,dt$$
$$= \frac{2}{\pi}\int_0^\pi F_{n-1}(t)t^\alpha\,dt = \frac{2^{\alpha+1}}{\pi n}\int_0^{\pi/2} t^\alpha \frac{\sin^2 nt}{\sin^2 t}\,dt.$$

Using the boundness of the function $(\sin t)^{-2} - t^{-2}$ on $(0, \pi/2)$ we can write

$$e(H^\alpha, \mathcal{F}_{n-1})_C = \frac{2^{\alpha+1}}{\pi n}\int_0^{\pi/2} \frac{\sin^2 nt}{t^{2-\alpha}}\,dt + O(n^{-1})$$
$$= \frac{2^{\alpha+1}}{\pi n^\alpha}\int_0^{n\pi/2} \frac{\sin^2 t}{t^{2-\alpha}}\,dt + O(n^{-1}) = \frac{2^{\alpha+1}}{\pi n^\alpha}\int_0^\infty \frac{\sin^2 t}{t^{2-\alpha}}\,dt + O(n^{-1}). \tag{3.55}$$

Integration by parts gives for $0 < \alpha < 1$

$$\int_0^\infty \frac{\sin^2 t}{t^{2-\alpha}}\,dt = \frac{1}{(1-\alpha)2^\alpha}\int_0^\infty t^{\alpha-1}\sin t\,dt = \frac{\Gamma(\alpha)}{(1-\alpha)2^\alpha}\sin\frac{\alpha\pi}{2},$$

where $\Gamma(\alpha)$ is the gamma-function. Replacing this in (3.55) and combining it with (3.54) we get:

Proposition 4.3.7 We have the following asymptotic relations for approximating the classes H^α by Fejér sums (3.52)

$$e(H^\alpha, \mathcal{F}_{n-1})_C = \begin{cases} \dfrac{2\ln n}{\pi n} + O(n^{-1}), & \alpha = 1, \\ \dfrac{2\Gamma(\alpha)\sin(\alpha\pi/2)}{\pi(1-\alpha)n^\alpha} + o(n^{-\alpha}), & 0 < \alpha < 1. \end{cases}$$

4.3.6 On approximation in L_1 There is a close relationship between the problem of approximating the classes W^r_∞ in the C metric and the classes W^r_1 in the L_1 metric by trigonometric polynomials (in particular by partial sums of Fourier series and their means). This relationship, based on the duality of the extremal problems, was discovered by Nikol'skii [6].

Theorem 4.3.8 Let

$$F(\varphi, t) = \int_0^{2\pi} K(t-u)\,\varphi(u)\,du,$$

where K is a fixed L_1 function. Then

$$M_1 =: \sup_{\|\varphi\|_1 \le 1, \varphi \perp 1} \|F(\varphi)\|_1 \le \sup_{\|\varphi\|_\infty \le 1, \varphi \perp 1} \|F(\varphi)\|_C =: M_\infty. \tag{3.56}$$

A necessary and sufficient condition for (3.56) to be an equality is the existence of an u_0 such that for almost all t

$$[K(t) + c_*][K(t+u_0) + c_*] \le 0, \tag{3.57}$$

where the number c_* is defined by

$$\min_c \int_0^{2\pi} |K(t) + c|\,dt = \int_0^{2\pi} |K(t) + c_*|\,dt. \tag{3.58}$$

Proof From Corollary 1.4.3 with $p = 1$ and from the periodicity of the kernel K we obtain

$$M_\infty = \int_0^{2\pi} |K(t) + c_*|\,dt. \tag{3.59}$$

The same corollary with $p = \infty$ gives for $f \in C$

$$\sup_{\|h\|_1 \le 1, h \perp 1} \int_0^{2\pi} f(t)h(t)\,dt = \frac{1}{2} \max_{t_1, t_2} [f(t_1) - f(t_2)]. \tag{3.60}$$

Now using (3.60), the well-known extremal relationship (see Section

A1), Fubini's theorem for interchanging the order of integration and the continuity of a convolution if one of the functions in it is bounded (see Proposition 1.5.5) we obtain

$$
\begin{aligned}
M_1 &= \sup_{\|\varphi\|_1 \le 1, \varphi \perp 1} \left\{ \sup_{\|\psi\|_\infty \le 1} \int_0^{2\pi} \psi(t) \int_0^{2\pi} K(t-u)\varphi(u)\, du\, dt \right\} \\
&= \sup_{\psi} \left\{ \sup_{\varphi} \int_0^{2\pi} \varphi(u) \int_0^{2\pi} K(t-u)\psi(t)\, dt\, du \right\} \\
&= \sup_{\psi} \left\{ \frac{1}{2} \max_{u_1,u_2} \int_0^{2\pi} [K(t-u_1) - K(t-u_2)]\psi(t)\, dt \right\} \\
&= \frac{1}{2} \max_{u_1,u_2} \left\{ \sup_{\|\psi\|_\infty \le 1} \int_0^{2\pi} [K(t-u_1) - K(t-u_2)]\psi(t)\, dt \right\} \\
&= \frac{1}{2} \max_{u_1,u_2} \int_0^{2\pi} |K(t-u_1) - K(t-u_2)|\, dt \\
&= \frac{1}{2} \max_{u} \int_0^{2\pi} |K(t) - K(t+u)|\, dt. \qquad (3.61)
\end{aligned}
$$

But we have for every number u and c

$$
\frac{1}{2} \int_0^{2\pi} |K(t) - K(t+u)|\, dt \le \int_0^{2\pi} |K(t) + c|\, dt, \qquad (3.62)
$$

which in view of (3.59) proves (3.56).

We have an equality in (3.62) with $c = c_*$ if and only if for some $u = u_0$ (3.57) holds almost everywhere. The theorem is provided. □

Going back to (3.61) and taking into account that (3.58) and (3.59) imply $M_\infty \le \|K\|_1$ we obtain

$$
\begin{aligned}
M_1 &\ge \frac{1}{2} \int_0^{2\pi} |K(t) - K(t+\pi)|\, dt = \int_{-\pi/2}^{\pi/2} |K(t) - K(t+\pi)|\, dt \\
&\ge \int_{-\pi/2}^{\pi/2} [|K(t)| - |K(t+\pi)|]\, dt = \int_{-\pi/2}^{\pi/2} |K(t)|\, dt - \int_{\pi/2}^{3\pi/2} |K(t)|\, dt \\
&= \int_{-\pi}^{\pi} |K(t)|\, dt - 2 \int_{\pi/2 \le |t| \le \pi} |K(t)|\, dt \ge M_\infty - 2 \int_{\pi/2 \le |t| \le \pi} |K(t)|\, dt.
\end{aligned}
$$

Therefore we have in the conditions of Theorem 4.3.8

$$
M_1 = M_\infty - 2\theta \int_{\pi/2 \le |t| \le \pi} |K(t)|\, dt, \qquad (3.63)
$$

where θ $(0 \le \theta \le 1)$ is a constant which depends on the kernel K.

Let us see what these general results give for the problems for approximation by λ-methods. In view of (1.12)

$$e_n(W_p^r, \lambda)_p =: \sup_{f\in W_p^r} \|f - U_n(f, \lambda)\|_p$$

$$= \frac{1}{\pi} \sup_{\|\varphi\|_p\le 1, \varphi\perp 1} \left\| \int_0^{2\pi} \Psi_{n,r}(\gamma, \lambda; \bullet - u)\varphi(u)\, du \right\|_p, \qquad (3.64)$$

where $\Psi_{n,r}(\gamma, \lambda; t)$ is the function defined in (1.11) and the parameter γ which is at our disposition can be assumed equal to zero because $\varphi \perp 1$.

Setting $p = 1$ and $p = \infty$ in (3.64) Theorem 4.3.8 and equality (3.63) allow us to write the following inequalities for every λ-method

$$e_n(W_\infty^r, \lambda)_C - \frac{2}{\pi}\int_{\pi/2\le|t|\le\pi} |\Psi_{n,r}(0, \lambda; t)\ |\, dt \le e_n(W_1^r, \lambda)_1 \le e_n(W_\infty^r, \lambda)_C. \qquad (3.65)$$

Inequalities (3.65) have a special value in the cases when the integral has an order higher than $e_n(W_p^r, \lambda)_C$ for $n \to \infty$. In particular this is the case with the partial sums of Fourier series where

$$\Psi_{n,r}(0, \lambda; t) = \Psi_{n,r}(0, 1; t) = D_{n,r}(t) = \sum_{k=n+1}^{\infty} \frac{\cos(kt - r\pi/2)}{k^r}.$$

Considering (3.8) and (3.10) more carefully we see that

$$\int_{\pi/2\le|t|\le\pi} |D_{n,r}(t)|\, dt = O(n^{-r})$$

and therefore for the partial sums of Fourier series (i.e. $\lambda_k^n = 1$) we have

$$e_n(W_1^r, 1)_1 = e_n(W_\infty^r, 1)_C + 0(n^{-r}).$$

Comparing this result with Theorem 4.3.1 we get

$$\sup_{f\in W_1^r} \|f - S_n(f)\|_1 = \frac{4}{\pi^2}\frac{\ln n}{n^r} + O(n^{-r}), \qquad r = 1, 2, \ldots. \qquad (3.66)$$

Let us note the following general result for positive λ-methods: we have for every λ^+-method

$$e_n(W_1^r, \lambda^+)_1 = e_n(W_\infty^r, 1^+)_C, \qquad r = 1, 2, \ldots. \qquad (3.67)$$

This equality automatically transfers all asymptotic relations for $e_n(W_1^r, \lambda^+)_1$ obtained in Sections 4.3.4 and 4.3.5 to $e_n(W_1^r, \lambda^+)_C$.

4.4 Approximation by algebraic polynomials on an interval

4.4.1 General considerations The exact results in the previous sections have been obtained by virtue of specific properties of the periodic case – the invariance of the functional classes with respect to translations of the argument, $2\pi/n$-periodicity and symmetries of the extremal functions. This specific character is very clear if we consider the 2π-periodic functions as defined on a circumference with unit radius: all points on this circle are in one and the same position and the extremal properties appear uniformly in every point.

The functions from $W_p^m[a,b]$ are determined by their derivative $f^{(m)}$ up to a polynomial of degree $m-1$. This property has two implications when we estimate the approximation error on this class. On one hand for $m \geq 2$ the effective construction of the extremal elements with a different behaviour at the end and in the middle of the interval $[a,b]$ becomes more difficult and even in the best case we are forced to restrict ourselves to asymptotically best results. On the other hand we can sometimes (see Section 4.4.2) improve the uniform estimate near the end-points of the interval substantially when approximating by algebraic polynomials.

For brevity we shall write

$$E_n(f)_{C[a,b]} = E(f, \mathcal{N}_n^A)_{C[a,b]}, \qquad E_n(\mathcal{M})_{C[a,b]} = \sup_{f \in \mathcal{M}} E_n(f)_{C[a,b]}.$$

Some of the asymptotically exact results in the problem of polynomial approximation on a finite interval are obtained by means of a reduction to the trigonometric case. Two well known reductions proved to be effective in some uniform approximation problems. One of them (namely the change of variables $x = \cos t$) was mentioned in Section 2.2.5 and will be considered in more detail later. The second one is based on a transformation to the infinite interval $(-\infty, \infty)$ and approximating there by entire functions of exponential types. Without going into details we just mention that Bernstein [5, 6] obtained the following result in this way.

Let $W^rH^\alpha[-1,1]$ be the class of all functions $f \in C^r[-1,1]$ such that we have for every $x', x'' \in [-1,1]$

$$|f^{(r)}(x') - f^{(r)}(x'')| \leq |x' - x''|^\alpha, \qquad 0 < \alpha \leq 1, \tag{4.1}$$

and let W^rH^α be the class of 2π-periodic functions $f \in C^r$ which satisfies (4.1) on the whole real line. Then

$$\lim_{n\to\infty} n^{r+\alpha} E_n(W^rH^\alpha[-1,1])_{C[-1,1]}$$
$$= \lim_{n\to\infty} n^{r+\alpha} E_n(W^rH^\alpha)_C, \qquad 0<\alpha\leq 1, \qquad r = 0,1,2,\ldots. \tag{4.2}$$

As usual $f^{(0)}$ denotes f and C^0 and $C^0[-1,1]$ stand for C and $C[-1,1]$ respectively.

Because $W^rH^1[-1,1] = W^{r+1}_\infty[-1,1]$ and $W^rH^1 = W^{r+1}_\infty$ relation (4.2) and the Favard result (2.5) give

$$\lim_{n\to\infty} n^r E_n(W^r_\infty[-1,1])_{C[-1,1]} = K_r, \qquad r = 1,2,3,\dots, \tag{4.3}$$

i.e. we have for $n \to \infty$

$$E_n(W^r_\infty[-1,1])_{C[-1,1]} = E_n(W^r_\infty)_C[1+o(1)] = K_r n^{-r} + o(n^{-r}). \tag{4.4}$$

In the case $0 < \alpha < 1$ it became possible to use (4.2) for solving extremal problems for approximation on the interval only after the obtaining of the supremum $E_n(W^rH^\alpha)_C$ in the trigonometric case (see Chapter 7).

4.4.2 An approximation estimate depending on the position of the point in the interval The specific character of the approximation on a finite interval was first demonstrated via reduction to the periodic case – a function $f \in C[-1,1]$ corresponds to a continuous 2π-periodic function $\varphi(t) = f(\cos t)$. In this manner one establishes a one-to-one correspondence between the space $\mathcal{N}^A_n$ of all algebraic polynomials of degree $n-1$ and the space of all even trigonometric polynomials of degree $n-1$.

In view of Proposition 2.1.6 we have

$$E(f, \mathcal{N}^A_n)_{C[-1,1]} = E(\varphi, \mathcal{N}^T_{2n-1})_C. \tag{4.5}$$

If $f \in W^1_\infty[-1,1]$ then $|\varphi'(t)| \le |f'(\cos t)| \le 1$, i.e. $\varphi \in W^1_\infty$ and from (4.5) and (2.5) we obtain

$$E_n(W^1_\infty[-1,1])_{C[-1,1]} \le \frac{\pi}{2n}, \qquad n = 1,2,\dots. \tag{4.6}$$

This is an easy consequence of Theorem 4.2.1 but there is a deeper result connected with this problem.

The change of variables $x = \cos t$ maps the points from the unit circle to the interval $[-1,1]$. A uniform partition of the half-circle $[0,\pi]$ is transformed to a non-uniform partition condensing to the end-points of $[-1,1]$. This geometric consideration explains to some extent one specific feature of the approximations by algebraic polynomials – preserving the best uniform approximation on the whole interval $[-1,1]$ one can essentially improve the error at the ends of the interval.

This effect was discovered in 1946 by Nikol'skii [7] where he showed that for the sequence of Fourier–Chebyshev λ^*-means (2.2.37)

$$V_{n-1}(f,\lambda^*,x) = \frac{c_0}{2} + \sum_{k=1}^{n-1} \frac{k\pi}{2n} \operatorname{ctg} \frac{k\pi}{2n} c_k \cos[k \arccos x] \tag{4.7}$$

the following asymptotic estimate holds uniformly on $x \in [-1, 1]$ for every $f \in W^1_\infty[-1, 1]$

$$|f(x) - V_{n-1}(f, \lambda^*, x)| \le \frac{\pi}{2n}[(1 - x^2)^{1/2} + o(1)]. \tag{4.8}$$

Soon after this it became clear that, with the help of Favard coefficients (1.40), (1.41) for every $r = 1, 2, \ldots$, one can construct a sequence of linear operators with values in $\mathcal{N}_n^A$ which provides a similar effect for the class $W^r_\infty[-1, 1]$. The proof is based on a lemma which should be compared with the following relation from Theorem 4.1.3

$$\frac{1}{\pi}\int_{-\pi}^{\pi} |B_r(t) - \tau_{m,r}(t)|\, dt = K_r\,(m + 1)^{-r}, \qquad m = 0, 1, \ldots, \tag{4.9}$$

Here B_r is the Bernoulli function (1.7) and $\tau_{m,r}$ is the trigonometric polynomial of degree m which interpolates B_r at the zeros of $\cos[(m + 1)t + \pi r/2]$ and which is the polynomial of best $L_1 = L_1[0, 2\pi]$ approximation to B_r.

Lemma 4.4.1 We have for every $\varepsilon \in (0, \pi)$

$$\lim_{n\to\infty} n^r \int_{\varepsilon}^{\pi} |B_r(t) - \tau_{n-1,r}(t)|\, dt = 0, \qquad r = 1, 2, \ldots.$$

Proof For definiteness let r be odd (the arguments are similar for even rs). Then $B_r(t) - \tau_{n-1,r}(t)$ is an odd function and

$$\operatorname{sgn}[B_r(t) - \tau_{n-1,r}(t)] = \pm\operatorname{sgn}\sin nt, \qquad n = 1, 2, \ldots. \tag{4.10}$$

In view of Proposition 4.1.2 and in view of (4.10) we have for every $\nu = 1, 2, \ldots,$

$$\left|\int_{-\pi}^{\pi} [B_r(t) - \tau_{n-1,r}(t)] \operatorname{sgn}\sin(n + \nu)t\, dt\right| = \int_{-\pi}^{\pi} |B_r(t) - \tau_{n-1+\nu,r}(t)|\, dt$$

and we obtain from the evenness of the integrands and (4.9)

$$\left|\int_0^{\pi} [B_r(t) - \tau_{n-1,r}(t)][\operatorname{sgn}\sin nt - \operatorname{sgn}\sin(n + \nu)t]\, dt\right|$$

$$= \frac{1}{2}\int_{-\pi}^{\pi} |B_r(t) - \tau_{n-1,r}(t)|\, dt - \frac{1}{2}\int_{-\pi}^{\pi} |B_r(t) - \tau_{n-1+\nu,r}(t)|\, dt$$

$$= \frac{\pi}{2} K_r \frac{r\nu}{n^{r+1}} + O(n^{-r-2}). \tag{4.11}$$

Now let us fix ε $(0 < \varepsilon < \pi)$ and set $N = [2\pi/\varepsilon] + 1$. The following considerations are based on the observation that: for every natural n, $n \geq N$, and for every $t \in [\varepsilon, \pi)$ $(t \neq k\pi/n)$ there is a natural $j \leq N$ such that

$$\operatorname{sgn} \sin nt = -\operatorname{sgn} \sin (n+j)t. \tag{4.12}$$

Indeed, let us write nt as

$$nt = k\pi + s, \tag{4.13}$$

with k natural and $0 \leq s < \pi$. For any j satisfying the equalities

$$(k+1)\pi < (n+j)t \leq (k+2)\pi \tag{4.14}$$

(4.12) is obviously fulfilled and we only have to see that there is such a j among the numbers $1, 2, \ldots, N$. But this becomes clear if, with the help of (4.13), we rewrite (4.14) as

$$\frac{\pi - s}{t} < j < \frac{2\pi - s}{t}$$

and if we remember that $0 \leq s < \pi$, $N = [2\pi/\varepsilon] + 1$, $\varepsilon \leq t < \pi$.

Denote by M_j^n the set of points $t \in [0, \pi]$ for which (4.12) holds $(j = 1, 2, \ldots, N)$. Then all but a finite number of points from $[0, \pi]$ belong to at least one of the sets $M_1^n, M_2^n, \ldots, M_N^n$. Then, using (4.11), (4.12) we can write

$$\begin{aligned}
&\int_\varepsilon^\pi |B_r(t) - \tau_{n-1,r}(t)|\,\mathrm{d}t \\
&\quad \leq \frac{1}{2}\sum_{j=1}^N \left| \int_{M_j^n} [B_r(t) - \tau_{n-1,r}(t)][\operatorname{sgn} \sin nt - \operatorname{sgn} \sin (n+j)t]\,\mathrm{d}t \right| \\
&\quad \leq \frac{1}{2}\sum_{j=1}^N \left| \int_0^\pi [B_r(t) - \tau_{n-1,r}(t)][\operatorname{sgn} \sin nt - \operatorname{sgn} \sin (n+j)t]\,\mathrm{d}t \right| \\
&\quad \leq \frac{1}{2}\sum_{j=1}^N \left[\frac{\pi}{2} K_r \frac{rj}{n^{r+1}} + O(n^{-r-2}) \right] \leq \frac{\pi}{2} K_r \frac{rN^2}{n^{r+1}} + O(Nn^{-r-2}),
\end{aligned}$$

which proves the lemma because N does not depend on n. □

Theorem 4.4.2 For every $f \in W_\infty^r[-1, 1]$ $(r = 1, 2, \ldots,)$ there is a sequence of algebraic polynomials $p_{n-1,r}(f, x)$ of degree $n - 1$ $(n = 1, 2, \ldots)$ linearly dependent on f such that we have uniformly on $x \in [-1, 1]$

$$\overline{\lim_{n\to\infty}}\, n^r |f(x) - p_{n-1,r}(f, x)| \leq K_r (1 - x^2)^{r/2}. \tag{4.15}$$

The (Favard) constant K_r cannot be decreased.

Proof First we shall consider the case $r = 1$ in detail where the essence is not hidden in technical details and we shall prove (4.8).

Let $f \in W^1_\infty[-1, 1]$, $\varphi(t) = f(\cos t)$. Then

$$\varphi(t) = \frac{a_0}{2} + \sum_{k=1}^{\infty} a_k \cos kt = \frac{a_0}{2} + \frac{1}{\pi}\int_0^{2\pi} B_1(u)\varphi'(t-u)\,du$$

and we have for the coefficients (1.42) (see (1.1), (1.45))

$$U_{n-1,1}(\varphi, \lambda^*, t) = \frac{a_0}{2} + \sum_{k=1}^{n-1} \frac{k\pi}{2n} \operatorname{ctg} \frac{k\pi}{2n} a_k \cos kt$$

$$= \frac{a_0}{2} + \frac{1}{\pi}\int_0^{2\pi} \tau_{n-1,1}(u)\varphi'(t-u)\,du.$$

Therefore

$$\varphi(t) - U_{n-1,1}(\varphi, \lambda^*, t) = \frac{1}{\pi}\int_0^{2\pi} [B_1(u) - \tau_{n-1,1}(u)]\varphi'(t-u)\,du.$$

From $|f'(x)| \le 1$ for $|x| \le 1$ we have

$$|\varphi'(t-u)| = |f'(\cos(t-u))\sin(t-u)| \le |\sin t| + |\sin u|$$

and using (4.9) and $K_1 = \pi/2$ we obtain

$$|\varphi(t) - U_{n-1,1}(\varphi, \lambda^*, t)| \le \frac{\pi}{2n}|\sin t| + I_n,$$

where

$$I_n = \frac{2}{\pi}\int_0^{\pi} |B_1(u) - \tau_{n-1,1}(u)| \sin u\,du. \tag{4.16}$$

We shall show that $nI_n \to 0$ for $n \to \infty$. Given an ε $(0 < \varepsilon < \pi/2)$ we have using once more (4.9)

$$nI_n = n\left(\frac{2}{\pi}\int_0^{\varepsilon} |B_1(u) - \tau_{n-1,1}(u)| \sin u\,du\right.$$

$$\left. + \frac{2}{\pi}\int_{\varepsilon}^{\pi} |B_1(u) - \tau_{n-1,1}(u)| \sin u\,du\right)$$

$$\le \varepsilon K_1 + n\frac{2}{\pi}\int_{\varepsilon}^{\pi} |B_1(u) - \tau_{n-1,1}(u)|\,du$$

and the last summand will be less than ε for $n \ge n_\varepsilon$ in view of Lemma 4.4.1. Therefore $nI_n < \varepsilon(\pi/2 + 1)$ for $n > n_\varepsilon$, i.e. $I_n = o(n^{-1})$ uniformly on t because of the arbitrariness of ε. Hence

$$|\varphi(t) - U_{n-1,1}(f, \lambda^*, t)| \le \frac{\pi}{2n}|\sin t| + o(n^{-1}). \tag{4.17}$$

Letting $t = \arccos x$ in (4.17) we get (4.8), where $V_{n-1}(f, \lambda^*, x)$ are the algebraic polynomials (4.7) and (4.8) holds uniformly for $x \in [-1, 1]$.

We shall sketch the proof for $r \ge 2$. Let $f \in W^r_\infty[-1, 1]$. Without loss of generality we assume that $f^{(\nu)}(0) = 0$ $(\nu = 1, 2, \dots, r-1)$. If $\varphi(t) = f(\cos t)$ then

$$\varphi^{(r)}(t) = (-1)^r f^{(r)}(\cos t)\sin^r t + \psi_r(t), \tag{4.18}$$

where

$$\psi_r(t) = \sum_{\nu=0}^{r-1} q_\nu(t) f^{(\nu)}(\cos t) \tag{4.19}$$

and q_ν are some trigonometric polynomials. Only the derivatives of f of order less than r are included in the sum (4.19) and hence ψ_r' exists a.e. and it is bounded. It is also essential that function ψ_r' is even for odd r and odd for even r.

Using the last observation and using the form of $\tau_{m,r}$ we easily see that the trigonometric polynomial

$$\frac{a_0}{2} + \frac{(-1)^r}{\pi}\int_0^{2\pi} \tau_{m,r}(t-u) f^{(r)}(\cos u)\sin^r u\, du + \frac{1}{\pi}\int_0^{2\pi} \tau_{m,r+1}(t-u)\psi_r'(u)\, du$$

of degree m which depends linearly on f is an even function for every $r = 1, 2, \dots$, and hence it can be represented as $p_{m,r}(\cos t)$, where $p_{m,r}(f, x)$ is an algebraic polynomial of degree m.

Recalling (4.18) we write φ as

$$\varphi(t) = f(\cos t) = \frac{a_0}{2} + \frac{(-1)^r}{\pi}\int_0^{2\pi} B_r(u) f^{(r)}(\cos(t-u))\sin^r(t-u)\, du$$
$$+ \frac{1}{\pi}\int_0^{2\pi} B_{r+1}(u)\psi_r'(t-u)\, du.$$

Therefore

$$f(\cos t) - p_{n-1,r}(f, \cos t)$$
$$= \frac{(-1)^r}{\pi}\int_0^{2\pi} [B_r(u) - \tau_{n-1,r}(u)] f^{(r)}(\cos(t-u))\sin^r(t-u)\, du$$
$$+ \frac{1}{\pi}\int_0^{2\pi} [B_{r+1}(u) - \tau_{n-1,r+1}(u)]\psi_r'(t-u)\, du =: A_n + A_n'.$$

Because $|\psi_r'(t)| \le M_r$ we get from (4.9) $|A_n'| \le M_r\, K_{r+1}\, n^{-r-1}$. For the other summand using $|f^{(r)}(x)| \le 1$ we can write

$$|A_n| \le \frac{|\sin^r t|}{\pi} \int_0^{2\pi} |B_r(u) - \tau_{n-1,r}(u)|\, du$$

$$+ \frac{2}{\pi} \int_0^{\pi} |B_r(u) - \tau_{n-1,r}(u)| \sum_{\nu=1}^{r} \binom{r}{\nu} |\sin u|^{r-\nu}\, du$$

$$= K_r n^{-r} |\sin^r t| + A_n''.$$

With the help of Lemma 4.4.1 we establish (see the estimate of integral (4.16) above) that $A_n'' = o(n^{-r})$ uniformly on t and hence

$$|f(\cos t) - p_{n-1,r}(f, \cos t)| \le K_r n^{-r} |\sin^r t| + o(n^{-r}).$$

Setting $\cos t = x$ we get the first part of the theorem. It follows from (4.3) that we cannot improve the constant K_r in (4.15). □

Comparing (4.15) and (4.3) we observe

(1) The polynomials $p_{n-1,r}(f, x)$ $(n = 2, 3, \ldots)$ defined for every $r = 1, 2, \ldots$, in uniform manner (coinciding with (4.7) for $r = 1$) form a sequence of asymptotically best methods of approximation to $f \in W_\infty^r[-1, 1]$ in $C[-1, 1]$.
(2) At every point $x \in [-1, 1]$ $(x \ne 0)$ we have for large enough ns $(n \ge N(x))$ and for every $f \in W_\infty^r[-1, 1]$

$$|f(x) - p_{n-1,r}(f, x)| < E_n(f, W_\infty^r[-1, 1])_{C[-1,1]}.$$

Finally let us remark that the approximation improvement of magnitude $O(n^{-r})$ of the approximation at the ends of the interval $[-1, 1]$ which polynomials $p_{n-1,r}(f, x)$ supply for every $f \in W_\infty^r[-1, 1]$ is at the expense of an increase in the error in a neighbourhood of the origin, but the increase is $o(n^{-r})$.

In comparison with relation (4.3) estimate (4.15) in Theorem 4.4.2 can be written as

$$|f(x) - p_{n-1,r}(f, x)| \le E_n(f, W_\infty^r[-1, 1])_{C[-1,1]}[(1 - x^2)^{r/2} + o(1)]. \quad (4.20)$$

4.4.3 On the extremal functions Let us go back to the asymptotic relation (4.3). It can be obtained independently on (4.2) if, using (4.15), we provide for every $n = 1, 2, \ldots$, a function $f_n \in W_\infty^r[-1, 1]$ such that $E_n(f_n)_{C[-1,1]} \ge K_r n^{-r} - \gamma_{nr}$, where $\gamma_{nr} = o(n^{-r})$ for $n \to \infty$. Unlike the trigonometric case, where the extremal functions $\varphi_{n,r}$ for the class W_∞^r were easily found because their polynomial of best approximation was the constant zero, obtaining extremals (even asymptotic ones) in the algebraic case is connected with overcoming some essential difficulties.

We shall give here the interesting construction of Nikol'skii [7] for the case $r = 1$ based on the interpolating polynomials on Chebyshev knots. Let n be even: $n = 2m$. For $x = 0$ the polynomial $\bar{A}_{n-1}(f, x)$ (see (2.2.39)) takes the value

$$\bar{A}_{n-1}(f, 0) = \frac{1}{n} \sum_{k=1}^{n} (-1)^{k+m} f(x_k) \tan t_k,$$

where $t_k = t_k^n = (2k-1)\pi/(2n)$, $x_k = x_k^n = \cos t_k$. Evaluating the norm of the functional $\bar{A}_{n-1}(f, 0)$ we obtain

$$M_n(0) =: \sup_{|f(x)| \le 1} |\bar{A}_{n-1}(f, 0)| = \frac{1}{n} \sum_{k=1}^{n} |\tan t_k|$$

$$= \frac{2}{\pi} \int_0^{t_m} \tan t \, \mathrm{d}t + O(1) = \frac{2}{\pi} \ln n + O(1). \tag{4.21}$$

If $f \in C[-1, 1]$ and if p is its polynomial of best approximation in $\mathcal{N}_n^A$ then

$$|\bar{A}_{n-1}(f - p, 0)| \le M_n(0) \, \|f - p\|_{C[-1,1]} = M_n(0) \, E_n(f)_{C[-1,1]},$$

and because $\bar{A}_{n-1}(f, x)$ is a linear projector on $\mathcal{N}_n^A$ we obtain

$$|\bar{A}_{n-1}(f, 0) - f(0)| \le |\bar{A}_{n-1}(f - p, 0) - [f(0) - p(0)]|$$

$$\le [1 + M_n(0)] \, E_n(f)_{C[-1,1]}. \tag{4.22}$$

For every $n = 1, 2, \ldots,$ we construct a continuous function f_n which is even on $[-1, 1]$ and which is linear between the points x_k and

$$f(0) = 0, \qquad f_n(x_k) = (-1)^{m+k} \tan \frac{\pi}{2n} \sin t_k, \qquad k = 0, 1, \ldots, m.$$

Then inequality (4.22) gives

$$|\bar{A}_{n-1}(f_n, 0) - f_n(0)| = \frac{2}{n} \tan \frac{\pi}{2n} \sum_{k=1}^{m} \frac{\sin^2 t_k}{\cos t_k}$$

$$= \frac{2}{\pi} \tan \frac{\pi}{2n} \sum_{k=1}^{m} \frac{1}{\cos t_k} + O(n^{-1})$$

$$= \frac{2}{n} \sum_{k=1}^{m} \frac{1}{2(m-k)+1} + O(n^{-1}) = \frac{\ln n}{n} + O(n^{-1}). \tag{4.23}$$

Now from (4.21), (4.22) and (4.23) we obtain

$$E_n(f)_{C[-1,1]} \geq \frac{|\bar{A}_{n-1}(f_n, 0) - f_n(0)|}{1 + M_n(0)} = \frac{\pi}{2n} + O\left(\frac{1}{n \ln n}\right), \qquad (4.24)$$

where as can be easily verified, $f \in W^r_\infty[-1, 1]$ ($n = 1, 2, \ldots,$).

For $n = 2m + 1$ the arguments are similar but we evaluate the norm of the functional $\bar{A}_{n-1}(f_n, x_*)$, where $x_* = x^n{}_* = \cos[(n-1)\pi/(2n)]$. We construct a polygon f_n such that

$$|\bar{A}_{n-1}(f_n, x_*) - f_n(x_*)| = \frac{\ln n}{n} + O(n^{-1}),$$

and which implies estimate (4.24) with x_* instead of 0.

Therefore for every $n = 1, 2, \ldots,$ there is a function f_n in $W^r_\infty[-1, 1]$ such that

$$E_n(f_n)_{C[-1,1]} \geq \frac{\pi}{2n} + O\left(\frac{1}{n \ln n}\right).$$

This fact, together with (4.6), gives

$$E_n(W^1_\infty[-1, 1])_{C[-1,1]} = \frac{\pi}{2n} - \varepsilon_n, \qquad \varepsilon_n \geq 0, \qquad \varepsilon_n = O\left(\frac{1}{n \ln n}\right),$$

which is an improvement of (4.6) for $r = 1$.

The functions f_n depend heavily on n and so it is interesting to know whether there is a function $f \in W^1_\infty[-1, 1]$ such that

$$\overline{\lim_{n\to\infty}}\, n\, E_n(f)_{C[-1,1]} = \pi/2. \qquad (4.25)$$

4.4.4 Approximation by Fourier–Chebyshev sums Let us note another exact result which takes the position of the point in the interval into account. We shall consider the approximation of $W^1_\infty[-1, 1] = H^1[-1, 1]$ by the polynomials (see Section 2.2.5)

$$Q_n(f, x) = \frac{c_0}{2} + \sum_{k=1}^{n} c_k \cos(k \arccos x),$$

$$c_k = \frac{2}{\pi} \int_{-1}^{1} f(x) \cos(k \arccos x)\, (1 - x^2)^{-1/2}\, dx. \qquad (4.26)$$

Theorem 4.4.3 The asymptotic equality

$$\sup_{f \in H^1[-1,1]} |f(x) - Q_{n-1}(f, x)| = \frac{4}{\pi^2} \frac{\ln n}{n} (1 - x^2)^{1/2} + O(n^{-1})$$

holds uniformly on $x \in [-1, 1]$.

We give only a sketch of the proof because all the arguments have been applied earlier in this chapter.

Let $f \in H^1[-1, 1]$, $\varphi(t) = f(\cos t)$. The change of variables $x = \cos t$ transfers $Q_n(f, x)$ into the partial sum $S_n(\varphi, t)$ of a Fourier series of φ and the class $H^1[-1, 1]$ into the class H_s^1 of all even 2π-periodic functions which are integrals of functions of the type $\psi(t) \sin t$, ψ even, $|\psi(t)| \leq 1$. Hence

$$\sup_{f \in H^1[-1,1]} |f(x) - Q_{n-1}(f, x)| = \sup_{\varphi \in H_s^1} |\varphi(t) - S_{n-1}(\varphi, t)|. \tag{4.27}$$

We have in view of (3.3)

$$S_{n-1}(\varphi, t) - \varphi(t) = \frac{1}{\pi} \int_0^{2\pi} D_{n-1,1}(u)\, \psi(t+u) \sin(t+u)\, du, \tag{4.28}$$

where

$$D_{m,1}(u) = \sum_{k=m+1}^{\infty} \frac{\sin kt}{k}. \tag{4.29}$$

Both φ and $S_{n-1}(\varphi)$ are even functions so we shall consider (4.28) only for $0 \leq t \leq \pi$. In view of (3.10) we can represent $D_{m,1}$ as

$$D_{m,1}(u) = \frac{1}{m+1} \frac{\cos[(2m+1)u/2]}{2 \sin(u/2)} + G_m(u),$$

where

$$\int_{-1/m}^{1/m} |D_{m,1}(u)|\, du = O\left(\frac{1}{m}\right), \qquad \int_{1/m}^{2\pi - 1/m} |G_m(u)|\, du = O\left(\frac{1}{m}\right).$$

Using these relations, the inequality $|\psi(t)| \leq 1$ and $\sin(t+u) = \sin t + O(u)$ for $u \to 0$ we obtain

$$S_{n-1}(\varphi, t) - \varphi(t) = \frac{1}{n} I_n(\psi, t) + O\left(\frac{1}{n}\right), \tag{4.30}$$

where

$$I_n(\psi, t) = \frac{\sin t}{\pi} \int_{1/n}^{2\pi - 1/n} \frac{\cos[(2n-1)u/2]}{2 \sin(u/2)} \psi(t+u)\, du. \tag{4.31}$$

For $0 \leq t \leq 1/n$ or $\pi - 1/n \leq t \leq \pi$ we obviously have $I_n(\psi, t) = O(1)$. Let $1/n \leq t \leq 1 - 1/n$. On the intervals $[\pi - t, \pi]$ and $[\pi, \pi + t]$ the integrand in (4.31) is bounded and hence (see the proof of Theorem 4.3.1)

$$|I_n(\psi, t)| \leq \frac{|\sin t|}{\pi} \left(\int_{-t}^{-1/n} \left| \frac{\cos[(2n-1)u/2]}{2 \sin(u/2)} \right| du \right.$$

$$\left. + \int_{1/n}^{\pi - t} \left| \frac{\cos[(2n-1)u/2]}{2 \sin(u/2)} \right| du \right) + O(1)$$

$$= \frac{2}{\pi^2} |\sin t| \, [\ln (nt) + \ln (\pi n - nt)] + O(1) = \frac{4}{\pi^2} |\sin t| \ln n + O(1). \tag{4.32}$$

Now we have to compare (4.32), (4.30) and (4.27) and also to remark that (4.32) holds as an equality if we set $\psi(u) = \psi_{n,t}(u)$ in (4.31), where $\psi_{n,t}$ is an even 2π-periodic function of u such that $|\psi_{n,t}(u)| \leq 1$ and such that

$$\psi_{n,t}(t+u) = \operatorname{sgn}\left(\sin t \frac{\cos [(2n-1)u/2]}{2 \sin (u/2)}\right), \qquad -t < u < \pi - t.$$

Finally let us remark that the functions $\psi_{n,t}(u)$ realizing the asymptotic equality in (4.32) depend on n and hence the following result is of importance: For every $x \in (-1, 1)$ there is $f \in H^1[-1, 1]$ depending on x but not on n such that

$$|f(x) - Q_{n-1}(f, x)| > \frac{4}{\pi^2} \frac{\ln n}{n} (1 - x^2)^{1/2} (1 - \varepsilon_n), \tag{4.33}$$

where $\varepsilon_n \to 0$ for some subsequence of values of n.

Comments

Sections 4.1 and 4.2 The problem of the best λ-method has been formulated by Favard [2, 3]. Starting from his paper [1], where in particular Theorem 4.1.3 is proved, Favard in [2, 3] establishes the validity of relations (2.5) of Theorem 4.2.1. The same relations have also been proved by Akhiezer and Krein [1] who also started with the Favard paper [1]. All these authors have also treated the conjugate case in C (see Theorem 4.2.6). These results were immediately followed by the fundamental paper of Szokefalvy-Nagy [1] in which the problem of the best uniform approximation of classes of periodic convolutions was studied and where more general linear means of the partial sums of Fourier series than λ-means (1.1) were used (see [9A]).

The first equality in (2.11) has been proved by Bohr [1] for $r = 1$ and Favard [1] for $r \geq 2$. The first exact result for the best one-sided approximation of classes of trigonometric functions is due to Ganelius [1].

Essentially new ideas in this field have been brought by the profound work of S. M. Nikol'skii [6] who applied duality relations for obtaining fundamental results for convolution classes of the general type. In particular he proved the validity of (2.6) and of (2.7) in Theorem 4.2.1, (2.12) in Theorem 4.2.3 and also the L_1 results in Theorem 4.2.6. Theorem 4.2.4 was proved by Taikov [2]; for $p = \infty$ it was established using other methods by Turovets [1].

There are several stages in solving the problem of the best approximation of classes W_p^r $(p = 1, \infty)$ with fractional r. The first exact result was obtained by Dzyadyk [2] for $0 < r < 1$. After that Stechkin [2] proved a similar fact for classes $W_\infty^{r,\beta}$ for $0 < r < 1$, $r < \beta < 2 - r$. For arbitrary $r > 0$, $\beta \in \mathbb{R}$ the problem was solved by Sun Yong-sheng [1] and Dzyadyk [3, 4]; the last paper contains the most general result.

Beginning with the works of Freud [1] and Ganelius [1] there are many exact estimates for the best one-sided approximation (in the L_1 metric) of classes of functions by polynomials; see V. G. Doronin [1], V. G. Doronin and Ligun [5] and also monograph [11A], where one can find a full citation of these problems.

Some generalizations of the results on best and best one-sided approximations by trigonometric polynomials of classes W_p^r to classes of convolutions with the functions from the unit ball of L_p are contained in the articles of Pinkus [1], Sun Yong-sheng and Hyang Daren [1] and V. F. Babenko [13].

Section 4.3 Kolmogorov's result (3.1) is a starting point for a large set of papers on approximation of classes of functions by Fourier means. It has been established that the asymptotic result (3.1) for a uniform metric remains valid if class W_∞^r ($r = 1, 2, \ldots,$) is replaced by W_∞^r with fractional $r > 0$ (Pinkevich [1]), by the classes $\tilde{W}_\infty^r$ of conjugate functions (S. M. Nikol'skii [5]) and finally by $W_\infty^{r,\beta}$ (Efimov [3] and Telyakovskii [1]). The refinement of the remainder in (3.1) was started by Sokolov [1]; the most exhaustive result in this direction is obtained by Telyakovskii [2], but for $r \to \infty$ a more precise relation is due to Stechkin [7] (see Exercise 4.3). Theorem 4.3.2 is due to Dzyadyk and Stepanets [1]; relation (3.31) was obtained by the author.

The bases of the linear positive operators (in particular λ^+-methods) are given in Korovkin's monograph [12A] (see also the survey of Garkavi [3]). Estimate (3.34) for the coefficients of λ^+-methods is from Egervary and Szasz [1]. Theorem 4.3.6 and relation (3.49) for $r = 2$ are proved by Korovkin [1], for $r \geq 3$ this relation is obtained by Davidchik [1, 2]. Asymptotically exact results for approximation by positive methods on similar functional classes can also be found in Petrov's work [1]. The asymptotic results in Proposition 4.3.7 are due to S. M. Nikol'skii [5]; the remaining ones have been refined by Lorch [1] and Telyakovskii [3].

After S. M. Nikol'skii [6] the relationship between the quantities (3.64) for $p = \infty$ and $p = 1$ were studied by Stechkin and Telyakovskii [1], who obtained (3.65) and gave an example when there is no asymptotic coincidence between these quantities. (3.67) is established by Motornyi [2]. For one-sided analogs see [11A].

Section 4.4 Lemma 4.4.1 has been formulated by S. M. Nikol'skii and the proof given in this book is due to Dzyadyk [1]. Theorem 4.4.2 for $r = 2, 3, \ldots,$ was proved by Timan [1].

(4.25) was proved by S. M. Nikol'skii [7]; in the same paper the proof of Theorem 4.4.3 and the construction of a function satisfying (4.33) can be found.

The best (α, β)-approximations and (as a consequence) the best one-sided approximations of classes $W_1^r[a, b]$ by algebraic polynomials in the $L_1[a, b]$ metric are studied in the paper by V. F. Babenko and Kofanov [1].

Other results and exercises

1. Let A_p^h be the class of functions $f \in C$ which can be analytically extended to the strip $\{z = t + iu: -h < u < h\}$ and $\|\mathrm{Re}\, f(\bullet + iu)\|_p \leq 1$ $(|u| < h)$. Then we have for $p = 1, \infty$

$$E_n(A_p^h)_p = \frac{4}{\pi} \sum_{\nu=0}^{\infty} \frac{(-1)^\nu}{(2\nu+1)\,\mathrm{ch}\,(2\nu+1)h}$$

(Akhiezer [1] ($p = \infty$), S. M. Nikol'skii [6] ($p = 1$), see also [2A, 11A, 25A]).

2. Let Γ_p^ρ be the class of functions $f \in C$ represented in the form $f = c + \chi_\rho * \varphi$, where $c \in \mathbb{R}$, $\|\varphi\|_p \leq 1$ and

$$\chi_\rho(t) = \frac{1}{2} + \sum_{k=1}^{\infty} \rho^k \cos kt = \frac{1-\rho}{2(1 - 2\rho\cos t + \rho^2)}, \qquad 0 < \rho < 1.$$

Then we have for $p = 1, \infty$

$$E_n(\Gamma_p^\rho)_p = \frac{4}{\pi}\,\mathrm{arctg}\,\rho^n, \qquad 0 < \rho < 1$$

(Krein [1], Szokefalvy-Nagy [1] ($p = 1$), see also [2A, 11A, 25A]). On the conjugate class to Γ_p^ρ a similar result is obtained by Szokefalvy-Nagy [1] and S. M. Nikol'skii [1].

3. For the approximations by partial sums of Fourier series we have

$$S_n(W_\infty^{r,\beta})_C = n^{-r}\left(\frac{4}{\pi^2}\ln\frac{n+r}{r+1} + \frac{2}{\pi r}\left|\sin\frac{\beta\pi}{2}\right| + O(1)\right)$$

uniformly with respect to $n = 1, 2, \ldots$, and $r > 0, \beta \in \mathbb{R}$ (Telyakovskii [2]) and also the following relation holds uniformly with respect of $n \geq 1, r \geq 1, \beta \in \mathbb{R}$

$$S_n(W_\infty^{r,\beta})_C = n^{-r}\left(\frac{8}{\pi^2}K(\exp(-r/n)) + O(r^{-1})\right),$$

where K is the elliptic integral of the first kind (Stechkin [7]).

4. If $f \in V^{r+1}[-1, 1]$ then

$$\lim_{n\to\infty} n^r E_n(f)_{C[-1,1]} = \frac{\mu(r)}{2r} \max_{-1<x<1} |f^{(r)}(x+0) - f^{(r)}(x-0)|\,(1-x^2)^{r/2},$$

where $\mu(r) = \lim_{n\to\infty} n^r E_n(x^{r-1}|x|)_{C[-1,1]}$ (Nikol'skii [10]). The constant $\mu(r)$ is introduced by Bernstein [3A, vol. 2, p. 262].

5. If A_n are linear operators from C to $\mathcal{N}_{2n-1}^T$ then

$$\lim_{n\to\infty} n^r\{\inf_{A_n}\sup_{f\in W_1^{r+1}} \|f - A_n f\|_C\} = \frac{\mu(r)}{2r!}, \qquad r = 0, 1, \ldots,$$

where $\mu(r)$ is the constant from 4 (V. F. Babenko and Pichugov [2]).

6. For functions $f \in L_\infty^r$ $(r = 2, 3, \ldots,)$ we have the exact inequalities

$$\left\{\frac{E_n(f^{(r-m)})_\infty}{E_n(f^{(r)})_\infty K_m}\right\}^{1/m} \leq \left\{\frac{E_n(f)_C}{E_n(f^{(r)})_\infty K_r}\right\}^{1/r}, \qquad m = 1, \ldots, r$$

(Zhuk [3] $(2 \leq r \leq 5)$ and Ligun [1]).

7. If A_n are linear operators from C to $\mathcal{N}_{2n-1}^T$ then we have for $n, r = 1, 2, \ldots,$

$$\inf_{A_n} \sup_{f \in W_p^r} \|f - A_n f\|_\infty = \frac{1}{\pi} E_n(B_r)_{p'}, \qquad 1 \le p \le \infty, \qquad \frac{1}{p} + \frac{1}{p'} = 1$$

(V. F. Babenko and Pichugov [1]).

8. Let A_n be linear operators from L_q to $\mathcal{N}_{2n-1}^T$. Then we have for $r = 1, 2, \ldots, 1 \le q \le \infty$

$$\inf_{A_n} \sup_{f \in W_2^r} E_1(f - A_n f)_q = \sup_{f \in W_2^r} E_1(f - S_n(f))_q = \sup_{g \in W_{q'}^r} \|g - S_n(g)\|_2$$

$$\frac{1}{q} + \frac{1}{q'} = 1$$

(V. F. Babenko and Pichugov [1]).

9. Let $\tilde{W}_\infty^r$ be the class of functions trigonometrically conjugated with $f \in W_\infty^r$. If $\alpha = \beta$ and $r = 1, 2, \ldots,$ or $\alpha \ne \beta$ and r is even then

$$E_n(\tilde{W}_\infty^r)_{1;\alpha,\beta} = \|\tilde{\varphi}_{n,r}(\alpha, \beta; \bullet)\|_1, \qquad n = 1, 2, \ldots,$$

(V. F. Babenko [5]).

10. For $f \in L_1$ set $\Pi(f, t) =: r(f_+, t) - r(f_-, 2\pi - t)$. If $f \in L_1^r$ then we have for all $n, r = 1, 2, \ldots,$ and $\alpha, \beta > 0$

$$E_n(f)_{1;\alpha,\beta} \le \int_0^{2\pi} \Pi(\varphi_{n,r}(\alpha, \beta; \bullet), t)\, \Pi(f^{(r)}, t)\, dt;$$

the inequality is exact (V. F. Babenko [5]).

11. The set $F \subset L_1$ is called Π-invariant if $f \in F$ and $\Pi(g, t) \equiv \Pi(f, t)$ (see 10) implies $g \in F$. For a given Π-invariant set $F \subset L_1$ set $W^r F = \{f; f \in L_1^r, f^{(r)} \in F\}$. Then

$$E_n(W^r F)_{1;\alpha,\beta} = \sup_{g \in F, g \perp 1} \int_0^{2\pi} \Pi(\varphi_{n,r}(\alpha, \beta; \bullet), t)\, \Pi(g, t)\, dt;$$

(V. F. Babenko [5]). In papers [8, 13] by the same author the results from 10 and 11 are extended to convolution classes with a non-increasing number of sign-changing kernels.

12. Using Theorem 4.4.2 prove that for every $f \in W_\infty^r[-1, 1]$ there is a sequence of polynomials $p_{n,r}(f, t)$ $(n = 1, 2, \ldots,)$ of degree $n - 1$ such that we have uniformly on $t \in [-1, 1]$

$$0 \le f(t) - p_{n,r}(f, t) \le 2 K_r n^{-r} (1 - t^2)^{r/2} + o(n^{-r}), \qquad n \to \infty$$

and the constant $2K_r$ cannot be improved.

13. Obtain an analog of Theorem 4.1.4 for conjugate functions [26A, p. 331].

14. Show that in (2.12) the class W_1^r can be replaced by W_V^{r-1}.

15. Prove that for every $n = 1, 2, \ldots,$ there is the best λ-method (Section 4.2.1) for the class $W_p^r (1 \le p \le \infty)$ when estimating the error in $L_q (1 \le q \le \infty)$ [9A, p. 87].

16. (Added by the translator.) For every $f \in W_p^1$ we have

$$(E_n^+(f)_p^p + E_n^-(f)_p^p)^{1/p} \le \frac{2\pi}{n} E_n(f')_p$$

(cf. Theorem 4.1.4) (Sendov and Popov [1, p. 243]). The inequality is exact (see e.g. [11A, p. 84]).

5

Spline approximation of classes of functions with a bounded rth derivative

In this chapter we consider polynomial spline approximation. Approximation by splines entered the theory relatively late and immediately acquired a large following; in particular because of their advantages over classical polynomials in the problem of the interpolation of functions. It turns out that, in addition to computational advantage due to the existence of bases with local supports, interpolating splines realize the minimal (for a fixed dimension) deviation for many classes of function. Interpolating polynomials do not even give the best order.

The results in this chapter will in general be connected with two aspects of spline approximation of classes of functions with the rth derivative bounded in L_p: (1) the best approximation by splines with minimal defects; (2) spline interpolation. In the first place we consider those situations when the best approximation is realized by interpolating splines and here it turns out that elementary analysis tools are sufficient if the specific properties of the polynomial splines are used. In Section 5.1 we give some basic broad-based facts to which many problems in spline interpolation can be reduced.

The methods of obtaining exact results in the estimation of the error in best approximation by splines with minimal defects (Section 5.4) are based on essentially different ideas – application of duality relations. These methods allow us to obtain solutions in situations in which the exact estimates for spline interpolation are not known.

Let us note that a better understanding of the significance of the results in Sections 5.2–5.4 will follow from the calculation of N-widths in Chapter 8.

5.1 Inequalities for functions with prescribed zeros

5.1.1 Functional classes connected with a partition In order to estimate the spline approximation error in the functional class $\mathcal{M}$ we investigate the difference $\delta(f, t) = f(t) - s(f, t)$ where $f \in \mathcal{M}$ and $s(f, t)$ is a spline with respect to a fixed partition $\Delta_N[a, b]$ interpolating the function f (and maybe some of its derivatives) at given points τ_k. Therefore $\delta(f, \tau_k) = 0$ and the problem of identifying the extremal properties of a new class $\mathcal{M}_1$ of functions containing $\delta(f, t)$ with zeros at the points τ_k arises. The class $\mathcal{M}_1$ contains, in particular, the functions $f \in \mathcal{M}$ for which $s(f, t) \equiv 0$ and it was found that in many important cases these are the functions which realize the supremum of the spline approximation error. Using the specifics of the splines we can sometimes appropriately describe the class $\mathcal{M}_1$ connected with a partition or estimate the norm exactly (or even the value at every point) of a function from $\mathcal{M}_1$ with the norm or values of the perfect spline with respect to the same partition.

We shall introduce such classes and shall prove some theorems which will be a base for obtaining exact estimates for spline interpolation in different particular cases. These are deep theorems of independent interest; they will be used in investigations on other extremal problems, in particular, those connected with the optimal recovery of a function with discrete information.

Let a fixed partition of $[a, b]$ be given by

$$\Delta_N[a, b]\colon a = t_0 < t_1 < \cdots < t_N = b. \tag{1.1}$$

Denote by $KH^1_{\Delta_N}[a, b]$ the class of functions f defined on the interval $[a, b]$ satisfying on every interval (t_{i-1}, t_i) $(i = 1, 2, \ldots, N)$ from partition (1.1) a Lipschitz condition with constant K:

$$|f(t') - f(t'')| \le K|t' - t''|, \qquad t_{i-1} < t', t'' < t_i; \tag{1.2}$$

at the points $t_1, t_2, \ldots, t_{N-1}$ the functions from $KH^1_{\Delta_N}[a, b]$ may be discontinuous. In other words $f \in KH^1_{\Delta_N}[a, b]$ if and only if $f = g + s_0$ on $[a, b]$ where $g \in KH^1[a, b]$ and s_0 is a step function with possible jumps at the points $t_1, t_2, \ldots, t_{N-1}$.

We denote the class of functions $f \in C^{m-1}[a, b]$ such that $f^{(m)}(t)$ exists for all $t \ne t_i$ $(i = 1, 2, \ldots, N-1)$ and $f^{(m)} \in KH^1_{\Delta_N}[a, b]$ by $W^m KH^1_{\Delta_N}[a, b]$ (m = 1, 2, . . .,). Hence $W^m KH^1_{\Delta_N}[a, b]$ is the set of all mth integrals of functions from $KH^1_{\Delta_N}[a, b]$ and so

$$W^m KH^1_{\Delta_N}[a, b] = W^m KH^1[a, b] \oplus S_m(\Delta_N[a, b]).$$

In relations of general type $W^0 KH^1_{\Delta_N}[a, b]$ will mean $KH^1_{\Delta_N}[a, b]$. In the case of the uniform partition we shall write $KH^1_N[a, b]$ and $W^m KH^1_N[a, b]$.

It is clear that for every partition (1.1) the class $W^m KH^1[a, b]$ introduced in Section 1.5 (coinciding with $KW^{m+1}_\infty[a, b]$) is contained in $W^m KH^1_{\Delta_N}[a, b]$. But the latter is essentially larger because it allows discontinuities in the mth derivative at the inner points from the partition.

In the periodic case we shall define similar classes of functions with respect to a partition on every interval with length 2π. Set

$$\Delta_N^t\colon t_0 < t_1 < \cdots < t_N = t_0 + 2\pi \tag{1.3}$$

and let $KH^1_{\Delta_N^t}$ be the class of 2π-periodic functions satisfying on every interval (t_{i-1}, t_i) of partition (1.3) condition (1.2). We denote by $W^m KH^1_{\Delta_N^t}$ $(m = 1, 2, \ldots,)$ the set of all 2π-periodic mth integrals of functions $f \in KH^1_{\Delta_N^t}$ with zero mean value over the period. In general relationships $W^0 KH^1_{\Delta_N^t}$ will mean $KH^1_{\Delta_N^t}$. Let us remark that for every partition (1.3) class $W^m KH^1_{\Delta_N^t}$ contains the class KW^{m+1}_∞ introduced in Section 1.5 but it is essentially larger because it is a direct sum of KW^{m+1}_∞ and the linear manifold $S_N(\Delta_N^t)$ of 2π-periodic splines of degree m defect 1 with respect to (1.3).

In the cases of a uniform partition, i.e. when $t_k = t_0 + 2\pi k/N$, we shall write $KH^1_{N,t}$ and $W^m KH^1_{N,t}$ and when $t_0 = 0$ we shall omit t in the notation and write $KH^1_{\Delta_N}$ and KH^1_N.

Because of the homogeneity with respect to K of the estimates proved below for functions in the introduced classes we can assume that $K = 1$ and we shall omit this 1 in the notation, e.g. $W^m 1 H^1_{\Delta_N^t} = W^m H^1_{\Delta_N^t}$.

5.1.2 Main theorems In the periodic case the following theorem is central.

Theorem 5.1.1 Let the function φ from the class $W^{m-1} H^1_{\Delta_N^t}$ determined by partition (1.3) be such that

$$\pm\varphi^{(m)}(t) = (-1)^i, \qquad t_{i-1} < t < t_i, \qquad i = 1, 2, \ldots, N, \tag{1.4}$$

and there are N points in the period at which φ vanishes:

$$\varphi(\tau_j) = 0, \qquad j = 1, 2, \ldots, N, \qquad \tau_0 < \tau_1 < \cdots < \tau_N < \tau_{N+1} = \tau_0 + 2\pi.$$

If $f \in W^{m-1} H^1_{\Delta_N^t}$ and $f(\tau_j) = 0$, $j = 1, 2, \ldots, N$, then for every t ($t \neq t_i$ for $m = 1$) we have

$$|f(t)| \leq |\varphi(t)|. \tag{1.5}$$

Proof If $m = 1$ then everything is simple because in this case on every interval (t_{i-1}, t_i) we have $|f'(t)| \leq 1$, φ is a linear function with slope ± 1 and $\varphi(\tau_k) = \mathrm{f}(\tau_k) = 0$ for $t_{i-1} \leq \tau_k \leq t_i$.

Assuming that $m \geq 2$ we shall argue by contradiction. Suppose that at the point ξ $(\tau_1 < \xi < \tau_{N+1}, \xi \neq \tau_j)$ we have $|f(\xi)| > |\varphi(\xi)|$. Choose λ $(0 < |\lambda| < 1)$ so that $\lambda f(\xi) = \varphi(\xi)$. Then the difference

$$\delta(t) = \varphi(t) - \lambda f(t) \tag{1.6}$$

vanishes at the points $\tau_1, \tau_2, \ldots, \tau_N$ and at ξ. Hence δ has $N + 1$ different zeros on the period $[\tau_1, \tau_N)$.

Consider the $(m-1)$th derivative

$$\delta^{(m-1)}(t) = \varphi^{(m-1)}(t) - \lambda f^{(m-1)}(t),$$

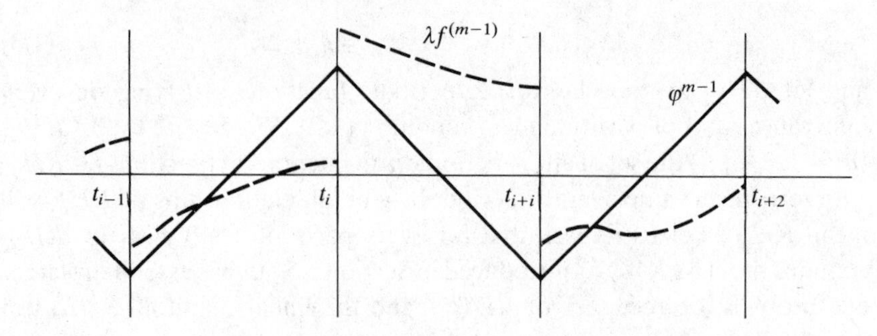

Fig. 5.1 To the proof of Theorem 5.1.1.

which exists for all $t \neq t_i$, i.e. everywhere except maybe at the points from partition (1.3). For $t_{i-1} < t' < t'' < t_i$ we have

$$|\varphi^{(m-1)}(t') - \varphi^{(m-1)}(t'')| = |t' - t''|$$

and

$$|\lambda f^{(m-1)}(t') - \lambda f^{(m-1)}(t'')| \leq |\lambda| \, |t' - t''| < |t' - t''|,$$

i.e.

$$|\lambda f^{(m-1)}(t') - \lambda f^{(m-1)}(t'')| < |\varphi^{(m-1)}(t') - \varphi^{(m-1}(t'')|.$$

From this fact and condition (1.4) we can draw two important conclusions.

(1) There is no interval $[\alpha, \beta]$ with a positive length on which $\delta^{(m-1)}$, and hence δ vanishes identically too. Therefore all zeros of δ are separated (see Section 2.4).
(2) On every interval (t_{i-1}, t_i) the function $\delta^{(m-1)}$ can change sign no more than once, moreover if $\delta^{(m-1)}$ essentially changes sign at t_i then at least in one of the intervals (t_{i-1}, t_i) or (t_i, t_{i+1}) the derivative $\delta^{(m-1)}$ does not change sign.

In view of (2) function $\delta^{(m-1)}$ has no more than N essential sign changes and now from Proposition 2.4.2 δ has no more than N separated zeros on the period – a contradiction. This proves (1.5). □

If $f \in W^m_\infty$ then the function $f(t + \alpha)$ belongs to $W^{m-1}H^1_{\Delta'_N}$ for every α and we obtain:

Corollary 5.1.2 If $f \in W^m_\infty$ and in the conditions of Theorem 5.1.1 $f(\tau_j + \alpha) = 0$ $(j = 1, 2, \ldots, N)$ for some α then $|f(t + \alpha)| \leq |\varphi(t)|$ for all t and $\|f\|_p \leq \|\varphi\|_p$ $(1 \leq p \leq \infty)$.

In the case of the uniform partition $t_i = i\pi/n$ ($i = 0, 1, \ldots, 2n$) the spline φ is the perfect Euler spline $\varphi_{n,m}$ and hence:

Proposition 5.1.3 Let $f \in W^{m-1}H^1_{2n}$, $f(\theta_{mj}) = 0$ ($j = 1, 2, \ldots, 2n$), where $\theta_{mj} = j\pi/n$ for even m and $\theta_{mj} = (2j-1)\pi/(2n)$ for odd m. Then for all t ($t \neq i\pi/n$ for $m = 1$) we have $|f(t)| \leq |\varphi_{n,m}(t)|$ and hence

$$\begin{aligned}\sup\{\|f\|_p: f \in W^{m-1}H^1_{2n}, f(\theta_{mj}) = 0, j = 1, 2, \ldots, 2n\}\\ = \sup\{\|f\|_p: f \in W^m_\infty, f(j\pi/n + \alpha) = 0, j = 1, 2, \ldots, 2n\}\\ = \|\varphi_{n,m}\|_p, \qquad 1 \leq p \leq \infty.\end{aligned}$$

Let us note that we obtain from this proposition and from Theorem 3.3.4

$$\sup\{\|f^{(k)}\|_p: f \in W^m_\infty, f(j\pi/n + \alpha) = 0, j = 1, 2, \ldots, 2n\} = \|\varphi_{n,m-k}\|_p,$$
$$1 \leq p \leq \infty, \qquad k = 1, 2, \ldots, m-1.$$

A similar result to these is

$$\sup\{\|f\|_1: f \in W^m_p, f(j\pi/n + \alpha) = 0, j = 1, 2, \ldots, 2n\} = \|\varphi_{n,m-k}\|_{p'}, \quad (1.7)$$
$$1 \leq p \leq \infty, \qquad 1/p + 1/p' = 1,$$

which will follow from Theorem 5.2.8 in the next section.

Analogs of Statements 5.1.1–5.1.3 can be obtained for functions defined on a finite interval $[a, b]$ assuming some zeros for the derivatives at the end-points a and b.

As in Section 2.4.2 for a fixed $m = 1, 2, \ldots$, we consider two subsets – I_a and I_b – of $Q_m = \{0, 1, \ldots, m-1\}$. By $|I_a|$ and $|I_b|$ we denote the number of elements in I_a and I_b respectively.

Theorem 5.1.4 Let the function φ be from the set $W^{m-1}H^1_{\Delta_N}[a, b]$ determined by partition (1.1) and let it satisfy

$$\pm\varphi^{(m)}(t) = (-1)^i, \qquad t_{i-1} < t < t_i, \qquad i = 1, 2, \ldots, N;$$
$$\varphi(\tau_j) = 0, \qquad j = 1, 2, \ldots, L, \qquad a < \tau_1 < \tau_2 < \cdots < \tau_N < b; \quad (1.8)$$
$$\varphi^{(\mu)}(a) = 0, \mu \in I_a; \qquad \varphi^{(\kappa)}(b) = 0, \kappa \in I_b;$$

and $L + |I_a| + |I_b| = N + m - 1$. If $f \in W^{m-1}H^1_{\Delta_N}[a, b]$, $f(\tau_j) = 0$, $j = 1, 2, \ldots, L$, and

$$f^{(\mu)}(a) = 0, \mu \in I_a; \qquad f^{(\kappa)}(b) = 0, \kappa \in I_b; \quad (1.9)$$

then for every $t \in [a, b]$ ($t \neq t_i$ for $m = 1$) we have $|f(t)| \leq |\varphi(t)|$.

The proof follows the same scheme as the periodic case but we use Proposition 2.4.1 instead of 2.4.2. If we assume that at point $\xi \in (a, b)$ $(\xi \neq \tau_i)$ the inequality $|f(\xi)| > |\varphi(\xi)|$ holds, then for some λ $(0 < |\lambda| < 1)$ the difference $\delta(t) = \varphi(t) - \lambda f(t)$ will have at least $L + 1$ different zeros on (a, b) $(\nu(\delta; (a, b)) \geq L + 1)$ which are separated because we have for $t', t'' \in (t_{i-1}, t_i)$

$$|\lambda f^{(m-1)}(t') - \lambda f^{(m-1)}(t'')| < |\varphi^{(m-1)}(t') - \varphi^{(m-1)}(t'')|.$$

From here and from (1.8), as in the proof Theorem 5.1.1, we can see that the function $\delta^{(m-1)}$ may essentially change sign on $[a, b]$ no more than N times: $\mu(\delta^{(m-1)}; [a, b]) \leq N$.

On the other hand, starting with the inequality $\nu(\delta; [a, b]) \geq L + 1$ we consecutively estimate, with the help of Proposition 2.4.1, the number of separated zeros of $\delta^{(k)}$ $(k = 1, 2, \ldots, m - 2)$ on $[a, b]$. It is essential to remember here that every differentiation may decrease this number by one while every one of conditions (1.9) will surely increase it by one. Finally, going from $\delta^{(m-2)}$ to $\delta^{(m-1)}$ (again with the help of Proposition 2.4.1) we obtain

$$\mu(\delta^{(m-1)}; [a, b]) \geq L + 1 - m + 1 + |I_a| + |I_b| = N + 1$$

which is a contradiction.

Let us note three partial cases corresponding to the following choices of boundary conditions: (1) I_a, I_b are empty; (2) I_a, I_b coincide with Q_m; and (3) I_a, I_b determined by Leedstone boundary conditions (see Section 2.4.2). The existence of corresponding perfect splines follows from Propositions 2.5.5–2.5.7.

Corollary 5.1.5 Let the perfect spline $\varphi \in S_m(\Delta_N[a, b])$ have $(N + m - 1)$ different zeros τ_j on $[a, b]$. If $f \in W^{m-1}H^1_{\Delta_N}[a, b]$ and $f(\tau_j) = 0$ $(j = 1, 2, \ldots, N + m - 1)$ then for every $t \in [a, b]$ we have $|f(t)| \leq |\varphi(t)|$.

Corollary 5.1.6 Let $N > m$ and let the perfect spline $\varphi \in S_m(\Delta_N[a, b])$ have $(N - m - 1)$ different zeros τ_j on (a, b) and $\varphi^{(\nu)}(a) = \varphi^{(\nu)}(b) = 0$ $(\nu = 0, 1, \ldots, m - 1)$. If $f \in W^{m-1}H^1_{\Delta_N}[a, b]$, $f(\tau_j) = 0$ $(j = 1, 2, \ldots, N - m - 1)$ and $f^{(\nu)}(a) = f^{(\nu)}(b) = 0$ $(\nu = 0, 1, \ldots, m - 1)$ then for every $t \in [a, b]$ we have $|f(t)| \leq |\varphi(t)|$.

For Leedstone boundary conditions we set $[a, b] = [0, \pi]$ and we consider only the case of uniform partition $t_i = i\pi/N$ when $\varphi = \varphi_{N,m}$ is a Euler perfect spline.

Corollary 5.1.7 Let $f \in W^{m-1}H^1_N[0, \pi]$ and for even m

$$f(j\pi/N) = 0 \ (j = 1, 2, \ldots, N), f^{(2\nu)}(0) = f^{(2\nu)}(\pi) = 0,$$
$$\nu = 0, 1, \ldots, m/2 - 1,$$

and for odd n

$$f((2j-1)\pi/(2N)) = 0 \ (j = 1, 2, \ldots, N), f^{(2\nu-1)}(0) = f^{(2\nu-1)}(\pi) = 0,$$
$$\nu = 0, 1, \ldots, (m-1)/2,$$

Then for every $t \in [0, \pi]$ we have the exact inequality $|f(t)| \leq |\varphi_{N,m}(t)|$.

5.1.3 The non-symmetric cases In the previous section we estimated the values and norms of a function f whose zeros coincide with the zeros of the function (in particular a perfect spline) φ such that $\varphi^{(m)} = \pm 1$. One can consider similar problems connected with the zeros of splines φ of degree m and defect 1 such that $\varphi^{(m)}$ takes only values α and $-\beta$ ($\alpha > 0, \beta > 0$); these splines can be called (α, β)-perfect splines. But in the non-symmetric case $\alpha \neq \beta$) we shall be satisfied by weaker statements than those ones from Section 5.1.2.

We shall consider the periodic case. In Section 3.3 we introduced the class $W^m_{\alpha,\beta}$ of 2π-periodic functions $f \in L^m_\infty$ such that $-\beta \leq f^{(m)}(t) \leq \alpha$ a.e. It is clear that $W^m_{\alpha,\beta}$ contains all 2π-periodic (α, β)-perfect splines of degree m with respect to every partition and, in particular, the standard splines $\varphi_{n,m}(\alpha, \beta; t)$ with period $2\pi/n$ (Section 3.1).

Proposition 5.1.8 Let the spline $\varphi(t) = \varphi_{n,m}(\alpha, \beta; t) + \lambda$ have (at the expense of the choice of λ) on the period $[0, 2\pi)$ $2n$ equidistant zeros τ_j ($j = 1, 2, \ldots, 2n$), $\tau_j - \tau_{j-1} = \pi/n$. *If* $f \in W^m_{\alpha,\beta}$ and $f(\tau_j + \gamma) = 0$ ($j = 1, 2, \ldots, 2n$) for some γ then for all t we have

$$\min\{\varphi(t), \varphi(t+\pi/n)\} \leq f(t+\gamma) \leq \max\{\varphi(t), \varphi(t+\pi/n)\}. \quad (1.10)$$

Proof Let one of the inequalities (1.10) be false at some point ξ ($\xi \neq \tau_j$); for definiteness assume that $0 < \varphi(\xi) < f(\xi + \gamma)$. Then the difference

$$\delta(t) = \varphi(t) - \lambda f(t+\gamma), \qquad \lambda = \varphi(\xi)/f(\xi+\gamma),$$

has at least $2n + 1$ separated zeros on the period. Hence the number of essential sign changes of $\delta^{(m)}$ has to be at least $2n + 1$. But $\delta^{(m)}(t) = \varphi^{(m)}(t) - \lambda f^{(m)}(t+\gamma)$ and because $0 < \lambda < 1$ we get that $\mu(\delta^{(m)}; 2\pi) = 2n$ – a contradiction. □

Corollary 5.1.9 Under the conditions of Proposition 5.1.8 we have $\|f\|_C \leq \|\varphi\|_C$, $E_1(f)_C \leq E_1(\varphi)_C$.

5.2 The interpolation error of the periodic splines with minimal defect on classes W_p^r

5.2.1 General estimates The space of periodic splines $\mathcal{N}$ is said to be *interpolating* by the spline s on a set of N points $\tau_1 < \tau_2 < \cdots < \tau_N < \tau_1 + 2\pi$ if for every $f \in C$ there exists a unique spline $s(t) = s(f, t)$ ($s(f, t) \in \mathcal{N}$) such that $s(f, \tau_j) = f(\tau_j)$ ($j = 1, 2, \ldots, N$).

Let us fix two independent partitions

$$\Delta_N^t\colon t_0 < t_1 < \cdots < t_N = t_0 + 2\pi, \tag{2.1}$$

$$\Delta_N^\tau\colon \tau_0 < \tau_1 < \cdots < \tau_N = \tau_0 + 2\pi. \tag{2.2}$$

We shall connect the classes of functions $W^m H^1_{\Delta_N^t}$ and $W^m H^1_{\Delta_N^\tau}$ defined in Section 5.1 and the subspaces $S_m(\Delta_N^t)$ and $S_m(\Delta_N^\tau)$ of splines of degree m defect 1 with them.

Theorem 5.2.1 Let the subspace $S_r(\Delta_N^t)$ be interpolating by the spline σ at the points τ_j ($j = 1, 2, \ldots, N$) of partition (2.2). If $f \in W^r H^1_{\Delta_N^t}$ then

$$|f(t) - \sigma(f, t)| \le |\varphi(t) - \sigma(\varphi, t)|, \qquad |t| < \infty, \tag{2.3}$$

where φ is the function from $W^r H^1_{\Delta_N^t}$ determined by

$$\pm\varphi^{(r+1)}(t) = (-1)^i, \qquad t_{i-1} < t < t_i, \qquad i = 1, 2, \ldots, N,$$

$$\varphi^{(r)}(t) = \int_{t_0}^{t} \varphi^{(r+1)}(u)\, du + c, \qquad t_0 < t < t_0 + 2\pi, \qquad \int_0^{2\pi} \varphi^{(r)}(u)\, du = 0.$$

If in addition $\varphi(\tau_j) = 0$ ($j = 1, 2, \ldots, N$) then

$$|f(t) - \sigma(f, t)| \le |\varphi(t)|, \qquad |t| < \infty. \tag{2.4}$$

Inequalities (2.3) and (2.4) cannot be improved at every point t on $W^r H^1_{\Delta_N^t}$ and also on W_∞^{r+1} provided that

$$(t_1 - t_0) - (t_2 - t_1) + (t_3 - t_2) - \cdots + (-1)^{N-1}(t_N - t_{N-1}) = 0. \tag{2.4'}$$

The proof consists of reducing the problem so that it becomes amenable to Theorem 5.1.1. Indeed, together with the functions f and φ the space $W^r H^1_{\Delta_N^t}$ contains the differences $\delta(t) = f(t) - \sigma(f, t)$ and $\psi(t) = \varphi(t) - \sigma(\varphi, t)$. But $\psi^{(r+1)}(t) = \delta^{(r+1)}(t) = \pm(-1)^i$ for $t \in (t_{i-1}, t_i)$ and Theorem 5.1.1 gives $|\delta(t)| \le |\psi(t)|$. In addition we have $\sigma(\phi, t) \equiv 0$ in the case $\phi(\tau_j) = 0$ ($j = 1, 2, \ldots, N$) because of the uniqueness of the interpolating spline and it remains to be noted that relation (2.4') implies $\varphi^{(r)} \in C$ and hence $\varphi \in W_\infty^{r+1}$. □

Theorem 5.2.1 allows us to obtain, in particular, the exact estimate for the spline interpolation error on the class W_∞^{r+1} in the L_p ($1 \le p \le \infty$) metric.

The corresponding estimates for the class W_p^{r+1} in the L_1 metric are obtained using a method based on the following lemma.

Lemma 5.2.2 Let the subspace $S_r(\Delta_N^t)$ be interpolating by the spline σ at the points τ_j from (2.2) and the subspace $S_r(\Delta_N^\tau)$ be interpolating by the spline s at the points t_j from (2.1). Then for every $f \in C^r$ there is a function $g(t) = g(f, t)$ from $W^r H^1_{\Delta_N^\tau}$ such that

$$\|f - \sigma(f)\|_1 = \left| \int_0^{2\pi} [g'(t) - s'(g, t)] f^{(r)}(t)\, dt \right|. \tag{2.5}$$

Proof Fix $f \in C^r$ and set

$$\delta(t) = f(t) - \sigma(f, t).$$

Hence $\delta(\tau_j) = 0$ $(j = 1, 2, \ldots, N)$ and so

$$\int_{\tau_{j-1}}^{\tau_j} \delta'(\mathrm{t})\, dt = 0, \qquad j = 1, 2, \ldots, N. \tag{2.6}$$

By integrating by parts on every interval (τ_{j-1}, τ_j) we obtain

$$\|\delta\|_1 = \int_{\tau_0}^{\tau_N} \delta(t) \operatorname{sgn} \delta(t)\, dt = \sum_{j=1}^{N} \int_{\tau_{j-1}}^{\tau_j} \delta(t) \operatorname{sgn} \delta(t)\, dt$$

$$= -\sum_{j=1}^{N} \int_{\tau_{j-1}}^{\tau_j} \delta'(t) l(t)\, dt = -\int_{\tau_0}^{\tau_N} \delta'(t) l(t)\, dt,$$

where the function l is such that $l'(t) = \operatorname{sgn} \delta(t)$ for $\tau_{j-1} < t < \tau_j$. Hence $l \in H^1_{\Delta_N^\tau}$ and, in view of (2.6), we may assume that $l \perp 1$.

Let g be the rth periodic antiderivative of l; then $g \in W^r H^1_{\Delta_N^\tau}$ and if we set

$$\eta(t) = g(t) - s(g, t),$$

we have $\eta(t_i) = 0$ $(i = 1, 2, \ldots, N)$ and

$$\int_{t_{j-1}}^{t_j} \eta'(t)\, dt = 0, \qquad i = 1, 2, \ldots, N. \tag{2.7}$$

From $s^{(r)}(g, t) =$ constant for $t \in (t_{i-1}, t_i)$ and (2.6) we get

$$\|\delta\|_1 = -\int_{\tau_0}^{\tau_N} \delta'(t) g^{(r)}(t)\, dt = -\int_{\tau_0}^{\tau_N} \delta'(t) \eta^{(r)}(t)\, dt = -\int_{t_0}^{t_N} \delta'(t) \eta^{(r)}(t)\, dt.$$

Integrating by parts $r - 1$ times (the non integral terms disappear in view of the periodicity) we obtain

$$\|\delta\|_1 = (-1)^r \int_{t_0}^{t_N} \delta^{(r)}(t) \eta'(t)\, dt,$$

and using (2.7) we have

$$\|\delta\|_1 = (-1)^r \int_{t_0}^{t_N} f^{(r)}(t)\eta'(t)\,dt = \left|\int_{t_0}^{t_N} f^{(r)}(t)\eta'(t)\,dt\right|.$$

The lemma is proved. □

Theorem 5.2.3 Let subspace $S_r(\Delta_N^t)$ be interpolating by the spline σ at the points τ_j of partition (2.2) and let subspace $S_r(\Delta_N^\tau)$ be interpolating by the spline s at the points t_j of partition (2.1).

If $f \in W_p^{r+1}$ $(1 \le p \le \infty)$ then

$$\|f - \sigma(f)\|_1 \le \|\varphi - \sigma(\varphi)\|_{p'}, \qquad 1/p + 1/p' = 1, \tag{2.8}$$

where $\varphi \in W^r H^1_{\Delta_N^\tau}$ determined by

$$\pm\varphi^{(r+1)}(t) = (-1)^j, \qquad \tau_{j-1} < t < \tau_j, \qquad j = 1, 2, \ldots, N;$$

$$\varphi^{(r)}(t) = \int_{\tau_0}^{t} \varphi^{(r+1)}(u)\,du + c, \qquad \tau_0 < t < \tau_0 + 2\pi, \qquad \int_0^{2\pi} \varphi^{(r)}(u)\,du = 0.$$

If in addition $\varphi(t_i) = 0$ $(i = 1, 2, \ldots, N)$ and

$$\begin{aligned} |\varphi(t)|^{p'-1}\,\mathrm{sgn}\,\varphi(t) \perp 1, &\qquad 1 \le p' < \infty, \\ \max \varphi(t) = -\min \varphi(t), &\qquad p' = \infty, \end{aligned} \tag{2.9}$$

then

$$\sup_{f \in W_p^{r+1}} \|f - \sigma(f)\|_1 = \|\varphi\|_{p'}, \qquad 1 \le p \le \infty, \qquad 1/p + 1/p' = 1.$$

Proof For $f \in W_p^{r+1}$ we apply (2.5), integrate by parts once more and use Hölder's inequality:

$$\|f - \sigma(f)\|_1 = \left|\int_0^{2\pi} [g(t) - s(g,t)] f^{(r+1)}(t)\,dt\right|$$

$$= \|g - s(g)\|_{p'} \|f^{(r+1)}\|_p \le \|g - s(g)\|_{p'}.$$

Because $g \in W^r H^1_{\Delta_N^\tau}$ from Theorem 5.2.1 we get $\|g - s(g)\|_{p'} \le \|\varphi - s(\varphi)\|_{p'}$, where function φ is determined as in Theorem 5.2.1 but with respect to partition Δ_N^τ. This proves (2.8).

If $\varphi(t_i) = 0$ $(i = 1, 2, \ldots, N)$ then in view of the unicity we get $s(\varphi, t) \equiv 0$ and instead of (2.8) we have

$$\|f - \sigma(f)\|_1 \le \|\varphi\|_{p'}. \tag{2.10}$$

Let $1 \le p' < \infty$, $|\varphi(t)|^{p'-1}\,\mathrm{sgn}\,\varphi(t) \perp 1$ and f_0 be the function from W_p^{r+1} such that

$$f_0^{(r+1)}(t) = \|\varphi\|_p^{1-p'} |\varphi(t)|^{p'-1}\,\mathrm{sgn}\,\varphi(t), \qquad \|f^{(r+1)}\|_p = 1.$$

The function φ preserves sign on every interval (t_{i-1}, t_i) (this can be easily established by arguing by contradiction) and so $f_0^{(r)}$ is strictly monotonic on every interval (t_{i-1}, t_i). Hence the difference $\delta_0(t) = f_0(t) - \sigma(f_0, t)$ changes sign at the zeros τ_j and has no other zeros, i.e. sgn $\delta_0(t) = \pm$ sgn $\varphi^{(r+1)}(t)$ and $\sigma(f_0, t) \equiv 0$. But then

$$\|f_0 - \sigma(f_0)\|_1 = \|f_0\|_1 = \left|\int_0^{2\pi} f_0(t)\varphi^{(r+1)}(t)\,\mathrm{d}t\right|$$

$$= \left|\int_0^{2\pi} f_0^{(r+1)}(t)\varphi(t)\,\mathrm{d}t\right| = \|\varphi\|_{p'}.$$

The exactness of (2.10) for $p' = \infty$ provided $\max \varphi(t) = -\min \varphi(t)$ can be obtained using Stekhlov functions. □

Remark If the difference $\psi(t) = \varphi(t) - s(\varphi, t)$ satisfies conditions similar to (2.9) then the estimate (2.8) will also be exact on W_p^{r+1}.

5.2.2 The uniform partition If the partitions (2.1) and (2.2) are uniform and appropriately coordinated for $N = 2n$ the spline interpolation error in W_p^{r+1} is estimated by the norm of the Euler perfect splines. This case is of special importance because it turns out that the estimates obtained cannot be improved in the most general sense – with respect to any approximating subspaces of the same dimension (see Chapter 8).

As before (see Section 2.4.3) $S_{2n,m}$ denotes the linear manifold of 2π-periodic splines of degree m and defect 1 with respect to the uniform partition $\{k\pi/n\}$ $(k = 1, 2, \ldots, 2n)$. For a fixed $n = 1, 2, \ldots,$ and every $m = 0, 1, \ldots,$ we consider the system of points

$$\theta_{mk} = \frac{k\pi}{n} + [1 - (-1)^m]\frac{\pi}{4n}, \qquad k = 0, \pm 1, \pm 2, \ldots, \tag{2.11}$$

which are the zeros of the perfect spline $\varphi_{n,m}$. In other words $\theta_{mk} = k\pi/n$ for even m and $\theta_{mk} = (2k+1)\pi/(2n)$ for odd m. Because of $\theta_{m,k-1} < \theta_{m+1,k} < \theta_{mk}$ Corollary 2.4.11 implies that for every function f defined at the points $\theta_{m+1,k}$ there is a unique spline $\sigma_{2n,m}(f, t) \in S_{2n,m}$ satisfying

$$\sigma_{2n,m}(f, \theta_{m+1,k}) = f(\theta_{m+1,k}), \qquad k = 0, 1, \ldots, 2n-1.$$

Let $W^m H_{2n}^1$ be the class of functions $f \in C^{m-1}$ introduced in Section 5.1.1 such that

$$|f^{(m)}(t') - f^{(m)}(t'')| \le |t' - t''|, \qquad (k-1)\pi/n < t', t'' < k\pi/n,$$
$$k = 1, 2, \ldots, 2n.$$

Then Theorem 5.2.1 immediately implies:

Proposition 5.2.4 For every function $f \in W^r H^1_{2n}$ $(r = 0, 1, \ldots,)$ we have

$$|f(t) - \sigma_{2n,r}(f, t)| \le |\varphi_{n,r+1}(t)|, \qquad 0 \le t \le 2\pi,$$

which cannot be improved for every t even on the class W^{r+1}_∞.

Of course Proposition 5.2.4 implies that for every $f \in W^r H^1_{2n}$

$$\|f - \sigma_{2n,r}(f)\|_q \le \|\varphi_{n,r+1}\|_q, \qquad 1 \le q \le \infty, \tag{2.12}$$

and this estimate is exact because $\varphi_{n,r+1} \in W^{r+1}_\infty \subset W^r H^1_{2n}$. It is interesting to see whether estimate (2.12) can be improved if we replace the interpolation by a best approximation with functions from $S_{2n,r}$. We have:

Proposition 5.2.5 For every fixed $m, r = 0, 1, \ldots,$ the best approximation in the metric L_q $(1 \le q \le \infty)$ of the function

$$\psi(t) = \varphi_{n,r}(t - \beta), \qquad \beta = \beta_{r,m} = [1 + (-1)^{r+m}]\pi/(4n), \tag{2.13}$$

by splines from the subspace $S_{2n,m}$ is realized by the constant zero and we have

$$E(W^r_\infty, S_{2n,m})_q \ge E(\psi, S_{2n,m})_q = \|\psi\|_q = \|\varphi_{n,r}\|_q, \qquad n = 1, 2, \ldots, \tag{2.14}$$

Proof First let $1 \le q < \infty$ and

$$h(t) = |\psi(t)|^{q-1} \operatorname{sgn} \psi(t). \tag{2.15}$$

In view of Proposition 1.4.6 it is sufficient to establish that $h \perp S_{2n,m}$. But if $s \in S_{2n,m}$ and if h_m is the mth periodic integral of h $(h \perp 1)$ then

$$\int_0^{2\pi} h(t)s(t)\,dt = (-1)^m \int_0^{2\pi} h_m(t)s^{(m)}(t)\,dt = 0,$$

because (2.15) and the choice of β in (2.13) imply that h_m is $2\pi/n$-periodic and

$$\int_{t_{i-1}}^{t_i} h_m(t)\,dt = 0, \qquad i = 1, 2, \ldots, 2n, \qquad t_k = k\pi/n,$$

and $s^{(m)}(t) = c_i$ on (t_{i-1}, t_i).

For $q = \infty$ we can get the necessary result using the well known fact [18A, p. 21] that $\|g\|_q \to \|g\|_\infty$ for $q \to \infty$ provided that $g \in L_\infty$. If $s(\psi, t)$ is the spline of the best L_∞ approximation to ψ in $S_{2n,m}$ then for every $\varepsilon > 0$ there is a $q(\varepsilon)$ such that for every $q > q(\varepsilon)$ we have

$$\begin{aligned} E(\psi, S_{2n,m})_\infty &= \|\psi - s(\psi)\|_\infty > \|\psi - s(\psi)\|_q - \varepsilon/2 \\ &\ge E(\psi, S_{2n,m})_q - \varepsilon/2 = \|\psi\|_q - \varepsilon/2 = \|\psi\|_\infty - \varepsilon = \|\varphi_{n,r}\|_\infty - \varepsilon. \end{aligned}$$

Because $E(\psi, S_{2n,m})_\infty \le \|\psi\|_\infty$ inequality (2.14) is proved for $q = \infty$. □

From Propositions 5.2.4 and 5.2.5 we obtain:

Theorem 5.2.6 For all $n = 1, 2, \ldots$, and $r = 0, 1, \ldots$, we have

$$\begin{aligned} E(W^r H^1_{2n}, S_{2n,r})_q &= E(W^{r+1}_\infty, S_{2n,r})_q = \sup_{f \in W^r H^1_{2n}} \|f - \sigma_{2n,r}(f)\|_q \\ &= \sup_{f \in W^{r+1}_\infty} \|f - \sigma_{2n,r}(f)\|_q = \|\varphi_{n,r+1}\|_q \\ &= n^{-r-1}\|\varphi_{1,r+1}\|_q, \qquad 1 \le q \le \infty. \end{aligned} \tag{2.16}$$

Therefore the best L_q $(1 \le q \le \infty)$ approximation of the classes $W^r H^1_{2n}$ and W^{r+1}_∞ by the subspace $S_{2n,r}$ is realized by the interpolation splines $\sigma_{2n,r}(f, t)$. Let us remember in connection with (2.16) that $\|\varphi_{1,m}\|_\infty = K_m$, $\|\varphi_{1,m}\|_1 = 4K_{m+1}$.

If we take into account that $\sigma_{2n,r}(f, t)$ is a linear projector into $S_{2n,r}$ then in the notation of (1.2.2) and (1.2.5), we can write

$$\begin{aligned} \mathscr{E}(W^r_\infty, S_{2n,r-1})_q &= \mathscr{E}_\perp(W^r_\infty, S_{2n,r-1})_q \\ &= E(W^r_\infty, S_{2n,r-1})_q = \|\varphi_{n,r}\|_q, \qquad 1 \le q \le \infty. \end{aligned}$$

Similar equalities also hold for classes $W^{r-1}H^1_{2n}$.

Let us turn our attention to the dual situation – approximation of the classes W^{m+1}_p in L_1 by the splines $\sigma_{2n,m}(f, t)$. We begin with the estimates given below.

Let $1 < p \le \infty$ and for fixed $n = 1, 2, \ldots$, and $r = 0, 1, \ldots$, we set

$$f_0(t) = \|\varphi_{n,r}\|^{1-p'}_{p'} \, |\varphi_{n,r}(t)|^{p'-1} \operatorname{sgn} \varphi_{n,r}(t).$$

It is clear that $f_0 \perp 1$ and if f_r $(r > 0)$ denotes the rth periodic interval of f_0 such that $f_r \perp 1$, then function f_r is $2\pi/n$-periodic with simple zeros at $k\pi/n$. Moreover f_r is similar to the function $\pm\sin nt$ and hence

$$\operatorname{sgn} f_r(t) = \pm\varphi_{n,0}(t), \qquad r = 0, 1, \ldots,. \tag{2.17}$$

For $m = 0, 1, \ldots$, set

$$g(t) = f_r(t - \gamma_m), \qquad \gamma_m = [1 - (-1)^m]\pi/(4n), \tag{2.18}$$

and we remark that $g \in W^r_p$ because $\|g^{(r)}\|_p = \|f_0\|_p = 1$.

Proposition 5.2.7 For every $n, r = 1, 2, \ldots$, and $m = 0, 1, \ldots$, we have

$$E(W^r_p, S_{2n,m})_1 \ge \|\varphi_{n,r}\|_{p'}, \qquad 1 \le p \le \infty, \qquad 1/p + 1/p' = 1. \tag{2.19}$$

If $1 < p \le \infty$ then for function $g \in W^r_p$ defined by (2.18) we have

$$E(g, S_{2n,m})_1 = \|g\|_1 = \|\varphi_{n,r}\|_{p'}. \tag{2.20}$$

Proof Let $1 < p \leq \infty$. If ψ_m is the mth periodic integral of $\psi_0(t) = \operatorname{sgn} g(t)$ and $\psi_m \perp 1$ then the definition of g implies that the mean value of ψ_m on every interval $[(i-1)\pi/n, i\pi/n]$ is zero. Hence

$$\int_0^{2\pi} \psi_0(t)s(t)\,dt = (-1)^m \int_0^{2\pi} \psi_m(t)s^{(m)}(t)\,dt = 0 \qquad \forall\, s \in S_{2n,m}.$$

This means, in view of Proposition 1.4.6, that the best approximation in the L_1 metric of function g by the subspace $S_{2n,m}$ is realized by the constant zero and the first equality in (2.20). Taking (2.17) into account we have

$$\|g\|_1 = \|f_r\|_1 = \left|\int_0^{2\pi} f_0(t)\varphi_{n,r}(t)\,dt\right|$$

$$= \|\varphi_{n,r}\|_{p'}^{1-p'} \int_0^{2\pi} |\varphi_{n,r}(t)|^{p'}\,dt = \|\varphi_{n,r}\|_{p'}.$$

It remains to prove (2.19) for $p = 1$. Set

$$f(t) = \frac{1}{4n}\varphi_{n,r-1}(t - \gamma_{r-1}),$$

where γ_{r-1} is chosen by the condition $f(\pi/(2n)) = 0$. If f_h is the Stekhlov function for f then $f_h \in L_1^r$ and $\|f_h^{(r)}\|_1 \leq 1$, i.e. $f_h \in W_1^r$. Moreover f_h has a zero mean value on every interval $[(i-1)\pi/n, i\pi/n]$. Using Proposition 1.4.6 we prove similarly to the above that $E(f_h, S_{2n,m})_1 = \|f_h\|_1$. Because

$$\lim_{h\to\infty} \|f_h\|_1 = \frac{1}{4n}\|\varphi_{n,r-1}\|_1 = \|\varphi_{n,r}\|_C$$

for h small enough we have

$$E(f_h, S_{2n,m})_1 > \|\varphi_{n,r}\|_C - \varepsilon,$$

which in view of the arbitrariness of ε implies $E(f_h, S_{2n,m})_1 \geq \|\varphi_{n,r}\|_C$. □

Now let us go back to Theorem 5.2.3 and consider the partial case when, for every $r = 1, 2, \ldots$, partition $\Delta_N^t = \bar{\Delta}_{2n}$ is generated by the points $t_i = i\pi/n$ and partition $\Delta_N^\tau = \bar{\Delta}_{2n}^\tau$ is generated by the points $\tau_i = \theta_{r+1,j}$ defined in (2.11). It was mentioned that the space $S_{2n,r} = S_r(\bar{\Delta}_{2n})$ interpolates at the points $\theta_{r+1,j}$; let us note now that the space $S_{2n,r}^\tau = S_r(\bar{\Delta}_{2n}^\tau)$ interpolates at the points $t_i = i\pi/n$. For odd r this statement follows from the previous one because partitions $\bar{\Delta}_{2n}$ and $\bar{\Delta}_{2n}^\tau$ coincide here. For even r one should remark that a $\pi/(2n)$ translation takes points $j\pi/n$ to $\theta_{r+1,j}$ and space $S_{2n,r}^\tau$ to $S_{2n,r}$. Moreover the perfect spline

$$\varphi(t) = \varphi_{n,r+1}(t - \beta_r), \qquad \beta = [1 + (-1)^r]\pi/(4n),$$

vanishes at the points $i\pi/n$ and $\varphi^{(r+1)}(t) = \pm(-1)^j\ (\theta_{r+1,j-1} < t < \theta_{r+1,j})$.

In connection with the above relation (2.5) in this case can be written as

$$\|f - \sigma_{2n,r}(f)\|_1 = \left| \int_0^{2\pi} [g'(f,t) - \sigma'_{2n,r}(g,t)] f^{(r)}(t - \beta_r)\,\mathrm{d}t \right|, \qquad (2.21)$$

where $f \in C^r$, $g \in W^r H^1_{2n}$ and $\beta = [1 + (-1)^r]\pi/(4n)$. We shall use this relation later.

Therefore in view of Theorem 5.2.3 for $f \in W^r_p$ we have

$$\|f - \sigma_{2n,r-1}(f)\|_1 \le \|\varphi_{2n,r}\|_{p'}, \qquad p' = p/(p-1), \qquad (2.22)$$

and in the case $1 < p \le \infty$ for the function $f_0 \in W^r_p$ given by

$$f_0^{(r)}(t) = \|\varphi_{n,r}\|_{p'}^{1-p'}\ |\varphi_{n,r}(t)|^{p'-1}\ \mathrm{sgn}\ \varphi_{n,r}(t)$$

(2.22) is an equality.

Let us remark that for $p = 1$ estimate (2.22) holds on the class W^{r-1}_V which is larger than W^r_1. Indeed, for every $f \in W^{r-1}_V$ its Stekhlov function $f_h \in W^r_1$ and $\|f - f_h\|_1 \to 0$ for $h \to 0$. From the representation of $\sigma_{2n,m}(f,t)$ via the fundamental splines (see (2.4.30)) it is clear that $\|\sigma(f) - \sigma(f_h)\|_1 \to 0$ for $h \to 0$. It is easy to check that for $f_1(t) = (4n)^{-1}\varphi_{n,r-1}(t - \alpha)$, choosing α such that $f_1(\theta_{rj}) = 0$ $(j = 1, 2, \ldots, 2n)$, (2.22) with $p' = \infty$ is satisfied as an equality and we have $f_1 \in W^{r-1}_V$. So we have proved:

Theorem 5.2.8 For all $n, r = 1, 2, \ldots,$ we have

$$E(W^r_p, S_{2n,r-1})_1 = \sup_{f \in W^r_p} \|f - \sigma_{2n,r-1}(f)\|_1$$

$$= \|\varphi_{n,r}\|_{p'} = n^{-r}\|\varphi_{1,r}\|_{p'}, \qquad 1 \le p \le \infty, \qquad 1/p + 1/p' = 1. \qquad (2.23)$$

In particular we have for $p = 1$

$$E(W^r_1, S_{2n,r-1})_1 = E(W^{r-1}_V, S_{2n,r-1})_1$$

$$= \sup_{f \in W^{r-1}_V} \|f - \sigma_{2n,r-1}(f)\|_1 = \|\varphi_{n,r}\|_C = K_r n^{-r}. \qquad (2.24)$$

Of course we can also write $\mathscr{E}_\perp(W^r_p, S_{2n,r-1})_1$ and $\mathscr{E}_\perp(W^r_1, S_{2n,r-1})_1$ on the left-hand sides of (2.23) and (2.24).

From (2.23) we immediately get relation (1.7) stated in the previous section because the functions on which the unimprovability of (2.22) is checked vanish at the interpolation points.

5.2.3 One interpretation, generalizations and approximations of the derivative The results obtained in Sections 5.2.1 and 5.2.2 for classes W_p^r may have a slightly different interpretation if we consider the interpolation of an individual function $f \in L_p^r$ and the norm $\|f^{(r)}\|_p$ is included in the error estimate. Simply one should note that for $f \neq$ constant the function $f(t)/\|f^{(r)}\|_p \in W_p^r$ and the operator $\sigma_{2n,m}(f, t)$ is homogeneous.

For example, the following statement is equivalent to the first statement in Theorem 5.2.8. If $f \in L_p^r$ $(1 \leq p \leq \infty)$ then

$$E(f, S_{2n,r-1})_1 \leq \|f - \sigma_{2n,r-1}(f)\|_1 \leq \|\varphi_{n,r}\|_{p'} \|f^{(r)}\|_p \tag{2.25}$$

and the inequalities (2.25) cannot be improved on L_p^r.

Proposition 5.2.4 also contains the following result: If $f \in L_\infty^r$ then the inequality

$$|f(t) - \sigma_{2n,r-1}(f, t)| \leq |\varphi_{n,r}(t)| \, \|f^{(r)}\|_\infty, \qquad 0 \leq t \leq 2\pi,$$

holds and it cannot be improved on L_∞^r for every point t.

One can interpret the other results for approximation of W_p^r in a similar manner.

Let us consider some generalizations. In Sections 5.2.1 and 5.2.2 we have seen that the estimates for interpolation by splines from $S_r(\Delta_N)$ on the classes W_∞^{r+1} and $W^r H_{\Delta_N}^1$ coincide, i.e. the addition of any spline from $S_r(\Delta_N)$ (not belonging to C^r in general) to a function from W_∞^{r+1} does not increase the spline interpolation error. The reason is that the mapping $f \to \sigma(f)$, where $\sigma(f)$ is the interpolating spline from $S_r(\Delta_N)$, is a linear projection on this subspace. Hence if $\mathcal{M}$ is a class of functions from C and $f = g + s$, where $g \in \mathcal{M}$ and $s \in S_r(\Delta_N)$ then $\sigma(f) = \sigma(g) + s$ and

$$f(t) - \sigma(f, t) = g(t) - \sigma(g, t).$$

Therefore we have

$$\sup_{f \in \mathcal{M}} \|f - \sigma(f)\|_q = \sup_{f \in \mathcal{M} \oplus S_r(\Delta_N)} \|f - \sigma(f)\|_q, \qquad 1 \leq q \leq \infty,$$

and this allows us to extend all estimates for spline interpolation proved above for classes W_p^{r+1} to classes $W_p^{r+1} \oplus S_r(\Delta_N)$.

It is also clear that all results from Section 5.2.2 for estimating the spline interpolation error are automatically transferrable to the corresponding classes of $2l$-periodic functions by replacing the standard function $\varphi_{n,m}$ by the $2l/n$-periodic function $\varphi_{\lambda,n}$, where $\lambda = \pi n/l$. This remark also concerns the results in Section 5.4 about the best approximation for classes of periodic functions by means of splines with uniform nodes.

Finally we shall briefly consider the simultaneous approximation of the derivative. If $s(f, t)$ is some spline corresponding to a function f then together with the estimate for the error $f(t) - s(f, t)$ it is important to estimate the deviation of $s'(f, t)$ from the derivative $f'(t)$ of the approximating function. Although the order of simultaneous approximation by interpolating splines of a function and its derivative in many important cases was found some time ago (see e.g. [1A, 23A]) exact results have only begun to appear relatively recently.

We have seen that exact estimates for the spline interpolation error can easily be obtained in some cases. The problem of obtaining the exact estimates for the derivatives in these cases has turned out to be much more difficult and it seems that the shortest way to the final results has not yet been found. In connection with this we restrict ourselves here, and in Section 5.3, to the formulation of the results known to us.

As before let $\sigma_{2n,r}(f, t)$ be the 2π-periodic spline from $S_{2n,r}$ coinciding with $f \in C$ at points $j\pi/n$ $(j = 1, 2, \dots, 2n)$ for odd r and points $(2j-1)\pi/(2n)$ $(j = 1, 2, \dots, 2n)$ for even r.

Proposition 5.2.9 For all $r = 1, 2, \dots,$ we have

$$\sup_{f \in W^r H^1_{2n}} \|f' - \sigma'_{2n,r}(f)\|_q = \sup_{f \in W^{r+1}_\infty} \|f' - \sigma'_{2n,r}(f)\|_q$$
$$= \|\varphi_{n,r}\|_q = n^{-r}\|\varphi_{1,r+1}\|_q, \qquad 1 \le q \le \infty. \qquad (2.26)$$

Comparing (2.16) and (2.26) we see that spline $\sigma_{2n,r}(f, t)$ and its derivative $\sigma'_{2n,r}(f, t)$ approximate in the best way in the metric L_q $(1 \le q \le \infty)$ to the function $f \in W^{r+1}_\infty$ (in the subspace $S_{2n,r}$) and its derivative f' (in $S_{2n,r-1}$).

5.2.4 An integral representation of the error The proofs of the main results in the previous sections were based on geometric considerations. There is another method for estimating the spline interpolation error based on the analytic representation of the error. One may call such methods *analytic*. We give it here not only in order to examine another approach for obtaining known results but because this method will be more effective in solving the more complicated problems from Chapter 7 when the method from Sections 5.2.1–5.2.3 will not work.

Restricting ourselves to the periodic case, we start with representation (4.1.6) which is vlid for functions $f \in L^r_1$ where B_r are Bernoulli kernels (4.1.7).

Let

$$\Delta_N\colon 0 = t_0 < t_1 < \cdots < t_N = 2\pi$$

be an arbitrary partition and let the subspace $S_r(\Delta_N)$ $(r = 1, 2, \dots,)$ of splines of degree r defect 1 interpolate at the N points τ_j $(0 < \tau_0 < \tau_1 < \cdots < \tau_N \le 2\pi)$ on the period. The spline $s \in S_r(\Delta_N)$ interpolating the function f at the points τ_j can be written as

$$s(t) = s(f, t) = \sum_{i=1}^{N} f(\tau_i) s_i(t), \tag{2.27}$$

where s_i are the fundamental splines from $S_r(\Delta_N)$ determined by the equations $s_i(\tau_j) = \delta_{ij}$. If in particular $f \equiv 1$ then in view of the unicity we have

$$s_1(t) + s_2(t) + \cdots + s_N(t) \equiv 1. \tag{2.27'}$$

For $f \in L_1^r$ replacing in (2.27) the values of $f(\tau_j)$ from (4.1.6) and using (2.27′) we obtain

$$s(f, t) = \frac{a_0}{2} + \frac{1}{\pi} \int_0^{2\pi} \sum_{i=1}^{N} B_r(\tau_i - u) s_i(t) f^{(r)}(u)\, du.$$

Then we have the representation

$$\delta(f, t) = \frac{1}{\pi} \int_0^{2\pi} \left[B_r(t - u) - \sum_{i=1}^{N} B_r(\tau_i - u) s_i(t) \right] f^{(r)}(u)\, du \tag{2.28}$$

for the error $\delta(t) = f(t) - s(f, t)$.

The specific and the properties of the interpolating spline $s(f, t)$ become apparent in the properties of the kernel of the representation (2.28), i.e. the function

$$Q_r(t, u) = B_r(t - u) - \sum_{i=1}^{N} B_r(\tau_i - u) s_i(t), \tag{2.29}$$

which is 2π-periodic on every variable and

$$Q_r(\tau_j, u) = 0, \qquad j = 1, 2, \ldots, N,$$

because $s_i(\tau_j) = \delta_{ij}$. Moreover from

$$\frac{\partial^{r-1}}{\partial u^{r-1}} Q_r(t, u) = (-1)^{r-1} \left[B_1(t - u) - \sum_{i=1}^{N} B_1(\tau_i - u) s_i(t) \right]$$

from (2.27′) and from Lemma 2.3.4 for $t \neq \tau_j$ the function $Q_r(t, u)$ of the variable u is a spline of degree $r - 1$ and defect 1 with respect to partition Δ_N augmented with the point t.

Now let $N = 2n$ and let the partition $\Delta_{2n} = \bar{\Delta}_{2n}$ be the uniform one, i.e. $t_i = i\pi/n$ $(i = 1, 2, \ldots, 2n)$. $\sigma_{2n,r}(f, t)$ as in Section 5.2.2 is the spline from $S_{2n,r}$ interpolating f at the zeros of $\varphi_{n,r+1}$, i.e. at the points $\tau_{rj} = j\pi/n - \beta_r$, where $\beta_r = 0$ for odd r and $\beta_r = \pi/(2n)$ for even r. For $f \in L_1^r$ the error $\delta(f, t) = f(t) - \sigma_{2n,r}(f, t)$ can be represented as

$$\delta(f, t) = \frac{1}{\pi} \int_0^{2\pi} q_r(t, u) f^{(r)}(u + \beta_r)\, du, \tag{2.30}$$

where

$$q_r(t,u) = B_r(t-u-\beta_r) - \sum_{i=1}^{2n} B_r(\tau_{ri}-u-\beta_r)s_{r,i}(t),$$

and $s_{r,i}$ are the corresponding fundamental splines. If we set

$$F_r(t,u) = \int_{-\beta_r}^{u} q_r(t,v)\,dv, \tag{2.31}$$

then we have for $f \in L_1^{r+1}$

$$\delta(f,t) = -\frac{1}{\pi}\int_0^{2\pi} F_r(t,u)f^{(r+1)}(u+\beta_r)\,du \tag{2.32}$$

and

$$q_r(t,u) = B_r(t) - B_r(t-u-\beta_r) - \sum_{i=1}^{2n} [B_{r+1}(\tau_{ri}) - B_{r+1}(\tau_{ri}-u-\beta_r)]s_{r,i}(t). \tag{2.33}$$

Lemma 5.2.10 The function $F_r(t,u)$ has the following properties:

(1) $F_r(\tau_{ri},u) = F_r(t,\tau_{ri}) = 0 \qquad (i = 1, 2, \ldots, 2n)$.

(2) $F_r(t,u)$ is a spline of degree r and defect 1 with respect to each variable t, u with knots $\bar{\Delta}_{2n}$ augmented with the points $u+\beta_r$ or $t-\beta_r$ respectively. Moreover

$$\operatorname{sgn} F_r(t,u) = \pm\varphi_{n,0}(t+\beta_r), \qquad u \neq \tau_{rj}, \tag{2.34}$$

$$\operatorname{sgn} F_r(t,u) = \pm\varphi_{n,0}(u+\beta_r), \qquad t \neq \tau_{rj}. \tag{2.35}$$

(3) For all t and u we have

$$F_{2k-1}(t,u) = F_{2k-1}(u,t), \qquad F_{2k}(t,u) = F_{2k}(u,t-\pi/n), \qquad k = 1, 2, \ldots,. \tag{2.36}$$

(4) We have the identities

$$\frac{1}{\pi}\int_0^{2\pi} |F_r(t,u)|\,du = |\varphi_{n,r+1}(t)|, \tag{2.37}$$

$$\frac{1}{\pi}\int_0^{2\pi} |F_r(t,u)|\,dt = |\varphi_{n,r+1}(u)|. \tag{2.38}$$

Proof Property (1) If we set for a fixed u

$$g_u(t) = B_{r+1}(t) - B_{r+1}(t-u-\beta_r),$$

then (2.33) implies

$$F_r(t,u) = g_u(t) - \sigma_{2n,r}(g_u,t)$$

and hence $F_r(\tau_{ri},u) \equiv 0$ $(i = 1, 2, \ldots, 2n)$. Moreover $g_u^{(r)}(t) = B_1(t) -$

$B_1(t-u-\beta_r)$ and Lemma 2.3.4 shows that $g_u^{(r)}$ is a step function with jumps at the points $t=0$, $t=u+\beta_r$. Therefore if $u=\tau_{ri}=t_i-\beta_r$ then the jumps of $g_u^{(r)}$ are between the points from the partition $\bar{\Delta}_{2n}$ and hence $g_u \in S_{2n,r}$. But then the unicity of the interpolation gives $\sigma_{2n,r}(g_u, t) \equiv g_u(t)$, i.e. $F_r(t, \tau_{ri}) \equiv 0$.

Property (2) For a fixed u the function $F_r(t, u)$ (as a difference of the splines g_u and $\sigma_{2n,r}(g_u)$) is a spline of degree r defect 1 with respect to the partition $\bar{\Delta}_{2n} \cup \{u+\beta_r\}$. Moreover the step function (of the variable t)

$$\frac{\partial^r}{\partial t^r} F_r(t, u) = g_u^{(r)}(t) - \sigma_{2n,r}^{(r)}(g_u, t)$$

may have no more than $2n+1$ jumps on the period and hence, in view of the periodicity, no more than $2n$ essential sign changes. Using this fact and arguing by contradiction it is easy to obtain that for $u \neq \tau_{rj}$ the zeros $t=\tau_{ri}$ $(i=1,2,\ldots,2n)$ of the function $F_r(t,u)$ are simple, and $F_r(t,u)$ changes sign there; and there are no other zeros on the period. This proves (2.34).

Now let us fix $t \neq \tau_{rj}$. Because $\tau_{ri}-\beta_r$ coincide either with t_i or with t_{i-1} we have

$$\frac{\partial^r}{\partial u^r} F_r(t, u) = (-1)^{r+1}\left[B_1(t-u-\beta_r) - \sum_{i=1}^{2n} B_1(\tau_{ri}-u-\beta_r)s_{r,i}(t)\right]$$

$$= (-1)^r\left[B_1(u-t+\beta_r) - \sum_{i=1}^{2n} B_1(u-\tau_{ri}+\beta_r)s_{r,i}(t)\right]$$

and (2.27′) and Lemma 2.3.4 imply that $\partial^r F_r(t,u)/\partial u^r$ is a step function of u with possible jumps at the point $t_i = i\pi/n$ and an obligatory jump at $u=t-\beta_r$. Therefore $F_r(t,u)$ as a function of u is a spline of degree r and defect 1 with respect to the partition $\bar{\Delta}_{2n} \cup \{t-\beta_r\}$. Moreover $F_r(t,u)$ has $2n$ simple zeros τ_{ri} $(i=1,2,\ldots,2n)$ at which it changes sign, i.e. (2.35) holds.

Property (3) In order to prove equality (2.36) which can be written as

$$F_r(t,u) = F_r(u, t-2\beta_r), \qquad r=1,2,\ldots, \tag{2.39}$$

we note that in view of (1) it suffices to establish (2.39) for $t \neq \tau_{ri}$ and $u \neq \tau_{rj}$. (2) shows that for a fixed t the functions $\varphi(u) = F_r(t,u)$, $\psi(u) = F_r(u, t-2\beta_r)$ are splines of degree r and defect 1 with respect to one and the same partition $\bar{\Delta}_{2n} \cup \{t-\beta_r\}$. If we assume that $\varphi(u) \not\equiv \psi(u)$, e.g. $\varphi(u_*) > \psi(u_*) > 0$ for some $u_* \neq \tau_{ri}$, we get that the spline

$$\Phi(u) = \varphi(u) - \psi(u)\varphi(u_*)/\psi(u_*)$$

has $2n$ zeros τ_{ri} on the period and one additional zero u_*. This is impossible because the derivative $\Phi^{(r)}(u)$ has no more than $2n$ essential sign changes. Therefore (2.39) is an identity for every fixed t.

Property (4) Using (2.35), (2.33) and the orthogonality $\varphi_{n,0}(t+\alpha) \perp 1$ for every α ($\varphi_{n,0}(t) = \operatorname{sgn} \sin nt$) we can write

$$\int_0^{2\pi} |F_r(t,u)|\, du = \left| \int_0^{2\pi} F_r(t,u)\varphi_{n,0}(u+\beta_r)\, du \right|$$

$$= \left| \int_0^{2\pi} B_{r+1}(t-u-\beta_r)\varphi_{n,0}(u+\beta_r)\, du \right.$$

$$\left. + \sum_{i=1}^{2n} s_{r,i}(t) \int_0^{2\pi} B_{r+1}(\tau_{ri}-u-\beta_r)\varphi_{n,0}(u+\beta_r)\, du \right|.$$

But we have from the definition of the perfect spline $\varphi_{n,m}$ and the general formula (4.1.6) for every t

$$\frac{1}{\pi}\int_0^{2\pi} B_{r+1}(t-u-\beta_r)\varphi_{n,0}(u+\beta_r)\, du = \varphi_{n,r+1}(t),$$

which proves (2.37) because $\varphi_{n,r+1}(\tau_{ri}) = 0$. Now using (2.39) and the periodicity of $F_r(t,u)$ we obtain

$$\int_0^{2\pi} |F_r(t,u)|\, dt = \int_0^{2\pi} |F_r(u, t-2\beta_r)|\, dt = \int_0^{2\pi} |F_r(u,t)|\, dt = \pi|\varphi_{n,r+1}(u)|.$$

The lemma is proved. □

The representation (2.32) and the identities (2.37) and (2.38) allow us to get immediately some of the exact estimates obtained earlier in Section 5.2.2 with a different technique. If $f \in W_\infty^{r+1}$ then we have for every t

$$|f(t) - \sigma_{2n,r}(f,t)| = \left| \frac{1}{\pi}\int_0^{2\pi} F_r(t,u) f^{(r+1)}(u+\beta_r)\, du \right|$$

$$\leq \frac{1}{\pi}\int_0^{2\pi} |F_r(t,u)|\, du = |\varphi_{n,r+1}(t)|.$$

If $f \in W_p^{r+1}$ $(1 \leq p \leq \infty)$ then

$$\|f - \sigma_{2n,r}(f)\|_1 = \frac{1}{\pi}\int_0^{2\pi} \left| \int_0^{2\pi} F_r(t,u) f^{(r+1)}(u+\beta_r)\, du \right| dt$$

$$\leq \frac{1}{\pi}\int_0^{2\pi} \left(\int_0^{2\pi} |F_r(t,u)|\, dt \right) |f^{(r+1)}(u+\beta_r)|\, du$$

$$= \int_0^{2\pi} |\varphi_{n,r+1}(u)| f^{(r+1)}(u+\beta_r)\, du$$

$$\leq \|\varphi_{n,r+1}\|_{p'} \|f^{(r+1)}\|_p \leq \|\varphi_{n,r+1}\|_{p'}, \qquad p' = p/(p-1).$$

5.3 Estimates for spline interpolation on classes $W_p^r[a, b]$

5.3.1 General theorems In the spline interpolation on an interval the boundary conditions should also be considered and we begin with a case sufficiently general in this respect.

Let, as in Section 5.1.2, $Q_{r+1} = \{0, 1, \ldots, r\}$ for a fixed $r = 0, 1, \ldots$; I_a, I_b be subsets of Q_{r+1} determining the boundary conditions at the end-points of the interval $[a, b]$. We say that the space of splines $\mathcal{N}$ defined on $[a, b]$ is *interpolating* by the spline s at the points $\tau_j \in [a, b]$ $(j = 1, 2, \ldots, N)$ with boundary conditions (I_a, I_b) if for every $f \in C^r[a, b]$ there exists a unique spline $s \in \mathcal{N}$ $(s(t) = s(f, t))$ such that $s(f, \tau_j) = f(\tau_j)$ $(j = 1, 2, \ldots, N)$ and

$$s^{(\mu)}(f, a) = f^{(\mu)}(a), \mu \in I_a; \qquad s^{(\kappa)}(f, b) = f^{(\kappa)}(b), \kappa \in I_b.$$

Let us fix two partitions

$$\Delta_N^t[a, b]: a = t_0 < t_1 < \cdots < t_N = b, \tag{3.1}$$

$$\Delta_N^\tau[a, b]: a = \tau_0 < \tau_1 < \cdots < \tau_N = b. \tag{3.2}$$

We shall connect them with the subspaces $S_m(\Delta_N^t[a, b])$ and $S_m(\Delta_N^\tau[a, b])$ of splines of degree m and defect 1 and the classes of functions $W^m H^1_{\Delta_N^t}[a, b]$ and $W^m H^1_{\Delta_N^\tau}[a, b]$ defined in Section 5.1.1 with them.

Theorem 5.3.1 Let the subspace $S_r(\Delta_N^t[a, b])$ be interpolating by the spline σ at the inner points τ_j $(j = 1, 2, \ldots, M - 1)$ of partition (3.2) with boundary conditions $I = (I_a, I_b)$, where $M + |I_a| + |I_b| = N + r + 1$. If $f \in W^r H^1_{\Delta_N^t}[a, b]$ then

$$|f(t) - \sigma(f, t)| \le |\psi(t) - \sigma(\psi, t)|, \qquad a \le t \le b,$$

where ψ is a function from $W_\infty^{r+1}[a, b]$ such that

$$\pm\psi^{(r+1)}(t) = (-1)^i, \qquad t_{i-1} < t < t_i, \qquad i = 1, 2, \ldots, N.$$

In particular, if

$$\psi(\tau_j) = 0, \qquad j = 1, 2, \ldots, M - 1,$$

$$\psi^{(\mu)}(a) = \psi^{(\kappa)}(b) = 0, \qquad \mu \in I_a, \qquad \kappa \in I_b.$$

then

$$|f(t) - \sigma(f, t)| \le |\psi(t)|, \qquad a \le t \le b.$$

The validity of this theorem follows immediately from Theorem 5.1.4 if we note that both differences $\delta(t) = f(t) - \sigma(f, t)$ and $\eta(t) = \psi(t) - \sigma(\psi, t)$ belong to $W^r H^1_{\Delta_N^t}[a, b]$ and $\eta^{(r+1)}(t) = \psi^{(r+1)}(t)$ $(t \in (t_{i-1}, t_i))$.

The boundary conditions $J = (J_a, J_b)$ are called *complementary* to

$I = (I_a, I_b)$ if every set J_a, J_b contains exactly these elements ν of Q_{r+1} for which $r - \nu$ does not belong to I_a or I_b respectively, i.e. for example

$$J_a = \{\nu: \nu \in Q_{r+1}, \nu = r - n, n \in Q_{r+1} \backslash I_a\}.$$

Theorem 5.3.2 Let the subspace $S_r(\Delta_N^t[a, b])$ be interpolating by the spline σ at the points τ_j $(j = 1, 2, \ldots, M-1)$ of partition (3.2) with boundary conditions $I = (I_a, I_b)$, where $M + |I_a| + |I_b| = N + r + 1$ and $r \notin I_a \cup I_b$ and the subspace $S_r(\Delta_N^\tau[a, b])$ be interpolating by the spline s at the points t_i $(i = 1, 2, \ldots, N-1)$ of partition (3.1) with boundary conditions $J = (J_a, J_b)$ complementary to I. If $f \in W_p^{r+1}[a, b]$ $(1 \le p \le \infty)$ then

$$\|f - \sigma(f)\|_1 \le \|\psi - s(\psi)\|_{p'}, \qquad p' = p/(p-1), \tag{3.3}$$

where $\psi \in W_\infty^{r+1}[a, b]$ and

$$\pm\psi^{(r+1)}(t) = (-1)^i, \qquad \tau_{j-1} < t < \tau_j, \qquad j = 1, 2, \ldots, M.$$

In particular if $\psi(t_i) = 0$ $(i = 1, 2, \ldots, N-1)$ and $\psi^{(\mu)}(a) = \psi^{(\kappa)}(b) = 0$ $(\mu \in J_a, \kappa \in J_b)$ then $s(\psi, t) \equiv 0$ and

$$\|f - \sigma(f)\|_1 \le \|\psi\|_{p'}. \tag{3.4}$$

Estimates (3.3) and (3.4) are exact on the class $W_p^{r+1}[a, b]$.

The proof follows a known scheme (see Lemma 5.2.2 and Theorem 5.2.3) and we shall only sketch it pointing out the places connected with the boundary conditions. Let $f \in W_p^{r+1}[a, b]$, $\delta(t) = f(t) - \sigma(f, t)$. Then

$$\delta(\tau_j) = 0, \qquad j = 1, 2, \ldots, M-1,$$

$$\delta^{(\mu)}(a) = \psi^{(\kappa)}(b) = 0, \qquad \mu \in I_a, \qquad \kappa \in I_b. \tag{3.5}$$

Take a function $g \in W_\infty^{r+1}[a, b]$ such that $g^{(r+1)}(t) = \operatorname{sgn} \delta(t)$ $(a \le t \le b)$. For the difference $\eta(t) = g(t) - s(g, t)$ we have

$$\eta(t_i) = 0, \qquad i = 1, 2, \ldots, N-1,$$

$$\eta^{(\mu)}(a) = \eta^{(\kappa)}(b) = 0, \qquad \mu \in J_a, \kappa \in J_b. \tag{3.6}$$

moreover $r \notin I_a \cup I_b$ implies $0 \in J_a \cap J_b$, i.e. $\eta(a) = \eta(b) = 0$.

If $0 \in I_a \cap I_b$ then $\delta(a) = \delta(b) = 0$ and recalling the first equalities (3.5) and $s^{(r)}(g, t) =$ constant for $t \in (\tau_{j-1}, \tau_j)$ we obtain

$$\|\delta\|_1 = \int_a^b \delta(t) g^{(r+1)}(t)\,\mathrm{d}t = -\int_a^b \delta'(t) g^{(r)}(t)\,\mathrm{d}t = -\int_a^b \delta'(t)\eta^{(r)}(t)\,\mathrm{d}t.$$

If $0 \notin I_a$ $(0 \notin I_b)$ then $r \in J_a$ $(r \in J_b)$ and then $\eta^{(r)}(a) = 0$ $(\eta^{(r)}(b) = 0)$. Because of $\eta^{(r)}(t) = g^{(r)}(t)$ for $t \in (\tau_{j-1}, \tau_j)$ we obtain the same result integrating by parts on every interval (τ_{j-1}, τ_j):

$$\|\delta\|_1 = \sum_{j=1}^{M} \int_{\tau_{j-1}}^{\tau_j} \delta(t)\, d\eta^{(r)}(t) = -\int_a^b \delta'(t)\eta^{(r)}(t)\, dt.$$

Consecutively integrating $r-1$ times on $[a, b]$ by parts the right-hand side we get

$$\|\delta\|_1 = \sum_{\nu=1}^{r-1} (-1)^\nu[\delta^{(\nu)}(b)\eta^{(r-\nu)}(b) - \delta^{(\nu)}(a)\eta^{(r-\nu)}(a)]$$
$$+ (-1)^r \int_a^b \delta^{(r)}(t)\eta'(t)\, dt.$$

In view of the complementarity of the boundary conditions I and J (see (3.5) and (3.6)) all terms in the sum are zero. From $\eta(t_i) = 0$ $(i = 0, 1, \ldots, N)$ and $\sigma^{(r)}(f, t) =$ constant for $t \in (t_{i-1}, t_i)$ we obtain for $f \in W_p^{r+1}[a, b]$

$$\|\delta\|_1 = \left|\int_a^b f^{(r)}(t)\eta'(t)\, dt\right| = \left|\int_a^b f^{(r+1)}(t)\eta(t)\, dt\right| \le \|\eta\|_{p'}.$$

But $\eta \in W^r H^1_{\Delta_N^\tau}[a, b]$ and we have in view of Theorem 5.3.1

$$\|\eta\|_{p'} \le \|\psi - s(\psi)\|_{p'},$$

i.e. inequalities (3.3) and (3.4) are proved.

The exactness of these inequalities for $1 < p \le \infty$ can be easily obtained by considering a function $f_0 \in W_p^{r+1}[a, b]$ such that

$$f_0^{(r+1)}(t) = \|\eta_0\|_{p'}^{1-p'} |\eta_0(t)|^{p'-1} \operatorname{sgn} \eta_0(t),$$

where $\eta_0 = \psi - s(\psi)$. One should only note that $\operatorname{sgn}[f_0(t) - \sigma(f_0, t)] = \pm\operatorname{sgn} \psi^{(r+1)}(t)$. For $p = 1$ the proof of the exactness of (3.3), (3.4) uses Steklov functions as in the other similar cases. □

5.3.2 Some particular cases We consider three partial cases of Theorems 5.3.1 and 5.3.2, each being of importance by itself. The existence and uniqueness of the interpolating splines in these cases was established in Section 2.4.2.

Let us begin with the situation when, as in the periodic case, the exact estimates are given in terms of the Euler perfect spline $\varphi_{N,n}$. Let $[a, b] = [0, \pi]$; let $\bar{\Delta}_N[0, \pi]$ be the uniform partition of $[0, \pi)$ by the points $t_i = i\pi/N$; let $\sigma_r(f, t)$ be the spline from $S_r(\bar{\Delta}_N[0, \pi])$ interpolating $f \in C^{r-1}[0, \pi]$ at the zeros of $\varphi_{N,r+1}$ with Leedstone boundary conditions (Section 2.4.2). In other words we have for odd r

$$\sigma_r(f, j\pi/N) = f(j\pi/N), \qquad j = 1, 2, \ldots, N,$$
$$\sigma_r^{(2\nu)}(f, 0) = f^{(2\nu)}(0), \qquad \sigma_r^{(2\nu)}(f, \pi) = f^{(2\nu)}(\pi), \nu = 1, 2, \ldots, (r-1)/2, \tag{3.7}$$

and for even r

$$\sigma_r\left(f, \frac{(2j-1)\pi}{2N}\right) = f\left(\frac{(2j-1)\pi}{2N}\right), \qquad j = 1, 2, \ldots, N,$$

$$\sigma_r^{(2\nu-1)}(f, 0) = f^{(2\nu-1)}(0), \sigma_r^{(2\nu-1)}(f, \pi) = f^{(2\nu-1)}(\pi), \qquad \nu = 1, 2, \ldots, r/2. \tag{3.8}$$

The conditions of Theorem 5.3.1 now hold and hence:

Proposition 5.3.3 If $f \in W^r H_N^1[0, \pi]$ then

$$|f(t) - \sigma_r(f, t)| \le |\varphi_{N,r+1}(t)|, \qquad 0 \le t \le \pi,$$

and the estimate cannot be improved for every $t \in [0, \pi]$ even on $W_\infty^{r+1}[0, \pi]$.

From this proposition we obtain immediately ($\|\bullet\|_q = \|\bullet\|_{L_q[0,\pi]}$)

$$\sup_{f \in W^r H_N^1[0,\pi]} \|f - \sigma_r(f)\|_q = \sup_{f \in W_\infty^{r+1}[0,\pi]} \|f - \sigma_r(f)\|_q$$

$$= \|\varphi_{N,r+1}\|_q = N^{-r-1}\|\varphi_{1,r+1}\|_q, \qquad 1 \le q \le \infty.$$

Let us remark that relations similar to (2.26) for simultaneous approximation of the derivative hold for a spline $\sigma_r(f)$ interpolating f on $[0, \pi]$ at the zeros of $\varphi_{N,r+1}$ with boundary conditions (3.7) or (3.8):

$$\sup_{f \in W^r H_N^1[0,\pi]} \|f' - \sigma_r'(f)\|_q = \sup_{f \in W_\infty^{r+1}[0,\pi]} \|f' - \sigma_r'(f)\|_q = \|\varphi_{N,r}\|_q = N^{-r}\|\varphi_{1,r}\|_q,$$

$$1 \le q \le \infty, \qquad \|\bullet\|_q = \|\bullet\|_{L_q[0,\pi]}. \tag{3.9}$$

In order to obtain a corollary of Theorem 5.3.2 in our case we introduce partition $\bar{\Delta}^\tau[0, \pi]$ of interval $[0, \pi]$ by the points $\tau_j = j\pi/N$ $(j = 0, 1, \ldots, N)$ for odd r and the points $\tau_0 = 0$, $\tau_j = (2j-1)\pi/(2N)$ $(j = 1, 2, \ldots, N)$, $\tau_{N+1} = \pi$ for even r. The subspace $S_r(\bar{\Delta}^\tau\ [0, \pi])$ interpolates at the points $i\pi/N$ $(i = 0, 1, \ldots, N)$ with boundary conditions which are complementary to (3.7) or (3.8). Moreover the perfect spline $\psi(t) = \varphi_{N,r+1}(t + \beta_r)$, where β_r is chosen so that $\psi(i\pi/N) = 0$, has knots exactly at the points $j\pi/N$ for odd r and the points $(2j-1)\pi/(2N)$ for even r. Therefore Theorem 5.3.2. implies:

Proposition 5.3.4 For all $N = 1, 2, \ldots; r = 0, 1, \ldots,$ we have the equalities ($\|\bullet\|_q = \|\bullet\|_{L_q[0,\pi]}$)

$$\sup_{f \in W_p^{r+1}[0,\pi]} \|f - \sigma_r(f)\|_1 = \|\varphi_{N,r+1}\|_{p'} = N^{-r-1}\|\varphi_{1,r+1}\|_{p'},$$

$$1 \le p \le \infty, \qquad 1/p + 1/p' = 1.$$

For $p = 1$ class $W_1^{r+1}[0, \pi]$ can be replaced by $W_V^r[0, \pi]$.

Consider now a case when there are no boundary conditions in Theorem 5.3.1, i.e. the sets I_a and I_b are empty, and the partition $\Delta_N^t[a, b]$ and the interpolating points τ_j are chosen in a special way. In Section 2.5.3 we have introduced the set $\Gamma_{N,m}[a, b]$ of perfect splines of degree m with no more than $N - 1$ knots on (a, b). Let us denote $\Gamma_{N,m}^*[a, b]$ the set of these perfect splines from $\Gamma_{N,m}[a, b]$ which have exactly $N - 1$ knots on (a, b) and the maximal possible number of zeros, i.e. $N + m - 1$, on $[a, b]$. Obviously these zeros can only be simple. Proposition 2.5.5 implies that $\Gamma_{N,m}^*[a, b]$ is not empty.

For every $m = 1, 2, \ldots$, we consider the perfect spline $\Psi_{N,m}(t) = \Psi_{N,m}(q, t)$ from $\Gamma_{N,m}[a, b]$ with minimal $L_q[a, b]$ $(1 \leq q \leq \infty)$ norm. There is always such a spline; it belongs to $\Gamma_{N,m}^*[a, b]$ and all its simple $N + m - 1$ zeros are inside (a, b). Let, for a fixed q $(1 \leq q \leq \infty)$,

$$\Delta_N^*[a, b]\colon\ a = t_0 < t_1 < \cdots < t_N = b$$

denote the partition formed from the knots of $\Psi_{N,r}(t) = \Psi_{N,r}(q, t)$, i.e. $\Psi_{N,r}^{(r)}(t) = \pm(-1)^i$ for $t_{i-1} < t < t_i$. Recalling Theorem 2.4.4 and inequalities (2.5.21), and characterizing the disposition of the zeros of the perfect spline from Proposition 2.5.5, the subspace $S_{r-1}(\Delta_N^*[a, b])$ interpolates at the zeros τ_j $(j = 1, 2, \ldots, N + m - 1)$ of spline $\Psi_{N,r}(t)$ on the interval (a, b).

Proposition 5.3.5 Let $s(f, t)$ be the spline from $S_{r-1}(\Delta_N^*[a, b])$ satisfying for $f \in C[a, b]$ the interpolating conditions

$$s(f, \tau_j) = f(\tau_j), \qquad j = 1, 2, \ldots, N + r - 1,$$

where τ_j are the zeros of $\Psi_{N,r}$. If $f \in W^{r-1}H^1_{\Delta_N^*}[a, b]$ then

$$|f(t) - s(f, t)| \leq |\Psi_{N,r}(t)|, \qquad a \leq t \leq b,$$

and the estimate cannot be unproved on $W^r_\infty[a, b]$ for every $t \in [a, b]$. Therefore $(\|\bullet\|_q = \|\bullet\|_{L_q[0,\pi]})$,

$$\sup_{f \in W^{r-1}H^1_{\Delta_N^*}[a,b]} \|f - s(f)\|_q = \sup_{f \in W^r_\infty[a,b]} \|f - s(f)\|_q = \|\Psi_{N,r}(q)\|_q,$$

$$1 \leq q \leq \infty.$$

Now let us consider the case when the maximal number of boundary conditions is being imposed in Theorem 5.3.1. Let $\Gamma_{N,m}^0[a, b]$ be the set of perfect splines ψ of degree m with no more than $N + m - 1$ knots on (a, b) which also satisfy the conditions

$$\psi^{(\nu)}(a) = \psi^{(\nu)}(b) = 0, \qquad \nu = 0, 1, \ldots, m - 1.$$

A spline $\psi \in \Gamma_{N,m}^0[a, b]$ has no more than $N - 1$ zeros on $[a, b]$. Let us denote by $\Gamma_{N,m}^{0,*}[a, b]$ the set of those perfect splines from $\Gamma_{N,m}[a, b]$ which have the maximal possible number $N+m-1$ of knots on (a,b) and the maximal possible number of zeros, i.e. $N-1$, on $[a,b]$. Proposition 2.5.6 implies that $\Gamma^{0,*}_{N,m}[a,b]$ is not empty. For every $q \in [1,\infty]$ there exists a spline $\Psi^0_{N,m}(t) = \Psi^0_{N,m}(q,t)$ from $\Gamma^0_{N,m}[a,b]$ with minimal $L_q[a,b]$ norm, moreover it belongs to $\Gamma^{0,*}_{N,m}[a,b]$ and all its zeros are simple and inside (a,b).

Consider the partition

$$\Delta_{N+r}[a,b]: a = t_0 < t_1 < \cdots < t_{N+r} = b$$

formed by the knots of $\Psi^0_{N,r}(t)$, and let τ_j $(j=1,2,\ldots,N-1,$ $a<\tau_0<\tau_1<\cdots<\tau_{N-1}<b)$ be the zeros of $\Psi^0_{N,r}(t)$. In view of Theorem 2.4.4 and inequalities (2.5.23), the subspace $S_{r-1}(\Delta_{N+r}[a,b])$ interpolates at the points τ_j $(j=1,2,\ldots,N-1)$ with boundary conditions (I_a, I_b) determined by the sets $I_a = I_b = \{0,1,\ldots,r-1\}$.

In this case Theorem 5.3.1 implies:

Proposition 5.3.6 Let $s(f,t)$ be the spline from $S_{r-1}(\Delta_{N+r}[a,b])$ satisfying for $f \in C^{r-1}[a,b]$ the interpolating conditions

$$s(f,\tau_j) = f(\tau_j), \qquad j=1,2,\ldots,N-1,$$
$$s^{(\nu)}(f,a) = f^{(\nu)}(a), \qquad s^{(\nu)}(f,b) = f^{(\nu)}(b), \qquad \nu = 0,1,\ldots,r-1, \quad (3.10)$$

where τ_j are the zeros of $\Psi^0_{N,r}(t) = \Psi^0_{N,r}(q,t)$. If $f \in W^{r-1}H^1_{\Delta_{N+r}}[a,b]$ then $|f(t) - s(f,t)| \le |\Psi^0_{N,r}(t)|$ $(a \le t \le b)$ and the following relations hold $(\|\bullet\|_q = \|\bullet\|_{L_q[a,b]})$,

$$\sup_{f \in W^{r-1}H^1_{\Delta_{N+r}}[a,b]} \|f - s(f)\|_q = \sup_{f \in W^r_\infty[a,b]} \|f - s(f)\|_q = \|\Psi^0_{N,r}(q)\|_q, \qquad 1 \le q \le \infty.$$

Let us use an application of Theorem 5.3.2 in this case. Together with partition $\Delta_{N+r}[a,b]$ in which $t_1, t_2, \ldots, t_{N+r-1}$ are the knots of spline $\Psi^0_{N,r}(q,t)$ we consider the partition

$$\Delta^\tau_N[a,b]: a = \tau_0 < t_1 < \cdots < \tau_N = b,$$

where τ_j $(j=1,2,\ldots,N-1)$ are the zeros of the same spline $\Psi^0_{N,r}(q,t)$. The subspace $S_{r-1}(\Delta^\tau_N[a,b])$ now interpolates at points $t_1, t_2, \ldots, t_{N+r-1}$ – this follows from Theorem 2.4.4 and inequalities (2.5.22). Because the boundary conditions (3.10) and the conditions determined by the empty sets $J_a = J_b = \varnothing$ are complementary, we obtain the following result from Theorem 5.3.2:

Proposition 5.3.7 Let $s(f,t)$ be the spline from $S_{r-1}(\Delta^\tau_N[a,b])$ such that $s(f,t_i) = f(t_i)$ $(i=1,2,\ldots,N+r-1)$, where t_i are the knots of spline $\Psi^0_{N,r}(q,t)$. Then

$$\sup_{f\in W^r_{q'}[a,b]} \|f - s(f)\|_{L_1[a,b]} = \|\Psi^0_{N,r}(q)\|_{L_q[a,b]}, \qquad 1 \le q \le \infty,\ 1/q + 1/q' = 1.$$

We also note that, as in the periodic case (see Section 5.2.3), the results for estimating the error for interpolation by splines from $S_m(\Delta_N[a,b])$ on classes $W_p^{m+1}[a,b]$ are valid also for the larger sets $W_p^{m+1}[a,b] \oplus \mathrm{S_m}(\Delta_N[a,b])$.

5.3.3 Estimates for Hermite spline interpolation In Section 3.4.3 we introduced Hermite splines which are uniquely determined on every interval $[t_{i-1}, t_i]$ of the partition by the values of the spline and some of its derivatives at the points t_{i-1} and t_i. It was proved there that if $S^n_{2n-1}(\Delta_N[a,b])$ is the set of all splines of degree $2n-1$ defect n with respect to partition

$$\Delta_N[a,b]\colon\ a = t_0 < t_1 < \cdots < t_N = b,$$

then for every $f \in C^{n-1}[a,b]$ there is a unique spline $s \in S^n_{2n-1}(\Delta_N[a,b])$ satisfying

$$s^{(\nu)}(f, t_i) = f^{(\nu)}(t_i); \qquad \nu = 0, 1, \ldots, n-1; \qquad i = 0, 1, \ldots, N. \tag{3.11}$$

Let us remark that, unlike the splines of minimal defect (and degree at least 2), the value of the Hermite spline $s(f,t)$ at $t \in (t_{k-1}, t_k)$ does not depend on the interpolating conditions at the points t_i $(i \ne k-1, k)$.

The Hermite spline interpolation error for every function $f \in W^{2n}_\infty[a,b]$ at every point is bounded by the values of the standard function q given by

$$q(t) = \frac{1}{(2n)!}[(t - t_{i-1})(t_i - t)]^n, \qquad t_{i-1} \le t \le t_i,\ i = 1, 2, \ldots, N. \tag{3.12}$$

Note that $q^{(\nu)}(t_i) = 0$ $(\nu = 0, 1, \ldots, n-1)$ and $q^{(2n)}(t) = (-1)^n$.

Theorem 5.3.8 If the function $f \in W^{2n}_\infty[a,b]$ and the spline $s(f) \in S^n_{2n-1}(\Delta_N[a,b])$ satisfy conditions (3.11) then

$$|f(t) - s(f,t)| \le |q(t)|, \qquad a \le t \le b. \tag{3.13}$$

The estimate cannot be improved at every point $t \in [a,b]$.

The proof follows a known procedure (see e.g. Theorem 5.1.1) but is technically simpler, because each interval of the partition can be worked separately. If $\delta(t) = f(t) - s(f,t)$ then

$$\delta^{(\nu)}(t_i) = 0, \qquad \nu = 0, 1, \ldots, n-1, \qquad i = 0, 1, \ldots, N, \tag{3.14}$$

and we have to prove that $|\delta(t)| \le |q(t)|$, $(t \in [a,b])$. Assume that this is not so and that at some point $t = t_*$ $(t_{k-1} < t_* < t_k)$ the inequality $|\delta(t_*)| > |q(t_*)|$ holds. Choose λ $(0 < |\lambda| < 1)$ such that $\lambda\delta(t_*) = q(t_*)$. Then the difference $\eta(t) = q(t) - \lambda\delta(t)$ has at least three separated zeros on $[t_{k-1}, t_k]$. Because of

$$\eta^{(\nu)}(t_{k-1}) = \eta^{(\nu)}(t_k) = 0, \qquad \nu = 1, 2, \ldots, n-1,$$

the derivative $\eta^{(n-1)}$ has at least $n+2$ separated zeros inside (a, b) and hence $\eta^{(2n-1)}$ changes sign at least twice there. But this is impossible because $|q^{(2n)}(t)| = 1$ for $t \in (t_{k-1}, t_k)$ and $|\lambda\delta^{(2n)}(t)| = |\lambda f^{(2n)}(t)| < 1$. The exactness of estimate (3.13) at every point $t \in [a, b]$ follows from $q \in W_\infty^{2n}[a, b]$ and $s(q, t) \equiv 0$. □

Setting $h_i = t_i - t_{i-1}$ $(i = 1, 2, \ldots, N)$, $h = \max_i h_i$ and calculating the norm of function (3.12) we obtain:

Corollary 5.3.9 We have for the interpolating Hermite splines $s(f, t) \in S_{2n-1}^n(\Delta_N[a, b])$

$$\sup_{f \in W_\infty^{2n}[a,b]} \|f - s(f)\|_{C[a,b]} = \frac{h^{2n}}{(2n)!\, 2^{2n}}, \tag{3.15}$$

$$\sup_{f \in W_\infty^{2n}[a,b]} \|f - s(f)\|_{L_p[a,b]} = \frac{M_{np}}{(2n)!}\left(\sum_{i=1}^{N} h_i^{2np+1}\right)^{1/p}, \qquad 1 \le p < \infty, \tag{3.16}$$

where

$$\mathrm{M}_{np} = \left[\frac{\Gamma^2(np+1)}{\Gamma(2np+2)}\right]^{1/p},$$

and where $\Gamma(u)$ is a Euler gamma-function. For a fixed N the right-hand sides of (3.15) and (3.16) for the uniform partition $(h = h_i = (b-a)/N)$ take their minimal values

$$\frac{1}{(2n)!}\left(\frac{b-a}{2N}\right)^{2n} \quad \text{and} \quad \frac{M_{np}(b-a)^{1/p}}{(2n)!}\left(\frac{b-a}{N}\right)^{2n}$$

respectively.

Using relations (3.15) and (3.16) we can, as in the case of minimal defect spline interpolation, establish exact results on classes $W_p^r[a, b]$ in the $L_1[a, b]$ metric.

Theorem 5.3.10 For the interpolating Hermite splines $s(f, t) \in S_{2n-1}^n(\Delta_N[a, b])$ we have

$$\sup_{f \in W_1^{2n}[a,b]} \|f - s(f)\|_1 = \frac{h^{2n}}{(2n)!\, 2^{2n}}, \tag{3.17}$$

$$\sup_{f \in W_p^{2n}[a,b]} \|f - s(f)\|_1 = \frac{M_{np'}}{(2n)!} \left(\sum_{i=1}^{N} h_i^{2np'+1} \right)^{1/p'}, \qquad 1 \le p \le \infty, p' = p/(p-1). \tag{3.18}$$

For a fixed N the right-hand sides of (3.17) and (3.18) take their minimal values in the case $h = h_i = (b-a)/N$, which are

$$\frac{1}{(2n)!}\left(\frac{b-a}{2N}\right)^{2n}, \quad \text{and} \quad \frac{M_{np'}(b-a)^{1/p'}}{(2n)!}\left(\frac{b-a}{N}\right)^{2n},$$

respectively.

Proof For $\delta(t) = f(t) - s(f, t)$ $(f \in W_p^{2n}[a, b])$ equalities (3.14) hold. If g is a function from $W_\infty^{2n}[a, b]$ such that $g^{(2n)}(t) = \operatorname{sgn} \delta(t)$ $(a \le t \le b)$ and $\eta(t) = g(t) - s(g, t)$, then

$$\eta^{(j)}(t_i) = 0, \qquad j = 0, 1, \ldots, n-1; \qquad i = 0, 1, \ldots, N, \tag{3.19}$$

From (3.14) with $\nu = 0$ we obtain

$$\begin{aligned} \|\delta\|_{L_1[a,b]} &= \int_a^b \delta(t) g^{(2n)}(t)\, \mathrm{d}t = \left| \int_a^b \delta'(t) g^{(2n-1)}(t)\, \mathrm{d}t \right| \\ &= \left| \sum_{i=1}^{N} \int_{t_{i-1}}^{t_i} \delta'(t) \eta^{(2n-1)}(t)\, \mathrm{d}t \right|. \end{aligned}$$

Now we integrate by parts on every interval (t_{i-1}, t_i). The terms from the integrals vanish because of (3.14) or (3.19). We obtain (here and later $\|\bullet\|_q = \|\bullet\|_{L_q[a,b]}$)

$$\begin{aligned} \|\delta\|_1 &= \left| \sum_{i=1}^{N} \int_{t_{i-1}}^{t_i} \delta^{(2n-1)}(t) \eta'(t)\, \mathrm{d}t \right| \\ &= \left| \sum_{i=1}^{N} \int_{t_{i-1}}^{t_i} f^{(2n-1)}(t) \eta'(t)\, \mathrm{d}t \right| = \left| \int_a^b f^{(2n)}(t) \eta(t)\, \mathrm{d}t \right| \le \|f^{(2n)}\|_p \|\eta\|_{p'}. \end{aligned}$$

But $\|f^{(2n)}\|_p \le 1$ and Theorem 5.3.8 implies $\|\eta\|_{p'} \le \|q\|_{p'}$ because of $g \in W_\infty^{2n}[a, b]$. Therefore

$$\|f - s(f)\|_1 \le \|q\|_{p'} \qquad \forall f \in W_p^{2n}[a, b]. \tag{3.20}$$

For $1 < p \le \infty$ a function $f \in L_p^{2n}[a, b]$ with

$$f^{(2n)}(t) = \|q\|_p^{1-p'} \, |q(t)|^{p'-1} \operatorname{sgn} q(t)$$

belongs to $W_p^{2n}[a, b]$ and (3.20) is fulfilled as an equality for it. In the case

$p = 1$ one can easily get a function $f_\varepsilon \in W_1^{2n}[a, b]$ for every $\varepsilon > 0$ such that $\|f_\varepsilon - s(f_\varepsilon)\|_1 > \|q\|_{C[a,b]} - \varepsilon$.

We leave the proof of the last statement of Theorem 5.3.10 and Corollary 5.3.9 to the reader.

As in the problem of minimal defect spline interpolation here we also have another approach, based on the integral representation of the error, for obtaining exact results. In view of the local nature of Hermite splines we shall restrict our considerations to one interval from the partition $[\alpha, \beta]$, say.

Taylor's formula (1.5.11) implies for $f \in L_1^m[a, b]$

$$f(t) = p_m(t) + \int_\alpha^\beta \psi_m(t, u) f^{(m)}(u)\, du, \qquad \alpha \le t \le \beta, \qquad (3.21)$$

where

$$p_m(t) = \frac{1}{2} \sum_{k=0}^{m-1} \frac{1}{k!} [f^{(k)}(\alpha)(t-\beta)^k + f^{(k)}(\beta)(t-\alpha)^k],$$

$$\psi_m(t, u) = \begin{cases} \dfrac{1}{2} \dfrac{(t-u)^{m-1}}{(m-1)!}, & \alpha \le u \le t, \\ -\dfrac{1}{2} \dfrac{(t-u)^{m-1}}{(m-1)!}, & t \le u \le \beta. \end{cases}$$

The Hermite polynomial $H(f, t)$ of degree $2n - 1$ which is defined by the equalities

$$H^{(\nu)}(f, \alpha) = f^{(\nu)}(\alpha), \qquad H^{(\nu)}(f, \beta) = f^{(\nu)}(\beta), \qquad \nu = 0, 1, \ldots, n-1,$$

can be represented as

$$H(f, t) = \sum_{k=0}^{n-1} [f^{(k)}(\alpha) H_k(t-\alpha) + f^{(k)}(\beta) H_k(\beta - t)], \qquad (3.22)$$

where $H_k(u)$ are basic Hermite polynomials

$$H_k^{(\nu)}(0) = \delta_{k\nu}. \qquad H_k^{(\nu)}(\beta - \alpha) = 0, \qquad \nu = 0, 1, \ldots, n-1.$$

Recalling that $H(p, t) \equiv p(t)$ for every polynomial p of degree $2n - 1$, from (3.21) and (3.22) for every $f \in L_1^m[\alpha, \beta]$ $(n \le m \le 2n)$ we obtain

$$f(t) - H(f, t) = \int_\alpha^\beta [\psi_m(t, u) - H(\psi_m(\bullet, u), t)] f^{(m)}(u)\, du, \qquad \alpha \le t \le \beta, \qquad (3.23)$$

and the problem is reduced to investigating the function

$$Q_m(t,u) = \psi_m(t,u) - H(\psi_m(\bullet,u),t), \qquad \alpha \le t \le \beta, n \le m \le 2n,$$

which can be easily expressed by polynomials H_k.

The properties of Q_m which are important for us are:

$$\operatorname{sgn} Q_{2n}(t,u) = (-1)^n, \qquad \alpha \le t, u \le \beta,$$

$$\frac{\partial^\nu}{\partial u^\nu} Q_m(t,u)|_{u=\alpha,\beta} = 0, \qquad \nu = 0, 1, \ldots, n-1,$$

$$\max_{\alpha \le t, u \le \beta} |Q_{2n}(t,u)| = |Q_{2n}((\alpha+\beta)/2, (\alpha+\beta)/2)|$$

$$= (\beta-\alpha)^{2n-1}(2n-1)^{-1}[(2n-2)!!]^{-2}2^{-2n} =: \eta_n(\alpha,\beta), \tag{3.24}$$

$$\frac{\partial}{\partial u} Q_m(t,u) = -Q_{m-1}(t,u), \qquad n+1 \le m \le 2n. \tag{3.25}$$

Moreover, for a fixed $t \in (\alpha,\beta)$ function $Q_{2n-1}(t,u)$ changes sign once on (α,β).

Now if $f \in W_\infty^{2n-1}[\alpha,\beta]$ then we have for every $t \in [\alpha,\beta]$

$$|f(t) - H(f,t)| \le \int_\alpha^\beta |Q_{2n-1}(t,u)|\,\mathrm{d}u \le 2 \max_{\alpha \le u \le \beta} |Q_{2n}(t,u)| \le 2\eta_n(\alpha,\beta), \tag{3.26}$$

and in the case $f \in W_1^{2n}[\alpha,\beta]$

$$|f(t) - H(f,t)| \le \max_{\alpha \le u \le \beta} |Q_{2n}(t,u)| = \eta_n(\alpha,\beta). \tag{3.27}$$

For a function $f \in W_\infty^{2n-1}[\alpha,\beta]$ with a derivative $f^{(2n-1)}$ equal to 1 on $[\alpha, (\alpha+\beta)/2]$ and -1 on $[(\alpha+\beta)/2, b]$ we have equalities in (3.26). The exactness of (3.27) on $W_1^{2n}[\alpha,\beta]$ can be verified by constructing a function f such that $f^{(2n)}$ has a singularity in the neighborhood of point $(\alpha+\beta)/2$.

Therefore we have proved:

Theorem 5.3.11 For the interpolating Hermite splines $s(f,t) \in S_{2n-1}^n(\Delta_N[a,b])$ we have

$$\sup_{f \in W_\infty^{2n-1}[a,b]} \max_{t_{i-1} \le t \le t_i} |f(t) - s(f,t)| = 2 \sup_{f \in W_1^{2n}[a,b]} \max_{t_{i-1} \le t \le t_i} |f(t) - s(f,t)|$$

$$= \frac{h_i^{2n-1}}{(2n-1)[(2n-2)!!]^2 2^{2n-1}},$$

where $h_i = t_i - t_{i-1}$, t_k are the points from the partition $\Delta_N[a,b]$.

Up to now we have considered interpolation by Hermite splines of odd degree $2n-1$. In order to interpolate a function $f \in C^n[a,b]$ by a unique spline of even degree $2n$ in Section 2.4.2, we have added the knots $z_i = (t_{i-1}+t_i)/2$ $(i=1,2,\ldots,N)$ to the partition $\Delta_N[a,b]$ and we have denoted by $S_{2n}^n(\Delta'_N[a,b])$ the linear manifold of splines s of degree $2n$ with defect n at every knot t_i $(i=1,2,\ldots,N-1)$ and defect 1 at every knot z_i $(i=1,2,\ldots,N)$. The spline $s \in S_{2n}^n(\Delta'_N[a,b])$ is uniquely determined (see Proposition 2.4.7) by the conditions

$$s^{(\nu)}(f,t_i) = f^{(\nu)}(t_i), \qquad \nu = 0,1,\ldots,n; \qquad i=1,2,\ldots,N. \quad (3.28)$$

When interpolating functions from $W_\infty^{2n+1}[a,b]$ by Hermite splines from $S_{2n}^n(\Delta'_N[a,b])$ the error is estimated by the values of the standard function

$$\psi(t) = \psi_i(t), \qquad t_{i-1} \le t \le t_i, \qquad i=1,2,\ldots,N,$$

where ψ_i is a perfect spline of degree $2n+1$ defined on $[t_{i-1},t_i]$ such that $\psi_i^{(2n+1)}$ changes sign exactly at z_i and

$$\psi_i^{(\nu)}(t_{i-1}) = \psi_i^{(\nu)}(t_i) = 0, \qquad \nu = 0,1,\ldots,n.$$

Theorem 5.3.12 If $f \in W_\infty^{2n+1}[a,b]$ and if the spline $s(f) \in S_{2n}^n(\Delta'_N[a,b])$ satisfies conditions (3.28) then

$$|f(t) - s(f,t)| \le |\psi(t)|, \qquad a \le t \le b. \quad (3.29)$$

The estimate cannot be improved at every point $t \in [a,b]$.

The proof is similar to the proof of Theorem 5.3.8. Assuming that for some $t_* \in (t_{k-1},t_k)$ (3.29) does not hold we conclude that for some λ $(0<|\lambda|<1)$ the function $\psi_k^{(2n)}(t) - \lambda[f^{(2n)}(t) - s^{(2n)}(f,t)]$ changes sign on the interval (t_{k-1},t_k) at least three times. But this is impossible because the difference $\psi_k^{(2n+1)}(t) - \lambda f^{(2n+1)}(t)$ changes sign on this interval only once. The exactness of inequality (3.29) is checked on function ψ.

From Theorem 5.3.12 we immediately obtain

$$\sup_{f \in W_\infty^{2n+1}[a,b]} \|f - s(f)\|_{L_p[a,b]} = \|\psi\|_{L_p[a,b]}, \qquad 1 \le p \le \infty,$$

and, in particular

$$\sup_{f \in W_\infty^{2n+1}[a,b]} \|f - s(f)\|_{C[a,b]} = \frac{h^{2n+1}}{(2n+1)[(2n)!!]^2 2^{2n+1}} \qquad h = \max_i (t_i - t_{i-1}),$$

$$\sup_{f \in W_\infty^{2n+1}[a,b]} \|f - s(f)\|_{L_1[a,b]} = \sum_{i=1}^N h_i^{2n+2}(2^{2n+1}(2n+2)!)^{-1}, \qquad h_i = t_i - t_{i-1}.$$

The same technique which we have used for deriving Theorem 5.3.10 from Theorem 5.3.8 gives in this case

$$\sup_{f \in W_p^{2n+1}[a,b]} \|f - s(f)\|_{L_1[a,b]} = \|\psi\|_{L_{p'}[a,b]}, \qquad 1 \le p \le \infty, 1/p + 1/p' = 1.$$

If we compare the estimates for the errors of interpolation of functions for a fixed N from one and the same class $W_p^m[a, b]$ by Hermite splines and splines with minimal defect (e.g. with Leedstone boundary conditions) then we observe a greater precision for Hermite interpolation. But one should not forget that the price is, essentially, the increase in the dimension of the approximating subspace (e.g. $\dim S_{2n-1}^n(\Delta_N[a, b]) = (N + 1)n$) and hence the number of interpolating conditions.

Let us note that, since the Hermite spline is a projector on the corresponding subspace, the results for estimating the error for Hermite spline interpolation on the classes $W_p^{2n}[a, b]$ and $W_p^{2n+1}[a, b]$ are valid also for the largers sets $W_p^{2n} \oplus S_{2n-1}^n(\Delta_N[a, b])$ and $W_p^{2n+1}[a, b] \oplus S_{2n}^n(\Delta_N[a, b])$ (see Section 5.2.3).

5.3.4 Local splines of minimal defect We have mentioned in Section 2.4.4 that in order to compute an interpolating spline $s(f, t)$ from $S_m(\Delta_N[a, b])$ it is more convenient to represent it as a linear combination of B-splines which have local supports:

$$s(f, t) = \sum_{i=-m}^{N-1} c_i B_{m,i}(t). \tag{3.30}$$

But in order to determine the coefficients $c_i = c_i(f)$ in this case we have to solve a system of equations involving the values of f at all points of interpolation τ_j. It was noted that the dependence of the value of $s(f, t)$ at a fixed point t on the values $f(\tau_j)$ decreases with moving τ_j from t.

In this context one considers local splines

$$l(f, t) = \sum_{i=1}^{s} d_i(f) B_{m,i}(t), \qquad 1 \le s \le N, \tag{3.31}$$

in which the value of $d_i(f)$ is a linear combination of the values $f(\tau_j)$ only for τ_j near to t.

We assume that the partition of $[a, b] = [0, 1]$ is the uniform one: $t_i = i/N$ $(i = 0, 1, \ldots, N)$. We also assume that the values of function f at points $\tau_j = j/N - [1 + (-1)^m]/(4N)$ are known and that B-splines $B_{m,i}$ are normalized by the conditions

$$\max_t B_{m,i}(t) = B_{m,i}(\tau_i), \qquad \sum_i B_{m,i}(t) \equiv 1.$$

In the simple case when $d_i(f) = f(\tau_i)$ in (3.31) we obtain a linear positive operator. Such operators (see Section 4.3.4) cannot provide an order of the approximation error better than $O(N^{-2})$.

If we set

$$l_m(f, t) = \sum_{i=1}^{N} \sum_{k=-n}^{n} \gamma_k f(\tau_{i+k}) B_{m,i}(t), \qquad n = \left[\frac{m}{2}\right], \tag{3.32}$$

and if coefficients γ_k are chosen from the condition $l_m(f, t) \equiv f(t)$ with f being a polynomial of degree m then it turns out that the operator (3.32) provides the best order of approximation on the classes $W_\infty^{m+1}[0, 1]$, $W_\infty^m[0, 1]$. It is also clear that the calculation of the coefficients of the spline (3.32) is much simpler than the calculation of the coefficients of the interpolation spline (3.30) when m is much smaller than N. But in order to determine the splines (3.30) and (3.32) on the whole interval $[0, 1]$ we have to extend the system of points $\{t_i\}$ and $\{\tau_j\}$ slightly beyond this interval and somehow to continue function f on an interval $[a, b]$ containing the supports of all B-splines from (3.30) or (3.32).

From Taylor's formula

$$f(t) = \sum_{\nu=0}^{m} a_\nu t^\nu + \frac{1}{m!} \int_a^b (t-u)_+^m f^{(m+1)}(u)\, du, \qquad f \in L_1^{m+1}[a, b],$$

and the choice of γ_k in (3.32) we obtain

$$f(t) - l_m(f, t) = \int_a^b K_m(t, u) f^{(m+1)}(u)\, du, \qquad 0 \le t \le 1, \tag{3.33}$$

where

$$K_m(t, u) = \frac{1}{m!}\left[(t-u)_+^m - \sum_{i=1}^{N} \sum_{k=-n}^{n} \gamma_k (\tau_{i+k} - u)_+^m B_{m,i}(t)\right].$$

Without studying in detail the properties of the kernel $K_m(t, u)$ (see [10A, p. 318]) we give the general relation

$$\sup_{f \in W_p^{m+1}[0,1]} \|f - l_m(f)\|_{C[0,1]} = \max_t \|K_m(t, \bullet)\|_{L_{p'}[0,1]}, \qquad 1/p + 1/p' = 1,$$

which is obtained from (3.33) in the usual way, and consider some particular cases.

Case 1 $m = 3$. Then $n = 1$, $\tau_j = j/N$ in (3.32) and the coefficients γ_k which provide exactness for the cubic polynomials are $\gamma_{-1} = \gamma_1 = -\frac{1}{6}$, $\gamma_0 = \frac{4}{3}$. The calculation of the maximum on t of the norm $\|K_3(t, \bullet)\|_{p'}$ for $p' = 1, \infty$ gives the following estimates ($\|\bullet\|_X = \|\bullet\|_{X[0,1]}$)

$$\|f - l_3(f)\|_C \le \frac{35}{1152}\|f^{(4)}\|_C N^{-4}, \qquad f \in C^4[0, 1],$$

$$\|f - l_3(f)\|_C \le \frac{185}{6912}\|f^{(4)}\|_{L_1} N^{-3}, \qquad f \in L_1^4[0, 1], \tag{3.34}$$

$$\|f - l_3(f)\|_C \le \frac{185}{3456}\|f^{(3)}\|_C N^{-3}, \qquad f \in C^3[0, 1],$$

each of which is exact. Let us note that estimate (3.34) is slightly worse than the exact estimate for the best approximation by the same subspaces of splines with periodic boundary conditions: in the last case the constant on the right-hand side is 5/384 (see Section 5.2).

Case 2 $m = 2$. In this case $n = 1$ too, but now $\tau_j = (2j - 1)/N$ and the coefficients γ_k which provide exactness for the quadratic polynomials are $\gamma_{-1} = \gamma_1 = -\frac{1}{8}$, $\gamma_0 = \frac{5}{4}$. We have the following exact estimates ($\|\bullet\|_X = \|\bullet\|_{X[0,1]}$):

$$\|f - l_2(f)\|_C \le \frac{3}{64}\|f^{(3)}\|_C N^{-3}, \qquad f \in C^3[0, 1],$$

$$\|f - l_2(f)\|_C \le \frac{9}{128}\|f^{(3)}\|_{L_1} N^{-2}, \qquad f \in L_1^3[0, 1], \tag{3.35}$$

$$\|f - l_2(f)\|_C \le \frac{9}{32}\|f^{(2)}\|_C N^{-2}, \qquad f \in C^2[0, 1].$$

The exact estimate corresponding to (3.35) for the best approximation in the periodic case has a constant 1/24 on the right-hand side.

5.4 Best approximation by splines of minimal defect

5.4.1 The duality problem in the periodic case In Section 5.2 we have used splines of minimal defect with respect to a fixed partition as an approximation tool. The estimates obtained there for the interpolation error of splines $\sigma_{2n,r-1}(f, t) \in S_{2n,r-1}$ on classes W_∞^r in the L_q metric and on classes W_p^r in the L_1 metric cannot be improved even in the sense of best spline approximation. In other words we have found the exact values of $E(W_\infty^r, S_{2n,m})_q$ and $E(W_p^r, S_{2n,m})_1$ for $m = r - 1$.

What is the behaviour of these quantities when we increase the degree m for fixed p and r? It will follow from the results of Chapter 8 that they cannot decrease. We shall show here that in many cases they do not increase either. It turns out that for every $m \ge r - 1$ the best approximation by subspace

$S_{2n,m}$ of class W_∞^r in C and of class W_p^r in L_1 is the same. One can establish this fact using the duality relations from Section 1.5 for classes W_p^r of periodic functions.

If

$$\Delta_N = \Delta_N[0, 2\pi):\ 0 = t_0 < t_1 < \cdots < t_N = 2\pi \tag{4.1}$$

is a fixed partition then we have by Proposition 1.5.4

$$E(W_p^r, S_m(\Delta_N))_q = \sup\{E_1(g)_{p'}: g \in W_{q'}^r[S_m(\Delta_N)]\}, \tag{4.2}$$

where

$$E_1(g)_p = \inf_c \|g - c\|_p,$$

$$W_p^r[S_m(\Delta_N)] = \{g: g \in W_p^r, g^{(r)} \perp S_m(\Delta_N)\}. \tag{4.3}$$

The specific features of the splines allow us to derive from the orthogonality condition $g^{(r)} \perp S_m(\Delta_N)$ explicit information for the properties of the functions in $W_p^r[S_m(\Delta_N)]$ connected with the partition Δ_N.

Lemma 5.4.1 Let partition (4.1) be fixed. A necessary and sufficient condition for the orthogonality of a function $f \in L_1$ to the set $S_0(\Delta_N)$ is

$$\int_{t_{i-1}}^{t_i} g(t)\,\mathrm{d}t = 0, \qquad i = 1, 2, \ldots, N. \tag{4.4}$$

Indeed, if we denote the left-hand side in (4.4) by α_i then condition $g \perp S_0(\Delta_N)$ means $c_1\alpha_1 + c_2\alpha_2 + \cdots + c_N\alpha_N = 0$ for every c_i. In particular for $c_i = \alpha_i$ we obtain $\alpha_1^2 + \alpha_2^2 + \cdots + \alpha_N^2 = 0$, i.e. $\alpha_i = 0$ $(i = 1, 2, \ldots, N)$. The sufficiency is obvious.

Lemma 5.4.2 A sufficient condition for $g^{(m)} \perp S_m(\Delta_N)$, $g \in L_1^m$ $(m = 1, 2, \ldots)$, is condition (4.4). If $g \perp 1$ then this condition is necessary too.

Indeed if $\psi \in S_m(\Delta_N)$ then integration by parts gives

$$\int_0^{2\pi} g^{(m)}(t)\psi(t)\,\mathrm{d}t = (-1)^m \int_0^{2\pi} g(t)\psi^{(m)}(t)\,\mathrm{d}t,$$

and hence condition $g^{(m)} \perp S_m(\Delta_N)$ is equivalent to $g \perp \psi^{(m)}\ \forall\, \psi \in S_m(\Delta_N)$. Because $\psi^{(m)} \in S_0(\Delta_N)$ it remains to apply Lemma 5.4.1 and to take into account that $\psi^{(m)} \perp 1$.

If $g \in W_p^r[S_m(\Delta_N)]$ and $g \perp 1$ then for

$$g_1(t) = \int_0^t g(u)\,\mathrm{d}u \tag{4.5}$$

we have the equalities $g_1(t_i) = 0$ $(i = 0, 1, \ldots, N)$ in view of Lemma 5.4.2. This motivates us to introduce the classes

$$W_p^r(\Delta_N) = \{g: g \in W_p^r, g(t_i) = 0, i = 0, 1, \ldots, N\},$$
$$m = 1, 2, \ldots; \qquad 1 \le p \le \infty, \tag{4.6}$$

determined by partition (4.1).

From (4.6) we obtain directly:

Proposition 5.4.3 If $g \in W_p^m(\Delta_N)$ then $g^{(k)} \perp S_{k-1}(\Delta_N)$ $(k = 1, 2, \ldots, m)$.

Indeed we get $g' \perp S_0(\Delta_N)$ from $g \in W_p^m(\Delta_N)$ which, combined with $\psi^{(k-1)} \in S_0(\Delta_N)$, implied by $\psi \in S_{k-1}(\Delta_N)$, gives

$$\int_0^{2\pi} g^{(k)}(t)\psi(t)\,\mathrm{d}t = (-1)^{k-1}\int_o^{2\pi} g'(t)\psi^{(k-1)}(t)\,\mathrm{d}t = 0.$$

Let us investigate the connection between classes (4.3) and (4.6) for $m \ge r - 1$. First let $m = r - 1$. If $g \in W_p^r(\Delta_N)$ then in view of Lemma 5.4.2, $g' \in W_p^{r-1}[S_{r-1}(\Delta_N)]$ and hence $g \in W_p^r[S_{r-1}(\Delta_N)]$. On the other hand we get from $g \in W_p^r[S_{r-1}(\Delta_N)]$ and Lemma 5.4.2 that $\int_{t_{i-1}}^{t_i} g'(t)\,\mathrm{d}t = 0$ $(i = 1, 2, \ldots, N)$ which implies $g \in W_p^r(\Delta_N)$ as soon as $g(0) = 0$. Therefore we have the following equality (in the sense of coincidence of functional sets):

$$W_p^r(\Delta_N) = \{g: g \in W_p^r[S_{r-1}(\Delta_N)], g(0) = 0\}, \qquad r = 1, 2, \ldots. \tag{4.7}$$

We shall show that the equality

$$\{\psi: \psi \in W_p^r[S_m(\Delta_N)], \psi \perp 1\} = \{\psi: \psi = g^{(m-r+1)}, g \in W_p^{m+1}(\Delta_N)\}, \tag{4.8}$$

holds for $m \ge r$ (in the same sense).

If $\psi \in W_p^r[S_m(\Delta_N)]$, $\psi \perp 1$ then ψ can be considered as the $(m - r)$th derivative of a function $g \in W_p^{m+1}(\Delta_N)$ $(g \perp 1)$. Moreover $g^{(m)} = \psi^{(r)}$ and hence $g^{(m)} \perp S_m(\Delta_N)$. Lemma 5.4.2 implies that the function g satisfies equalities (4.4) and hence its integral (4.5) belongs to the class $W_p^{m+1}(\Delta_N)$. On the other hand, if $g \in W_p^{m+1}(\Delta_N)$ then the function $\psi = g^{(m-r+1)}$ belongs to W_p^r and according to Proposition 5.4.3 its derivative $\psi^{(r)} = g^{(m+1)}$ is orthogonal to the subspace $S_m(\Delta_N)$. This means $\psi \in W_p^r[S_m(\Delta_N)]$ and also $\psi \perp 1$.

Taking (4.7) and (4.8), into account we can write (4.2) as

$$E(W_p^r, S_m(\Delta_N))_q = \sup\{E_1(g^{(m-r+1)})_{p'}: g \in W_{q'}^{m+1}(\Delta_N)\}, \tag{4.9}$$

$$r = 1, 2, \ldots,; \quad m \ge r - 1; \quad 1 \le p, q \le \infty; \quad 1/p + 1/p' = 1/q + 1/q' = 1.$$

5.4.2 Basic results First we identify some extremal properties of the class (4.6). We require that N is even: $N = 2n$.

For every partition

$$\Delta_{2n}: 0 = t_o < t_1 < \cdots < t_{2n} = 2\pi \tag{4.10}$$

there is, in view of Theorem 2.5.2, a unique (up to a sign) 2π-periodic perfect spline $\varphi_m(t) = \varphi_m(t; \Delta_{2n})$ of degree $m \geq 1$ with $2n$ knots on the period satisfying the conditions $\varphi_m(t_i) = 0$, where t_i $(i = 1, 2, \ldots, 2n)$ are the points from partition Δ_{2n}. Moreover these zeros are simple and the spline has no other zeros on $[0, 2\pi]$. An application of Theorem 5.1.1 gives:

Theorem 5.4.4 If $g \in W^m_\infty(\Delta_{2n})$ then $|g(t)| \leq |\varphi_m(t)|$ for all t and hence

$$\sup_{g \in W^m_\infty(\Delta_{2n})} \|g\|_q = \|\varphi_m\|_{q'} \qquad 1 \leq q \leq \infty, \qquad m = 1, 2, \ldots.$$

We can further advance in the case when partition $\Delta_{2n} = \bar{\Delta}_{2n}$ is the uniform one, i.e. $t_i = i\pi/n$. In this case the perfect spline $\varphi_m(t, \bar{\Delta}_{2n})$ is $\varphi_{n,m}(t - \gamma_m)$ where γ_m is chosen from conditions $\varphi_{n,m}(i\pi/n - \gamma_m) = 0$, i.e. $\gamma_m = 0$ for even m and $\gamma_m = \pi/(2n)$ for odd m. Obviously $\varphi_{n,m}(t - \gamma_m)$ belongs to $W^m_\infty(\bar{\Delta}_{2n})$.

Proposition 5.4.5 If $g \in W^m_\infty(\bar{\Delta}_{2n})$ then $|g(t)| \leq |\varphi_{n,m}(t - \gamma_m)|$ for all t and

$$\sup_{g \in W^m_\infty(\bar{\Delta}_{2n})} \|g^{(k)}\|_q = \|\varphi_{n,m-k}\|_q, \qquad k = 0, 1, \ldots, m - 1; \qquad 1 \leq q \leq \infty. \tag{4.11}$$

Indeed the first part of the statement is contained in Proposition 5.4.4 and because of $\|g\|_C \leq \|\varphi_{n,m}\|_C$ the Theorem 3.3.4, together with the definition of the class $W^m_\infty(\bar{\Delta}_{2n})$, immediately implies (4.11).

Proposition 5.4.6 We have

$$\sup_{g \in W^m_p(\bar{\Delta}_{2n})} \|g\|_1 = \|\varphi_{n,m}\|_{p'}, \qquad 1 \leq p \leq \infty, \qquad p' = p/(p-1). \tag{4.12}$$

Proof Let $g \in W^m_p(\bar{\Delta}_{2n})$, $\psi_0(t) = \operatorname{sgn} g(t)$ and

$$\psi_1(t) = \int_0^t \psi_0(t)\,dt + c, \qquad 0 \leq t \leq 2\pi, \qquad \psi_1(t + 2\pi) = \psi_1(t),$$

where the constant c is chosen so that $\psi_1 \perp 1$. We have

$$\|g\|_1 = \int_0^{2\pi} g(t)\psi_0(t)\,dt = (-1)^m \int_0^{2\pi} g^{(m)}(t)\,\psi_m(t)\,dt,$$

where ψ_m is the $(m-1)$st periodic integral of ψ_1. Because of Proposition 5.4.3 $g^{(m)} \perp S_{2n,m-1}$ and for every $s \in S_{2n,m-1}$ we have

$$\|g\|_1 = (-1)^m \int_0^{2\pi} g^{(m)}(t)[\psi_m(t) - s(t)]\, dt. \tag{4.13}$$

Let us choose the spline s from the conditions $s(\theta_{m,j}) = \psi_m(\theta_{m,j})$ where $\theta_{m,j} = j\pi/n$ for even m and $\theta_{m,j} = (2j-1)\pi/n$ for odd m; this spline exists in view of Corollary 2.4.11. If $\delta(t) = \psi_m(t) - s(t)$ then $\delta \in C^{m-2}$, $\delta(\theta_{m,j}) = 0$ and $\delta^{(m-1)}(t) = \psi_1(t) - c_i$ $(t_{i-1} \le t \le t_i)$, i.e. $\delta \in W^{m-1}H^1_{2n}$. From $g \in W^r_p$, from Hölder's inequality applied to the right-hand side of (4.13) and from Corollary 5.1.2 we get the estimate

$$\|g\|_1 \le \|\varphi_{n,m}\|_{p'}, \tag{4.14}$$

and it remains to obtain its exactness.

For $1 < p \le \infty$ let g_0 be a function from L_p such that

$$g_0^{(m)}(t) = \|\varphi_{n,m}\|_p^{1-p'} |\varphi_{n,m}(t)|^{p'-1} \operatorname{sgn} \varphi_{n,m}(t)$$

and $g_0 \perp 1$. It is easy to see that $\|g_0^{(m)}\|_p = 1$, $\operatorname{sgn} g_0(t) = \pm\varphi_{n,0}(t)$ and hence $g_0 \in W^m_p(\bar{\Delta}_{2n})$. Moreover

$$\|g\|_1 = \left|\int_0^{2\pi} g_0(t)\psi_{n,0}(t)\, dt\right| = \left|\int_0^{2\pi} g_0^{(m)}(t)\psi_{n,m}(t)\, dt\right| = \|\varphi_{n,m}\|_{p'}.$$

In the case $p = 1$ the exactness of the estimate (4.14) for $g \in W^m_1(\bar{\Delta}_{2n})$ is obtained by considering the Stekhlov function of $\varphi_{n,m-1}(t)/(4n)$ for small h. □

Concerning the norm $\|g^{(k)}\|_1$ of the derivatives of a function from $W^m_p(\bar{\Delta}_{2n})$ unlike (4.11) we can get exact estimates (except in the case $p = \infty$) only for $p = 1$.

Proposition 5.4.7 For $g \in W^m_1(\bar{\Delta}_{2n})$ we have inequalities

$$\|g^{(k)}\|_1 = \|\varphi^{(k)}_{n,m}\|_C = K_{m-k} n^{k-m}, \qquad k = 1, 2, \ldots, m-1. \tag{4.15}$$

Proof Fix k and set $\psi(t) = \operatorname{sgn} g^{(k)}(t)$ and

$$\Phi(t) = \int_0^{2\pi} g(t+u)\psi(u)\, du.$$

Then we obtain from (4.14)

$$\|\Phi\|_C \le \|g\|_1 \le \|\varphi_{n,m}\|_C. \tag{4.16}$$

Because of

$$\Phi^{(j)}(t) = \int_0^{2\pi} g^{(j)}(t+u)\psi(u)\,du, \qquad j = 1, 2, \ldots, m,$$

we have

$$\|\Phi^{(k)}\|_C \geq \Phi^{(k)}(0) = \|g^{(k)}\|_1 \tag{4.17}$$

and

$$\|\Phi^{(m)}\|_C \leq \|g^{(m)}\|_1 \, \|\psi\|_\infty \leq 1. \tag{4.18}$$

Now we apply Kolmogorov's inequality (3.3.12) and we use estimates (4.16)–(4.18):

$$\|g^{(k)}\|_1 \leq \|\Phi^{(k)}\|_C \leq K_{m-k} K_m^{k/m-1} \|\varphi_{n,m}\|_C^{1-k/m} = K_{m-k} n^{k-m} = \|\varphi_{n,m}^{(k)}\|_C.$$

The exactness of the inequalities (4.15) is checked on the Stekhlov functions of $\varphi_{n,m-1}(t)/(4n)$. □

Proposition 5.4.5 together with relation (4.9) allows us to get new exact estimates for the best approximation by subspaces $S_{2n,m}$ of classes W_p^r. It seems that these estimates cannot be obtained using spline interpolation.

Theorem 5.4.8 For all $n, r = 1, 2, \ldots,$ and $m \geq r-1$ we have

$$E(W_p^r, S_{2n,m})_1 = \|\varphi_{n,r}\|_{p'} = n^{-r}\|\varphi_{1,r}\|_{p'}, \qquad 1 \leq p \leq \infty, \qquad 1/p + 1/p' = 1; \tag{4.19}$$

$$E(W_\infty^r, S_{2n,m})_C = \|\varphi_{n,r}\|_C = K_r n^{-r}. \tag{4.20}$$

Proof Starting with (4.9) and using (4.11) for $m \geq r-1$ we have

$$\begin{aligned} E(W_p^r, S_{2n,m})_1 &\leq \sup\{\|g^{(m-r+1)}\|_{p'} : g \in W_\infty^{m+1}(\bar{\Delta}_{2n})\} \\ &= \|\varphi_{n,m+1}^{(m-r+1)}\|_{p'} = \|\varphi_{n,r}\|_{p'}, \end{aligned}$$

which together with Proposition 5.2.7 gives (4.19).

Also from (4.9) and from Proposition 5.4.7 follows for $m \geq r-1$ that

$$E(W_\infty^r, S_{2n,m})_C \leq \sup\{\|g^{(m-r+1)}\|_1 : g \in W_1^{m+1}(\bar{\Delta}_{2n})\} = \|\varphi_{n,r}\|_C = K_r n^{-r},$$

which gives (4.20) together with Proposition 5.2.5. □

Let us remark that (4.9), (4.12) and Proposition 5.2.5 imply the equality

$$E(W_\infty^r, S_{2n,r-1})_q = \|\varphi_{n,r}\|_q, \qquad 1 \leq q \leq \infty, \tag{4.21}$$

which was obtained in Section 5.2 via spline interpolation. It seems that (4.21) remains valid if we replace $S_{2n,r-1}$ with every subspace $S_{2n,m}$ $(m \geq r-1)$ but we have no proof of this fact.

The results (4.19) and (4.20) have an equivalent form in the terms of the best approximation of individual functions from L_p^r, namely: if $f \in L_p^r$ $(1 \le p \le \infty)$ then

$$E(f, S_{2n,m})_1 \le \|\varphi_{1,r}\|_{p'} n^{-r} \|f^{(r)}\|_{p'} \qquad m \ge r-1, \tag{4.22}$$

and in the case $p = \infty$, i.e. when $f \in L_\infty^r$, we also have

$$E(f, S_{2n,m})_C \le K_r n^{-r} \|f^{(r)}\|_\infty, \qquad m \ge r-1. \tag{4.23}$$

Inequalities (4.22) and (4.23) cannot be improved, the second one not even on C^r.

Using the orthogonality relation in the definition of the class $W_p^r[S_m(\Delta_N)]$ (see (4.3)) we can get better estimates in the same situation in which the norm $\|f^{(r)}\|_p$ is replaced by the best approximation of $f^{(r)}$ by splines from $S_{2n,m-r}$.

We have, in view of (1.5.25) for arbitrary partition (4.1) and for every $f \in L_1^r$

$$E(f, S_m(\Delta_N))_q = \sup\left\{\int_0^{2\pi} f^{(r)}(t)g(t)\,\mathrm{d}t \colon g \in W_{q'}^r[S_m(\Delta_N)]\right\}, \tag{4.24}$$

where we can additionally assume that $g \perp 1$ because $f^{(r)} \perp 1$. For $m \ge r$ let ψ be an arbitrary spline from $S_{m-r}(\Delta_N)$, let $\psi_0(t) = \psi(t) + c$, where c is chosen from $\psi_0 \perp 1$, and let ψ_r be the rth periodic integral of ψ_0. Then $\psi_r \in S_m(\Delta_N)$ and we have for $g \in W_{q'}^r[S_m(\Delta_N)]$, $g \perp 1$

$$\int_0^{2\pi} g(t)\psi(t)\,\mathrm{d}t = \int_0^{2\pi} g(t)\psi_0(t)\,\mathrm{d}t = (-1)^r \int_0^{2\pi} g^{(r)}(t)\psi_r\,\mathrm{d}t = 0.$$

This means that the function g in (4.24) can be assumed orthogonal to the subspace $S_{m-r}(\Delta_N)$ and hence

$$\int_0^{2\pi} f^{(r)}(t)g(t)\,\mathrm{d}t = \int_0^{2\pi} [f^{(r)}(t) - s(t)]g(t)\,\mathrm{d}t,$$

where s is any spline from $S_{m-r}(\Delta_N)$. For $f \in L_p^r$ we choose $s(t) = s(f, t)$ such that

$$\|f^{(r)} - s\|_p = E(f^{(r)}, S_{m-r}(\Delta_N))_p.$$

Then Hölder's inequality gives

$$\int_0^{2\pi} f^{(r)}(t)g(t)\,\mathrm{d}t \le \|g\|_{p'} E(f^{(r)}, S_{m-r}(\Delta_N))_p. \tag{4.25}$$

From (4.24) and (4.25) for $f \in L_p^r$ and $m \ge r$ we obtain

$$E(f, S_m(\Delta_N))_q \le E(f^{(r)}, S_{m-r}(\Delta_N))_p \sup\{\|g^{(m-r+1)}\|_{p'} \colon g \in W_{q'}^{m+1}(\Delta_N)\}. \tag{4.26}$$

We have for the uniform partition:

Proposition 5.4.9 For $m \geq r \geq 1$ and $f \in L_p^r$ $(1 \leq p \leq \infty)$ we have

$$E(f, S_{2n,m})_1 \leq \|\varphi_{1,r}\|_{p'} n^{-r} E(f^{(r)}, S_{2n,m-r})_p, \qquad p' = p/(p-1). \tag{4.27}$$

Moreover if $f \in L_\infty^r$ then we also have

$$E(f, S_{2n,m})_C \leq K_r n^{-r} E(f^{(r)}, S_{2n,m-r})_\infty. \tag{4.28}$$

Inequalities (4.27), (4.28) are exact.

Indeed inequalities (4.27) and (4.28) follow immediately from (4.26) and Propositions 5.4.5 and 5.4.7 respectively. The exactness as in Theorem 5.4.8, follows from Propositions 5.2.7 and 5.2.5.

5.4.3 Best approximation on the interval Let us consider the best approximation of the classes $W_p^r[a, b]$ by subspaces $S_m(\Delta_N[a, b])$ of splines defined on $[a, b]$ of degree m and defect 1 with respect to the partition

$$\Delta_N[a, b]\colon a = t_0 < t_1 < \cdots t_N = b. \tag{4.29}$$

If, as before, P_ν is the set of all algebraic polynomials of degree ν, then obviously $P_{r-1} \subset W_p^r[a, b]$ and $P_m \subset S_m(\Delta_N[a, b])$. The subspace $S_m(\Delta_N[a, b])$ has dimension $N + m$ and hence, unlike the periodic case, an increase in the degree of the splines increases the dimension.

The usage of duality relation (1.5.20) in Proposition 1.5.3 and definition (1.5.17) led us to the necessity of investigating the properties of the functions from the set

$$W_p^r[S_m(\Delta_N[a, b])]_0 = \{g\colon g \in W_p^r[a, b], g^{(r)} \perp S_m(\Delta_N[a, b]),$$
$$g^{(k)}(a) = g^{(k)}(b) = 0, k = 0, 1, \ldots, r-1\},$$

where because of the arguments given at the end of Section 1.5.1 we may assume for our purposes that $m \geq r - 1$.

Using integration by parts as in Lemmas 5.4.1 and 5.4.2 we can prove:

Lemma 5.4.10 Let $g \in L_1^m[a, b]$ and let it satisfy

$$g^{(k)}(a) = g^{(k)}(b) = 0, \qquad k = 0, 1, \ldots, m-1.$$

A necessary and sufficient condition for the orthogonality relation $g^{(m)} \perp S_m(\Delta_N[a, b])$ is

$$\int_{t_{i-1}}^{t_i} g(t)\, \mathrm{d}t = 0, \qquad i = 1, 2, \ldots, N,$$

where t_j are the points from partition (4.29).

Let us introduce the set

$$W_p^r(\Delta_N[a,b])_0 = \{g\colon g \in W_p^r[a,b],\, g^{(k)}(a) = g^{(k)}(b) = 0,$$
$$k = 0, 1, \ldots, r-1,\, g(t_i) = 0,\, i = 1, 2, \ldots, N-1\}.$$

With the help of Lemma 5.4.10 we easily verify the equalities (in the sense of coincidence of sets)

$$W_p^{m+1}[S_m(\Delta_N[a,b])]_0 = W_p^{m+1}(\Delta_N[a,b])_0. \tag{4.30}$$

From (4.30) and (1.5.20) we have

$$E(W_p^r[a,b], S_m(\Delta_N[a,b]))_{L_q[a,b]}$$
$$= \sup\{\|g^{(m-r+1)}\|_{L_{p'}[a,b]}\colon g \in W_{q'}^{m+1}(\Delta_N[a,b])_0\},$$
$$r = 1, 2, \ldots; \qquad m \geq r-1; \qquad 1 \leq p, q \leq \infty. \tag{4.31}$$

Finding the supremum in (4.31) for an arbitrary partition and for $m \geq 2$ is a technically difficult problem to solve. By analogy with the proof of theorem 5.4.4 we can show that the solution is provided for $q = 1$ $(q' = \infty)$ by the perfect spline with zero boundary conditions. Its existence follows from Proposition 2.5.6.

If $[a, b] = [0, \pi]$ and if the partition $\Delta_N[0, \pi] = \bar{\Delta}_N[0, \pi]$ is the uniform one, i.e. $t_j = j\pi/N$, then the right-hand side of (4.31) can be bound from above by the norm of the perfect spline $\varphi_{N,r}$ using a reduction to the periodic case.

Let $g \in W_p^r(\bar{\Delta}_N[0, \pi])_0$. In particular we have $\int_0^\pi g^{(r)}(t)\,dt = 0$. Let us continue function g as an even function on $[-\pi, \pi]$ and then that as a 2π-periodic on the real line. The new function, f say, belongs to W_p^r, $f(t) = g(t)$ $(0 \leq t \leq \pi)$ and

$$\int_0^\pi |g^{(k)}(t)|^q\,dt = \frac{1}{2}\int_0^\pi |f^{(k)}(t)|^q\,dt, \qquad k = 0, 1, \ldots, r-1; \qquad q \geq 0. \tag{4.32}$$

Applying Propositions 5.4.5–5.4.7 to f we obtain:

Proposition 5.4.11 If $g \in W_\infty^r(\bar{\Delta}_N[0, \pi])_0$ then ($\|\bullet\|_q = \|\bullet\|_{L_q[0,\pi]}$)

$$\|g^{(k)}\|_q \leq \|\varphi_{N,r-k}\|_q = \|\varphi_{1,r-k}\|_q N^{-r+k}, \qquad k = 0, 1, \ldots, r; \qquad 1 \leq q \leq \infty.$$

If $g \in W_p^r(\bar{\Delta}_N[0, \pi])_0$ $(1 \leq p < \infty)$ then

$$\|g\|_1 \leq 2^{-1/p}\,\|\varphi_{N,r}\|_{p'}, \qquad p' = p/(p-1),$$

and in the case $p = 1$ we also have

$$\|g^{(k)}\|_1 \leq \tfrac{1}{2}\|\varphi_{N,r-k}\|_C = \tfrac{1}{2}K_{r-k}N^{-r+k},\ k = 0, 1, \ldots, r.$$

It will follow from the results in Section 8.1 that we have for every partition $\Delta_N[0, \pi]$ and for every $m = 0, 1, \ldots,; r = 1, 2, \ldots,$

$$E(W^r_\infty, S_m(\Delta_N[0, \pi]))_q \geq (N+m)^{-r}\|\varphi_{1,r}\|_{L_q[0,\pi]}, \tag{4.33}$$

$$E(W^r_p, S_m(\Delta_N[0,\pi]))_1 \geq (N+m)^{-r}\|\varphi_{1,r}\|_{L_{p'}[0,\pi]}. \tag{4.34}$$

Bounding the right-hand side of (4.31) from above in the case of a uniform partition with the help of Proposition 5.4.11, of into account (4.33), and of (4.34), we obtain:

Theorem 5.4.12 We have

$$E(W^r_p[0, \pi], S_{N,m}[0, \pi])_1 = \|\varphi_{1,r}\|_{L_{p'}[0,\pi]}N^{-r}(1-\eta_1),$$

$$m \geq r-1, 1 \leq p \leq \infty;$$

$$E(W^r_\infty[0, \pi], S_{N,r-1}[0, \pi])_q = 2^{1/q-1}\|\varphi_{1,r}\|_{L_q[0,\pi]}N^{-r}(1-\eta_2),$$

$$1 \leq q < \infty;$$

$$E(W^r_\infty[0, \pi], S_{N,m}[0, \pi])_C = 2^{-1}K_r N^{-r}(1-\eta_3),$$

$$m \geq r-1,$$

where quantities η_1, η_2, η_3 satisfy the inequalities

$$0 < \eta_1, \eta_3 < 1 - \left(\frac{N}{N+m}\right)^r, \quad \text{and} \quad 0 < \eta_2 < 1 - \left(\frac{N}{N+r}\right)^r,$$

and hence $\eta_i = o(1)$ $(i = 1, 2, 3)$ for $n \to \infty$ and fixed m and r.

We shall also get an asymptotic analog of Proposition 5.4.9 on the interval. Let $f \in L^r_1[a, b]$. Then we have from relation (1.5.18) for $m \geq r$

$$E(f, S_m(\Delta_N[a, b]))_{L_q[a,b]}$$

$$= \sup\left\{\int_a^b f^{(r)}(t)g^{(m-r+1)}(t)\,dt\colon g \in W^{m+1}_{q'}[S_m(\Delta_N[a, b])]_0\right\}. \tag{4.35}$$

If $\psi \in S_{m-r}(\Delta_N[a, b])$ and ψ_r is the rth integral of ψ then

$$\int_a^b g^{(m-r+1)}(t)\psi(t)\,dt = (-1)^r \int_a^b g^{(m+1)}(t)\psi_r(t)\,dt = 0$$

because of $g^{(m+1)} \perp S_m(\Delta_N[a, b])$, i.e. $g^{(m-r+1)} \perp S_{m-r}(\Delta_N[a, b])$. Hence we can subtract under the integral sign in (4.35) from $f^{(r)}$ the spline $s \in S_{m-r}(\Delta_N[a, b])$ of best approximation to $f^{(r)}$ in the $L_p[a, b]$ metric.

Estimating the integral later in the usual way and taking (4.30) into account we get

$$E(f, S_m(\Delta_N[a,b]))_{L_q[a,b]}$$
$$\leq E(f^{(r)}, S_{m-r}(\Delta_N[a,b]))_{L_p[a,b]}$$
$$\times \sup\{\|g^{(m-r+1)}\|_{L_{p'}[a,b]} : g \in W_{q'}^{m+1}(\Delta_N[a,b])_0\}, \qquad m \geq r \geq 1. \quad (4.36)$$

Going to the uniform partition for $[a,b] = [0,\pi]$ and using Proposition 5.4.11, and inequalities (4.33) and (4.34) we get

$$\sup_{f \in L_p^r[0,\pi]} \frac{E(f, S_{N,m}[0,\pi])_1}{E(f^{(r)}, S_{N,m-r}[0,\pi])_p} = \|\varphi_{1,r}\|_{L_{p'}[0,\pi]} N^{-r}(1-\eta_1), \quad m \geq r, \quad (4.37)$$

$$\sup_{f \in L_\infty^r[0,\pi]} \frac{E(f, S_{N,m}[0,\pi])_C}{E(f^{(r)}, S_{N,m-r}[0,\pi])_\infty} = K_r N^{-r}(1-\eta_2), \qquad m \geq r, \quad (4.38)$$

where η_1, η_2 are the same as in Theorem 5.4.12.

5.3.4 Non-symmetric approximations Let $r = 1, 2, \ldots;\ 1 \leq p \leq \infty;$ $\gamma, \delta > 0$. Set $W_{p;\gamma,\delta}^r = \{f : f \in L_p^r, \|f^{(r)}\|_{p;\gamma,\delta} \leq 1\}$ (see Section 1.4.5). It is clear that $W_{p;1,1}^r = W_p^r$. For $p > 1$ and $\gamma \neq \delta$ the classes $W_{p;\gamma,\delta}^r$ are non-symmetric.

The Theorems 4.2.4 and 5.4.8 which are about the best approximation by trigonometric polynomials and splines of the classes W_p^r in L_1 metric have analogs in the cases of best (α,β)-approximation of classes $W_{p;\gamma,\delta}^r$ in L_1.

Theorem 5.4.13 Let $n, r = 1, 2, \ldots;\ 1 \leq p \leq \infty;\ \gamma, \delta > 0;\ \alpha, \beta > 0$, $\mathcal{N}$ being either $\mathcal{N}_{2n-1}^{\mathrm{T}}$ or $S_{2n,m}$ $(m \geq r)$. Then

$$E(W_{p;\gamma,\delta}^r, \mathcal{N})_{1;\alpha,\beta} = n^{-r} \inf_{\lambda \in \mathbb{R}} \|\varphi_{1,r}(\alpha,\beta;\bullet) - \lambda\|_{p';\gamma^{-1},\delta^{-1}}, \qquad 1/p + 1/p' = 1. \quad (4.39)$$

Letting α or β go to infinity in (4.39) and having Proposition 1.5.9 for the best one-sided approximations in mind we get:

Theorem 5.4.14 Let $n, r, m, \gamma, \delta, \mathcal{N}$ be as in Theorem 5.4.13. Then

$$E^{\pm}(W_{p;\gamma,\delta}^r, \mathcal{N})_1 = 2n^{-r} \inf_{\lambda \in \mathbb{R}} \|B_r - \lambda\|_{p';\gamma^{-1},\delta^{-1}}.$$

The scheme of the proof of Theorem 5.4.13 is as follows. First we establish that for

$$W_{p;\gamma,\delta}^r(\mathcal{N}) = \{g : g \in W_{p;\gamma,\delta}^r, g \perp 1, g^{(r)} \perp \mathcal{N}\}$$

and $f \in W^r_{\infty;\alpha^{-1},\beta^{-1}}$ we have

$$\|f_\pm\|_\infty \le \|(\varphi_{n,r}(\alpha,\beta;\bullet))_\pm\|_\infty. \tag{4.40}$$

In the case when $\mathcal{N} = \mathcal{N}^{\mathrm{T}}_{2n-1}$ this is a generalization of one of the statements of theorem 4.2.3 and it can be proved by calculating the best (α,β)-approximation of Bernoulli kernels B_r. In the case $\mathcal{N} = S_{2n,m}$ this is a generalization of (4.11) and it can be proved by taking the specific properties of the splines and Theorem 3.3.9 into account.

In view of (4.40) we can apply Theorem 3.3.10 for functions f and $\varphi_{n,r}(\alpha,\beta)$ and obtain:

Theorem 5.4.15 If $g \in W^r_{\infty;\alpha^{-1},\beta^{-1}}(\mathcal{N})$, where $\mathcal{N}$ is either $\mathcal{N}^{\mathrm{T}}_{2n-1}$ or $S_{2n,m}$ $(m \ge r)$, then for every $\lambda \in \mathbb{R}$ and $x \in [0, 2\pi]$ we have

$$\int_0^x r((g-\lambda)_\pm, t)\,\mathrm{d}t \le \int_0^x r((\varphi_{n,r}(\alpha,\beta;\bullet)-\lambda)_\pm, t)\,\mathrm{d}t.$$

The inequality is exact.

From Theorem 5.4.15 we easily get

$$\sup\{\inf_{\lambda\in\mathbb{R}} \|g-\lambda\|_{p';\gamma^{-1},\delta^{-1}}: g \in W^r_{\infty;\alpha^{-1},\beta^{-1}}(\mathcal{N})\}$$
$$= \inf_{\lambda\in\mathbb{R}} \|\varphi_{n,r}(\alpha,\beta;\bullet)-\lambda\|_{p';\gamma^{-1},\delta^{-1}},$$

and using Proposition 1.4.9 we complete the proof of (4.39). □

Comments

Sections 5.1 and 5.2 Theorem 5.1.1 has a basic rôle for obtaining exact results for the errors of the spline interpolation on classes W^r_p $(p = 1, \infty)$. In the proof we use the scheme from Tikhomirov's paper [5], in which relation (2.12) for the classes W^{r+1}_∞ and $q = \infty$ is obtained. Proposition 5.2.4 for the class W^{r+1}_∞ (in a slightly different form) has appeared in the work of Zhensykbaev [2], who used the integral representation of the error (see Section 5.2.4) for its proof.

Theorem 5.2.8 is proved by the author [17] on the basis of (2.32) with the help of Lemma 5.2.10. Relation (2.26) is also due to the author [25, 28–30] (see also [10A]).

Section 5.3 Theorem 5.3.1 is from von Golitschek paper [1], Theorem 5.3.2 and Propositions 5.3.4 and 5.3.7 which follow directly from it are obtained by Korneichuk and Ligun [1]. Equality (3.9) is proved by Hall and Meyer [1] for $r = 2$ and by the author [25, 28–30] in the other cases. On the existence of perfect splines with minimal L_q norm on the interval see e.g. the papers of Miccelli and Pinkus [1] and Pinkus [1].

Equality (3.15) is established by Ciarlet, Schultz and Varga [1], relation (3.16) and Theorems 5.3.10 and 5.3.11 have been proved by Velikin [3] using the integral representation of the error. Theorem 5.3.12 was obtained by Nazarenko and Pereverzev [1]. For the approximation and extremal properties of Hermite splines see also the monographs [1A, 10A, 23A, 28A]. The results in Section 5.3.4 were established by the author [10A, 26]. Some other exact estimates for approximation by local splines of minimal defect are contained in the papers of Avakyan [1, 2]; much attention is paid to these splines in [28A].

Exact estimates for approximation by bivariate local (including Hermitian) splines can be found in the papers of Birkhoff, Schultz and Varga [1], Storchai [2], Pereverzev [1, 2] and Avakyan [2]; there are some exact estimates for blending splines in Shabozov [1]. Results with exact constants for approximation by Hermitian and other local splines on parametrically given curves and surfaces are contained in Martynyuk and Storchai [1], Nazarenko [1, 3, 4], Vakarchuk [1, 2], Martynyuk [1, 2]; it turns out that for such questions the Hausdorff metric is a natural one. Its approximation sides are studied in Sendov's monograph [21A].

Section 5.4 The results in Sections 5.4.1 and 5.4.2 are taken from the author's papers [16, 18, 21]. Relation (4.20) for $m = r - 1$ was obtained by Tikhomirov [5]; the other results of Theorem 5.4.8 were proved by Ligun [4]. Proposition 5.4.9 was established by Korneichuk [18, 21]; the results for best approximation on the interval in Section 5.4.3 are also due to him [20].

Exact results for best one-sided approximation by splines of classes W_1^r were obtained in Doronin and Ligun's paper [1], see also [11A]. The results in Section 5.4.4 are due to V. F. Babenko [4, 5]. More general results on the best and best one-sided approximation by splines of convolution classes are contained in the articles of Pinkus [1], Sun Yong-sheng and Hyang Daren [1].

For limit relations between estimates for trigonometric and spline approximation on classes of functions see Velikin [5].

A very extensive review of the results on the best approximation of classes W_p^r and $W_p^r[a, b]$ by polynomials and splines is given in Tikhomirov [6].

Other results and exercises

1. For all $r = 1, 2, \ldots$; $m \geq r$: $1 \leq p \leq \infty$ we have for the one-sided approximation

$$E_n^+(W_p^r)_1 = E^+(W_p^r, S_{2n,m})_1 = 2n^{-r} \inf_{\lambda \in \mathbb{R}} \|B_r - \lambda\|_{p'},$$

where $p' = p/(p-1)$ (V. G. Doronin and Ligun [4]).

2. If $f \in L_1^r$; $\alpha, \beta > 0$; $n, r, = 1, 2, \ldots$; $m \geq r$ then

$$E(f, S_{2n,m})_{1;\alpha,\beta} \leq \int_0^{2\pi} \Pi(\varphi_{n,r}(\alpha, \beta; \bullet), t)\Pi(f^{(r)}, t)\, dt,$$

where $\Pi(g, t)$ is defined in Exercise 4.10. The inequality is exact. For a Π-invariant set $F \subset L_1$ we have (for the notation see Exercise 4.11)

$$E(W^rF, S_{2n,m})_{1;\alpha,\beta} = \sup_{g\in F, g\perp 1} \int_0^{2\pi} \Pi(g,t)\Pi(\varphi_{n,r}(\alpha,\beta;\bullet),t)\,dt$$

(V. F. Babenko [5]; in [8, 13] these results are extended to convolutional classes with kernels which do not increase the oscillation).

3. Under the conditions of Exercise 4.9, we have

$$E(\tilde{W}^r_\infty, S_{2n,m})_{1;\alpha,\beta} = \|\tilde{\varphi}_{n,r}(\alpha,\beta;\bullet)\|_1, \qquad m \ge r$$

(V. F. Babenko [5]).

4. If $\sigma_{2n,r}(f,t)$ is the spline from $S_{2n,r}$ interpolating f at the zeros of $\varphi_{n,r+1}$ then

$$\sup_{g\in W^{r+1}_\infty} \|\tilde{f} - \tilde{\sigma}_{2n,r}\|_1 = \|\tilde{\varphi}_{n,r+1}\|_1,$$

where $\tilde{g}$ is the conjugate function to g (V. F. Babenko [9]).

5. Prove that

$$E^+(W^1_p, S_{2n,0})_1 = \frac{\pi}{n}\left(\frac{2p-2}{2p-1}\pi\right)^{1-1/p}, \qquad p \ge 1.$$

6. Prove the exactness of estimate (2.25) for $p = \infty$ on C^r.

7. Under the conditions of Theorem 5.1.4 show

$$|f^{(\nu)}(a)| \le |\varphi^{(\nu)}(a)|, |f^{(\nu)}(b)| \le |\varphi^{(\nu)}(b)|, \qquad \nu = 1, 2, \ldots, m.$$

8. Check that for every polynomial of the type

$$p(t) = [(t-\alpha)(t-\beta)]^n/(2n)!$$

we have an equality in (3.13).

9. Let $l(f,t)$ ($l \in S_{2n,1}$) be the polygon interpolating $f \in C$ at the points $k\pi/n$ ($k = 0, \pm 1, \pm 2, \ldots,$) and $\alpha, \beta > 0$. Evaluate

$$\sup_{f\in W^1_{\infty;\alpha,\beta}} \|f - l(f)\|_C.$$

6

Exact constants in Jackson inequalities

Jackson inequalities (or inequalities of Jackson's type) are expressions in which the approximation error of an individual function is estimated using the modulus of continuity of the function or some of its derivatives. In this context we can consider best approximations either from fixed subspaces or by given methods, and here in particular by linear ones. The only requirement for the approximating function is that the modulus used to estimate the error makes sense for it. The problem of obtaining inequalities which are exact on certain sets can also be posed.

The modulus of continuity is a better characteristic for a function than, for example, its norm in C or L_p and hence obtaining the exact constant in Jackson inequalities needs methods of investigation essentially different from those used in Chapters 4 and 5.

The proofs for exact inequalities of the Jackson type for polynomial and spline approximation are the main content of this chapter. The problem of finding the smallest constants in such inequalities with respect to the whole class of approximating subspaces of fixed dimension will be considered in Section 8.3.

6.1 Modulus of continuity and general statement of the problems

6.1.1 Modulus of continuity The notion of the modulus of continuity $\omega(f, \delta)$ of a continuous function f characterizing the maximal oscillations of f on the intervals of length δ was introduced in Section 3.2 in connection with the comparison theorems. Now we shall approach this definition from a more general point of view.

Let X be a normed space of (in particular periodic) functions f defined on the real line. For every $f \in X$ we set

$$\omega(f, \delta)_X = \sup_{0 \le u \le \delta} \|f(\bullet + u) - f(\bullet)\|_X, \qquad \delta \ge 0. \tag{1.1}$$

The function $\omega(f, \delta)_X$ is called the *modulus of continuity of the function f* in the space X. If X is the space of functions defined on a finite interval $[a, b]$ then the norm in definition (1.1) is taken on the interval $[a, b - u]$; in this case $0 \le \delta \le b - a$. We have in particular for $X = C[a, b]$

$$\omega(f, \delta)_{C[a,b]} = \sup \{|f(t') - f(t'')|: |t' - t''| \le \delta, t', t'' \in [a, b]\}. \tag{1.2}$$

We shall denote the modulus of continuity of functions $f \in C(-\infty, \infty)$ in the metric of this space by $\omega(f, \delta)_C$; this is also valid for 2π-periodic functions, i.e. $f \in C$.

In the case $X = L_p$ or $X = L_p[a, b]$ $(1 \le p < \infty)$ the function (1.1) is called sometimes the *integral modulus of continuity*.

The modulus of continuity as a function of δ has the following properties:

(1) $\omega(f, 0)_X = 0$;

(2) $\omega(f, \delta)_X$ is non-decreasing;

(3) $\omega(f, \delta)_X$ is a sub-additive function, i.e.

$$\omega(f, \delta_1 + \delta_2)_X \le \omega(f, \delta_1)_X + \omega(f, \delta_2)_X; \tag{1.3}$$

(4) if X is $C[a, b]$ (or C), $L_p[a, b]$ (or L_p) $(1 \le p < \infty)$ then $\omega(f, \delta)_X$ is continuous on the interval $[0, b - a]$ (or on the half-axis $[0, \infty)$) function.

Properties (1) and (2) follow immediately from definition (1.1). In order to prove (1.3) it suffices to fix δ_1 and δ_2 and to establish that we have for every $u \ge 0$, $u \le \delta_1 + \delta_2$,

$$\|f(\bullet + u) - f(\bullet)\|_X \le \omega(f, \delta_1)_X + \omega(f, \delta_2)_X.$$

This is obvious if $0 \le u \le \delta_1$ or $0 \le u \le \delta_2$. Let $\max\{\delta_1, \delta_2\} < u \le \delta_1 + \delta_2$. Then $|u - \delta_2| \le \delta_1$ and

$$\|f(\bullet + u) - f(\bullet)\|_X \le \|f(\bullet + u) - f(\bullet + \delta_2)\|_X + \|f(\bullet + \delta_2) - f(\bullet)\|_X$$
$$\le \omega(f, \delta_1)_X + \omega(f, \delta_2)_X.$$

The proof of Property (4) reduces to establishing

$$\lim_{\delta \to 0+0} \omega(f, \delta)_X = 0, \tag{1.4}$$

because in view of (1.3) for $\delta + h \ge 0$ independently of the sign of h we have

$$|\omega(f, \delta + h)_X - \omega(f, \delta)_X| \le \omega(f, |h|)_X.$$

Moreover, if (1.4) holds then $\omega(f, \delta + h)_X \to \omega(f, \delta)_X$ for $h \to 0$. The limit (1.4) follows the case $X = C[a, b]$ (or $X = C$) from the uniform continuity of the function $f \in C[a, b]$ ($f \in C$) on $[a, b]$ (on $\mathbb{R}$). If $X = L_p[a, b]$ $(p \ge 1)$ then (1.4) is equivalent to the relation (see e.g. [10B, p. 500])

$$\lim_{u\to 0}\|f(\bullet+u)-f(\bullet)\|_{L_p[a,b]}=0, \qquad f\in L_p[a,b], \qquad 1\le p<\infty. \quad (1.5)$$

(1.5) follows from the well known fact that the set of the functions which are continuous on $[a, b]$ (for which (1.5) is trivial) is dense in $L_p[a, b]$ $(1 \le p < \infty)$.

Remark For $X = L_\infty[a, b]$ the function $\omega(f, \delta)_X$ is not, in general, continuous: e.g. for $f(t) = \operatorname{sgn} t\,(-1 \le t \le 1)$ we have $\omega(f, 0+0)_{L_\infty[-1,1]} = 2$. Hence consideration of the modulus of continuity in $L_\infty[a, b]$ is of limited interest.

Let us note that if X is a space of functions of period $2l$ then $\omega(f, \delta)_X = \omega(f, l)_X$ for $\delta \ge l$, because for every u $(|u| > l)$ there always exists an integer m such that $|2ml - u| \le l$ and hence

$$\|f(\bullet+u)-f(\bullet)\|_X=\|f(\bullet+u)-f(\bullet+2ml)\|_X\le\omega(f,l)_X.$$

It is easy to deduce from Property (3) that

$$\omega(f, n\delta)_X\le n\,\omega(f,\delta)_X, \qquad n=1,2,\ldots, \quad (1.6)$$

if point $n\delta$ belongs to the domain of function $\omega(f, \delta)_X$. For $n = 1$ this is trivial and if it holds for $n = k$ then

$$\omega(f,(k+1)\delta)_X\le\omega(f,k\delta)_X+\omega(f,\delta)_X\le(k+1)\omega(f,\delta)_X.$$

If we take an arbitrary positive number λ instead of n then we can only assert

$$\omega(f,\lambda\delta)_X\le([\lambda]+1)\,\omega(f,\delta)_X \quad (1.7)$$

($[\lambda]$ is the integral part of λ) because we have in view of (1.6)

$$\omega(f,\lambda\delta)_X\le\omega(f,([\lambda]+1)\delta)_X\le([\lambda]+1)\,\omega(f,\delta)_X.$$

For $X = C[a, b]$ Properties (1)–(4) of the modulus of continuity (1.1) are characteristic in the sense that, as we shall see later, every function $\omega(\delta)$ defined on $[0, b - a]$ possessing these properties is a modulus of continuity $\omega(f, \delta)_{C[a,b]}$ of some function $f \in C[a, b]$. This is also true for $X = C$ provided $\omega(\delta)$ also satisfies $\omega(\delta) = \omega(\pi)$ for $\delta \ge \pi$. In view of this it is appropriate for the concept of an independent modulus of continuity, i.e. one not connected with any particular function, to be introduced.

The *modulus of continuity* is the function $\omega(t)$ which is everywhere continuous on the interval $[0, d]$ or on the half-axis $[0, \infty)$ and which is non-decreasing, sub-additive and vanishes at zero. All these conditions are contained in the following conditions:

$$\lim_{t\to 0+0}\omega(t)=\omega(0)=0,$$

$$0\le\omega(t_2)-\omega(t_1)\le\omega(t_2-t_1), \qquad 0\le t_1\le t_2.$$

If ω is a modulus of continuity on $[0, d]$ $(d \geq b - a)$ then the function $f(t - a) = \theta(t - a)$ $(a \leq t \leq b)$ belongs to $C[a, b]$ and it is easy to check that $\omega(f, \delta)_{C[a,b]} = \delta(\delta)$. By setting $f_*(t) = \omega(|t|)$ $(-\pi \leq t \leq \pi)$ and $f_*(t + 2\pi) = f_*(t)$ for $d \geq \pi$ we get a 2π-periodic function f_* continuous on the real line such that $\omega(f_*, \delta)_C = \omega(\delta)$ $(0 \leq \delta \leq \pi)$.

In view of the sub-additivity property of the modulus of continuity $\omega(t)$ we can get, as above for $\omega(f, \delta)_X$, the inequalities:

$$\omega(nt) \leq n\omega(t), \qquad n = 1, 2, \ldots; \qquad \omega(\lambda t) \leq ([\lambda] + 1)\omega(t), \qquad \lambda > 0. \tag{1.8}$$

In order to check the sub-additivity of ω easily we remark that this property follows if the ratio $\omega(t)/t$ is non-increasing. Indeed we have in this case for $t_1 > 0$, $t_2 > 0$

$$\omega(t_1 + t_2) = t_1 \frac{\omega(t_1 + t_2)}{t_1 + t_2} + t_2 \frac{\omega(t_1 + t_2)}{t_1 + t_2}$$

$$\leq t_1 \frac{\omega(t_1)}{t_1} + t_2 \frac{\omega(t_2)}{t_2} = \omega(t_1) + \omega(t_2).$$

If ω is concave and $\omega(0) = 0$ then the ratio $\omega(t)/t$ is non-increasing and hence we have:

Proposition 6.1.1 Every function ω which is continuous at 0, non-increasing and concave on $[0, d]$ or on $[0, \infty)$ such that $\omega(0) = 0$ is a modulus of continuity.

An important example is $\omega(t) = Kt^\alpha(0 < \alpha \leq 1, K > 0)$, where for $t \geq l > 0$ we can set $\omega(t) = Kl^\alpha$.

In Chapters 7 and 8 we shall primarily be dealing with the concave moduli of continuity and so we shall immediately give some properties that follow from the concavity.

Lemma 6.1.2 The function ψ concave on $[a, b]$ has at every inner point $t \in (a, b)$ finite one-sided derivatives $\psi'_-(t)$, $\psi'_+(t)$ which are non-increasing on (a, b) and

$$\psi'_+(t) \leq \psi'_-(t). \tag{1.9}$$

Indeed from the definition of concavity for $h > 0$ we obtain

$$\frac{\psi(t + h) - \psi(t)}{h} \leq \frac{\psi(t) - \psi(t - h)}{h}. \tag{1.10}$$

For $h \to 0$ the ratio on the left-hand side of (1.10) does not decrease and the ratio on the right-hand side does not increase. Hence the finite limits $\psi'_+(t)$, $\psi'_-(t)$ exist and are connected with (1.9). If $a < t_1 < t_2 < b$ then we have for $0 < h < t_2 - t_1$

$$\psi(t_1 + h) + \psi(t_2 - h) \geq \psi(t_1) + \psi(t_2)$$

and hence

$$\frac{1}{h}[\psi(t_1 + h) - \psi(t_1)] \geq \frac{1}{h}[\psi(t_2) - \psi(t_2 - h)].$$

Taking the limit for $h \to 0$ we get $\psi'_+(t_1) \geq \psi'_-(t_2)$ which together with (1.9) gives $\psi'_+(t_1) \geq \psi'_+(t_2)$ and $\psi'_-(t_1) \geq \psi'_-(t_2)$.

Proposition 6.1.3 If ω is a concave modulus of continuity on $[0, d]$ then: (1) on the interval $(0, d)$ there exist finite non-increasing non-negative one-sided derivatives $\omega'_-(t), \omega'_+(t)$ such that $\omega'_+(t) \leq \omega'_-(t)$; (2) ω is absolutely continuous on $[0, d]$.

Proof Property (1) follows from Lemma 6.1.2 and the monotonicity of ω. In order to prove (2) take $\varepsilon > 0$ and let $\{(\alpha_k, b_k)\}$ $(k = 1, 2, \ldots, n)$ be a system of disjoint intervals in $[0, d]$ such that the sum of their lengths is less than δ, where $\delta = \delta(\varepsilon)$ is determined from $\omega(\delta) = \varepsilon$. We can assume that $0 \leq \alpha_1 < \beta_1 \leq \alpha_2 < \beta_2 \leq \ldots \leq \alpha_n < \beta_n \leq d$. From the concavity for $0 \leq t' \leq t'' \leq d$ and $h > 0$ we have

$$\omega(t'' + h) - \omega(t'') \leq \omega(t' + h) - \omega(t')$$

and hence

$$\begin{aligned}
&\omega(\beta_1) - \omega(\alpha_1) \leq \omega(\beta_1 - \alpha_1),\\
&\omega(\beta_2) - \omega(\alpha_2) \leq \omega[(\beta_2 - \alpha_2) + (\beta_1 - \alpha_1)] - \omega(\beta_1 - \alpha_1),\\
&\ldots\ldots\ldots\ldots\ldots\ldots\ldots\ldots\ldots\ldots\ldots\ldots\\
&\omega(\beta_n) - \omega(\alpha_n) \leq \omega\left[\sum_{k=1}^{n} (\beta_k - \alpha_k)\right] - \omega\left[\sum_{k=1}^{n-1} (\beta_k - \alpha_k)\right],
\end{aligned}$$

Summing these inequalities we obtain

$$\sum_{k=1}^{n}[\omega(\beta_n) - \omega(\alpha_n)] \leq \omega\left[\sum_{k=1}^{n} (\beta_k - \alpha_k)\right] \leq \omega(\delta) = \varepsilon.$$

This proves the proposition. □

The absolute continuity of the modulus of continuity ω concave on $[0, d]$ means that

$$\omega(t) = \int_0^t \omega'(u)\,du,$$

where the derivative ω' exists almost everywhere in $[0, d]$ and is non-increasing and summable in the Lebesgue sense. If we set

$$\omega'(t) = [\omega'_+(t) + \omega'_-(t)]/2, \qquad 0 < t < d,$$

then the non-increasing function $\omega'(t)$ will be defined for every $t \in (0, d)$.

It is easy to give an example of a modulus of continuity which is not concave. Let $0 < h < 1$ and

$$\omega(t) = \begin{cases} t/h, & 0 \le t \le h, \\ 1, & h \le t \le 1, \end{cases}$$
$$\omega(t) = \omega(t-1) + 1, \qquad t \ge 1. \tag{1.11}$$

One can easily check that this function although not concave, satisfies all requirements defining a modulus of continuity on $[0, \infty)$.

Later we shall need the following lemma for the concave majorant of an arbitrary modulus of continuity.

Lemma 6.1.4 For every modulus of continuity $\omega(t) \not\equiv 0$ $(0 \le t \le d)$ there exists a concave modulus of continuity ω_* such that

$$\omega(t) \le \omega_*(t) < 2\omega(t), \qquad 0 \le t \le d. \tag{1.12}$$

The constant 2 on the right-hand side cannot be decreased.

Proof For every fixed $t \in [0, d]$ we set

$$\omega_*(t) = \sup_{0 \le x \le t \le z \le d} \frac{(z-t)\omega(x) + (t-x)\omega(z)}{z-x}. \tag{1.13}$$

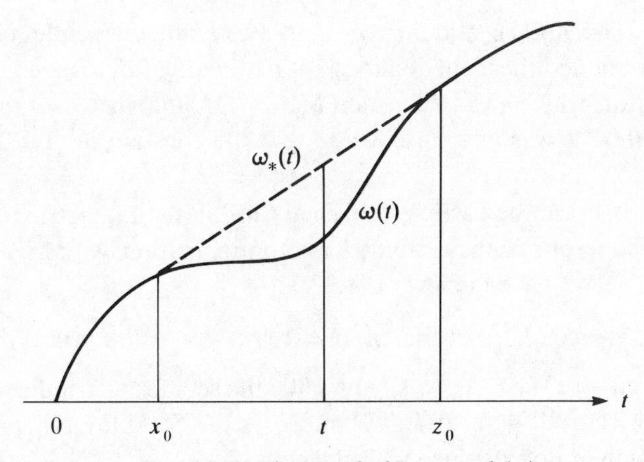

Fig. 6.1 To the proof of Lemma 6.1.4.

The graph of the function (1.13) is the upper boundary of the convex hull of the set $\{(t, y): 0 \le t \le d, 0 \le y \le \omega(t)\}$. It is clear that $\omega_*(0) = 0$, that ω_* is a continuous, non-decreasing, concave on $[0, d]$ function and that $\omega_*(t) \ge \omega(t)$.

In order to prove the second inequality in (1.12) we may assume that for some $t \in (0, d)$ we have $\omega_*(t) > \omega(t)$. Then there are points x_0, z_0 $(0 \le x_0 < t < z_0 \le d)$ such that

$$\omega_*(t) = [(z_0 - t)\omega(x_0) + (t - x_0)\omega(z_0)]/(z_0 - x_0).$$

Taking into account that $\omega(x_0) \le \omega(t)$ and recalling that (1.8) implies $\omega(z_0) \le (1 + z_0/t)\omega(t)$ we get

$$(z_0 - t)\omega(x_0) + (t - x_0)\omega(z_0) \le \left[z_0 - t + (t - x_0)\left(\frac{z_0}{t} + 1\right)\right]\omega(t)$$

$$= (z_0 - x_0)\left(1 + \frac{1 - x_0/t}{1 - x_0/z_0}\right)\omega(t) < 2\,(z_0 - x_0)\,\omega(t)$$

and hence $\omega_*(t) < 2\omega(t)$. For the modulus of continuity (1.11) $\omega_*(1) = (2 - h)\,\omega(1)$ and because h can be taken arbitrary close to zero we cannot decrease the constant 2 in (1.12). □

6.1.2 Jackson inequalities In 1911 Jackson [1] proved the following inequalities for the best uniform approximation of 2π-periodic functions by trigonometric polynomials from the subspace $\mathcal{N}^T_{2n-1}$:

$$E_n(f)_C \le M\,\omega(f, 1/n), \qquad n = 1, 2, \ldots, \qquad f \in C, \tag{1.14}$$

$$E_n(f)_C \le M_r n^{-r}\,\omega(f^{(r)}, 1/n), \; n, r = 1, 2, \ldots, \qquad f \in C^r, \tag{1.15}$$

where $\omega(f, \delta) = \omega(f, \delta)_C$ and the constants M, M_r are independent on f and n. Jackson obtained these inequalities approximating f by a special kind of linear operators which gives a constant M_r in (1.15) that increases to infinity together with r. It was later discovered that the constant in (1.15) can be absolute.

The latter fact can be easily established (also in the L_p metric too) with the help of the results from Section 4.2 for approximation by Favard means: for every $N > 0$ we have in view of (4.2.2)

$$E_n(NW^r_p)_p \le N\,K_m\,n^{-m}, \qquad n, m = 1, 2, \ldots; \qquad 1 \le p \le \infty. \tag{1.16}$$

Let X be either C or L_p $(p \ge 1)$ and X^r be the set of rth periodic integrals of functions from $X^0 = X$. If $f \in X^r$ and f_h is the Stekhlov function for f (Section A2) then Proposition A2.2 gives

$$\|f^{(r)} - f_h{}^{(r)}\|_X \le \omega(f^{(r)}, h)_X, \qquad \|f_h{}^{(r+1)}\|_X \le h^{-1}\,\omega(f^{(r)}, h)_X.$$

Hence $f - f_h \in N_1 W_p^r$ ($p = \infty$ if $X = C$) with $N_1 = \omega(f^{(r)}, h)_X$, $f_h \in N_2 W_p^{r+1}$ with $N_2 = h^{-1}\,\omega(f^{(r)}, h)_X$. But then (see (1.16))

$$E_n(f - f_h)_X \le K_r n^{-r}\,\omega(f^{(r)}, h)_X, \qquad E_n(f_h)_X \le K_{r+1} n^{-r-1} h^{-1} \omega(f^{(r)}, h)_X.$$

In view of Proposition 1.1.1 we have $E_n(f) \le E_n(f - f_h) + E_n(f_h)$ and hence for $h = \pi/n$ and $f \in X^r$ we obtain

$$E_n(f)_X \le (K_r + \pi^{-1} K_{r+1})\, n^{-r}\,\omega(f^{(r)}, \pi/n)_X. \tag{1.17}$$

In particular for $f \in X$ ($r = 0$)

$$E_n(f)_X \le \frac{3}{2}\,\omega\left(f, \frac{\pi}{n}\right)_X. \tag{1.18}$$

In view of (1.7) we can multiply the coefficients on the right-hand sides of (1.17) and (1.18) by $(\pi + 1)$ and replace $\omega(f^{(r)}, \pi/n)_X$ and $\omega(f, \pi/n)_X$ by $\omega(f^{(r)}, 1/n)_X$ and $\omega(f, 1/n)_X$. The constants in (1.17) and (1.18) are not the best possible. In Section 6.2 we shall be able to find their smallest possible values in some cases with the help of advanced methods.

For the best approximation of functions on an interval by algebraic polynomials inequalities similar to (1.14) and (1.15) can be obtained by a reduction to the periodic case as in Section 4.4. For example let $f \in C[-1, 1]$. Then the function $\varphi(t) = f(\cos t)$ belongs to C. Moreover $\omega(\varphi, \delta)_C \le \omega(f, \delta)_{C[-1,1]}$. This follows directly from definition (1.2) and the inequality $|\cos t' - \cos t''| \le |t' - t''|$. But the inequality $E_n(\varphi)_C \le M\omega(\varphi, \beta/n)_C$ together with (4.4.5) implies

$$E(f, \mathcal{N}_{n-1}^A)_{C[-1,1]} \le M\omega(f, \beta/n)_{C[-1,1]}, \qquad f \in C[-1, 1].$$

Of course the problem of obtaining inequalities similar to (1.14) and (1.15) with exact constants can also be posed in the case of approximation by splines or some other subspace. The variety of such problems can also be enlarged by going to linear methods or to metrics different from the metric in the modulus of continuity.

6.1.3 General statements of the problem Let both X and Y be either of the normed functional spaces for $C[a, b]$, C, $L_p[a, b]$ or L_p ($1 \le p < \infty$) and let as before X^r ($r = 1, 2, \ldots$) be the set of rth integrals of functions from X. Here X^0 denotes X in general relationships. If $X^r \subset Y$ and $\mathcal{N}_N$ is an N-dimensional subspace of Y then *Jackson inequalities* can be described by estimates of the following type

$$E(f, \mathcal{N}_N)_Y \le M\,\omega(f^{(r)}, \gamma)_X, \qquad f \in X^r, \tag{1.19}$$

and also

$$\|f - Af\|_Y \leq M' \, \omega(f^{(r)}, \gamma)_X, \qquad f \in X^r, \tag{1.20}$$

where A is some operator from Y to $\mathcal{N}_N$. The constants M, M', for given spaces X, Y, cannot depend on f but they may depend on r, γ, on the subspace $\mathcal{N}_N$ (in particular from its dimension N), and M' may also depend on the operator A.

If we fix not only X, Y, but also $\mathcal{N}_N, A, r, \gamma$, then the problem of finding the smallest possible constants in (1.19) and (1.20) is equivalent to evaluating the quantities

$$\kappa_\gamma(X^r, Y, \mathcal{N}_N) = \sup_{f \in X^r} \frac{E(f, \mathcal{N}_N)_Y}{\omega(f^{(r)}, \gamma)_X}, \tag{1.21}$$

and

$$\kappa'_\gamma(X^r, Y, \mathcal{N}_N, A) = \sup_{f \in X^r} \frac{\|f - Af\|_Y}{\omega(f^{(r)}, \gamma)_X}, \tag{1.22}$$

respectively.

One should pay attention to the possibility of the denominators on the right-hand sides of (1.21) and (1.22) vanishing. This may happen (for $\gamma > 0$) if $f^{(r)} =$ constant and, following the reasoning given at the beginning of Section 1.5, we shall assume that $\mathcal{N}_N$ contains the polynomials of degree r when approximating on an interval or when $1 \in \mathcal{N}_N$ in the periodic case. Then the denominators in (1.21) and in (1.22) may vanish only together with the numerators and we make the convention that 0/0 be interpreted as 0.

Obviously we always have

$$\kappa'_\gamma(X^r, Y, \mathcal{N}_N, A) \geq \kappa_\gamma(X^r, Y, \mathcal{N}_N). \tag{1.23}$$

The following problem naturally arises: find

$$\kappa'_\gamma(X^r, Y, \mathcal{N}_N) = \inf_{A \in \mathcal{L}(Y, \mathcal{N}_N)} \kappa'_\gamma(X^r, Y, \mathcal{N}_N, A), \tag{1.24}$$

where $\mathcal{L}(Y, \mathcal{N}_N)$ is the set of all linear operators from Y to $\mathcal{N}_N$, and determine an operator $A_* \in \mathcal{L}(Y, \mathcal{N}_N)$ realizing the infimum in (1.24) and defining the best method in this sense. Of course A will be such an operator if it turns out that it gives equality in (1.23).

Remark On the right-hand sides of (1.21) and (1.22) we can calculate the supremums on the smaller set X^r_γ of functions $f \in X^r$ for which $\omega(f^{(r)}, \gamma)_X = 1$. This follows from the positive homogeneity of the quantities in the numerators and the denominators: the positive homogeneity of $E(f, \mathcal{N}_N)_X$ is proved in Proposition 1.1.1 and the equality

$$\omega(\lambda f, \delta)_X = |\lambda| \omega(f, \delta)_X, \qquad \delta \geq 0, \qquad \lambda \in \mathbb{R},$$

is implied by definition (1.1)

We have stated the problems in a general setting, but we should immediately note that exact results are only known to us (and they are given later) for a few cases connected with polynomial and spline approximations. Nevertheless these results, as we shall see in Chapter 8, are at the root of the solution of the problem of finding the smallest possible constant in (1.21) and (1.22) with respect to all subspaces $\mathcal{N}_N$ of fixed dimension N.

6.2 Jackson inequalities for polynomial approximations

6.2.1 An intermediate approximation In order to bound from above the best approximation $E_n(f)_C =: E(f, \mathcal{N}^T_{2n-1})_C$ of function $f \in C$ by trigonometric polynomials from above using the modulus of continuity $\omega(f, \delta)_C$ we employ the following idea: writing (in view of Proposition 1.1.1)

$$E_n(f)_C \le E_n(f-\varphi)_C + E_n(\varphi)_C \le \|f-\varphi\|_C + E_n(\varphi)_C, \qquad (2.1)$$

where $\varphi \in KH^1$, we look for the exact estimate for $\|f-\varphi\|_C$, i.e. the best approximation of a fixed function $f \in C$ by means of functions from KH^1. Let us note that class KH^1 is not a subspace but it is a convex set.

Proposition 6.2.1 For every function $f \in C$ and every $K > 0$ we have

$$E(f, KH^1)_C =: \inf_{\varphi \in KH^1} \|f-\varphi\|_C = \frac{1}{2} \max_{0 \le t \le \pi} [\omega(f,t) - Kt], \qquad (2.2)$$

where $\omega(f,\delta) = \omega(f,\delta)_C$. There is a function $\varphi(f,t) = \varphi(f,K,t)$ in KH^1 such that $\|f - \varphi(f)\|_C = E(f, KH^1)_C$.

Proof Let $K > 0$, $f \in C \backslash KH^1$ (if $f \in KH^1$ then $\varphi(f) \equiv f$ and the assertion is trivial). Let

$$\mu = \frac{1}{2} \max_{0 \le t \le \pi} [\omega(f,t) - Kt] \qquad (2.3)$$

and noticing that $\mu > 0$ we set

$$\psi(t, x) = f(t) + \mu + K|x - t|, \qquad (2.4)$$

$$\varphi_0(x) = \inf_t \psi(t, x). \qquad (2.5)$$

For every x', x'', assuming for definiteness $\varphi_0(x') \le \varphi_0(x'')$, $\varphi_0(x') = \psi(t_1, x')$, we have

$$\begin{aligned}|\varphi_0(x'') - \varphi_0(x')| &= \inf_t \psi(t, x'') - \psi(t_1, x') \le \psi(t_1, x'') - \psi(t_1, x') \\ &= f(t_1) + \mu + K|x'' - t_1| - f(t_1) - \mu - K|x' - t_1| \\ &= K(|x'' - t_1| - |x' - t_1|) \le K|x'' - x'|.\end{aligned}$$

Therefore $\varphi_0 \in KH^1$. We have also from definitions (2.4) and (2.5) that $\varphi_0(x) \le \psi(x, x) = f(x) + \mu$, i.e.

$$\varphi_0(x) - f(x) \le \mu. \tag{2.6}$$

On the other hand let for a fixed x

$$\varphi_0(x) = \psi(t_0, x) = f(t_0) + \mu + K|x - t_0|.$$

Then

$$\begin{aligned} f(x) - \varphi_0(x) &= f(x) - f(t_0) - \mu - K|x - t_0| \\ &\le \omega(f, |x - t_0|) - K|x - t_0| - \mu \\ &\le \max_{0 \le t \le \pi} [\omega(f, t) - Kt] - \mu \end{aligned}$$

(remember that $\omega(f, t) = \omega(f, \pi)$ for $\delta \ge \pi$). Therefore (see (2.3)) $f(x) - \varphi_0(x) \le \mu$, and this together with (2.6) gives $\|f - \varphi\|_C \le \mu$ and because of $\varphi_0 \in KH^1$ we have proved

$$E(f, KH^1)_C \le \mu = \frac{1}{2} \max_{0 \le t \le \pi} [\omega(f, t) - Kt]. \tag{2.7}$$

Let us show that we actually have an equality in (2.7). Let $2\mu = \omega(f, t_0) - Kt_0$ $(0 < t_0 < \pi)$. Let α and β be such points that $|\alpha - \beta| = t_0$ and $|f(\alpha) - f(\beta)| = \omega(f, t_0)$. Then for every function $\varphi \in KH^1$ we have

$$\begin{aligned} 2\|f - \varphi\|_C &\ge |f(\alpha) - \varphi(\alpha)| + |f(\beta) - \varphi(\beta)| \\ &\ge |f(\alpha) - f(\beta) - \varphi(\alpha) + \varphi(\beta)| \ge |f(\alpha) - f(\beta)| - |\varphi(\alpha) - \varphi(\beta)| \\ &\ge \omega(f, t_0) - Kt_0 = 2\mu. \end{aligned}$$

So we have established $E(f, KH^1)_C \ge \mu$ which together with (2.7) proves (2.2).

The existence of a function φ of best approximation to f in the class KH^1 follows from Proposition 1.1.3 because KH^1 is a convex and a locally compact set in C.§

6.2.2 Inequalities for $f \in C$ In the space C of the continuous 2π-periodic functions we select the subspace C_* of functions for which the modulus of continuity $\omega(f, \delta)$ is concave on $[0, \pi]$ (and hence on $[0, \infty)$).

Theorem 6.2.2 For every $f \in C$ ($f \not\equiv$ constant) we have

$$E_n(f)_C < \omega(f, \pi/n), \qquad n = 1, 2, \ldots, \tag{2.8}$$

§ *Translator's note* The function φ_0 constructed in the proof is the function of best approximation to f.

and the constant 1, which is independent of f and n, on the right-hand side of (2.8) cannot be decreased.

If $f \in C_*$ then

$$E_n(f)_C \le \frac{1}{2}\omega(f, \frac{\pi}{n}), \qquad n = 1, 2, \ldots, \tag{2.9}$$

and this inequality is also exact for every $n = 1, 2, \ldots$.

Proof Let $f \in C$ and n be a fixed integer. If for some $M > 0$ $\omega(f, t) = Mt$ for $0 \le t \le \pi/n$ then $f \in MH^1$ and inequality (2.9) follows from (1.16) with $p = \infty$ because $MH^1 = MW^r_\infty$ and $K_1 = \pi/2$. Hence we assume that $\omega(f, \delta) \not\equiv Mt$ on $[0, \pi/n]$ for any M.

For every $M > 0$ in class MH^1 there is a function $\varphi_0(t) = \varphi_0(f, M, t)$ such that $\|f - \varphi_0\|_C = E(f, MH^1)_C$. Using Proposition 6.2.1 and inequality (1.16) in (2.1) we get

$$E_n(f)_C \le \|f - \varphi_0\|_C + E_n(MH^1)_C \le \frac{1}{2}\max_{0\le t\le\pi}[\omega(f, t) - Mt] + \frac{MK_1}{n}. \tag{2.10}$$

If $f \in C_*$ and hence $\omega(f, t)$ is concave, then Lemma 6.1.2 implies $\omega'_+(f, t) \le \omega'_-(f, t)$ for every $t > 0$ and we can choose $M = M_n$ from the condition

$$\omega'_+(f, \pi/n) \le M \le \omega'_-(f, \pi/n). \tag{2.11}$$

For this choice of M we have

$$\max_{0\le t\le\pi}[\omega(f, t) - Mt] = \omega(f, \pi/n) - M\pi/n, \tag{2.12}$$

Replacing (2.12) in (2.10) we obtain (2.9).

Equality holds in (2.9) for the function

$$f(t) = \omega(|t|), \qquad -\pi/n \le t \le \pi/n, \qquad f(t + 2\pi/n) = f(t), \tag{2.13}$$

where ω is any (in particular concave) modulus of continuity. Indeed $\omega(f, t) = \omega(t)$ $(0 \le t \le \pi/n)$ for function (2.13) and in view of Chebyshev's Theorem 2.1.3 the best approximation to this function by polynomials from $\mathcal{N}^T_{2n-1}$ is realized by the constant $c = \omega(\pi/n)/2$.

Now let f be any function from C for which $\omega(f, \delta) \not\equiv Mt$ on $[0, \pi/n]$ according to the assumptions made at the beginning of the proof. In view of Lemma 6.1.4 there exists a concave modulus of continuity ω_* such that

$$\omega(f, t) \le \omega_*(t) < 2\omega(f, t), \qquad 0 \le t \le \pi. \tag{2.14}$$

The first inequality in (2.14) allows us to continue to estimate (2.10) and write for every $M > 0$

$$E_n(f)_C \le \frac{1}{2}\max_{0\le t\le\pi}[\omega_*(f, t) - Mt] + \frac{M\pi}{2n}. \tag{2.15}$$

If we choose $M = M_n$ from a condition of the type (2.11) for ω_* then the maximum in (2.15) is attained at the point $t = \pi/n$ and then using the second inequality in (2.14) we obtain

$$E_n(f)_C \leq \frac{1}{2}\omega_*\left(\frac{\pi}{n}\right) < \omega\left(f, \frac{\pi}{n}\right).$$

It remains to prove that the constant 1 on the right-hand side of (2.8) cannot be improved. This is established by the following lemma.

Lemma 6.2.3 For every fixed $n = 1, 2, \ldots,$ there exists a function $f_\varepsilon \in C$ ($f_\varepsilon \not\equiv$ constant) for every $\varepsilon > 0$ such that

$$E_n(f_\varepsilon)_C > \left(1 - \frac{1}{2n} - \varepsilon\right)\omega\left(f_\varepsilon, \frac{\pi}{n}\right). \tag{2.16}$$

Proof Let $0 < \varepsilon < \frac{1}{2}$. Set $h = \pi/n$ and

$$x_0 = 0, \qquad x_k = kh - (n - k)\beta, \qquad k = 1, 2, \ldots, n,$$

where $0 < \beta < 2\varepsilon/n^2$. It is easy to check that $0 = x_0 < x_1 < \ldots < x_n = \pi$. Noticing that $x_{k+1} - x_k = h + \beta$ $(k = 1, 2, \ldots, n - 1)$ we define a 2π-periodic even polygon f_ε by setting for $t \in [0, \pi]$

$$f_\varepsilon(x_k) = (-1)^{k+1}, \qquad k = 1, 2, \ldots, n,$$

$$f_\varepsilon(t) = 0, \qquad 0 \leq t \leq x_1 - \beta, x_k + \beta \leq t \leq x_{k+1} - \beta, \qquad k = 1, 2, \ldots, n - 1,$$

f_ε to be linear on the intervals $x_k - \beta \leq t \leq x_k$ $(k = 1, 2, \ldots, n)$ and $x_k \leq t \leq x_k + \beta$ $(k = 1, 2, \ldots, n - 1)$. It is easy to see (by drawing a diagram) that the oscillation of f_ε on every interval of length not greater than h does not exceed 1. Hence from $f_\varepsilon(x_1) - f_\varepsilon(0)$ and $x_1 < h$ we obtain

$$\omega(f_\varepsilon, h) = \omega(f_\varepsilon, \pi/n) = 1. \tag{2.17}$$

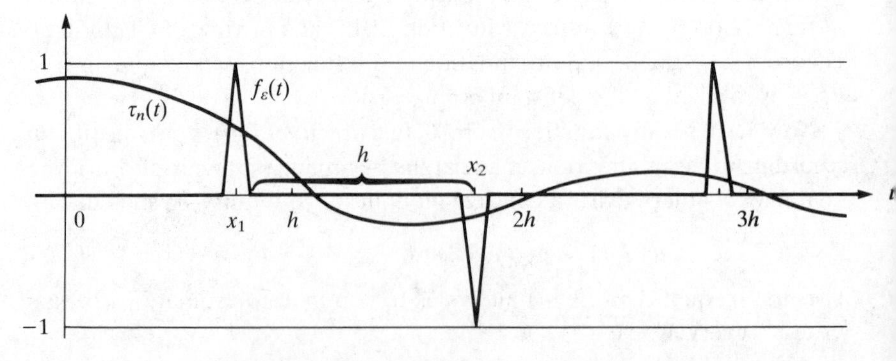

Fig. 6.2 To the proof of Lemma 6.2.3.

Consider the trigonometric polynomial

$$\tau_n(t) = \frac{1}{n}\left(\frac{1}{2} + \cos t + \cos 2t + \ldots + \cos (n-1)t\right) = \frac{1}{2n}\frac{\sin (n-\frac{1}{2})t}{\sin (t/2)}. \quad (2.18)$$

From (2.18) we get

$$\tau_n(0) = 1 - 1/(2n), \quad \tau_n(kh) = (-1)^{k+1}/(2n), \quad k = 1, 2, \ldots, n. \quad (2.19)$$

Because $|x_k - kh| = (n-k)\beta$, $|\tau_n'(t)| < (n-1)/2$ we have

$$|\tau_n(x_k) - \tau_n(kh)| < \max_t |\tau_n'(t)|\,|x_k - kh| < (n-1)(n-k)\beta/2 < \varepsilon,$$

$$k = 1, 2, \ldots, n-1. \quad (2.20)$$

Taking (2.19) and (2.20) into account together with the values of f_ε at points x_k $(k = 0, 1, \ldots, n)$ we obtain

$$\begin{aligned} f(x_k) - \tau_n(x_k) &= [f(x_k) - \tau_n(kh)] + [\tau_n(kh) - \tau_n(x_k)] \\ &= (-1)^{k+1}(1 - 1/(2n)) + \mu_k, \quad k = 0, 1, \ldots, n, \end{aligned}$$

where $0 \le |\mu_n| < \varepsilon < \frac{1}{2}$. Therefore at n points on $[0, \pi]$ the difference $(f - \tau_n)$ takes values which are greater than $1 - 1/(2n) - \varepsilon$ in modulus and which alternate in sign. Taking the evenness of f and of τ_n into account the de la Vallée-Poussin Theorem 2.1.5 gives

$$E_n(f_\varepsilon)_C > 1 - 1/(2n) - \varepsilon,$$

which together with (2.17) proves (2.16). This completes the proof of Theorem 6.2.2 because $\varepsilon > 0$ in lemma 6.2.3 was arbitrary. □

Let us note that in Theorem 6.2.2 we have proved the exactness of the constant in (2.8) when assumed to be independent of n. Considering the exact constant in (2.8) for each $n = 1, 2, \ldots$, we deduce from (2.8) and Lemma 6.2.3 that it is between $1 - 1/(2n)$ and 1. In other words

$$1 - \frac{1}{2n} \le \sup_{f \in C} \frac{E_n(f)_C}{\omega(f, \pi/n)} < 1, \quad n = 1, 2, \ldots,$$

or in the notation of (1.21)

$$1 - 1/(2n) \le \kappa_{\pi/n}(C, C, \mathcal{N}_{2n-1}^T) < 1, \quad n = 1, 2, \ldots.$$

The proof of the more general result

$$\frac{k+1}{2}\left(1 - \frac{1}{2n}\right) \le \kappa_{\pi/(kn)}(C, C, \mathcal{N}_{2n-1}^T) \le \frac{k+1}{2}, \quad n, k = 1, 2, \ldots, \quad (2.21)$$

is based on the same ideas.

6.2.3 Inequalities for $f \in L_2$ In inequality (1.18) with $X = L_2$ we can find the exact constant using the specific features of the metric in L_2. As before we set $\|f\|_p = \|f\|_{L_p}$, $E_n(f)_p = E_n(f)_{L_p}$ and $\omega(f, \delta)_p = \omega(f, \delta)_{L_p}$ for $f \in L_p$.

Theorem 6.2.4 If $f \in L_2$ and $f \not\equiv$ constant then

$$E_n(f)_2 < \frac{1}{2^{1/2}}\,\omega\left(f, \frac{\pi}{n}\right)_2, \qquad n = 1, 2, \ldots, \tag{2.22}$$

Constant $1/2^{1/2}$ is exact for every n.

Proof Let $f \in L_2$ and let

$$\frac{a_0}{2} + \sum_{k=1}^{\infty} (a_k \cos kt + b_k \sin kt) \tag{2.23}$$

be the Fourier series for f. If $S_m(f, t)$ denotes the mth partial sum of this series then Parseval's equality gives

$$E_n(f)_2 = \|f - S_{n-1}(f)\|_2 = \left[\pi \sum_{k=n}^{\infty} (a_k^2 + b_k^2)\right]^{1/2} \tag{2.24}$$

and we have to estimate the last expression by $\omega(f, \pi/n)_2$.

From (2.23) we see that the Fourier series for $f(\bullet + u) - f(\bullet)$ is

$$\sum_{k=1}^{\infty} \{[a_k(\cos ku - 1) + b_k \sin ku] \cos kt + [b_k (\cos ku - 1) - a_k \sin ku] \sin kt\},$$

and for brevity setting $a_k^2 + b_k^2 = \rho_k^2$ we get by another application of Parseval's equality

$$\|f(\bullet + u) - f(\bullet)\|_2^2 = 2\pi \sum_{k=1}^{\infty} \rho_k^2 (1 - \cos ku). \tag{2.25}$$

Now using (2.24) for every $\delta \geq 0$ we can write

$$\omega^2(f, \delta)_2 \geq \|f(\bullet + \delta) - f(\bullet)\|_2^2$$

$$= 2\pi \sum_{k=n}^{\infty} \rho_k^2 (1 - \cos k\delta) = 2E_n^2(f)_2 - 2\pi \sum_{k=n}^{\infty} \rho_k^2 \cos k\delta.$$

Changing δ to t we can write

$$E_n^2(f)_2 \leq \tfrac{1}{2}\,\omega^2(f, t)_2 + \pi \sum_{k=n}^{\infty} \rho_k^2 \cos kt, \qquad t \geq 0, \tag{2.26}$$

in which both the left- and right-hand sides are non-negative.

Multiplying both sides of (2.26) by sin nt and integrating with respect to t from 0 to π/n we obtain

$$\frac{2}{n}E_n^2(f)_2 \leq \frac{1}{2}\int_0^{\pi/n} \omega^2(f,t)_2 \sin nt \, \mathrm{d}t + \pi \sum_{k=n}^{\infty} \rho_k^2 c_k,$$

where

$$c_k = \int_0^{\pi/n} \sin nt \cos kt \, \mathrm{d}t.$$

It is easy to see that $c_n = 0$, $c_k \leq 0$ $(k > n)$ and hence

$$E_n(f)_2 \leq \frac{1}{2}\left\{n \int_0^{\pi/n} \omega^2(f,t)_2 \sin nt \, \mathrm{d}t\right\}^{1/2}. \tag{2.27}$$

The last inequality is of independent interest. Moreover it cannot be improved (in this form) as we shall see later. In order to deduce (2.22) from (2.27) we notice that $\omega(f, \pi/n)_2 > 0$ and hence

$$\int_0^{\pi/n} \omega^2(f,t)_2 \sin nt \, \mathrm{d}t < \omega^2\left(f, \frac{\pi}{n}\right)_2 \int_0^{\pi/n} \sin nt \, \mathrm{d}t = \frac{2}{n}\omega^2\left(f, \frac{\pi}{n}\right)_2.$$

We shall prove the exactness of (2.22) which in turn will imply the unimprovability of (2.27). For fixed n and $0 < \varepsilon < \pi/(2n)$ let $g(\varepsilon, t)$ be an even $2\pi/n$-periodic function equal to $1 - t/(2\varepsilon)$ for $0 \leq t \leq 2\varepsilon$ and zero for $2\varepsilon \leq t \leq \pi/n$.

The Fourier series for g is

$$g(\varepsilon, t) = \frac{\varepsilon}{\pi} + \frac{2}{\pi\varepsilon}\sum_{\nu=1}^{\infty}\left(\frac{\sin n\nu\varepsilon}{\nu n}\right)^2 \cos n\nu t.$$

Set

$$f_\varepsilon(t) = \left(\frac{2}{\pi\varepsilon}\right)^{1/2}\sum_{\nu=1}^{\infty}\frac{\sin n\nu\varepsilon}{\nu n}\cos n\nu t.$$

From the general expression (2.25) we obtain

$$\begin{aligned}\|f_\varepsilon(\bullet + u) - f_\varepsilon(\bullet)\|_2^2 &= \frac{4}{\varepsilon}\sum_{k=1}^{\infty}\left(\frac{\sin n\nu\varepsilon}{\nu n}\right)^2(1 - \cos n\nu u)\\ &= 2\pi\,[g(\varepsilon, 0) - g(\varepsilon, u)]\end{aligned}$$

and because g is non-increasing on $[0, \pi/n]$ we have

$$\omega^2(f_\varepsilon, \pi/n)_2 = 2\pi\,[g(\varepsilon, 0) - g(\varepsilon, \pi/n)] = 2\pi.$$

On the other hand (2.24) implies

$$E_n^2(f_\varepsilon)_2 = \frac{2}{\varepsilon}\sum_{\nu=1}^{\infty}\left(\frac{\sin n\nu\varepsilon}{\nu n}\right)^2 = \pi\,[g(\varepsilon, 0) - \varepsilon/\pi] = \pi - \varepsilon,$$

and hence

$$E_n^2(f_\varepsilon)_2 = \frac{\pi - \varepsilon}{2\pi}\,\omega^2\left(f_\varepsilon, \frac{\pi}{n}\right)_2,$$

which completes the proof of the theorem because ε was arbitrarily positive and $f_\varepsilon \in L_2$. □

In the notations of (2.21) and (2.24) this result can be written as

$$\kappa_{\pi/n}(L_2, L_2, \mathcal{N}_{2n-1}^T) = \kappa'_{\pi/n}(L_2, L_2, \mathcal{N}_{2n-1}^T) = 2^{-1/2}, \qquad n = 1, 2, \ldots,$$

where the exact constant is realized by Fourier means.

When the function $\omega^2(f, \delta)_2$ is concave on $[0, \pi/n]$ the estimate (2.22) can be improved. Indeed in this case there is a linear function l such that

$$l\left(\frac{\pi}{2n}\right) = \omega^2\left(f, \frac{\pi}{2n}\right)_2, \qquad \omega^2(f, \delta)_2 \le l(\delta), \qquad 0 \le \delta \le \pi/n.$$

Writing the integral in (2.27) as

$$\begin{aligned}\int_0^{\pi/n} \omega^2(f, t)_2 \sin nt\, \mathrm{d}t &= \int_0^{\pi/n} [\omega^2(f, t)_2 - l(t)] \sin nt\, \mathrm{d}t \\ &\quad + \int_0^{\pi/n} \left[l(t) - \omega^2\left(f, \frac{\pi}{2n}\right)_2\right] \sin nt\, \mathrm{d}t \\ &\quad + \omega^2\left(f, \frac{\pi}{2n}\right)_2 \int_0^{\pi/n} \sin nt\, \mathrm{d}t,\end{aligned}$$

we see that the first integral on the right-hand side is non-positive and the second one is zero. Hence

$$\int_0^{\pi/n} \omega^2(f, t)_2 \sin nt\, \mathrm{d}t \le \frac{2}{n}\,\omega^2\left(f, \frac{\pi}{2n}\right)_2$$

and substituting this estimate into (2.27) we see that under the assumption for concavity of $\omega^2(f, \delta)_2$ the inequality

$$E_n(f)_2 = \|f - S_{n-1}(f)\|_2 \le \frac{1}{2^{1/2}}\,\omega\left(f, \frac{\pi}{2n}\right)_2, \qquad n = 1, 2, \ldots, \quad (2.28)$$

holds. Estimate (2.28) cannot be improved on the set of functions $f \in L_2$ for which $\omega^2(f, \delta)_2$ is concave (see Exercise 10).

6.2.4 Inequalities for differentiable functions Let us go back to the inequality (1.17) which we obtained by general arguments and which is not exact, and let us show that for $X = C$ or $X = L_1$ and for odd r we can find the exact constant in it. Moreover this constant is realized by a linear method.

Let us fix $h > 0$. We define a sequence of functions $g_r(t) = g_r(h, t)$ $(r = 0, 1, \ldots,)$ on $[-h, h]$:

$$g_0(t) = -\frac{1}{2h}, \qquad -h \leq t \leq h;$$

$$g_1(t) = \frac{1}{2}\left(1 - \frac{t}{h}\right), 0 \leq t \leq h; \qquad g_1(t) = -g_1(-t), -h \leq t < 0;$$

$$g_r(t) = \int_{c_r}^{t} g_{r-1}(u)\, du, \qquad -h \leq t \leq h, \qquad r = 2, 3, \ldots,$$

where number c_r is chosen so that

$$\int_{-h}^{h} g_r(u)\, du = 0.$$

Let as before X^r $(r = 1, 2, \ldots,)$ denote the set of rth periodic integrals of functions from $X = C, L_p$, let f_h be the Stekhlov function for f (see Section A2) and let $U_{n-1,r}(f, \lambda^*, t)$ be the Favard means (Section 4.1.3). As we know (see (4.1.47)) if $f \in X^r$ then

$$\|f - U_{n-1,r}(f, \lambda^*)\|_X \leq K_r\, n^{-r}\, \|f^{(r)}\|_X. \tag{2.29}$$

By setting

$$\Phi_{n,r}(f, t) = -\frac{\pi}{n} \sum_{\nu=0}^{m_r} g_{2\nu}(h_n)\, U_{n-1,r-2\nu+1}(f_{h_n}^{(2\nu)}, \lambda^*, t), \tag{2.30}$$

where $m_r = [(r-1)/2]$, $h_n = \pi/(2n)$, we define a linear operator $\Phi_{n,r}$ to assign a trigonometric polynomial $\Phi_{n,r}(f) \in \mathcal{N}_{2n-1}^T$ to every function $f \in X^r$.

Theorem 6.2.5 If $f \in X^r$, where X is C or L_p $(1 \leq p < \infty)$, then we have for all odd $r = 1, 3, 5, \ldots,$

$$\|f - \Phi_{n,r}(f)\|_X \leq \frac{K_r}{2n^r}\, \omega\left(f^{(r)}, \frac{\omega}{n}\right)_X, \qquad n = 1, 2, \ldots, \tag{2.31}$$

where K_r are Favard constants.

Proof Setting

$$A_{h,r}(f, t) = f(t) + 2h \sum_{\nu=0}^{m_r} g_{2\nu}(h)\, f_h^{(2\nu)}(t), \tag{2.32}$$

we represent the error as

$$f(t) - \Phi_{n,r}(f,t) = A_{h_n,r}(f,t) - \frac{\pi}{n}\sum_{\nu=0}^{m_r} g_{2\nu}(h_n)$$
$$\times [f_{h_n}^{(2\nu)}(t) - U_{n-1,r-2\nu+1}(f_{h_n}^{(2\nu)}, \lambda^*, t)],$$

and hence

$$\|f - \Phi_{n,r}(f)\|_X \leq \|A_{h_n,r}(f)\|_X + \frac{\pi}{n}\sum_{\nu=0}^{m_r} |g_{2\nu}(h_n)|$$
$$\times \|f_{h_n}^{(2\nu)} - U_{n-1,r-2\nu+1}(f_{h_n}^{(2\nu)}, \lambda^*)\|_X. \tag{2.33}$$

But $f_h \in X^{r+1}$ and hence $f_h^{(2\nu)} \in X^{r-2\nu+1}$, We have in view of (2.29)

$$\|f_h^{(2\nu)} - U_{n-1,r-2\nu+1}(f_h^{(2\nu)}, \lambda^*)\|_X \leq \frac{K_{r-2\nu+1}}{n^{r-2\nu+1}}\|f_h^{(r+1)}\|_X.$$

But (see (A2.2))

$$\|f_h^{(r+1)}\|_X \leq \frac{1}{2h}\omega(f^{(r)}, 2h)_X$$

and hence

$$\|f_{h_n}^{(2\nu)} - U_{n-1,r-2\nu+1}(f_{h_n}^{(2\nu)}, \lambda^*)\|_X \leq \frac{n}{\pi}\frac{K_{r-2\nu+1}}{n^{r-2\nu+1}}\,\omega\left(f^{(r)}, \frac{\pi}{n}\right)_X. \tag{2.34}$$

In order to estimate the first summand on the right-hand side of (2.33) using $\omega(f^{(r)}, \pi/n)_X$ we shall represent the function (2.32) in a slightly different form, namely

$$A_{h,r}(f,t) = \int_{-h}^{h} G_r(u) f^{(r)}(t-u)\,du, \qquad r = 1, 2, \ldots, \tag{2.35}$$

where

$$G_r(t) = G_r(h,t) = g_r(t) - \frac{1 + (-1)^r}{2} g_r(h), \qquad -h \leq t \leq h. \tag{2.36}$$

First let $r = 1$. Then $G_1(t) = g_1(t)$ and the definition of g_1 and integration by parts on every interval $[-h, 0]$, $[0, h]$ give

$$\int_{-h}^{h} G_r(u) f'(t-u)\,du = f(t) - f_h(t) = A_{h,1}(f,t).$$

Let, by induction, equality (2.35) be valid for $r = 1, 2, \ldots, k$. If k is odd then $A_{h,k+1}(f,t) = A_{h,k}(f,t)$ and integration by parts and $G_r(-h) = G_r(h) = 0$ give

$$A_{h,k+1}(f,t) = \int_{-h}^{h} G_k(u)\, f^{(k)}(t-u)\,\mathrm{d}u = \int_{-h}^{h} G_{k+1}(u)\, f^{(k+1)}(t-u)\,\mathrm{d}u.$$

If k is even then

$$\begin{aligned} A_{h,k+1}(f,t) &= A_{h,k}(f,t) + 2hg_k(h)\, f_h^{(k)}(t) \\ &= \int_{-h}^{h} G_k(u)\, f^{(k)}(t-u)\,\mathrm{d}u + g_k(h)\int_{-h}^{h} f^{(k)}(t-u)\,\mathrm{d}u \\ &= \int_{-h}^{h} g_k(u)\, f^{(k)}(t-u)\,\mathrm{d}u = \int_{-h}^{h} G_{k+1}(u)\, f^{(k+1)}(t-u)\,\mathrm{d}u, \end{aligned}$$

and this proves (2.35). We also need:

Lemma 6.2.6 We have

$$\int_{-h_n}^{h_n} |G_r(h_n t)|\,\mathrm{d}t = \frac{K_r}{n^r} - 2\sum_{\nu=0}^{m_r} |g_{2\nu}(h_n)|\,\frac{K_{r-2\nu+1}}{n^{r-2\nu+1}}, \qquad n, r = 1, 2, \ldots, \tag{2.37}$$

where $h_n = \pi/(2n)$, $m_r = [(r-1)/2]$.

Indeed with odd r setting $f = \varphi_{n,r}$ (see Section 3.1) and $t = 0$ in the representations (2.32) and (2.35) we obtain

$$\begin{aligned} \varphi_{n,r}(0) + \sum_{\nu=0}^{m_r} g_{2\nu}(h_n)\int_{-h_n}^{h_n} \varphi_{n,r-2\nu}(t)\,\mathrm{d}t &= \int_{-h_n}^{h_n} G_r(t)\,\varphi_{n,0}(-t)\,\mathrm{d}t \\ &= (-1)^{m_r+1}\int_{-h_n}^{h_n} |G_r(t)|\,\mathrm{d}t, \end{aligned}$$

because $\operatorname{sgn} G_{2j-1}(t) = (-1)^{j-1}\varphi_{n,0}(t)$ for $|t| < h_n$. It remains to note that $g_{2\nu}(h_n) = (-1)^{\nu-1}\,|g_{2\nu}(h_n)|$ and for $r = 2j-1$

$$\varphi_{n,r}(0) = (-1)^j \frac{K_r}{n^r}; \qquad \int_{-h_n}^{h_n} \varphi_{n,r-2\nu}(t)\,\mathrm{d}t = (-1)^{j-\nu}\, 2\,\frac{K_{r-2\nu+1}}{n^{r-2\nu+1}}.$$

If r is even, $r = 2j$, then we get (2.37) by setting $f(t) = \varphi_{n,r}(t+h_n)$, $t = 0$ in (2.32) and (2.35), and taking into account that $\operatorname{sgn} G_{2j}(t) = (-1)^j\varphi_{n,0}(h_n - t)$ for $|t| < h_n$.

Now we can finish the proof of Theorem 6.2.5. Let r be odd. Then $G_r(-t) = -G_r(t)$ $(0 \le t \le h_n)$ and in view of (2.35)

$$\|A_{h_n,r}(f)\|_X = \left\| \int_0^{h_n} G_r(u)[f^{(r)}(t-u) - f^{(r)}(t+u)]\,\mathrm{d}u \right\|_X. \tag{2.38}$$

If $X = L_p$ $(1 \le p < \infty)$ then the generalized Minkowski inequality (see (A1.19)) gives

$$\|A_{h_n,r}(f)\|_p = \int_0^{h_n} |G_r(u)| \left\{ \int_0^{2\pi} |f^{(r)}(t-u) - f^{(r)}(t+u)|^p \, dt \right\}^{1/p} du$$

$$\leq \omega(f^{(r)}, 2h_n)_p \int_0^{h_n} |G_r(t)| \, dt = \frac{1}{2} \omega\left(f^{(r)}, \frac{\pi}{n} \right)_p \int_{-h_n}^{h_n} |G_r(t)| \, dt.$$

In the case $X = C$ we get the corresponding estimate from (2.38) in an obvious way. We have for $f \in X^r$ and $r = 1, 3, 5, \ldots,$

$$\|A_{h_n,r}(f)\|_X = \frac{1}{2} \omega\left(f^{(r)}, \frac{\pi}{n} \right)_X \int_{-h_n}^{h_n} |G_r(h_n, t)| \, dt. \tag{2.39}$$

Replacing (2.34) and (2.39) in (2.33) and using Lemma 6.2.6 for odd r we obtain

$$\|f - \Phi_{n,r}(f)\|_X \leq \omega\left(f^{(r)}, \frac{\pi}{n} \right)_X \left[\frac{1}{2} \int_{-h_n}^{h_n} |G_r(h_n, t)| \, dt + \sum_{\nu=0}^{m_r} |g_{2\nu}(h_n)| \frac{K_{r-2\nu+1}}{n^{r-2\nu+1}} \right]$$

$$= \frac{K_r}{2n^r} \omega\left(f^{(r)}, \frac{\pi}{n} \right)_X,$$

which proves the theorem. □

As a consequence we get for every $f \in X^r$ (X is either L_p ($p \geq 1$) or C) and $r = 1, 3, 5, \ldots,$

$$E_n(f)_X \leq \frac{K_r}{2n^r} \omega\left(f^{(r)}, \frac{\pi}{n} \right)_X, \qquad n = 1, 2, \ldots. \tag{2.40}$$

We shall show that, in the cases $X = C$ and $X = L_1$, inequality (2.40) cannot be improved (for every $r = 1, 2, \ldots,$) even if we replace $\omega(f^{(r)}, \pi/n)_X$ by $\omega(f^{(r)}, \gamma)_X$ for every $\gamma \in [\pi/n, \pi]$. Because $\omega(g, \delta)_X \leq 2\|g\|_X$ and assuming in the following that $f^{(r)} \not\equiv$ constant for every $\gamma > 0$ we have

$$\sup_{f \in X^r} \frac{E_n(f)_X}{\omega(f^{(r)}, \gamma)_X} \geq \frac{1}{2} \sup_{f \in X^r} \frac{E_n(f)_X}{\|f^{(r)}\|_X}. \tag{2.41}$$

The Stekhlov function $\varphi_{n,r,h}$ for the Euler spline $\varphi_{n,r}$ obviously belongs to C^r and for small enough h we have

$$\|\varphi_{n,r,h}^{(r)}\|_C = \|\varphi_{n,0,h}\|_C = 1.$$

From (2.41) with $X = C$ and the continuity of the functional of best approximation we obtain

$$\sup_{f \in C^r} \frac{E_n(f)_C}{\omega(f^{(r)}, \gamma)_C} \geq \tfrac{1}{2} E_n(\varphi_{n,r,h})_C = \tfrac{1}{2} E_n(\varphi_{n,r})_C + \varepsilon(h) = \frac{K_r}{2n^r} + \varepsilon(h),$$

where $\varepsilon(h) \to 0$ when $h \to 0$ and hence

$$\sup_{f \in C^r} \frac{E_n(f)_C}{\omega(f^{(r)}, \gamma)_C} \geq \frac{1}{2} \sup_{f \in C^r} \frac{E_n(f)_C}{\|f^{(r)}\|_C} \geq \frac{K_r}{2n^r}, \qquad n, r = 1, 2, \ldots. \tag{2.42}$$

Arguing in the same way using the Stekhlov function for $g_{n,r-1} = \varphi_{n,r-1}/(4n)$ we see that the right-hand side of (2.41) for $X = L_1$ is not less than $E_n(g_{n,r-1})_1/2 = \|g_{n,r-1}\|_1/2$ and hence

$$\sup_{f \in L_1^r} \frac{E_n(f)_1}{\omega(f^{(r)}, \gamma)_1} \geq \frac{1}{2} \sup_{f \in L_1^r} \frac{E_n(f)_1}{\|f^{(r)}\|_1} \geq \frac{K_r}{2n^r}, \qquad n, r = 1, 2, \ldots. \tag{2.43}$$

We see from (2.42) and (2.43) that the constant $K_r/2$ in (2.40) cannot be decreased for odd r and $X = C$ or $X = L_1$. Taking into account that estimate (2.40) is given by the linear method for approximation (2.30) and using the notation from (1.21) and (1.24) we can state:

Theorem 6.2.7 For all $\gamma \geq \pi/n$, $r = 1, 3, 5, \ldots$, we have

$$\kappa_\gamma(C^r, C, \mathcal{N}_{2n-1}^T) = \kappa'_\gamma(C^r, C, \mathcal{N}_{2n-1}^T) = K_r/(2n^r),$$
$$\kappa_\gamma(L_1^r, L_1, \mathcal{N}_{2n-1}^T) = \kappa'_\gamma(L_1^r, L_1, \mathcal{N}_{2n-1}^T) = K_r/(2n^r).$$

The situation for even r is slightly different. For $X = C$ and $r = 2, 4, 6, \ldots$, instead of (2.40) we can write (see the comments for references) the exact inequality

$$E_n(f)_C \leq \frac{K_r}{2n^r} \omega\left(f^{(r)}, \frac{2\pi}{n}\right)_C. \tag{2.44}$$

Let us note that (2.44) is deduced from a general inequality for the best approximation by an arbitrary subspace $\mathcal{N}$ containing the constants:

$$E(f, \mathcal{N})_C = \frac{K_r}{2} a_{r+1}^r \, \omega(f^{(r)}, \gamma)_C, \qquad r = 1, 2, \ldots, \tag{2.45}$$

where

$$a_m = [E(W_\infty^r, \mathcal{N})_C / K_m]^{1/m}$$

and $\gamma \geq \pi a_{r+1}$ for odd r and $\gamma \geq 2\pi a_{r+1}$ for even r.

Therefore equality $\kappa_\gamma(C^r, C, \mathcal{N}_{2n-1}^T) = K_r/(2n^r)$ for $r = 2, 4, \ldots$, holds for $\gamma \geq 2\pi/n$. We do not know the quantity $\kappa'_\gamma(C^r, C, \mathcal{N}_{2n-1}^T)$ ($r = 2, 4, \ldots$,) even for $\gamma \geq 2\pi/n$. We also do not know the exact constants for even r in the inequalities with the integral modulus of continuity $\omega(f^{(r)}, \gamma)_1$. But the following exact inequality holds:

$$E_n(f)_1 \geq \frac{K_r}{2n^r} \omega(f^{(r)}, \gamma)_C, \qquad r = 1, 2, \ldots, \ \gamma \geq \frac{2\pi}{n}, \ f \in C^r. \tag{2.46}$$

Finally let us note the validity of the exact inequality

$$E_n(f)_C \le \frac{K_r}{2n^r}\,\omega\left(f^{(r)}, \frac{\pi}{n}\right)_C, \qquad n = 1, 2, \ldots,$$

which can be guaranteed for all $r = 0, 1, \ldots,$ for all functions $f \in C_*^r$ for which the modulus of continuity $\omega(f^{(r)}, \delta)_C$ is concave. In the case $r = 0$ this fact was established in Section 2. For $r \ge 1$ it immediately follows from Theorem 4.1.4 and from inequality (2.9).

6.2.5 Approximation by algebraic polynomials on a finite interval Here there are only a few results which can be considered as exact in one sense or another. Let $E_n(f)_{C[a,b]}$ be the best approximation of a function $f \in C[a, b]$ by algebraic polynomials of degree $n - 1$. If $f \in C[-1, 1]$ then the function $\varphi(\theta) = f(\cos\theta)$ belongs to C and $E_n(\varphi)_C = E_n(f)_{C[-1,1]}$ in view of the results in Section 2.1. Because $\omega(\varphi, \delta)_C \le \omega(f, \delta)_{C[-1,1]}$ $(0 \le \delta \le 2)$ from Theorem 6.2.2 we immediately get the inequality

$$E_n(f)_{C[-1,1]} \le \omega(f, \pi/n)_{C[-1,1]}, \qquad n = 2, 3, \ldots, \tag{2.47}$$

which is valid for every $f \in C[-1, 1]$.

The specific properties of the approximation by algebraic polynomials on an interval allow (as in Section 4.4) for an improvement of estimate (2.47) by considering the position of the point in the interval $[-1, 1]$.

Proposition 6.2.8 For every function $f \in C[-1, 1]$ there exists a sequence of algebraic polynomials $p_n(f, t)$ $(n = 1, 2, \ldots,)$ of degree n with the property that for $n \to \infty$ we have uniformly on $t \in [-1, 1]$

$$|f(t) - p_n(f, t)| \le \omega\left(f, \frac{\pi}{n}(1 - t^2)^{1/2}\right) + o\left(\omega\left(f, \frac{1}{n}\right)\right), \tag{2.48}$$

where $\omega(f, \delta) = \omega(f, \delta)_{C[-1,1]}$. In the case of a concave modulus of continuity $\omega(f, \delta)$ the constant 1 in front of $\omega(f, \pi(1 - t^2)^{1/2}/n)$ can be replaced by $\frac{1}{2}$.

According to the exactness of inequalities (2.47) and (2.48) one can state the following: for every $\delta > 0$ there exists a sequence f_n $(n = 1, 2, \ldots,)$ of functions continuous on $[-1, 1]$ such that

$$E_n(f_n)_{C[-1,1]} \le (1 - \varepsilon_n)\,\omega(f_n, (\pi - \delta)/n)_{C[-1,1]},$$

where $\varepsilon_n \ge 0$, $\varepsilon_n = o(1)$.

With respect to the Jackson inequality for $f \in C^r[a, b]$ the following algebraic analog of Theorem 6.2.5 is known:

Theorem 6.2.9 Let $r = 1, 3, 5, \ldots,$. Then for every function $f \in C^r[-1, 1]$ there exists a sequence of algebraic polynomials $p_{n,r}(f, t)$ of degree

$n = 1, 2, \ldots$, depending linearly on f with the property that for $n \to \infty$ we have uniformly on $t \in [-1, 1]$

$$|f(t) - p_{n-1,r}(f, t)| \leq \frac{K_r(1-t^2)^{r/2}}{2n^r}\,\omega\!\left(f^{(r)}, \frac{\pi}{n}(1-t^2)^{1/2}\right)$$
$$+ o[n^{-r}\omega(f^{(r)}, n^{-1})], \tag{2.49}$$

where $\omega(g, \delta) = \omega(g, \delta)_{C[-1,1]}$. The constant $K_r/2$ on the right-hand side of (2.49) cannot be improved.

6.3 Jackson-type inequalities for spline approximation

6.3.1 Approximation by step functions A set of splines of some smoothness with respect to a fixed partition

$$\Delta_N[a, b]: \qquad a = t_0 < t_1 < \ldots < t_N = b, \tag{3.1}$$

is not invariant with respect to translations of the argument, unlike the subspaces of polynomials of a given degree. Hence the error in spline approximation of an individual function will depend on the partition $\Delta_N[a, b]$. Here, together with the modulus of continuity $\omega(f, \delta)_{C[a.b]}$ of a function $f \in C[a, b]$ we shall also consider the quantity

$$\Omega(f, \Delta_N[a, b]) = \max_{1 \leq i \leq N} \Omega_i(f, \Delta_N[a, b]),$$

where

$$\Omega_i(f, \Delta_N[a, b]) = \max_{t_{i-1} \leq t \leq t_i} f(t) - \min_{t_{i-1} \leq t \leq t_i} f(t), \qquad i = 1, 2, \ldots, N.$$

If we define

$$|\Delta_N| = \max_{1 \leq i \leq N} (t_i - t_{i-1}),$$

then clearly

$$\Omega(f, \Delta_N[a, b]) \leq \omega(f, |\Delta_N|)_{C[a,b]}.$$

Hence we have $\Omega(f, \Delta_N[a, b]) \to 0$ as $|\Delta_N| \to 0$ for $f \in C[a, b]$.

As in Section 2.4 let $S_0(\Delta_N[a, b])$ denote the set of step functions s on $[a, b]$ with possible discontinuities only at the points t_i from partition (3.1). For definiteness we set $s(t_i) = [s(t_i - 0) + s(t_i + 0)]/2$ $(i = 1, 2, \ldots, N-1)$. When there is no risk of ambiguity we shall write C, L_p instead of $C[a, b], L_p[a, b]$.

If $f \in C[a, b]$ then the function

$$s(f, t) = \frac{1}{2}\left[\max_{t_{i-1} \leq t \leq t_i} f(t) + \min_{t_{i-1} \leq t \leq t_i} f(t)\right], \qquad t_{i-1} < t < t_i, \qquad i = 1, 2, \ldots, N, \tag{3.2}$$

has the least deviation from f in the $L_\infty[a, b]$ metric among all functions from $S_0(\Delta_N[a, b])$. Specifically

$$\|f - s(f)\|_\infty = \tfrac{1}{2}\,\Omega(f, \Delta_N[a, b]) \le \tfrac{1}{2}\,\omega(f, |\Delta_N|)_C.$$

Moreover, we have for $p > 0$

$$\int_a^b |f(t) - s(f, t)|^p \,\mathrm{d}t \le 2^{-p} \sum_{i=1}^{N} (t_i - t_{i-1})\, \Omega_i^p(f, \Delta_N[a, b])$$

$$\le 2^{-p}\,(b - a)\, \Omega^p(f, \Delta_N[a, b]).$$

Therefore we have for $f \in C[a, b]$

$$E(f, S_0(\Delta_N[a, b]))_\infty = \tfrac{1}{2}\,\Omega(f, \Delta_N[a, b]) \le \tfrac{1}{2}\,\omega(f, |\Delta_N|)_C, \tag{3.3}$$

$$E(f, S_0(\Delta_N[a, b]))_p = \frac{(b - a)^{1/p}}{2}\,\Omega(f, \Delta_N[a, b]) \le \frac{(b - a)^{1/p}}{2}\,\omega(f, |\Delta_N|)_C. \tag{3.4}$$

In the case of uniform partition from (3.3) and (3.4) we obtain

$$E(f, S_{N,0}[a, b])_\infty \le \frac{1}{2}\,\omega\!\left(f, \frac{b - a}{N}\right)_C, \tag{3.5}$$

$$E(f, S_{N,0}[a, b])_p \le \frac{(b - a)^{1/p}}{2}\,\omega\!\left(f, \frac{b - a}{N}\right)_C. \tag{3.6}$$

Inequalities (3.3)–(3.6) cannot be improved on $C[a, b]$: to see this it is sufficient to consider the Stekhlov function for the function f taking only the values ± 1 on $[a, b]$ and changing sign at the middle of every interval (t_{i-1}, t_i) $(i = 1, 2, \ldots, N)$.

Using the notation of (1.21) we can write the results for the uniform partition as

$$\kappa_{(b-a)/N}(C[a, b], L_p[a, b], S_{N,0}[a, b]) = \frac{(b - a)^{1/p}}{2}, \qquad 1 \le p \le \infty. \tag{3.7}$$

The function $s(f, t)$ defined in (3.2) depends non-linearly on f and we may also look for linear methods. If $\sigma_0(f, t)$ is the spline from $S_0(\Delta_N[a, b])$ which coincides with f at the points $\tau_i = (t_i + t_{i-1})/2$ $(i = 1, 2, \ldots, N)$ then we have the exact estimates for $f \in C[a, b]$

$$\|f - \sigma_0(f)\|_\infty \le \omega(f, |\Delta_N|/2)_C,$$

$$\|f - \sigma_0(f)\|_p \le (b - a)^{1/p}\,\omega(f, |\Delta_N|/2)_C, \qquad 1 \le p < \infty, \tag{3.8}$$

which should be compared with (3.3) and (3.4). For every partition (3.1) it is easy to find a function $f_0 \in C[a, b]$ for which the right-hand sides in (3.8) are strictly less than those in (3.3) and (3.4).

Another linear method for approximation by step functions is provided by the function $\psi(f, t) \in S_0(\Delta_N[a, b])$ defined by

$$\psi(f, t) = \frac{1}{t_{i-1} - t_i} \int_{t_{i-1}}^{t_i} f(t)\, \mathrm{d}t, \qquad t_{i-1} < t < t_i, \qquad i = 1, 2, \ldots, N.$$

In the case $t_i = a + i(b - a)/N$ we write $\psi_N(f, t)$ instead of $\psi(f, t)$. If $t \in (t_{i-1}, t_i)$ then

$$|f(t) - \psi(f, t)| = \frac{1}{t_{i-1} - t_i} \int_{t_{i-1}}^{t_i} |f(t) - f(u)|\, \mathrm{d}u \le \Omega_i(f, \Delta_N[a, b])$$

and hence

$$\|f - \psi(f)\|_\infty \le \Omega(f, \Delta_N[a, b]) \le \omega(f, |\Delta_N|)_C. \tag{3.9}$$

Although the estimates in (3.9) are exact, the linear method using $\psi(f, t)$ as the interpolating splines $\sigma_0(f, t)$ does not realize the best inequalities (3.3) and (3.5). In the $L_p[a, b]$ metric for $1 \le p \le 3$ the step function $\psi_N(f, t)$ (for the uniform partition) provides the best estimate

$$\|f - \psi_N(f)\|_p \le \frac{(b - a)^{1/p}}{2} \omega\left(f, \frac{b - a}{N}\right)_C, \qquad 1 \le p \le 3, \tag{3.10}$$

and hence (see (1.24))

$$\kappa'_{(b-a)/N}(C[a, b], L_p[a, b], S_{N,0}[a, b]) = \frac{(b - a)^{1/p}}{2}, \qquad 1 \le p \le 3. \tag{3.11}$$

This fact is far from trivial unlike the other results in this section. It will be proved in Section 7.3.4 when solving the problem of the best approximation for functional classes determined by a majorant of the modulus of continuity.

Considering the periodic functions, all the estimates (with $[a, b] = [0, 2\pi]$) are also valid and for $N = 2n$ they cannot be improved on the class C. In particular for $f \in C$ the following exact inequality holds

$$E(f, S_{2n,0})_p \le \frac{(2\pi)^{1/p}}{2} \omega\left(f, \frac{\pi}{n}\right)_C, \qquad 1 \le p \le \infty.$$

For $1 \le p \le 3$ the above inequality is realized by the linear method $\psi_{2n}(f, t) \in S_{2n,0}$ which coincides with the mean value of f on every interval $((i - 1)\pi/n, i\pi/n)$.

Hence we have

$$\kappa_{\pi/n}(C, L_p, S_{2n,0}) = 2^{-1} (2\pi)^{1/p}, \qquad 1 \le p \le \infty, \tag{3.12}$$

$$\kappa'_{\pi/n}(C, L_p, S_{2n,0}) = 2^{-1} (2\pi)^{1/p}, \qquad 1 \le p \le 3. \tag{3.13}$$

6.3.2 Polygonal interpolation As in the case of step function approximation we can treat the problem locally on every interval of the partition. This allows us to obtain exact estimates both in C and L_p. Moreover in many cases the interpolating polygons are the best possible approximation tools. Such situations occur when we interpolate continuously differentiable functions and estimate the error using $\omega(f', \delta)_C$.

We begin with a lemma which will also be used in the next chapter.

Lemma 6.3.1 If $f \in C[a, b]$, $\int_a^b f(t)\,dt = 0$ and

$$F(x) = \int_a^x f(t)\,dt, \qquad a \le x \le b,$$

then we have $|r'(F, x)| \le \omega(f, x)/4$ a.e. on $[0, b-a]$, where $\omega(g, \delta) = \omega(g, \delta)_{C[a,b]}$ and where $r(g, x) = r(g; a, b; x)$ is the non-increasing rearrangement of g.

Proof The function $r(F, x)$ is absolutely continuous on $[0, b-a]$ (see Section 3.2.2) and hence it is differentiable a.e. Fix $x_0 \in (0, b-a)$ such that the derivative $r'(F, x_0)$ exists, $r'(F, x_0) < 0$ and let $r(F, x_0) = y_0$. From the definitions of the function F and its rearrangement it follows that there are points x_1, x_2 $(a < x_1 < x_2 < b)$ such that $|F(x_1)| = |F(x_2)| = r(F, x_0)$, $|F(x)| > y_0$ $(x_1 < x < x_2)$ and also $x_2 - x_1 \le x_0$, sgn $F(x_1) =$ sgn $F(x_2)$.

If $h > 0$ is small enough then there are $h_1 > 0$, $h_2 < 0$ such that

$$x_1 < x_1 + h_1 < x_2 + h_2 < x_2, \qquad |h_1 - h_2| \le h,$$

and (here drawing a diagram is advisable)

$$|r(F, x_0 + h) - r(F, x_0)| = |F(x_1 + h_1) - F(x_1)| = |F(x_2 + h_2) - F(x_2)|.$$

In view of the assumptions on h, h_1, h_2 we have

$$\left|\frac{1}{h_1} - \frac{1}{h_2}\right| \ge \left|\frac{4}{h}\right|$$

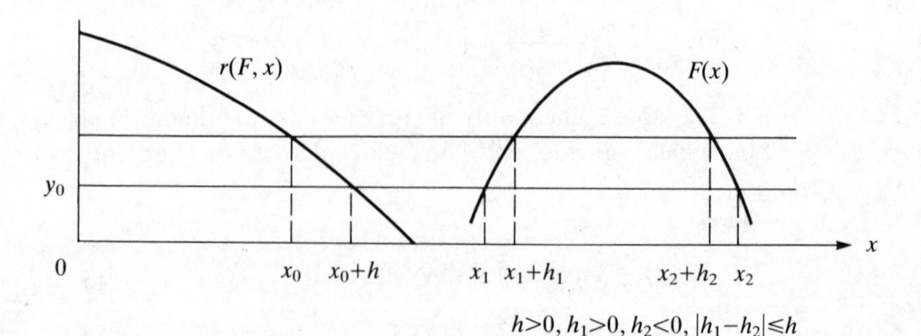

Fig. 6.3 To the proof of Lemma 6.3.1.

and hence

$$\left|\frac{r(F, x_0 + h) - r(F, x_0)}{h}\right| \leq \frac{1}{4}\left|\frac{F(x_1 + h_1) - F(x_1)}{h_1} - \frac{F(x_2 + h_2) - F(x_2)}{h_2}\right|.$$

Letting $h \to 0$ we obtain

$$\begin{aligned} 4|r'(F, x_0)| &\leq |F'(x_1) - F'(x_2)| = |f(x_1) - f(x_2)| \\ &\leq \omega(f, x_2 - x_1) \leq \omega(f, x_0) \end{aligned}$$

and the lemma is proved. □

Let us associate a function

$$f_0(t) = \begin{cases} \omega(f, a + b - 2t)/2, & a \leq t \leq (a + b)/2, \\ -\omega(f, 2t - a - b)/2, & (a + b)/2 \leq t \leq b, \end{cases}$$

with a function f satisfying the conditions of Lemma 6.3.1 and set

$$F_0(x) = \int_a^x f_0(t)\, \mathrm{d}t, \qquad a \leq x \leq b.$$

It is easy to check that $F_0(b) = 0$, $r(F_0, t) = F_0((a + b + t)/2)$ and

$$\omega(f, \delta) \leq \omega(f_0, \delta) = -4\, r'(F_0, \delta), \qquad 0 < \delta < b - a. \tag{3.14}$$

Lemma 6.3.2 Under the conditions of Lemma 6.3.1 we have

$$r(F, x) \leq r(F_0, x), \qquad 0 \leq x \leq b - a, \tag{3.15}$$

$$\int_a^b |F(x)|^p\, \mathrm{d}x \leq \int_a^b |F_0(x)|^p\, \mathrm{d}x \leq \frac{(b - a)^{p+1}}{p + 1}\left[\frac{1}{p}\,\omega(f, b - a)\right]^p, \qquad p > 0. \tag{3.16}$$

Indeed Lemma 6.3.1 and (3.14) imply that a.e. on $(0, b - a)$ we have $|r'(F, x)| \leq |r'(F_0, x)|$ which proves (3.15) because $r(F, b - a) = r(F_0, b - a) = 0$. Using (3.15) and (3.2.12) for all $p > 0$ we have

$$\int_a^b |F(x)|^p\, \mathrm{d}x = \int_0^{b-a} r^p(F, x)\, \mathrm{d}x \leq \int_0^{b-a} r^p(F_0, x)\, \mathrm{d}x = \int_a^b |F_0(x)|^p\, \mathrm{d}x.$$

The final relation in (3.16) is obtained by estimating the last integral above and recalling the definitions of F_0 and f_0 and the inequality $|f_0(x)| \leq \omega(f, b - a)/2$ $(a \leq x \leq b)$.

Now, as in Section 2.4, let $S_{N,1}[a, b]$ denote the set of all splines of degree 1 and defect 1 with respect to the partition $a + (b - a)k/N$ $(k = 0, 1, \ldots, N)$ of $[a, b]$. Lemma 6.3.2 allows an exact estimate of the interpolation error of a function $f \in C^1[a, b]$ by polygons $\sigma_1(f, t) \in S_{N,1}[a, b]$ defined by the equations

$$\sigma_1(f, t_k) = f(t_k), \qquad t_k = a + (b-a)k/N, \qquad k = 0, 1, \ldots, N.$$

Theorem 6.3.3 If $f \in C^1[a, b]$ then for all $p > 0$ we have

$$\int_a^b |f(t) - \sigma_1(f, t)|^p \, \mathrm{d}t \le \frac{b-a}{p+1}\left[\frac{b-a}{4N}\,\omega\left(f', \frac{b-a}{N}\right)\right]^p, \tag{3.17}$$

where $\omega(f', \delta) = \omega(f', \delta)_{C[a,b]}$, and hence

$$\|f - \sigma_1(f)\|_{L_p[a,b]} \le \left(\frac{b-a}{p+1}\right)^{1/p} \frac{b-a}{4N}\,\omega\left(f', \frac{b-a}{N}\right), \qquad p \ge 1 \tag{3.18}$$

$$\|f - \sigma_1(f)\|_{C[a,b]} \le \frac{b-a}{4N}\,\omega\left(f', \frac{b-a}{N}\right). \tag{3.19}$$

Inequalities (3.17)–(3.19) cannot be improved on $C^1[a, b]$ nor for even N on the set of $b-a$-periodic functions from $C^1(-\infty, \infty)$.

In order to obtain (3.17) we apply Lemma 6.3.2 on every interval $[t_{k-1}, t_k]$ $(k = 1, 2, \ldots, N)$ to the difference $\eta(f, t) = f(t) - \sigma_1(f, t)$ and we then sum over k. We observe that if we consider η on $[t_{k-1}, t_k]$ we may assume that the derivative $\sigma_1'(f, t)$ is a constant on this interval and hence

$$\omega(\eta'(f), \delta)_{C[t_{k-1}, t_k]} = \omega(f', \delta)_{C[t_{k-1}, t_k]}, \qquad 0 \le \delta \le t_k - t_{k-1}.$$

Estimate (3.19) follows from (3.18) by letting $p \to \infty$. The exactness of inequalities (3.17)–(3.19) on $C^1[a, b]$ may be checked by employing the Stekhlov function (for small h) for

$$f(t) = \int_a^t \operatorname{sgn} \cos \frac{\pi N}{b-a}(u-a)\, \mathrm{d}u, \qquad a \le t \le b,$$

where we also remember that $\sigma_1(f_h, t) \equiv 0$ and that f_h satisfies $f_h(a) = f_h(b)$ for even N.

In the periodic case when $[a, b] = [0, 2\pi]$ using the notation introduced for polygonal interpolation in Section 5.2 for $N = 2n$ we have:

Corollary 6.3.4 If $f \in C^1$ then

$$\int_0^{2\pi} |f(t) - \sigma_{2n,1}(f, t)|^p \, \mathrm{d}t \le \frac{2\pi}{p+1}\left[\frac{\pi}{4n}\,\omega\left(f', \frac{\pi}{n}\right)\right]^p, \qquad p > 0, \tag{3.20}$$

$$\|f - \sigma_{2n,1}(f)\|_C \le \frac{\pi}{4n}\,\omega\left(f', \frac{\pi}{n}\right). \tag{3.21}$$

The inequalities are exact.

Considering the Stekhlov function for the polygon $\varphi_{n,1}(t+\pi/(2n))$ in the place of f we see that (3.20) (for $p \geq 1$) and (3.21) cannot be improved even if we consider the best approximation by subspace $S_{2n,1}$ of splines of degree 1 on the uniform partition $\{k\pi/n\}$. Therefore

$$\kappa_{\pi/n}(C^1, C, S_{2n,1}) = \kappa'_{\pi/n}(C^1, C, S_{2n,1}) = \frac{\pi}{4n}, \tag{3.22}$$

$$\kappa_{\pi/n}(C^1, L_p, S_{2n,1}) = \kappa'_{\pi/n}(C^1, L_p, S_{2n,1}) = \left(\frac{2\pi}{p+1}\right)^{1/p} \frac{\pi}{4n}, \qquad 1 \leq p < \infty. \tag{3.23}$$

We shall speak in Section 8.3 on the exactness of estimates (3.20), (3.21) in a more general context.

Let us identify the approximation properties of the interpolation polygons on the set $C[a, b]$. A spline $l(f, t) \in S_1(\Delta_N[a, b])$ which coincides with $f \in C[a, b]$ at the points $t_0, t_1, \ldots, t_N$ from partition (3.1) can be written on every interval $[t_{k-1}, t_k]$ as

$$l(f, t) = \frac{1}{t_k - t_{k-1}} [f(t_{k-1})(t_k - t) + f(t_k)(t - t_{k-1})].$$

Fix $t \in (t_{k-1}, t_k)$ and assume for definiteness $t_{k-1} < t \leq t_{k-1} + h/2$, where $h = t_k - t_{k-1}$. For brevity setting $t - t_{k-1} = \theta h$ $(0 < \theta \leq \frac{1}{2})$ and $\omega(f, \delta)_{C[a,b]} = \omega(f, \delta)$ we have

$$\begin{aligned} |f(t) - l(f, t)| &= |[f(t) - f(t_{k-1})]\,(1-\theta) + [f(t) - f(t_k)]\,\theta| \\ &\leq (1-\theta)\,\omega(f, \theta h) + \theta\,\omega(f, (1-\theta)h) \\ &\leq (1-\theta)\,\omega(f, h/2) + \theta\,\omega(f, h) \\ &\leq (1-\theta)\,\omega(f, h/2) + 2\theta\,\omega(f, h/2) \\ &\leq (3/2)\,\omega(f, h/2). \end{aligned}$$

Therefore we have for every $f \in C[a, b]$

$$\|f - l(f)\|_{C[a,b]} \leq \frac{3}{2}\omega\left(f, \frac{|\Delta_N|}{2}\right)_{C[a,b]}, \tag{3.24}$$

where as before $|\Delta_N| = \max_{1 \leq k \leq N} (t_k - t_{k-1})$. As we can see the estimate (3.24) is worse than the corresponding estimate (3.8) for interpolating splines of zero degree. Nevertheless inequality (3.24) is exact on $C[a, b]$ (see Exercise 6.12).

Let us also give the trivial estimate

$$\|f - l(f)\|_{C[a,b]} \leq \omega(f, |\Delta_N|)_{C[a,b]}, \qquad f \in C[a, b],$$

which is also exact. In particular for the uniform partition we have the exact estimates for $f \in C[a, b]$

$$\|f - l(f)\|_C \le \min\left\{\omega\left(f, \frac{b-a}{N}\right)_C, \frac{3}{2}\,\omega\left(f, \frac{b-a}{2N}\right)_C\right\},$$

where $C = C[a, b]$.

6.3.3 Best approximation of differentiable functions In the periodic case one can obtain the Jackson inequalities for the best approximation of a function $f \in C^r$ by splines of minimal defect with respect to the uniform partition $\bar{\Delta}_{2n} = \{k\pi/n\}$ by starting with the inequalities proved in Proposition 5.4.9, which have the form

$$E(f, S_{2n,r})_C \le K_r\, n^{-r}\, E(f^{(r)}, S_{2n,0})_\infty, \qquad f \in L^r_\infty, \tag{3.25}$$

$$E(f, S_{2n,r})_1 \le \|\varphi_{n,r}\|_{p'}\, E(f^{(r)}, S_{2n,0})_p, \qquad f \in L^r_p, \tag{3.26}$$

for $m = r = 1, 2, \ldots$.

Let, in the notation of Section 6.3.1, $\Omega(g, \bar{\Delta}_{2n})$ be the largest oscillation of $g \in C$ on the intervals $((i-1)\pi/n, i\pi/n)$ $(i = 1, 2, \ldots, 2n)$.

Theorem 6.3.5 If $f \in C^r$ $(r = 0, 1, \ldots)$ then

$$E(f, S_{2n,r})_\infty \le \frac{K_r}{2n^r}\,\Omega(f^{(r)}, \bar{\Delta}_{2n}) \le \frac{K_r}{2n^r}\,\omega\left(f^{(r)}, \frac{\pi}{n}\right)_C, \tag{3.27}$$

$$E(f, S_{2n,r})_1 \le \frac{2K_{r+1}}{n^r}\,\Omega(f^{(r)}, \bar{\Delta}_{2n}) \le \frac{2K_{r+1}}{n^r}\,\omega\left(f^{(r)}, \frac{\pi}{n}\right)_C. \tag{3.28}$$

All inequalities are exact.

Here the inequalities (3.27) and (3.28) were obtained in Section 6.3.1 (see (3.5) and (3.6)) for $r = 0$. If $r \ge 1$ then (3.27) follows immediately from (3.25) and (3.5) and (3.28) follows from (3.26) (with $p = \infty$), (3.5) and $\|\varphi_{n,r}\|_1 = 4\, K_{r+1}\, n^{-r}$.

According to the exactness of estimate (3.27) and (3.28), this is true even for the less accurate inequalities

$$E(f, S_{2n,r})_C \le K_r\, n^{-r}\, \|f^{(r)}\|_C, \tag{3.29}$$

$$E(f, S_{2n,r})_1 \le 4\, K_{r+1}\, n^{-r}\, \|f^{(r)}\|_C. \tag{3.30}$$

Indeed Proposition 5.2.5 for function $\psi(t) = \varphi_{n,r}(t - \pi/(2n))$ implies $E(\psi, S_{2n,r})_q = \|\psi\|_q$. The Stekhlov function ψ_h of ψ belongs to C^r and $\|\psi_h\|_q \to \|\psi\|_q = \|\varphi_{n,r}\|_q$. Because $\omega(g, \delta) \le 2\,\|g\|_C$ the exactness of (3.27) and (3.28) follows from the exactness of (3.29) and (3.30).

Using the notation of (1.21) we can reformulate the statement of Theorem 6.3.5 as

$$\kappa_{\pi/n}(C^r, C, S_{2n,r}) = 2^{-1} K_r n^{-r}, \qquad r = 0, 1, \ldots, \tag{3.31}$$

$$\kappa_{\pi/n}(C^r, L_1, S_{2n,r}) = 2 K_{r+1} n^{-r}, \qquad r = 0, 1, \ldots. \tag{3.32}$$

When considering the similar problem of the best spline approximation of differentiable functions on a finite interval we shall be satisfied by an asymptotically exact estimate which can easily be derived from the results in Sections 5.3.3 and 6.3.1.

If $S_{N,m}[0, \pi]$ is the subspace of the splines of degree m and defect 1 defined on $[0, \pi]$ with respect to the uniform partition $t_k = k\pi/N$ $(k = 0, 1, \ldots, N)$ then (5.4.37) (with $p = \infty$) and (5.4.38) for $f \in C^r[0, \pi]$ $(r = 1, 2, \ldots)$ imply

$$E(f, S_{N,r}[0, \pi])_{C[0,\pi]} \leq K_r N^{-r} E(f^{(r)}, S_{N,0}[0, \pi])_\infty (1 - \eta_1),$$

$$E(f, S_{N,r}[0, \pi])_{L_1[0,\pi]} \leq 4 K_{r+1} N^{-r} E(f^{(r)}, S_{N,0}[0, \pi])_\infty (1 - \eta_2),$$

where $\eta_1 > 0$, $\eta_2 > 0$, and $\eta_i \to 0$ $(i = 1, 2)$ with $N \to \infty$.

Recalling estimate (3.3) for $f \in C^r[0, \pi]$ $(r = 0, 1, \ldots)$ we obtain

$$\begin{aligned} E(f, S_{N,r}[0, \pi])_{C[0,\pi]} &\leq \frac{K_r}{2N^r} \Omega(f^{(r)}, \bar{\Delta}_N[0, \pi]) \\ &\leq \frac{K_r}{2N^r} \omega\left(f^{(r)}, \frac{\pi}{N}\right)_{C[0,\pi]}, \end{aligned} \tag{3.33}$$

$$\begin{aligned} E(f, S_{N,r}[0,\pi])_{L_1[0,\pi]} &\leq \frac{2K_{r+1}}{N^r} \Omega(f^{(r)}, \bar{\Delta}_N[0,\pi]) \\ &\leq \frac{2K_{r+1}}{N^r} \omega\left(f^{(r)}, \frac{\pi}{N}\right)_{C[0,\pi]}. \end{aligned} \tag{3.34}$$

These inequalities are, of course, exact for $r = 0$ (see Section 6.3.1). If $r \geq 1$ then they are only asymptotically exact in the sense that we have for example for $N \to \infty$

$$\sup_{f \in C^r[0,\pi]} \frac{E(f, S_{N,r}[0, \pi])_{C[0,\pi]}}{\omega(f^{(r)}, \pi/N)_{C[0,\pi]}} = 2^{-1} K_r N^{-r} (1 + o(1)),$$

i.e.

$$\lim_{N \to \infty} N^r \kappa_{\pi/n}(C^r[0, \pi], C[0, \pi], S_{N,r}[0, \pi]) = 2^{-1} K_r.$$

6.3.4 Inequalities for interpolating splines Exact constants in Jackson-type inequalities for interpolation by splines of degree $r = 0, 1$ were considered in

Sections 6.3.1 and 6.3.2. In particular, we have established that, in the periodic case, splines $\sigma_{2n,1}(f, t)$ realize the best constants for estimating the approximation of a function $f \in C^1$ in L_p $(1 \le p \le \infty)$. The situation is different for the splines $\sigma_{2n,r}(f, t)$ for $r \ge 2$. Let us recall that the splines $\sigma_{2n,r}(f, t) \in S_{2n,r}$ have knots at points $t_i = i\pi/n$ $(i = 1, 2, \dots, 2n)$ and they are determined by the equations

$$\sigma_{2n,r}(f, \tau_j) = f(\tau_j), \qquad j = 1, 2, \dots, 2n,$$

where $\tau_j = \tau_{rj}$ are the zeros of the spline $\varphi_{n,r+1}(t)$, i.e. $\tau_j = j\pi/n + \beta_r$ $(\beta_r = [1 + (-1)^r]\pi/(4n))$.

If $f \in C^r$, $\delta(f, t) = f(t) - \sigma_{2n,r}(f, t)$ then we have in view of (5.2.30)

$$\delta(f, t) = \frac{1}{\pi}\sum_{j=1}^{2n}\int_{\tau_{j-1}}^{\tau_j} q_r(t, u) f^{(r)}(u + \beta_r)\, du, \qquad r = 1, 2, \dots,$$

where $q_r(t, u) = B_r(t - u - \beta_r) - \sigma_{2n,r}(B_r(\bullet - u - \beta_r), t)$. By statement (1) of Lemma 5.2.10 we have

$$\int_{\tau_{j-1}}^{\tau_j} q_r(t, u)\, du = 0, \qquad j = 1, 2, \dots, 2n,$$

and hence for every constant c_j we obtain

$$|\delta(f, t)| = \frac{1}{\pi}\sum_{j=1}^{2n}\int_{\tau_{j-1}}^{\tau_j} |q_r(t, u)|\, |f^{(r)}(u + \beta_r) - c_j|\, du.$$

Set

$$c_j = \left[\sup_{t_{i-1} \le t \le t_i} f^{(r)}(t) + \inf_{t_{i-1} \le t \le t_i} f^{(r)}(t)\right]/2. \tag{3.35}$$

Then

$$|f^{(r)}(u + \beta_r) - c_j| \le \Omega(f^{(r)}, \bar{\Delta}_{2n})/2, \qquad u \in (\tau_{j-1}, \tau_j),$$

and hence

$$|\delta(f, t)| = \frac{1}{2\pi}\Omega(f^{(r)}, \bar{\Delta}_{2n})\int_0^{2\pi} |q_r(t, u)|\, du, \qquad \forall f \in C^r. \tag{3.36}$$

Let f be a function from L^r_∞, $f^{(r)}(t) = [\operatorname{sgn} q_r(t, u - \beta_r)]/2 + d$, where the constant d is chosen from $f^{(r)} \perp 1$. Then the Stekhlov function f_h belongs to C^r, $\Omega(f_h^{(r)}, \bar{\Delta}_{2n}) = \omega(f^{(r)}, \pi/n) = 1$ for small enough h and

$$\lim_{h \to 0} \delta(f_h, t) = \delta(f, t) = \frac{1}{2\pi}\omega\left(f^{(r)}, \frac{\pi}{n}\right)\int_0^{2\pi} |q_r(t, u)|\, du.$$

Hence we have proved:

Proposition 6.3.6 For every t the following inequalities which cannot be improved on C^r $(r = 1, 2, \ldots)$ hold:

$$|f(t) - \sigma_{2n,r}(f, t)| \leq M_{n,r}(t)\Omega(f^{(r)}, \bar{\Delta}_{2n}) = M_{n,r}(t)\omega\left(f^{(r)}, \frac{\pi}{n}\right), \quad (3.37)$$

where

$$M_{n,r}(t) = \frac{1}{2\pi}\int_0^{2\pi} |B_r(t-u) - \sigma_{2n,r}(B_r(\bullet - u), t)|\, du,$$

and B_m is the Bernoulli function.

Comparing (3.37) and (3.27) we see that $\|M_{n,r}\|_C \geq 2^{-1}K_r n^{-r}$, but the investigation of function $M_{n,r}$ shows that we have a strict inequality for $r \geq 2$. Hence for $r = 2, 3, \ldots$, the interpolating splines $\sigma_{2n,r}(f, t)$ do not realize the constant in inequality (3.27). Nevertheless the estimate (3.37) has some advantages in comparison with (3.27). Specifically it is provided by an effectively constructed linear method and the error is estimated at any point t. Moreover quantity $\|M_{n,r}\|_C$ is slightly larger than $2^{-1}\, K_r n^{-r}$.

The situation is different when we estimate the error in L_1. Setting $g(f, t) - \sigma_{2n,r}(g(f), t) = \eta(t)$ in (5.2.21) and taking $\eta(\tau_j) = 0$ into account we can rewrite this relation as

$$\|f - \sigma_{2n,r}(f)\|_1 = \left|\int_0^{2\pi} \eta'(t + \beta_r)[f^{(r)}(t) - s_0(t)]\, dt\right|,$$

where $s_0(t) = s_0(f, t)$ is the step function defined on the intervals (t_{j-1}, t_j) by the values (3.35). Then obviously

$$\|f - \sigma_{2n,r}(f)\|_1 \leq \frac{1}{2}\|\eta'\|_1\, \Omega(f^{(r)}, \bar{\Delta}_{2n}) \leq \frac{1}{2}\|\eta'\|_1\, \omega\left(f^{(r)}, \frac{\pi}{n}\right).$$

But the function $g(f)$ in (5.2.21) belongs to $W^r H^1_{2n}$ and hence (5.2.26) implies

$$\|\eta'\|_1 \leq \|\varphi_{n,r}\|_1 = 4\, K_{r+1}\, n^{-r}.$$

Therefore we have the exact (see (3.28)) estimates

$$\|f - \sigma_{2n,r}(f)\|_1 \leq 2\, K_{r+1}\, n^{-r}\, \Omega(f^{(r)}, \bar{\Delta}_{2n}) = 2\, K_{r+1}\, n^{-r}\, \omega\left(f^{(r)}, \frac{\pi}{n}\right)$$

for $f \in C^r$ and we have established:

Proposition 6.3.7 We have for all $n, r = 1, 2, \ldots,$

$$\kappa'_{\pi/n}(C^r, L_1, S_{2n,r}) = \kappa_{\pi/n}(C^r, L_1, S_{2n,r}) = \sup_{f \in C^r} \frac{\|f - \sigma_{2n,r}(f)\|_1}{\omega(f^{(r)}, \pi/n)} = \frac{2K_{r+1}}{n^r}.$$

Using a reduction in the periodic case (or directly from the results in Section 5.3.2) we can establish the following. Let $\sigma_{N,r}(f, t)$ be the spline from $S_{N,r}[0, \pi]$ interpolating the function $f \in C^r[0, \pi]$ at the points $k\pi/N$ ($k = 1, 2, \ldots, N-1$) if r is odd or at the points $k\pi/N - \pi/(2N)$ ($k = 1, 2, \ldots, N$) if r is even with the boundary conditions

$$\sigma_{N,r}^{(\mu)}(f, 0) = f^{(\mu)}(0), \qquad \sigma_{N,r}^{(\mu)}(f, \pi) = f^{(\mu)}(\pi).$$

Here $\mu = 0, 2, 4, \ldots, r-1$ for odd r and $\mu = 1, 3, 5, \ldots, r-1$ for even r. Then we have for $f \in C^r[0, \pi]$

$$\|f - \sigma_{N,r}(f)\|_1 \leq 2\, K_{r+1}\, n^{-r}\, \Omega(f^{(r)}, \bar{\Delta}_N[0, \pi]) = 2\, K_{r+1}\, n^{-r}\, \omega\left(f^{(r)}, \frac{\pi}{N}\right)$$

and the inequalities cannot be improved on the set $C^r[0, \pi]$.

6.3.5 Estimates for Hermite splines Here the most convenient approach is to use the integral representation of the error (see Section 5.3.3). If $f \in C^{2n-1}[a, b]$ and the spline $s(f, t) \in S_{2n-1}^{n}(\Delta_N[a, b])$ is determined by the equalities

$$S^{(\nu)}(f, t_i) = f_i^{(\nu)}(t_i), \qquad \nu = 0, 1, \ldots, n-1; \qquad i = 0, 1, \ldots, N,$$

where t_k are the points of partition $\Delta_N[a, b]$, then in view of (5.3.23)

$$f(t) - s(f, t) = \int_{t_{i-1}}^{t_i} Q_{2n-1}(t, u)\, f^{(2n-1)}(u)\, du, \qquad t \in (t_{i-1}, t_i),$$

where $Q_m(t, u)$ is defined and studied (for $t_{i-1} = \alpha$, $t_i = \beta$) in Section 5.3.3.

Here it is important that

$$\int_{t_{i-1}}^{t_i} Q_{2n-1}(t, u)\, du = 0, \qquad t \in (t_{i-1}, t_i),$$

and that $Q_{2n-1}(t, u)$ changes sign exactly once on (t_{i-1}, t_i) as a function of u, say at x_i. But then if number c_i is defined as in (3.35) (with $r = 2n-1$) then we have for $t \in (t_{i-1}, t_i)$

$$\begin{aligned} |f(t) - s(f, t)| &= \left| \int_{t_{i-1}}^{t_i} Q_{2n-1}(t, u)[f^{(2n-1)}(u) - c_i]\, du \right| \\ &\leq \tfrac{1}{2}\, \Omega_i(f^{(2n-1}, \Delta_N[a, b]) \\ &\quad \times \left| \int_{t_{i-1}}^{x_i} Q_{2n-1}(t, u)\, du - \int_{x_i}^{t_i} Q_{2n-1}(t, u)\, du \right| \\ &= \tfrac{1}{2}\, \Omega_i(f^{(2n-1)}, \Delta_N[a, b]) \int_{t_{i-1}}^{t_i} |Q_{2n-1}(t, u)|\, du. \end{aligned}$$

Hence using (5.3.25) we can write

$$|f(t) - s(f, t)| \le \max_{t_{i-1} \le u \le t_i} |Q_{2n}(t, u)| \, \Omega_i(f^{(2n-1)}, \Delta_N[a, b]), \qquad t \in (t_{i-1}, t_i), \tag{3.38}$$

where

$$Q_{2n}(t, u) = -\int_{t_{i-1}}^{u} Q_{2n-1}(t, v)\, \mathrm{d}v, \qquad t \in [t_{i-1}, t_i].$$

The exactness of estimate (3.38) can easily be established by constructing a function $f \in C^{2n-1}[a, b]$ such that $f^{(2n-1)}(t) = -1$ for $t_{i-1} \le t \le x_i - \varepsilon$, $f^{(2n-1)}(t) = 1$ for $x_i + \varepsilon \le t \le t_i$, $|f^{(2n-1)}(t)| \le 1$ for $a \le t \le b$.

Therefore using relation (5.3.24) as well we can state:

Theorem 6.3.8 If $f \in C^{2n-1}[a, b]$ and $s(f, t)$ is the Hermite spline in $S^n_{2n-1}(\Delta_N[a, b])$ then

$$\max_{t_{i-1} \le t \le t_i} |f(t) - s(f, t)| \le \eta_{ni} \Omega_i(f^{(2n-1)}, \Delta_N[a, b]) \le \eta_{ni} \omega_i(f^{(2n-1)}, h_i),$$

where $h_i = t_i - t_{i-1}$ and

$$\eta_{ni} = \frac{h_i^{2n-1}}{(2n-1)[(2n-2)!!]^2 2^{2n}}.$$

The inequalities are exact.

Comments

Section 6.1 The notion of the modulus of continuity $\omega(f, \delta)$ for a function f was introduced by Lebesgue (for $f \in C$) and since then has been extensively used in the theory of functions and especially in approximation theory. The modulus of continuity $\omega(\delta)$ as an independent function defined by the properties of monotonicity and sub-additivity was considered for the first time by S. M. Nikol'skii [9].

The proofs of inequalities (1.14), (1.15) (with inexact constants) with the help of the Jackson operator [1] can be found in many monographs (see e.g. [14A, 16A, 26A]). The Stekhlov functions for sharpening the constants (see (1.17) and (1.18)) have been used by Akhiezer [2A], some estimates are contained in the papers of Stechkin [4, 6], Berdyshev [1] and Gavrilyuk [1].

Section 6.2 Proposition 6.2.1 is due to the author [3, 7], the current proof is from Pokrovskii's paper [1]. Theorem 6.2.2 and Lemma 6.2.3 have also been proved by the author [6], inequalities (2.21) were also obtained by him [27]. Theorem 6.2.4 and relation (2.28) was established by Chernykh [1, 2]. Inequality (2.40) and Theorem

6.2.7 for $r = 1$ was obtained by Zhuk [1, 2] and in the general case, i.e. for all odd r, by Ligun [2], who has also proved in [11–13] relations (2.44)–(2.46) for all $r = 1, 2, \ldots,$. Proposition 6.2.8 is proved in the papers of Korneichuk and Polovina [1–3], Theorem 6.2.9 was established by Ligun [9].

There are many papers devoted to finding the exact constants in inequality (1.20) for a variety of linear polynomial methods. From this point of view Fourier means (Gavrilyuk [1, 2], Gavrilyuk and Stechkin [1]), Favard sums (Stechkin [6]), Jackson polynomials (Wang Hing-hua [1], Bugaets and Martynyuk [1]), Rogosinski means and λ-methods (Gavrilyuk [1, 2]), interpolation polynomials and convolution methods (V. F. Babenko and Shalaev [1]) have been studied. Exact estimates for the best polynomial approximation of a function $f \in C^r$ by the modulus of smoothness and for $f \in C^r$ by a linear combination of the modulus of continuity and modulus of smoothness of $f^{(r)}$ have been obtained by Shalaev [1, 2].

Section 6.3 Relations (3.10)–(3.13) and the results in Section 6.3.2 on approximation by polygons of $f \in C^1$ are contained in the author's paper [23]. Estimate (3.24) was proved (with different arguments) by Loginov [1]. Inequalities (3.27) in Theorem 6.3.5 are established by Ligun [4] and inequalities (3.28) – by the author [18, 21]. In his work [20] relations (3.33) and (3.34) are obtained. Proposition 6.3.7 is from the author's paper [25] (see also [29, 30]). Theorem 6.3.8 was proved by Velikin [3].

Exact constants in Jackson-type inequalities for the best one-sided approximation are found in the paper of V. G. Doronin and Ligun [2] and in [11A].

Added by the translator. A characterization of the best one-sided approximations in spaces L_p $(1 \le p < \infty)$ can be obtained in the terms of averaged moduli of continuity and smoothness (see e.g. Sendov and Popov [1, Chapter 8]). The problem of determining the exact constants in the corresponding Jackson-type inequalities is still open.

Other results and exercises

1. If $f \in C^r$ $(r = 1, 2, \ldots,)$ and the subspace $\mathcal{N} \subset L_1$ contains the constants then

$$E(f, \mathcal{N})_1 \le 2\, K_{r+1}\, b_{r+2}^r\, \omega(f^{(r)}, \delta)_C, \qquad \delta \ge 2\pi B_{r+2},$$

where $b_k = (E(W_1^k, \mathcal{N})_1/K_k)^{1/k}$. For $\mathcal{N} = \mathcal{N}_{2n-1}^T$ or $S_{2n,m}$ $(m \ge r-1)$ the inequality is exact (Ligun [13]).

2. Let $\mathcal{L}_{2n-1}$ be the class of linear operators from C to $\mathcal{N}_{2n-1}^T$ and let $\mathcal{L}_{2n-1}^+$ be the class of positive operators from $\mathcal{L}_{2n-1}$. Then

$$\inf_{A \in \mathcal{L}_{2n-1}} \sup_{f \in C} \frac{\|f - Af\|_C}{\omega_*(f, \pi/n)_C} = \inf_{A \in \mathcal{L}_{2n-1}^+} \sup_{f \in C} \frac{\|f - Af\|_C}{\omega_*(f, \pi/n)_C} = 1,$$

where $\omega_*(f, \delta)_C$ is the smallest concave majorant of the modulus of continuity $\omega(f, \delta)_C$ (Davidchik and Ligun [2]).

3. For all $n, r = 1, 2, \ldots,$ we have

$$\sup_{f \in C^r} \frac{E^+(f, S_{2n,r})_1}{\omega(f^{(r)}, \pi/n)_C} = \frac{\|B_r\|_1}{n^r}$$

(V. G. Doronin and Ligun [2]). A generalization of this relation to (α, β)-approximation is obtained by V. F. Babenko [11].

4. For $\gamma \geq \dfrac{3\pi}{3(n + \frac{1}{2})}$ we have the equality

$$\sup_{f \in C, f \not\equiv \text{const}} \frac{\|f - S_n(f)\|_C}{\omega(f, \gamma)_C} = \tfrac{1}{2}\left(\|S_n\|_{(C)} + 1\right).$$

It is no longer true if $\gamma < \left(\dfrac{2\pi}{3} - \dfrac{\pi}{2n+1}\right)\left(n + \dfrac{1}{2}\right)^{-1}$ (Gavrilyuk and Stechkin [1]).

5. Let $p_n(f, t)$ be the algebraic polynomial of degree $n-1$ interpolating $f \in C[a, b]$ at the points from the set $\Delta = \{x_1, x_2, \ldots, x_n\}$, $a \leq x_1 < x_2 < \ldots < x_n \leq b$; let ψ_k be the fundamental interpolation polynomials ($\psi_k(x_i) = \delta_{ki}$) and let $\lambda_n(x) = \sum_{k=1}^n |\psi_k(t)|$ be the Lebesgue function. Then we have for every $x \in [a, b]$

$$|f(x) - p_n(f, x)| \leq \frac{1 + \lambda_n(x)}{2} \omega(f, \|\Delta_x\|)_{C[a,b]},$$

where $\Delta_x = \Delta \cup \{x\}$ and $\|\Delta_x\|$ is the largest distance between two consecutive points in Δ_x; the inequality is exact. For $f \in C^1[a, b]$ the following exact inequality holds

$$|f(x) - p_n(f, x)| \leq M_x\, \omega(f', \|\Delta_x\|)_{C[a,b]},$$

where

$$M_x = \int_a^b \left| \theta_x(t) - \sum_{k=1}^n \theta_{x_k}(t)\, \psi_k(t) \right| dt$$

and $\theta_x(t) = 0$ if $a \leq t < x$ and $\theta_x(t) = 1$ if $x \leq t \leq b$ (V. F. Babenko and Shalaev [1]).

6. Show that in the inequality $E_n(f)_C \leq M\, \omega(f, \gamma/n)_C$ for $f \in C$ the exact constant $M = M(\gamma)$ satisfies the bounds

$$[\pi/\gamma] + 1 \leq 2\, M(\gamma) \leq \pi/\gamma + 1, \qquad \gamma \neq \pi/k;$$

$$(\pi/\gamma + 1)(1 - 1/(2n)) \leq 2\, M(\gamma) \leq \pi/\gamma + 1, \qquad \gamma = \pi/k, \qquad k = 1, 2, \ldots,$$

(Korneichuk [27]). A similar problem for the space L_2 has been studied by A. G. Babenko [1].

7. Let $\Delta_N[0, 1]$: $0 = t_0 < t_1 < \ldots < t_N = 1$, $f \in C^1[0, 1]$ and $s(f, t)$ is the polygon from $S_1(\Delta_N[0, 1])$ for which $s(f, t_k) = f(t_k)$ $(k = 0, 1, \ldots, N)$. Prove that for every $t \in (t_{k-1}, t_k)$ the inequality

$$|f(t) - s(f, t)| \leq (t_k - t)(t - t_{k-1}) h_k^{-2} \int_0^{h_k} \omega(f', u)\, du, \qquad h_k = t_k - t_{k-1},$$

holds and is exact on $C^1[0, 1]$.

8. Using Theorem 6.2.9 prove that for every $f \in C^r[-1, 1]$ there is a sequence of polynomials $q_{n,r}(f, t)$ $(n = 1, 2, \ldots,)$ of degree $n-1$ such that we have $(\omega(g, \delta)_C = \omega(g, \delta)_{C[-1,1]})$

$$0 \le f(t) - q_{n,r}(f,t) \le K_r n^{-r} (1-t^2)^{r/2}\, \omega\left(f^{(r)}, \frac{\pi}{n}(1-t^2)^{1/2}\right)_C + o\left(n^{-r}\omega\left(f^{(r)}, \frac{\pi}{n}\right)_C\right),$$

uniformly on $t \in [-1, 1]$ for $n \to \infty$ where the constant K_r cannot be improved (see [11A]).

9. Let C^r_* denote the set of all functions from C^r for which $\omega(f^{(r)}, \delta)_C$ is a concave function. Using relations (7.2.3) and (7.2.4) (Chapter 7) prove that

$$\sup_{f \in C^r_*} \frac{n^r E_n(f)_C}{\omega(f^{(r)}, \eta/n)_C} = \begin{cases} K_{r+1}/\eta, & 0 < \eta < \eta_r, \\ K_r/2, & \eta_r < \eta < \pi, \end{cases}$$

where $\eta_r = 2\,K_{r+1}\,K_r^{-1}$. Obtain similar inequalities for $E_n(f)_1$.

10. Prove the exactness of estimate (2.28) on the set of all functions $f \in L_2$ for which $\omega^2(f, \delta)_2$ is concave.

11. For $f \in L_2^r$ prove the estimate

$$E_n(f)_2 \le \frac{1}{2n^r}\left\{ n \int_0^{\pi/n} \omega^2(f^{(r)}, t)_2 \sin nt\, dt \right\}^{1/2}.$$

Hint. Use (2.27).

12. Check the exactness of estimate (3.24) on $C[a, b]$.

13. Prove that for modulus of continuity $\omega(\delta)$ ($\delta \ge 0$) and its smallest majorant $\omega_*(\delta)$ for all $\lambda > 0$ the inequality $\omega_*(\lambda\delta) \le (\lambda + 1)\omega(\delta)$ holds and cannot be improved for $\lambda = 1, 2, \ldots$.

14. Obtain an analog of Theorem 6.2.5 for the approximation of $f \in X^r(-\infty, \infty)$ by entire functions of exponential type.

Hint. Use the results in Chapter 5 in [2A] and the scheme and the intermediate facts in the proof of Theorem 6.2.5.

7

Approximation of classes of functions determined by modulus of continuity

When considering the approximation of functions by polynomials and splines in Chapters 4 and 5 our aim was to obtain error estimates which could not be improved on the class W_p^r (or $W_p^r[a, b]$) of functions f which were defined by bounding the norm $\|f^{(r)}\|_p$. It was remarked that this is equivalent to bounding the error by the norm $\|f^{(r)}\|_p$ of the rth derivative of the approximating function. In Chapter 6 the error was estimated not by the norm but by values of the modulus of continuity $\omega(f^{(r)}, \delta)_X$ at some points δ and, as we have seen, sometimes it is possible to solve such problems using known methods.

A similar problem is that of approximating classes of functions f for which $\omega(f^{(r)}, \delta)_C \leq \omega(\delta)$, where $\omega(\delta)$ is a given modulus of continuity. Solving these problems exactly, normally requires essentially new ideas and has to be approached in a different way.

For the best approximation of the functional classes mentioned above by polynomials and splines a suitable approach was found which employs duality relations and a special type of rearrangement (Σ-rearrangements) of functions preserving both the L_1-norm and the variation of the function. Using Σ-rearrangements is not simple but during the last twenty years since the development of this method no other approach has been offered. Moreover it turns out that Σ-rearrangements are also a suitable tool for solving other extremal problems in approximation theory. The method is based on the so-called comparison theorems for Σ-rearrangements which are crucial to it. Their proofs need long computations, if only by elementary analysis and in order not to interrupt our considerations we have put them into the Appendices (see Section A3).

The last section is devoted to the approximation of functional classes defined by the modulus of continuity by linear operators constructed on the basis of polynomials or splines. Here the methods of investigation depend substantially on the specific functional classes and also on the peculiarities of the operator. This seems to be the reason for the small number of exact results in the linear approximation of the classes considered. Moreover the majority of them are only asymptotically exact.

7.1 General facts and Σ-rearrangements

7.1.1 Classes W^mH^ω, $W^mH^\omega[a, b]$ It has been noted in Section 1.5 that for a function $f \in L_\infty^{m+1}$ the condition $\|f^{(m+1)}\|_\infty \leq K$ is equivalent to:

$$|f^{(m)}(t) - f^{(m)}(t+\delta)| \leq K\delta, \qquad \delta \geq 0, \tag{1.1}$$

and so the classes KW_∞^{m+1} and W^mKH^1 coincide. We obtain the wider classes W^mKH^α if we replace $K\delta$ on the right-hand side of (1.1) by $K\delta^\alpha$ $(0 < \alpha \leq 1)$, i.e. if we are going to Hölder's condition of order α. The variety of classes can be further extended by making it both more accurate and more flexible if $K\delta^\alpha$ is replaced by an arbitrary modulus of continuity $\omega(\delta)$ (see Section 6.1).

The class of 2π-periodic functions $f \in C^m$ which are such that for all t

$$|f^{(m)}(t) - f^{(m)}(t+\delta)| \leq \omega(\delta), \qquad \delta \geq 0,$$

will be denoted by W^mH^ω $(m = 0, 1, \ldots)$. In particular $H^\omega = W^0H^\omega$ is the class of functions $f \in C$ for which

$$|f(t) - f(t+\delta)| \leq \omega(\delta), \qquad \delta \geq 0.$$

In the terms of the modulus of continuity $\omega(f, \delta) = \omega(f, \delta)_C$ of the function f or its mth derivative we can define the class W^mH^ω as

$$W^mH^\omega = \{f: f \in C^m, \omega(f^{(m)}, \delta) \leq \omega(\delta), \delta \geq 0\}.$$

In a similar way the classes $H^\omega[a, b]$, $W^mH^\omega[a, b]$ of a function defined on $[a, b]$ can be determined:

$$W^mH^\omega[a, b] = \{f: f \in C^m[a, b], \omega(f^{(m)}, \delta)_{C[a,b]} \leq \omega(\delta), 0 \leq \delta \leq b - a\};$$

here we also set $W^0H^\omega[a, b] = H^\omega[a, b]$.

When $\omega(\delta) = K\delta^\alpha$ $(0 < \alpha \leq 1)$ we get classes W^mKH^α and $W^mKH^\alpha[a, b]$.

Let us introduce the standard function which, as we shall see, will realize many extremal properties of the functions from W^mH^ω for concave $\omega(\delta)$. For every fixed $n = 1, 2, \ldots$, we set

$$f_{n,0}(t) = f_{n,0}(\omega, t) = \begin{cases} -\dfrac{1}{2}\omega\left(\dfrac{\pi}{n} - 2t\right), & 0 \leq t \leq \dfrac{\pi}{2n} \\ \dfrac{1}{2}\omega\left(2t - \dfrac{\pi}{n}\right), & \dfrac{\pi}{2n} \leq t \leq \dfrac{\pi}{n}; \end{cases} \tag{1.2}$$

$$f_{n,0}(t) = -f_{n,0}\left(t - \frac{\pi}{n}\right), \qquad \frac{\pi}{n} \le t \le 2\pi;$$

$$f_{n,0}(t + 2\pi) = f_{n,0}(t).$$

Obviously the function $f_{n,0}(t)$ is $2\pi/n$-periodic and monotonic on every interval $[(k-1)\pi/n, k\pi/n]$ and $f_{n,0}((2k-1)\pi/(2n)) = 0$ (the shape of the graph of $f_{n,0}(t)$ is similar to that of $-\cos nt$). Let us remark that in the case $\omega(\delta) = \delta$ $(0 \le \delta \le \pi/n)$ the function $f_{n,0}(\omega, t)$ becomes the Euler perfect spline $\varphi_{n,1}(t)$.

Further we set $f_{n,m}(t) = f_{n,m}(\omega, t)$ $(m = 0, 1, \dots,)$ to be the mth periodic integral of $f_{n,0}(t)$ with zero mean value on the period, i.e.

$$f_{n,m}(t) = \int_{\beta_m}^{t} f_{n,m-1}(u)\, du, \qquad m = 1, 2, \dots,$$

where $\beta_m = 0$ for odd m and $\beta_m = \pi/(2n)$ for even m. Therefore $f_{n,m}^{(k)}(t) = f_{n,m-k}(t)$ $(k = 1, 2, \dots, m)$ (the graph of $f_{n,m}$ is similar to the graph of $\varphi_{n,m+1}$ and coincides with it when $\omega(\delta) = \delta$).

Proposition 7.1.1 If $\omega(\delta)$ is a concave modulus of continuity then $f_{n,m} \in W^m H^\omega$ $(m = 1, 2, \dots,)$ and

$$\omega(f_{n,0}, \delta) \equiv \omega(\delta), \qquad 0 \le \delta \le \pi/n. \tag{1.3}$$

Proof If $0 \le t' < t'' \le \pi/(2n)$ then using the properties of the modulus of continuity (see Section 6.1) we have

$$|f_{n,0}(t') - f_{n,0}(t'')| = \frac{1}{2}\,\omega\left(\frac{\pi}{n} - 2t''\right) - \frac{1}{2}\,\omega\left(\frac{\pi}{n} - 2t'\right)$$

$$\le \frac{1}{2}\,\omega(2t'' - 2t') \le \omega(|t' - t''|).$$

The case $\pi/(2n) \le t' < t'' \le \pi/n$ is similar. If $0 \le t' < \pi/(2n) < t'' \le \pi/n$ then using the concavity of $\omega(\delta)$ we have

$$|f_{n,0}(t') - f_{n,0}(t'')| = \frac{1}{2}\,\omega\left(2t'' - \frac{\pi}{n}\right) + \frac{1}{2}\,\omega\left(\frac{\pi}{n} - 2t'\right) \le \omega(|t' - t''|).$$

In general for every point t', t'' we can always find points τ', τ'' on $[0,\pi/n]$ such that $|\tau' - \tau''| \le |t' - t''|$, $f_{n,0}(\tau') = f_{n,0}(t')$ and $f_{n,0}(\tau'') = f_{n,0}(t'')$. Hence in view of the above we have

$$|f_{n,0}(t') - f_{n,0}(t'')| = |f_{n,0}(\tau') - f_{n,0}(\tau'')| \le \omega(|\tau' - \tau''|) \le \omega(|t' - t''|).$$

This proves that $\omega(f_{n,0}, \delta) \leq \omega(\delta)$ and hence $f_{n,0} \in H^\omega$. If on the other hand $0 < \delta \leq \pi/n$ then we obtain from definition (1.2)

$$f_{n,0}\left(\frac{\pi}{2n} + \frac{\delta}{2}\right) - f_{n,0}\left(\frac{\pi}{2n} - \frac{\delta}{2}\right) = \frac{1}{2}\omega(\delta) + \frac{1}{2}\omega(\delta) = \omega(\delta),$$

i.e. (1.3) holds. The inclusion $f_{n,m} \in W^m H^\omega$ for $m = 1, 2, \ldots,$ follows from the validity of this fact for $m = 0$, the definition of $W^m H^\omega$ and the orthogonality $f_{n,0} \perp 1$. □

7.1.2 Duality and the main lemma When considering the problem of the best approximation in the L_q metric of class $W^r H^\omega$ by 2π-periodic functions of a fixed finite dimensional subspace $\mathcal{N} \subset L_q$, we shall begin with the general duality relation (1.4.10) which we write as

$$E(f, \mathcal{N})_q = \sup\left\{\int_0^{2\pi} f(t)h(t)\, dt: h \in W^0_{q'}(\mathcal{N})\right\},$$

where $W^0_p(\mathcal{N}) = \{\psi: \psi \in L_p, \|\psi\|_p \leq 1, \psi \perp \mathcal{N}\}$. If $1 \in \mathcal{N}$ and $W^r_p(\mathcal{N})$ is the set of all rth periodic integrals of functions $\psi \in W^0_p(\mathcal{N})$ then we get by r-fold integrating by part for $f \in C^r$

$$E(f, \mathcal{N})_q = \sup\left\{\int_0^{2\pi} f^{(r)}(t)g(t)\, dt: g \in W^r_{q'}(\mathcal{N})\right\}.$$

Assuming $f \in W^r H^\omega$ $(r = 1, 2, \ldots,)$ and hence $f^{(r)} \in H^\omega$, $f^{(r)} \perp 1$, we have

$$E(W^r H^\omega, \mathcal{N})_q =: \sup_{f \in W^r H^\omega} E(f, \mathcal{N})_q = \sup_{\varphi \in H^\omega, \varphi \perp 1} \; \sup_{g \in W^r_{q'}(\mathcal{N})} \int_0^{2\pi} \varphi(t)g(t)\, dt. \quad (1.4)$$

Because of the zero mean value of the function φ in the last integral, we may, when calculating the supremum in (1.4) on $g \in W^r_{q'}(\mathcal{N})$ consider just those g from $W^r_{q'}(\mathcal{N})$ for which $g \perp 1$ which simultaneously removes the orthogonality condition for $\varphi \in H^\omega$. Therefore

$$E(W^r H^\omega, \mathcal{N})_q = \sup_{g \in W^r_{q'}(\mathcal{N})_*} \; \sup_{\varphi \in H^\omega} \int_0^{2\pi} \varphi(t)g(t)\, dt, \quad (1.5)$$

where $W^r_p(\mathcal{N})_* = \{g: g \in W^r_{q'}(\mathcal{N}), g \perp 1\}$, and the problem is reduced to the investigating of the extremal properties of the functional

$$F_\omega(g) = \sup\left\{\int_0^{2\pi} f(t)g(t)\, dt: f \in H^\omega\right\} \quad (1.6)$$

on the set of functions $W^r_{q'}(\mathcal{N})_*$.

The first step in this direction is:

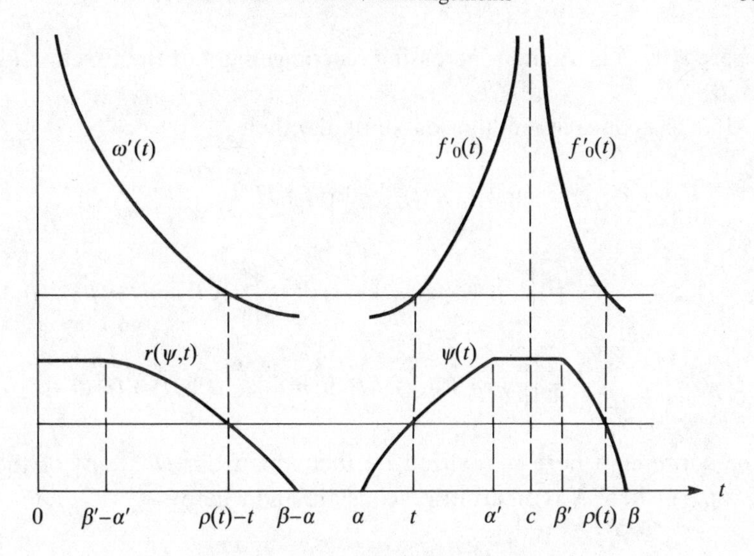

Fig. 7.1 To Lemma 7.1.2.

Lemma 7.1.2 Let the function ψ be summable on $[\alpha, \beta]$ and let it be such that

(1) $\psi(t) > 0$ a.e. on (α, α') and $\psi(t) < 0$ a.e. on (β', β) (or $\psi(t) < 0$ a.e. on (α, α') and $\psi(t) > 0$ a.e. on (β', β)), where $\alpha < \alpha' \leq \beta' < \beta$;
(2) if $\alpha' < \beta'$ then $\varphi(t) = 0$ a.e. on (α', β');
(3) $\int_\alpha^\beta \psi(t)\,\mathrm{d}t = 0$.

Moreover let

$$\Psi(t) = \int_\alpha^t \psi(u)\,\mathrm{d}u, \qquad \alpha \leq t \leq \beta, \tag{1.7}$$

and hence $\psi(\beta) = 0$; let ρ be a function defined for $\alpha \leq t \leq c = (\alpha' + \beta')/2$ by the equalities

$$\Psi(t) = \Psi(\rho(t)), \qquad \alpha \leq t \leq \alpha', \qquad \beta' \leq \rho(t) \leq \beta, \tag{1.8}$$

$$\rho(t) = \alpha' + \beta' - t, \qquad \alpha' < t \leq c, \tag{1.8}$$

and let ρ^{-1} be the inverse function of ρ.

Then we have for $f \in C[\alpha, \beta]$

$$\left|\int_\alpha^\beta \psi(t)f(t)\,\mathrm{d}t\right| \leq \int_\alpha^{\alpha'} |\psi(t)|\omega(f, \rho(t) - t)\,\mathrm{d}t$$

$$= \int_{\beta'}^\beta |\psi(t)|\omega(f, t - \rho^{-1}(t))\,\mathrm{d}t = \left|\int_0^{\beta-\alpha} r'(\Psi, t)\omega(f, t)\,\mathrm{d}t\right|, \tag{1.9}$$

where $r(\Psi, t)$ is the non-increasing rearrangement of the function (1.7) on $[\alpha, \beta]$.

If ω is a concave modulus of continuity then

$$F_\omega(\psi: \alpha, \beta) = \sup \left\{ \left| \int_\alpha^\beta \psi(t) f(t) \,\mathrm{d}t \right| : f \in H^\omega[\alpha, \beta] \right\}$$

$$= \int_\alpha^{\alpha'} |\psi(t)| \omega(\rho(t) - t) \,\mathrm{d}t = \int_{\beta'}^\beta |\psi(t)| \omega(t - \rho^{-1}(t)) \,\mathrm{d}t$$

$$= \left| \int_0^{\beta - \alpha} r'(\Psi, t) \omega(t) \,\mathrm{d}t \right| = \int_0^{\beta - \alpha} r(\Psi, t) \omega'(t) \,\mathrm{d}t. \tag{1.10}$$

The supremum here is realized by functions from $H^\omega[\alpha, \beta]$ of the type $K \pm f_0(t)$ where K is an arbitrary constant and where

$$f_0(t) = \begin{cases} -\int_t^c \omega'(\rho(u) - u) \,\mathrm{d}u, & \alpha \le t \le c, \\ \int_c^t \omega'(u - \rho^{-1}(u)) \,\mathrm{d}u, & c \le t \le \beta. \end{cases} \tag{1.11}$$

Proof Notice that the function ρ defined in (1.8) strictly monotonically and continuously maps $[\alpha, c]$ into $[c, \beta]$ and $\rho(\alpha) = \beta$, $\rho(c) = c$. The conditions imposed on ψ guarantee the absolute continuity of the functions ρ and ρ^{-1} on $[\alpha, c]$ into $[c, \beta]$ respectively (see [9A, Section 7.4]) and also offers the possibility for a formal differentiation of equality (1.8) and its equivalent expression:

$$\Psi(\rho^{-1}(t)) = \Psi(t), \qquad \beta' \le t \le \beta, \qquad \alpha \le \rho^{-1}(t) \le \alpha'.$$

Therefore we have a.e.

$$\psi(t) = \psi(\rho(t))\rho'(t), \qquad \alpha \le t \le c, \tag{1.12}$$

$$\psi(\rho^{-1}(t))(\rho^{-1}(t))' = \psi(t), \qquad c \le t \le \beta. \tag{1.13}$$

The change of the variables $t = \rho(z)$ gives for every function $f \in C[\alpha, \beta]$

$$\int_{\beta'}^\beta \psi(t) f(t) \,\mathrm{d}t = -\int_\alpha^{\alpha'} \psi(\rho(z)) f(\rho(z)) \rho'(z) \,\mathrm{d}z,$$

and because of $\psi(t) = 0$ a.e. on (α', β') we get, using (1.12),

$$\left| \int_\alpha^\beta \psi(t) f(t) \,\mathrm{d}t \right| = \left| \int_\alpha^{\alpha'} \psi(t)[f(t) - f(\rho(t))] \,\mathrm{d}t \right|$$

$$\le \int_\alpha^{\alpha'} |\psi(t)| \omega(f, \rho(t) - t) \,\mathrm{d}t. \tag{1.14}$$

Similarly setting $t = \rho^{-1}(z)$ on $[\alpha, \alpha']$ and using (1.13) we come to

$$\left|\int_\alpha^\beta \psi(t) f(t)\,\mathrm{d}t\right| \le \int_{\beta'}^\beta |\psi(t)|\omega(f, t - \rho^{-1}(t))\,\mathrm{d}t. \tag{1.14'}$$

The coincidence of the right-hand sides of (1.14) and (1.14′) is checked by one of the above changes of variables.

This proves the first two relationships in (1.9). The last equality in (1.9) will be established with the help of identity

$$|\Psi(t)| = r(\Psi, \rho(t) - t), \qquad \alpha \le t \le c, \tag{1.15}$$

whose validity on $[\alpha, \alpha']$ follows from the definitions of functions $r(\Psi, t)$ and $\rho(t)$ and whose validity on $[\alpha', c]$ follows from the fact that we have for $t \in [\alpha', c]$

$$|\Psi(t)| = \max_{\alpha \le u \le \beta} |\Psi(u)| = r(\Psi, \rho(t) - t), \qquad \rho(t) - t \le \beta' - \alpha'.$$

Using (1.15) and then setting $\rho(t) - t = z$ we obtain

$$\int_\alpha^c |\psi(t)|\omega(f, \rho(t) - t)\,\mathrm{d}t = \left|\int_\alpha^c \Psi'(t)\omega(f, \rho(t) - t)\,\mathrm{d}t\right|$$

$$\left|\int_\alpha^c r'(\Psi, \rho(t) - t)\omega(f, \rho(t) - t)(\rho'(t) - 1)\,\mathrm{d}t\right| = \left|\int_0^{\beta-\alpha} r'(\psi, t)\omega(f, t)\,\mathrm{d}t\right|.$$

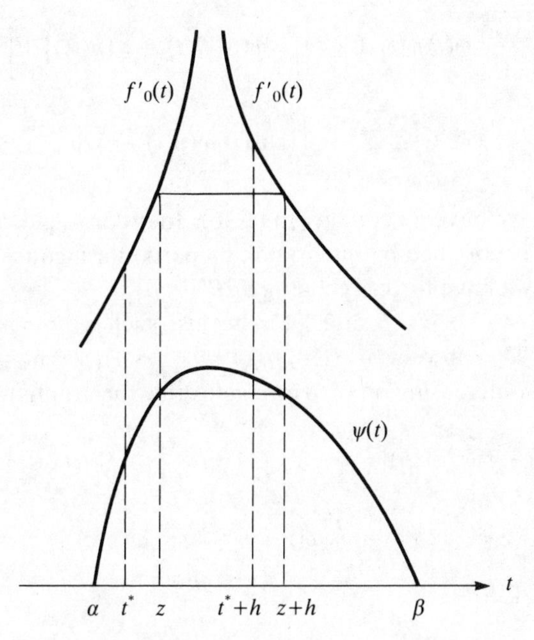

Fig. 7.2 To the proof that $f_0 \in H^\omega[\alpha, \beta]$ in Lemma 7.1.2.

It follows from (1.9) that for every modulus of continuity ω and for every $f \in H^\omega[\alpha, \beta]$ we have

$$\left|\int_\alpha^\beta \psi(t) f(t)\, dt\right| \le \int_\alpha^{\alpha'} |\psi(t)| \omega(\rho(t) - t)\, dt = \left|\int_0^{\beta-\alpha} r'(\Psi, t)\omega(t)\, dt\right|. \quad (1.16)$$

If ω is a concave modulus of continuity then the derivative $\omega'(\delta)$ exists a.e. and is non-increasing on $[0, \beta - \alpha]$ provided we define it as $\omega'(\delta) = [\omega'(\delta - 0) + \omega'(\delta + 0)]/2$ at the points of discontinuity. From definition (1.11) of function f_0 it follows that

$$f_0(\rho(t)) - f_0(t) = \omega(\rho(t) - t), \quad (1.17)$$

$$f_0'(\rho(t)) = f_0'(t), \qquad \alpha < t < c, \qquad c < \rho(t) < \beta. \quad (1.18)$$

Indeed if $\alpha < t < c$ then

$$f_0(\rho(t)) = \int_c^{\rho(t)} \omega'(u - \rho^{-1}(u))\, du = -\int_t^c \omega'(\rho(u) - u)\rho'(u)\, du$$

and hence

$$f_0(\rho(t)) - f_0(t) = -\int_t^c \omega'(\rho(u) - u)[\rho'(u) - 1]\, du = \omega(\rho(t) - t).$$

Relationship (1.18) is checked directly.

Now using the first equality in (1.14) and then (1.17) we obtain

$$\left|\int_\alpha^\beta \psi(t) f_0(t)\, dt\right| = \left|\int_\alpha^{\alpha'} \psi(t)[f_0(t) - f_0(\rho(t))]\, dt\right|$$

$$= \int_\alpha^{\alpha'} |\psi(t)| \omega(\rho(t) - t)\, dt,$$

i.e. for $f = f_0$ we have an equality in (1.16). In order to get relations (1.10) (the last one is obtained by integration by parts) for a concave modulus of continuity ω we have to verify that $f_0 \in H^\omega[\alpha, \beta]$.

Let $\alpha < t^* < t^* + h < \beta$ and let z be just such a point on (α, c) that $z + h = \rho(z)$. Then in view of (1.18) $f_0'(z) = f_0'(z + h)$, f_0' is non-decreasing on (α, c) and non-increasing on (c, β) we get (draw the graphs)

$$|f_0(t^*) - f_0(t^* + h)| = \int_{t^*}^{t^*+h} f_0'(u)\, du \le \int_z^{z+h} f_0'(u)\, du$$

$$= f_0(\rho(z)) - f_0(z) = \omega(\rho(z) - z) = \omega(h).$$

The lemma is proved. □

Remark If the graph of the function ψ in the statement of Lemma 7.1.2 is symmetric with respect to the mid-point $c = (\alpha + \beta)/2$ of the interval (α, β), i.e. $\psi(c-u) = -\psi(c+u)$ $(0 \le u < (\beta - \alpha)/2)$, then

$$\Psi(c-u) = \Psi(c+u), \qquad \rho(t) - t = 2(c-t),$$

and all calculations become much simpler. The extremal function f_0 will also be symmetric with respect to c:

$$f_0(t) = \begin{cases} -\omega(2c-2t)/2, & \alpha \le t \le c, \\ \omega(2t-2c)/2, & c \le t \le \beta. \end{cases}$$

Moreover in this case

$$r(\Psi, t) = \left|\Psi\left(\frac{\alpha+\beta-t}{2}\right)\right| = \left|\Psi\left(\frac{\alpha+\beta+t}{2}\right)\right|, \qquad 0 \le t \le \beta - \alpha.$$

7.1.3 Σ-rearrangements Lemma 7.1.2 allows us to estimate functional (1.6) if g changes sign once on $(0, 2\pi)$. In the general case the problem can be solved with the help of Σ-rearrangement which reduces the case to the simplest one described in Lemma 7.1.2. We shall describe the construction of this operator without paying attention to the proofs of some auxiliary facts; the reader can reproduce the fairly simple calculations by himself. A detailed and strict presentation of this material is given in [9A].

A function φ which is continuous on the interval $[\alpha, \beta]$ will be called *simple* if:

(1) $\varphi(\alpha) = \varphi(\beta) = 0, \qquad |\varphi(t)| > 0\ (\alpha < t < \beta)$;
(2) for every y $(0 < y < \max_t |\varphi(t)|)$ the equation $|\varphi(t)| = y$ has exactly two roots on the interval (α, β).

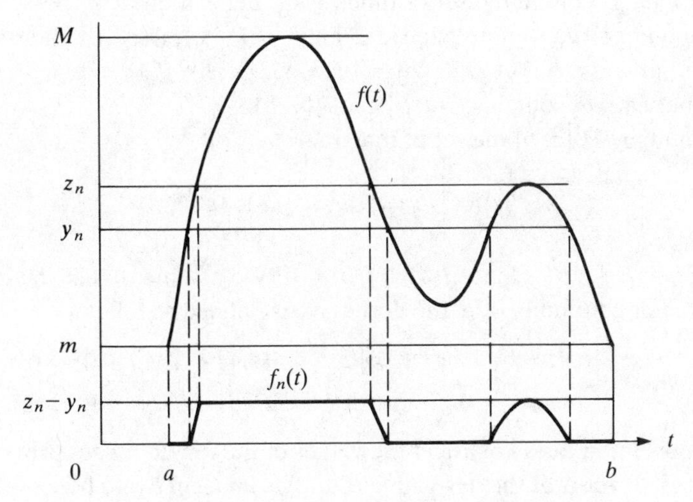

Fig. 7.3 The construction of Σ-expansion.

Sometimes it will be convenient to assume that φ is extended as zero to the real line: $\varphi(t) = 0$ $(t \leq \alpha, \beta \leq t)$ and we shall then describe φ as a finite function with support $[\alpha, \beta]$. Together with the support $[\alpha, \beta]$ we shall also relate the simple function φ to the intervals of strict monotonicity $[\alpha, \alpha']$ and $[\beta', \beta]$ and the interval $[\alpha', \beta'] \subset [\alpha, \beta]$ on which $|\varphi(t)|$ takes its maximum, allowing for the fact that it may degenerate to a point $(\alpha' = \beta')$.

As in Section 3.2.2 we denote the non-increasing rearrangement of $|\varphi|$ by $r(\varphi, t) = r(\varphi; \alpha, \beta; t)$ and we assume that $r(\varphi, t)$ is defined on the whole half-axis $[0, \infty)$, i.e. $r(\varphi, t) = 0$ for $t \geq \beta - \alpha$. Using inverse functions on the intervals of strict monotonicity $[\alpha, \alpha']$ and $[\beta', \beta]$ we can easily prove [9A, p. 135]:

Proposition 7.1.3 Let the function φ be simple and absolutely continuous on $[\alpha, \beta]$ and

$$|\varphi(t_1)| = |\varphi(t_2)| = r(\varphi t_0), \qquad t_0 = t_2 - t_1,$$

$$\alpha < t_1 < a' \leq \beta' < t_2 < \beta, \qquad 0 < t_0 < \beta - \alpha.$$

If φ is differentiable at t_1 and t_2 then

$$|r'(\varphi, t_0)| \leq \tfrac{1}{4}|\varphi'(t_1) - \varphi'(t_2)|$$

and an equality holds for simple functions for which $\varphi(\alpha + u) = \varphi(\beta - u)$ $(u > 0)$.

We are now going to show how under mild conditions a function f defined on $[a, b]$ can be represented as a finite or countable sum of simple functions so that the $L_1[a, b]$ norm and the variation on $[a, b]$ of f are exactly equal to the sums of the corresponding quantities for these simple functions.

Let $D[a, b]$ denote the set of functions ψ defined on $[a, b]$ possessing at every point $t \in (a, b)$ finite one-sided limits $\psi(t+0)$, $\psi(t-0)$ and only one of them at a and b. Set $D_0[a, b] = \{\psi: \psi \in D[a, b], \int_a^b \psi(t)\, dt = 0\}$. Let us note that $D[a, b]$ contains $C[a, b]$ and $V[a, b]$.

Denote by $D_0^1[a, b]$ the set of integrals

$$f(t) = \int_a^t \psi(u)\, du, \qquad a \leq t \leq b,$$

where $\psi \in D_0[a, b]$. Hence $f(a) = f(b) = 0$. We assume that at the interior points of discontinuity of f' the derivative is defined as follows:

$$f'(t) = \psi(t) = \begin{cases} [\psi(t-0) + \psi(t+0)]/2, & \text{if } \psi(t-0)\psi(t+0) > 0; \\ 0, & \text{if } \psi(t-0)\psi(t+0) \leq 0. \end{cases} \tag{1.19}$$

This convention does not affect the values of the function f but provides for the set F_0 of zeros of the derivative f' and its image $f(F_0)$ to be closed.

It is easy to see that meas $f(F_0) = 0$. Hence if

$$M = \max_{a \le t \le b} f(t), \qquad m = \min_{a \le t \le b} f(t)$$

and if $\{(y_n, z_n)\}$ is a system of disjoint intervals on the ordinate axis such that

$$\bigcup_n (y_n, z_n) = [m, M] \setminus (f(F_0) \cup \{0\})$$

(the zero is necessarily an end-point of one of the intervals (y_n, z_n)) then

$$\sum_n (z_n - y_n) = M - m.$$

Each interval (y_n, z_n) determines a strip parallel to the axis $0t$ which cuts the function f_n defined as follows from the graph of f: if $y_n \ge 0$ then

$$f_n(t) = \begin{cases} f(t) - y_n, & y_n \le f(t) < z_n, \\ z_n - y_n, & f(t) \ge z_n, \\ 0, & f(t) \le y_n; \end{cases}$$

if $z_n \le 0$ then

$$f_n(t) = \begin{cases} f(t) - z_n, & y_n < f(t) \le z_n, \\ y_n - z_n, & f(t) \le y_n, \\ 0, & f(t) \ge z_n; \end{cases}$$

Obviously

$$f(t) = \sum_n f_n(t), \qquad a \le t \le b,$$

and

$$\bigvee_a^b f = \sum_n \bigvee_a^b f_n, \qquad \int_a^b |f(t)| \, dt = \sum_n \int_a^b |f_n(t)| \, dt.$$

Every function f_n is a finite sum of simple functions ψ_{ni} ($i = 1, 2, \ldots, p_n$) with supports $[\alpha_{ni}, \beta_{ni}] \subset [a, b]$ and for every fixed n the intervals $(\alpha_{ni}, \beta_{ni})$ are disjointed. Hence

$$\bigvee_a^b f_n = \sum_{i=1}^{p_n} \bigvee_{\alpha_{ni}}^{\beta_{ni}} \psi_{ni}, \qquad \int_a^b |f_n(t)| \, dt = \sum_{i=1}^{p_n} \int_{\alpha_{ni}}^{\beta_{ni}} |\psi_{ni}(t)| \, dt.$$

Renumbering the functions ψ_{ni} with one index we get a finite or countable system $\{\varphi_k\}$ of simple functions with supports $[\alpha_k, \beta_k] \subset [a, b]$.

Hence for every $f \in D_0^1[a, b]$ we have a decomposition

$$f(t) = \sum_k \varphi_k(t), \qquad a \le t \le b, \tag{1.20}$$

which will be called the *Σ-representation* of the function f. Here are some properties of the Σ-representation which follow immediately from the construction:

(1) $$|f(t)| = \sum_k |\varphi_k(t)|, \qquad a \le t \le b;$$

(2) the intervals (α_k, α_k'), (β_k', β_k), on which the functions φ_k from representation (1.20) are strictly monotonic, are disjoint and on every one of them $\varphi_k(t) = f(t) + c_k$ ($c_k \in \mathbb{R}$) and hence $f'(t) = \Sigma_k \varphi_k'(t)$ a.e. on $[a, b]$;

(3) $$\int_a^b |f(t)|\,\mathrm{d}t = \sum_k \int_{\alpha_k}^{\beta_k} |\varphi_k(t)|\,\mathrm{d}t; \tag{1.21}$$

(4) $$\bigvee_a^b f = \sum_k \bigvee_{\alpha_k}^{\beta_k} \varphi_k = 2 \sum_k \max_t |\varphi_k(t)|. \tag{1.22}$$

Now starting with the Σ-representation (1.20) we introduce the operator $R(f)$ setting for every $f \in D_0^1[a, b]$

$$R(f; a, b; t) = \sum_k r(\varphi_k, t), \qquad 0 \le t \le b - a, \tag{1.23}$$

where $r(\varphi_k, t)$ is the non-increasing rearrangement of function $|\varphi_k|$ from representation (1.20). We call function $R(f; a, b; t)$ the *Σ-rearrangement* of function f. From its definition and relations (1.21) and (1.22) it follows that

(1) the function $R(f; a, b; t)$ is non-increasing and absolutely continuous on $[0, b - a]$ and

$$R(f; a, b; t) > 0, \qquad 0 < t < \Delta =: \sup_k (\beta_k - \alpha_k),$$

$$R(f; a, b; t) = 0, \qquad \Delta \le t \le b - a;$$

(2) $$\int_0^{b-a} R(f; a, b; t)\,\mathrm{d}t = \int_a^b |f(t)|\,\mathrm{d}t; \tag{1.24}$$

(3) $$R(f; a, b; 0) = \frac{1}{2} \bigvee_a^b f = \frac{1}{2} \|f'\|_{L_1[a,b]}. \tag{1.25}$$

Let us carry the definition of Σ-rearrangements over to 2π-periodic functions. Let D be the set of 2π-periodic functions ψ possessing one-sided limits $\psi(t+0)$, $\psi(t-0)$ at every point t. Hence D contains C and V. By D_0^1 we denote the set of periodic integrals of the type

$$f(t) = \int_a^t \psi(u)\,du,$$

where $\psi \in D_0$, $D_0 = \{\psi \in D, \psi \perp 1\}$.

If $f \in D_0^1$ then we have for some $f(a) = f(a+2\pi) = 0$ and the procedure above gives the Σ-rearrangement $R(f; a, a+2\pi; t)$ of the function f on an interval of length 2π. It is clear that the Σ-rearrangements of $f(t)$ and $f(t+\gamma)$ coincide and in the following we shall write $R(f, t)$ instead of $R(f; a, a+2\pi; t)$ in the periodic case. The properties (1)–(3) of $R(f; a, b; t)$ also hold for $R(f, t)$ with $[a, b] = [0, \pi]$.

7.1.4 Standard Σ-rearrangements and comparison theorems Comparison functions for Σ-rearrangements $R(f, t)$ defined for $f \in D_0^1$ will be defined by Σ-rearrangements of the standard functions $\varphi_{n,m}$ and $g_{n,m} = \varphi_{n,m}/(4n)$ introduced in Section 3.1.1. It is easy to see that

$$R(\varphi_{n,m}, t) = 2nr(\delta_{n,m}, t), \qquad R(g_{n,m}, t) = r(\delta_{n,m}, t)/2, \qquad (1.26)$$

where $\delta_{n,m}$ is the simple function (a 'hat') with a support of length π/n coinciding with $\varphi_{n,m}$ on an interval $[\alpha, \alpha + \pi/n]$ between two consecutive zeros of $\varphi_{n,m}$. Moreover

$$r(\delta_{n,m}, t) = \left|\delta_{n,m}\left(\alpha + \frac{\pi}{2n} \pm \frac{t}{2}\right)\right|, \qquad 0 \le t \le \frac{\pi}{n}. \qquad (1.27)$$

Let us also introduce the standard Σ-rearrangements depending on a continuous parameter a:

$$R_{a,0}(t) = \begin{cases} \frac{1}{2}, & 0 \le t < a, \\ 0, & t \ge a; \end{cases}$$

$$R_{a,m}(t) = \begin{cases} -\displaystyle\int_0^{a-t} R_{a,m-1}(u)\,du, & 0 \le t \le a, \\ 0, & t \ge a; \end{cases} \qquad m = 1, 2, \ldots \qquad (1.28)$$

One can easily see by induction that we have for all $\lambda > 0$

$$R_{a,m}(\lambda, t) = \lambda^m R_{a/\lambda,m}(t), \qquad m = 0, 1, \ldots, \qquad (1.29)$$

and using also (1.26) and (1.27) we obtain

$$R_{\pi/n,m}(t) = R(g_{n,m}, t) = \frac{1}{4n} R(\varphi_{n,m}, t), \qquad 0 \le t \le 2\pi, \qquad (1.30)$$

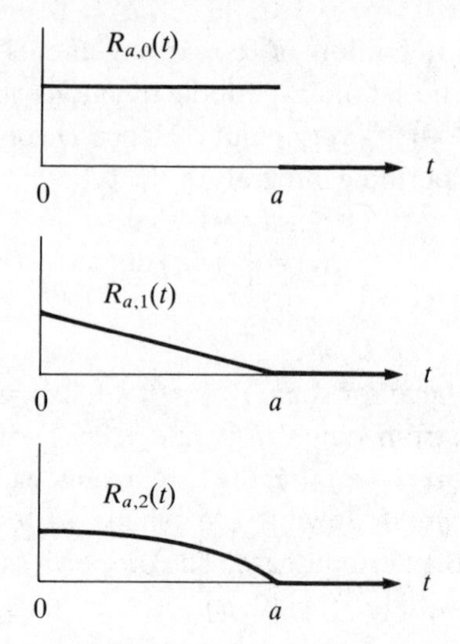

Fig. 7.4 Standard Σ-rearrangement.

$$\int_0^t R(g_{n,m}, u)\,\mathrm{d}u = \int_0^t R_{\pi/n,m}(u)\,\mathrm{d}u = \int_0^{2nt} r(g_{n,m}, u)\,\mathrm{d}u, \qquad 0 \le t \le \frac{\pi}{n}. \quad (1.31)$$

From definition (1.28) it is clear that, for every fixed $m = 0, 1, \ldots,$ the quantity

$$\|R_{a,m}\|_1 =: \int_0^a R_{a,m}(t)\,\mathrm{d}t = 2R_{a,m+1}(0)$$

depends continuously on a, increases monotonically with a and $\|R_{0+0,m}\|_1 = 0$. From (1.30), (1.31) and general properties (1.24), (1.25) we get

$$R_{\pi/n,m}(0) = \frac{1}{2}\|g_{n,m-1}\|_1 = \frac{K_m}{2n^m}, \qquad n, m = 1, 2, \ldots, \quad (1.32)$$

$$\|R_{\pi/n,m}\|_1 = \|g_{n,m}\|_1 = \frac{K_{m+1}}{n^{m+1}}, \qquad n = 1, 2, \ldots; m = 1, 2, \ldots. \quad (1.33)$$

We shall establish a representation of the norms in C and L_1 of the function $f_{n,m}(\omega, t)$ (see Section 7.1.1) by the standard rearrangements and the modulus of continuity $\omega(\delta)$. Setting $f_{n,m}(\omega, t) = f_{n,m}(t)$ for brevity and employing the structure of this function we can write

$$\|f_{n,m}\|_C = \frac{1}{4n}\|f_{n,m-1}\|_1 = \frac{1}{4n}\int_0^{2\pi} f_{n,m-1}(t)\psi_0(t)\,\mathrm{d}t,$$

where

$$\psi_0(t) = \operatorname{sgn} f_{n,m-1}(t) = \pm\varphi_{n,0}(t - \gamma_m), \qquad \gamma_m = [1 - (-1)^m]\pi/(4n).$$

If ω is a concave modulus of continuity then m-fold integration by parts and a suitable choice of arbitrary constants yield

$$\|f_{n,m}\|_C = \frac{1}{4n}\left|\int_0^{2\pi} f'_{n,0}(t)\psi_m(t)\,\mathrm{d}t\right| = \left|\int_0^{\pi/(2n)} f'_{n,0}(t)\psi_m(t)\,\mathrm{d}t\right|,$$

where ψ_m is the mth periodic integral of ψ_o with zero mean value on the period, i.e. $\psi_m(t) = \pm\varphi_{n,m}(t - \gamma_m)$. But (1.2) gives $f'_{n,0}(t) = \omega'(\pi/n - 2t)$ a.e. for $t \in (0, \pi/(2n))$ and hence

$$\|f_{n,m}\|_C = \int_0^{\pi/(2n)} \omega'\left(\frac{\pi}{n} - 2t\right)|\psi_m(t)|\,\mathrm{d}t = \frac{1}{2}\int_0^{\pi/n} \omega'(t)\left|\psi_m\left(\frac{\pi}{2n} - \frac{t}{2}\right)\right|\mathrm{d}t.$$

Because $\varphi_m(0) = 0$ we have

$$\left|\psi_m\left(\frac{\pi}{2n} - \frac{t}{2}\right)\right| = r(\delta_{nm}, t), \qquad 0 \le t \le \frac{\pi}{n},$$

and now relations (1.26) and (1.29) give

$$\|f_{n,m}\|_C = \int_0^{\pi/n} R_{\pi/n,m}(t)\omega'(t)\,\mathrm{d}t$$

$$= \frac{1}{n^{m+1}}\int_0^{\pi} R_{\pi,m}(t)\omega'\left(\frac{t}{n}\right)\mathrm{d}t = \frac{1}{2n^m}\int_0^{\pi} R_{\pi,m-1}(\pi - t)\omega\left(\frac{t}{n}\right)\mathrm{d}t. \tag{1.34}$$

The last equality is obtained by integration by part and using (1.28). Because $\|f_{n,m}\|_1 = 4n\|f_{n,m+1}\|_C$ (1.34) implies

$$\|f_{n,m}\|_1 = \frac{4}{n^{m+1}}\int_0^{\pi} R_{\pi,m+1}(t)\omega'\left(\frac{t}{n}\right)\mathrm{d}t = \frac{2}{n^m}\int_0^{\pi} R_{\pi,m}(\pi - t)\omega\left(\frac{t}{n}\right)\mathrm{d}t. \tag{1.35}$$

Now we state three comparison theorems for Σ-rearrangements which are at the root of the most important results in Chapters 7 and 8, and, in particular, the exact estimates for functional (1.6). When writing a relationship for the derivative $R'(f, t)$ we assume that it holds at the points where the derivative exists, i.e. almost everywhere. But it is possible in all these relationships to think of $R'(f, t)$ as either the right or the left derivative.

Theorem 7.1.4 Let $g \in W_V^m$ $(m = 1, 2, \ldots,)$, $\min_t |g(t)| = 0$ and the number a be chosen so that $R(g, 0) = R_{a,m}(0)$. Then

$$R(g, t) \ge R_{a,m}(t), \qquad 0 \le t \le 2\pi.$$

Theorem 7.1.5 If $g \in W_V^m$ $(m = 1, 2, \ldots,)$, $\min_t |g(t)| = 0$ and $\|g\|_1 \le \|R_{a,m}\|_1$ for some $a > 0$ then

$$R(g,0) \le R_{a,m}(0), \tag{1.36}$$

$$|R'(g,t)| \le |R'_{a,m}(t)|, \qquad 0<t<a. \tag{1.37}$$

Theorem 7.1.6 If $g \in W_\infty^m \cap D_0^1$ $(m = 1, 2, \ldots)$ and $\|g\|_C \le \|g_{n,m}\|_C$, then for $m \ge 3$ we have

$$|R'(g,t)| \le |R'(\varphi_{n,m},t)|, \qquad 0<t<\pi/n. \tag{1.38}$$

For $m = 1, 2$ this inequality may not be true but then

$$R(g,t) \le R(\varphi_{n,m},t), \qquad 0 \le t \le 2\pi. \tag{1.39}$$

The proofs of Theorems 7.1.4–7.1.6 can be found in the Appendices (see Section A3). Now we shall obtain two consequences which shall be used for the estimation of the functional (1.6) on classes of functions.

Proposition 7.1.7 If g satisfies the conditions of Theorem 7.1.5 then we have for every non-negative, non-increasing function μ, summable on $[0, 2\pi]$

$$\int_0^{2\pi} R(g,t)\mu(t)\,\mathrm{d}t \le \int_0^{2\pi} R_{a,m}(t)\mu(t)\,\mathrm{d}t.$$

Proof Functions $R(g,t)$ and $R_{a,m}(t)$ are non-increasing on $(0, 2\pi)$ and $R_{a,m}(t) = 0$ for $a \le t \le 2\pi$. Hence (1.36) and (1.37) imply that the difference

$$\delta(t) = R_{a,m}(t) - R(g,t)$$

is either non-negative on $(0, 2\pi)$ or changes sign once from '+' to '−'. In the first case everything is clear. Let

$$\delta(t) \ge 0 \; (0 \le t \le \eta), \qquad \delta(t) \le 0 \; (\eta \le t \le 2\pi).$$

Then

$$\int_0^{2\pi} \delta(t)\mu(t)\,\mathrm{d}t = \int_0^{\eta} \delta(t)\mu(t)\,\mathrm{d}t + \int_\eta^{2\pi} \delta(t)\mu(t)\,\mathrm{d}t$$

$$\ge \mu(\eta)\int_0^{\eta} \delta(t)\,\mathrm{d}t + \mu(\eta)\int_\eta^{2\pi} \delta(t)\,\mathrm{d}t = \mu(\eta)\int_0^{2\pi} \delta(t)\,\mathrm{d}t \ge 0,$$

because by assumption $\|R(g)\|_1 = \|g\|_1 \le \|R_{a,m}\|_1$ □

Proposition 7.1.8 Let g satisfy the conditions of Theorem 7.1.6 and

$$\max_{a,b} \left| \int_a^b g(t)\,\mathrm{d}t \right| \le 2\|\varphi_{n,m+1}\|_C. \tag{1.40}$$

Then we have for every non-negative, non-increasing function μ, which is summable on $[0, 2\pi]$

$$\int_0^{2\pi} R(g,t)\mu(t)\,dt \le \int_0^{2\pi} R(\varphi_{n,m},t)\mu(t)\,dt. \tag{1.41}$$

Proof The case (1.39) of Theorem 7.1.6 is trivial for (1.41). Let us consider the case

$$|R'(g,t)| \le |R'(\varphi_{n,m},t)|, \qquad 0<t<\pi/n. \tag{1.42}$$

The conditions of Theorem 7.1.6 and (1.40) allow us to use Theorem 3.3.4 (the case $k=0$) and we obtain $\|g\|_1 \le \|\varphi_{n,m}\|_1$. This fact and (1.42) imply $R(g,0)\le R(\varphi_{n,m},0)$ and the difference $R(\varphi_{n,m},t)-R(g,t)$ may change sign (from positive to negative) no more than once on $(0,2\pi)$. But then repeating the arguments from the previous proof we get (1.41). □

7.1.5 General estimates for functional $F_\omega(g)$ A starting point for considering the problem of the best approximation classes W^rH^ω by different subspaces is:

Theorem 7.1.9 Let $g \in D_0[a,b]$. For every modulus of continuity ω we have

$$\sup_{f\in H^\omega[a,b]} \left|\int_a^b g(t)f(t)\,dt\right| \le \int_0^{b-a} |R'(G;a,b;t)|\omega(t)\,dt, \tag{1.43}$$

where

$$G(t) = \int_a^t g(u)\,du, \qquad a\le t\le b, \qquad G(b)=0. \tag{1.44}$$

If $[a,b]=[0,2\pi]$ and g is continued as 2π-periodic to the real line then

$$F_\omega(g) = \sup_{f\in H^\omega[a,b]} \left|\int_0^{2\pi} g(t)f(t)\,dt\right| \le \min_c \int_0^{2\pi} |R'(G_c,t)|\omega(t)\,dt, \tag{1.45}$$

where

$$G_c(t) = \int_c^t g(u)\,du. \tag{1.46}$$

In the case of a concave modulus of continuity ω inequalities (1.43) and (1.45) cannot be improved in the sense that for some functions $g\in D_0[a,b]$ ($g \in D_0$) they become equalities.

Proof Function (1.44) belongs to $D_0^1[a,b]$ and (see Section 7.1.3) its Σ-representation is

$$G(t) = \sum_k \varphi_k(t), \qquad a\le t\le b,$$

where φ_k are simple functions. Moreover (see Property (2)) of representation (1.20)) we have a.e. on $[a, b]$

$$g(t) = G'(t) = \sum_k \varphi_k'(t).$$

But then we have for every $f \in C[a, b]$

$$\int_a^b f(t)g(t)\,dt = \sum_k \int_{\alpha_k}^{\beta_k} \varphi_k'(t)f(t)\,dt,$$

where $[\alpha_k, \beta_k]$ is the support of φ_k in $[a, b]$.

From the definition of a simple function it follows that the derivative φ_k' satisfies on $[\alpha_k, \beta_k]$ the conditions of Lemma 7.1.2 and hence

$$\left|\int_{\alpha_k}^{\beta_k} \varphi_k'(t)f(t)\,dt\right| \le \int_0^{\beta_k-\alpha_k} |r'(\varphi_k, t)|\omega(f, t)\,dt.$$

Therefore

$$\left|\int_a^b f(t)g(t)\,dt\right| \le \int_0^{b-a} \left|\sum_k r'(\varphi_k, t)\right|\omega(f, t)\,dt.$$

Taking the definition and the properties of the Σ-rearrangement for every $f \in H^\omega[a, b]$ (i.e. $\omega(f, t) \le \omega(t)$, $0 \le t \le b - a$) into account we have

$$\left|\int_a^b f(t)g(t)\,dt\right| \le \int_0^{b-a} |R'(G; a, b; t)|\omega(t)\,dt,$$

i.e. relation (1.43) is proved.

The proof of the periodic case is similar. Let us only note that beginning with the Σ-rearrangement of function G_c on the interval $[c, c + 2\pi]$ we get the inequality

$$F_\omega(g) \le \int_0^{2\pi} |R'(G_c, t)|\omega(t)\,dt$$

for every $c \in \mathbb{R}$.

The exactness of estimates (1.43) and (1.45) follows from Lemma 7.1.2 when G or G_c is only one simple function. Later on we shall see other cases of equality in (1.43) and (1.45). □

If we require g to satisfy the conditions of Comparison Theorems 7.1.5 and 7.1.6 for Σ-rearrangements of periodic functions we can estimate the right-hand side of (1.45) by standard Σ-rearrangements.

Theorem 7.1.10 Let $g \in W_V^m$ $(m = 0, 1, \ldots,)$, $g \perp 1$,

$$G_c(t) = \int_c^t g(u)\,du \tag{1.47}$$

and for some $a > 0$

$$\min_c \|G_c\|_1 \le \|R_{a,m+1}\|_1. \tag{1.48}$$

Then for every concave modulus of continuity ω we have

$$F_\omega(g) \le \int_0^a R_{a,m+1}(t)\omega'(t)\,dt = \frac{1}{2}\int_0^a R_{a,m}(a-t)\omega(t)\,dt. \tag{1.49}$$

For $a = \pi/n$ $(n = 1, 2, \ldots,)$ estimate (1.49) is exact; the equality holds for function $g = g_{n,m}$.

Proof The derivative ω' of a concave modulus of continuity ω is non-increasing and non-negative on $[0, 2\pi]$. By the conditions of the theorem $G_c \in W_V^{m+1}$. Moreover if c is such a number that $\|G_c\|_1 \le \|R_{a,m+1}\|_1$ then we can use Proposition 7.1.7 for estimating the right-hand side of (1.45). We have

$$\int_0^{2\pi} |R'(G_c, t)|\omega(t)\,dt = \int_0^{2\pi} R(G_c, t)\omega'(t)\,dt \tag{1.50}$$

$$\le \int_0^a R_{a,m+1}(t)\omega'(t)\,dt = \frac{1}{2}\int_0^a R_{a,m}(a-t)\omega(t)\,dt.$$

Only the last statement of the theorem remains to be proved.

Let $g = g_{n,m}$, $g_{n,m+1}(c) = 0$, $G_c(t) = g_{n,m+1}(t)$. Then for $a = \pi/n$ condition (1.48) is fulfilled. If ω is a concave modulus of continuity then for every β function $f_{n,0}(\omega, t+\beta)$ (see Section 7.1.1) belongs to H^ω. Let us choose $\beta = \beta_m$ so that the zeros of $g_{n,m}(t)$ and $f_{n,0}(\omega, t+\beta_m)$ coincide and for brevity set $\psi(t) = f_{n,0}(\omega, t+\beta_m)$. If $g_{n,m}(\alpha) = \psi(\alpha) = 0$ then $g_{n,m}(\alpha - u) = -g_{n,m}(\alpha + u)$ and the same for ψ. Moreover in view of the definition of $f_{n,0}(\omega, t)$ (see (1.2))

$$\psi(t) = \pm\tfrac{1}{2}\omega(2t - 2\alpha), \qquad \alpha \le t \le \alpha + \pi/(2n).$$

Therefore

$$F_\omega(g_{n,m}) \ge \left|\int_{\alpha-\pi}^{\alpha+\pi} \psi(t) g_{n,m}(t)\,dt\right| = 4n\left|\int_\alpha^{\alpha+\pi/(2n)} \psi(t) g_{n,m}(t)\,dt\right|$$

$$= 2n\left|\int_\alpha^{\alpha+\pi/(2n)} g_{n,m}(t)\omega(2t-2\alpha)\,dt\right| = n\int_0^{\pi/n} r\left(g_{n,m}, \frac{\pi}{n} - t\right)\omega(t)\,dt$$

$$= \frac{1}{2}\int_0^{\pi/n} R_{\pi/n,m}\left(\frac{\pi}{n} - t\right)\omega(t)\,dt.$$

Hence we have in view of the general inequality (1.49)

$$F_\omega(g_{n,m}) = \frac{1}{2}\int_0^{\pi/n} R_{\pi/n,m}\left(\frac{\pi}{n} - t\right)\omega(t)\,dt. \qquad (1.51) \quad \square$$

Theorem 7.1.11 Let $g \in W_\infty^m \cap D_0$ $(m = 0, 1, \ldots,)$ and let for some α the integral

$$G(t) = \int_\alpha^t g(u)\,du \qquad (1.52)$$

satisfy the relations

$$\|G\|_C \le \|\varphi_{n,m+1}\|_C, \qquad \max_{a,b}\left|\int_a^b G(t)\,dt\right| \le 2\|\varphi_{n,m+2}\|_C.$$

Then for every concave modulus of continuity ω we have the exact inequality

$$F_\omega(g) \le 4n\int_0^{\pi/n} R_{\pi/n,m+1}(t)\omega'(t)\,dt = 2n\int_0^{\pi/n} R_{\pi/n,m}\left(\frac{\pi}{n} - t\right)\omega(t)\,dt. \quad (1.53)$$

Equality holds for $g = \varphi_{n,m}$.

Proof The function (1.52) satisfies the conditions of Theorem 7.1.6 and hence estimating the right-hand side of (1.45) and using Proposition 7.1.8 and (1.30) we get

$$\begin{aligned} F_\omega(g) &= \int_0^{2\pi} |R'(G,t)|\,\omega(t)\,dt = \int_0^{2\pi} R(G,t)\omega'(t)\,dt \\ &\le \int_0^{2\pi} R(\varphi_{n,m+1}, t)\omega'(t)\,dt = 4n\int_0^{\pi/n} R_{\pi/n,m+1}(t)\omega'(t)\,dt \\ &= 4n\int_0^{\pi/n} |R'_{\pi/n,m+1}(t)|\omega(t)\,dt = 2n\int_0^{\pi/n} R_{\pi/n,m}\left(\frac{\pi}{n} - t\right)\omega(t)\,dt. \quad (1.54) \end{aligned}$$

Finally setting $g = \varphi_{n,m}$ using (1.51) we obtain

$$F_\omega(\varphi_{n,m}) = 4nF_\omega(g_{n,m}) = 2n\int_0^{\pi/n} R_{\pi/n,m}\left(\frac{\pi}{n} - t\right)\omega(t)\,dt. \qquad \square$$

7.2 Exact results for the best approximation of classes W^rH^ω, $W^rH^\omega[a, b]$

7.2.1 Approximation by trigonometric polynomials As in Chapters 4 and 6 we set

$$E_n(f)_X = \inf_{\varphi \in \mathcal{N}_{2n-1}} \|f - \varphi\|_X, \qquad E_n(\mathcal{M})_X = \sup_{f \in \mathcal{M}} E_n(f)_X,$$

where $\mathcal{N}_{2n-1}^{T}$ is the space of the trigonometric polynomials of degree $n-1$, X is either C or L_p $(1 \le p \le \infty)$. We write $\|\bullet\|_p$ instead of $\|\bullet\|_{L_p}$ as before.

An estimate from below is given by:

Proposition 7.2.1 For all $n = 1, 2, \ldots; m = 0, 1, \ldots,$ we have

$$E_n(f_{n,m}(\omega))_p = \|f_{n,m}(\omega)\|_p, \qquad 1 \le p \le \infty,$$

where $f_{n,m}(\omega, t)$ is the function defined in Section 7.1.1.

Indeed the polynomial of best approximation for $f_{n,m}(\omega, t)$ from $\mathcal{N}_{2n-1}^{T}$ in the L_p metric is the constant zero. For $p = \infty$ this is implied by Chebyshev's Theorem 2.1.2 in view of the continuity of $f_{n,m}(\omega, t)$. If $1 \le p < \infty$ then the function

$$g(t) = |f_{n,m}(\omega, t)|^{p-1} \operatorname{sgn} f_{n,m}(\omega, t)$$

is $2\pi/n$-periodic, orthogonal to subspace $\mathcal{N}_{2n-1}^{T}$ (Proposition 4.1.2) and we can apply Proposition 1.4.6.

If ω is a concave modulus of continuity then $f_{n,m}(\omega, t)$ belongs to W^mH^ω and we immediately obtain

$$E_n(W^mH^\omega)_C \ge \|f_{n,m}(\omega)\|_C, \qquad n = 1, 2, \ldots; m = 0, 1, \ldots; \tag{2.1}$$

$$E_n(W^mH^\omega)_p \ge \|f_{n,m}(\omega)\|_p, \qquad n = 1, 2, \ldots; m = 0, 1, \ldots, \qquad p \ge 1. \tag{2.2}$$

The norms $\|f_{n,m}(\omega)\|_C$, $\|f_{n,m}(\omega)\|_1$ can be expressed by the standard Σ-rearrangements (see (1.34) and (1.35)). Now we are ready to establish the following assertions.

Theorem 7.2.2 For every concave modulus of continuity $\omega(\delta)$ and for all $n = 1, 2, \ldots,; r = 0, 1, \ldots,$ we have

$$E_n(W^rH^\omega)_C = \|f_{n,r}(\omega)\|_C = \frac{1}{n^{r+1}} \int_0^\pi R_{\pi,r}(t)\omega'\left(\frac{t}{n}\right) dt, \tag{2.3}$$

$$E_n(W^rH^\omega)_1 = \|f_{n,r}(\omega)\|_1 = \frac{4}{n^{r+1}} \int_0^\pi R_{\pi,r+1}(t)\omega'\left(\frac{t}{n}\right) dt,$$

$$= \frac{2}{n^r} \int_0^\pi R_{\pi,r}(\pi - t)\omega\left(\frac{t}{n}\right) dt. \tag{2.4}$$

For $r = 1, 2, \ldots,$ relation (2.3) can be supplemented with the equivalent equality

$$E_n(W^rH^\omega)_C = \frac{1}{2n^r} \int_0^\pi R_{\pi,r-1}(\pi - t)\omega\left(\frac{t}{n}\right) dt. \tag{2.5}$$

Proof Let us begin with relation (1.5) in which we set $\mathcal{N} = \mathcal{N}_{2n-1}^{\mathrm{T}}$. If $g \in W_p^r(\mathcal{N}_{2n-1}^{\mathrm{T}})_*$, i.e. $g \in W_p^r$, $g^{(r)} \perp \mathcal{N}_{2n-1}^{\mathrm{T}}$ and $g \perp 1$ then the Fourier series of g is of the type

$$g(t) = \sum_{k=n}^{\infty} (a_k \cos kx + b_k \sin kx)$$

and hence $g \perp \mathcal{N}_{2n-1}^{\mathrm{T}}$. This means that the set $W_p^r(\mathcal{N}_{2n-1}^{\mathrm{T}})$ coincides with the set $W_{p,n}^r$ considered in Section 4.2.3 and relation (1.5) for $\mathcal{N} = \mathcal{N}_{2n-1}^{\mathrm{T}}$ can be written as

$$E_n(W^m H^\omega)_q = \sup_{g \in W_{q',n}^r} \sup_{f \in H^\omega} \int_0^{2\pi} f(t) g(t)\, dt = \sup_{g \in W_{q',n}^r} F_\omega(g). \tag{2.6}$$

For the estimate of the norm $\|g\|_{q'}$ for $q' = 1, \infty$ we apply Theorem 4.2.3 which implies that if $\psi \in W_{\infty,n}^m$ then

$$\|\psi\|_C \leq K_m n^{-m} = \|\varphi_{n,m}\|_C, \qquad n = 1, 2, \ldots; \qquad m = 1, 2, \ldots, \tag{2.7}$$

and if $\psi \in W_{1,n}^m$ then

$$\|\psi\|_C \leq K_m n^{-m} = \|g_{n,m-1}\|_1, \qquad n, m = 1, 2, \ldots. \tag{2.8}$$

Let $g \in W_{1,n}^r$ and let G be the integral of g with zero mean value on the period. Then $G \in W_{1,n}^{r+1}$ and we have from Theorem 4.2.3 (see (2.8))

$$\|G\|_1 \leq \|g_{n,r}\|_1 = \|R_{\pi/n,r}\|_1.$$

From the obvious inclusion $W_{1,n}^m \subset W_V^{m-1}$ we conclude that $G \in W_V^m$ and Theorem 7.1.10 with $a = \pi/n$ and concave ω gives the estimate

$$F_\omega(g) \leq \int_0^{\pi/n} R_{\pi/n,r}(t)\omega'(t)\, dt, \qquad r = 1, 2, \ldots,$$

which together with (2.6) implies for $r \geq 1$

$$E_n(W^r H^\omega)_C \leq \int_0^{\pi/n} R_{\pi/n,r}(t)\omega'(\tau)\, dt. \tag{2.9}$$

In the case $r = 0$ this inequality takes the simple form

$$E_n(H^\omega)_C \leq \frac{1}{2}\omega\left(\frac{\pi}{n}\right) \tag{2.10}$$

and in this form it follows from Theorem 6.2.2. Inequality (2.10) also follows from Theorem 7.1.9.

Hence estimate (2.9) is valid for all $r = 0, 1, \ldots$ (for a concave ω) and hence (see (1.34))

$$E_n(W^r H^\omega)_C \leq \|f_{n,r}(\omega)\|_C, \qquad n = 1, 2, \ldots; \qquad r = 0, 1, \ldots;$$

which together with (2.1) and (1.34) gives (2.3) and also (2.5) for $r \geq 1$.

Let us give the partial cases of equalities (2.3) and (2.5) for $r = 0, 1, 2$ (under the assumption of the concavity of ω):

$$E_n(H^\omega)_C = \frac{1}{2}\,\omega\left(\frac{\pi}{n}\right), \tag{2.11}$$

$$E_n(W^1H^\omega)_C = \frac{1}{4}\int_0^{\pi/n} \omega(t)\,\mathrm{d}t = \frac{1}{4n}\int_0^{\pi} \omega\left(\frac{t}{n}\right)\mathrm{d}t, \tag{2.12}$$

$$E_n(W^2H^\omega)_C = \frac{1}{8}\int_0^{\pi/n} t\omega(t)\,\mathrm{d}t = \frac{1}{8n^2}\int_0^{\pi} t\omega\left(\frac{t}{n}\right)\mathrm{d}t. \tag{2.13}$$

We now proceed to proving of (2.4). In view of (2.6) we have to estimate exactly the value of functional $F_\omega(g)$ on $W^r_{\infty,n}$. If $g \in W^r_{\infty,n}$ ($r = 0, 1, \ldots,$) and G and Φ are the first and second periodic integrals of g respectively with zero mean values on the period then clearly $G \in W^{r+1}_{\infty,n}$, $\Phi \in W^{r+2}_{\infty,n}$ and, in view of Theorem 4.2.3 (see (2.7)),

$$\|G\|_C \le \|\varphi_{n,r+1}\|_C, \qquad \|\Phi\|_C \le \|\varphi_{n,r+2}\|_C.$$

The second inequality implies

$$\max_{a,b}\left|\int_a^b G(t)\,\mathrm{d}t\right| \le 2\|\Phi\|_C \le 2\|\varphi_{n,r+2}\|_C,$$

and now for $r \ge 1$ and $W^1_\infty \subset V$ we can use Theorem 7.1.11 which implies for a concave modulus of continuity ω and for every $g \in W^r_{\infty,n}$

$$F_\omega(g) \le 4n\int_0^{\pi/n} R_{\pi/n,r+1}(t)\omega'(t)\,\mathrm{d}t. \tag{2.14}$$

Using Stekhlov functions in the case $r = 0$ and noticing that $g_h \in C \subset D_0$ we can deduce from Theorem 7.1.11 that for every $f \in H^\omega$

$$\left|\int_0^{2\pi} g_h(t)f(t)\,\mathrm{d}t\right| \le 4n\int_0^{\pi/n} R_{\pi/n,1}(t)\omega'(t)\,\mathrm{d}t,$$

which gives (2.14) with $r = 0$ when $h \to 0$.

Therefore inequality (2.14) holds for $g \in W^r_{\infty,n}$ for all $r = 0, 1, \ldots,$ and hence (see (2.6))

$$E_n(W^mH^\omega)_1 = \sup_{g\in W^r_{\infty,n}} F_\omega(g) \le 4n\int_0^{\pi/n} R_{\pi/n,r+1}(t)\omega'(t)\,\mathrm{d}t.$$

Now (2.2) and (1.35) imply that we actually have equality here for a concave ω. Relations (2.4) are proved and, as in the uniform case, we shall give their form for small r:

$$E_n(H^\omega)_1 = \int_0^\pi \omega\left(\frac{t}{n}\right) dt, \tag{2.15}$$

$$E_n(W^1H^\omega)_1 = \frac{1}{2n}\int_0^\pi t\,\omega\left(\frac{t}{n}\right) dt. \tag{2.16}$$ □

Let us note the case $\omega(t) = Kt^\alpha$ $(0 \le t \le \pi,\ 0 < \alpha \le 1)$, i.e. when the class $W^rH^\omega = W^rKH^\alpha$ is defined by Hölder's condition.

Corollary 7.2.3 For all $0 < \alpha \le 1$ we have

$$E_n(W^rKH^\alpha)_C = \frac{K}{2n^{r+\alpha}}\int_0^\pi t^\alpha R_{\pi,r-1}(\pi - t)\, dt, \qquad n, r = 1, 2, \ldots, \tag{2.17}$$

$$E_n(KH^\alpha)_C = \frac{K\pi^\alpha}{2n^\alpha}, \qquad n = 1, 2, \ldots, \tag{2.18}$$

$$E_n(W^rKH^\alpha)_1 = \frac{2K}{n^{r+\alpha}}\int_0^\pi t^\alpha R_{\pi,r}(\pi - t)\, dt, \quad n = 1, 2, \ldots; \quad r = 0, 1, \ldots.$$

We have seen that the extremal functions realizing the supremums $E_n(W^rH^\omega)_C$ and $E_n(W^rH^\omega)_1$ depend essentially on n and as in Section 4.2 the question of the asymptotic behaviour of individual functions from W^rH^ω arises. It turns out that for every modulus of continuity ω in the class W^rH^ω $(r = 0, 1, \ldots,)$ there is a function f such that

$$\overline{\lim_{n\to\infty}} \frac{E_n(f)_C}{E_n(W^rH^\omega)_C} = 1. \tag{2.19}$$

Moreover for a concave ω the upper limit can be replaced by the usual limit.

7.2.2 Spline approximation; the periodic case Using duality relations and the comparison theorems of Σ-rearrangements we can also get exact results for the best approximation of classes W^rH^ω (once more ω is concave) by splines of minimal defect.

Let, as before, $S_m(\Delta_N)$ be the linear manifold of 2π-periodic splines of degree m and defect 1 with respect to the fixed partition

$$\Delta_N: 0 = t_0 < t_1 < \cdots < t_N = 2\pi. \tag{2.20}$$

If $f \in L_1^r$ then, in view of relations (1.5.25) and $f^{(r)} \perp 1$, we have

$$E(f, S_m(\Delta_N))_q = \sup\left\{\int_0^{2\pi} f^{(r)}(t)g(t)\, dt: g \in W_{q'}^r[S_m(\Delta_N)],\ g \perp 1\right\}, \tag{2.21}$$

where

$$W_p^r[S_m(\Delta_N)] = \{g: g \in L_p^r, \|g^{(r)}\|_p \le 1, g^{(r)} \perp S_m(\Delta_N)\}.$$

Using relation (5.4.8) we can rewrite equality (2.21) for $m \ge r$ as

$$E(f, S_m(\Delta_N))_q = \sup\left\{\int_0^{2\pi} f^{(r)}(t)\psi^{(m-r+1)}(t)\,dt: \psi \in W_{q'}^{m+1}(\Delta_N)\right\},$$

where

$$W_p^{m+1}(\Delta_N) = \{\psi: \psi \in W_p^{m+1}, \psi(t_i) = 0, i = 1, 2, \ldots, N\}. \quad (2.22)$$

Thus for the best approximation of class W^rH^ω we get the equality

$$E(W^rH^\omega, S_m(\Delta_N))_q = \sup\{F_\omega(\psi^{(m-r+1)}): \psi \in W_{q'}^{m+1}(\Delta_N)\}, \quad (2.23)$$

$$r = 0, 1, \ldots,; \qquad m \ge r; \qquad 1 \le q \le \infty; \qquad 1/q + 1/q' = 1,$$

where $F_\omega(g)$ is functional (1.6). Now we have to estimate exactly the values of this functional on the $(m - r + 1)$th derivative of functions from $W_{q'}^{m+1}(\Delta_N)$. In view of (1.45) for a concave modulus of continuity ω for $\psi \in W_{q'}^{m+1}(\Delta_N)$ we have

$$F_\omega(\psi^{(m-r+1)}) \le \int_0^{2\pi} R(\psi^{(m-r)}, t)\omega'(t)\,dt, \qquad m \ge r. \quad (2.24)$$

Referring to the uniform partition $\bar{\Delta}_{2n}$, i.e. in (2.20) letting $N = 2n$ and $t_i = i\pi/n$, we can use the Comparison Theorems 7.1.5 and 7.1.6 (more precisely their corollaries – Propositions 7.1.7 and 7.1.8) in order to estimate the last integral when $q' = 1, \infty$. For this partition extremal relations were established in Section 5.4 which guarantee that the conditions of Theorems 7.1.5 and 7.1.6 hold.

If $\psi \in W_1^{m+1}(\bar{\Delta}_{2n})$ we have, in view of Proposition 5.4.7,

$$\|\psi^{(\nu)}\|_1 \le \|\varphi_{n,m+1}^{(\nu)}\|_C = \|g_{n,m+1}^{(\nu+1)}\|_1 = \|g_{n,m-\nu}\|_1, \qquad \nu = 1, 2, \ldots, m.$$

Now let $g = \psi^{(m-r)}$ $(m \ge r)$; then using (1.33) we have

$$\|g\|_1 = \|\psi^{(m-r)}\|_1 \le \|g_{n,r}\|_1 = \|R_{\pi/n,r}\|_1.$$

Because of $g^{(r+1)} = \psi^{(m+1)}$ and $\|\psi^{(m+1)}\|_1 \le 1$ we have $g \in W_1^{r+1} \subset W_V^r$ and Proposition 7.1.7 gives for $\psi \in W_1^{m+1}(\bar{\Delta}_{2n})$ the inequality

$$\int_0^{2\pi} R(\psi^{(m-r)}, t)\omega'(t)\,dt \le \int_0^{2\pi} R_{\pi/n,r}(t)\omega'(t)\,dt. \quad (2.25)$$

Thus taking (2.23) and (2.24) into account we get for a concave modulus of continuity ω the inequality

$$E(W^rH^\omega, S_{2n,m})_C \leq \int_0^{2\pi} R_{\pi/n,r}(t)\omega'(t)\,dt, \qquad r = 1, 2, \ldots; \qquad m \geq r. \tag{2.26}$$

In order to get the corresponding estimate for the best approximation in the L_1 metric with the help of Proposition 7.1.8 we apply Proposition 5.4.5 which gives for $\psi \in W_1^{m+1}(\bar{\Delta}_{2n})$

$$\|\psi^{(\nu)}\|_C \leq \|\varphi_{n,m+1}^{(\nu)}\|_C, \qquad \nu = 0, 1, \ldots, m.$$

From these inequalities we have, setting $g = \psi^{(m-r)}$,

$$\|g\|_C \leq \|\varphi_{n,r+1}\|_C, \qquad \max_{a,b} \left| \int_a^b g(t)\,dt \right| \leq 2\|\varphi_{n,r+2}\|_C,$$

and also $g \in W_\infty^{r+1}$. Now we can apply Proposition 7.1.8 which gives the inequality

$$\int_0^{2\pi} R(\psi^{(m-r)}, t)\omega'(t)\,dt \leq \int_0^{2\pi} R(\varphi_{n,r+1}, t)\omega'(t)\,dt, \qquad m \geq r \tag{2.27}$$

for $\psi \in W_\infty^{m+1}(\bar{\Delta}_{2n})$. With the help of this estimate from the general relationships (2.23) and (2.24) for a concave ω we obtain

$$E(W^rH^\omega, S_{2n,m})_1 \leq \int_0^{2\pi} R(\varphi_{n,r+1}, t)\omega'(t)\,dt; \qquad r = 0, 1, \ldots; \qquad m \geq r. \tag{2.28}$$

We shall show that we have actually an equality in (2.26) and (2.28).

Proposition 7.2.4 If both numbers m and r are either even or odd then for all $1 \leq q \leq \infty$ we have

$$E(f_{n,r}(\omega), S_{2n,m})_q = \|f_{n,r}(\omega)\|_q; \tag{2.29}$$

if one of m and r is odd and the other is even then

$$E(f_{n,r}(\omega, \bullet + \pi/(2n)), S_{2n,m})_q = \|f_{n,r}(\omega)\|_q. \tag{2.29'}$$

The simplest way to check the above for $1 \leq q < \infty$ is to use Proposition 1.4.6. If, for brevity, we set

$$\psi(t) = |f_{n,r}(\omega, t)|^{q-1} \operatorname{sgn} f_{n,r}(\omega, t)$$

then we can easily conclude from the definition of $f_{n,r}(\omega, t)$ and Lemma 5.4.2 that $\psi(t + \alpha)$ is orthogonal to subspace $S_{2n,m}$, where $\alpha = 0$ if $m + r$ is even and $\alpha = \pi/(2n)$ if $m + r$ is odd. Thus Proposition 1.4.6 implies that the best approximation in the space $S_{2n,m}$ to the function

$$f_{n,r}(\omega, t + \alpha_{mr}), \qquad \alpha_{m,r} = [1 - (-1)^{m+r}]\pi/(4n),$$

in the L_q ($q \geq 1$) metric is the constant zero.

For the proof of (2.29) and (2.29′) in the uniform metric we use the duality relation (see Section 1.4)

$$E(f, S_{2n,m})_\infty = \sup\left\{\int_0^{2\pi} f(t)h(t)\,dt\colon \|h\|_1 \le 1, h \perp S_{2n,m}\right\}. \tag{2.30}$$

Let h_0 be the derivative g_h' of Stekhlov function g_h of $g(t) = (1/4n)\operatorname{sgn}\sin n(t+\beta)$, where β is chosen from the condition $h_0 \perp S_{2n,m}$ (this is possible in view of Lemma 5.4.2). It is clear that $h_0 \in L_1$, $\|h_0\|_1 \le 1$ and if $f_*(t) = \pm f_{n,r}(\omega, t+\alpha)$ then for an appropriate choice of sign and the parameter α we have $f_*(t)h_0(t) \ge 0$ for all t and

$$\int_0^{2\pi} f_*(t)h_0(t)\,dt = \int_0^{2\pi} f_*(t)g_h'(t)\,dt = \|f_{n,r}(\omega)\|_C - \varepsilon_h,$$

where $\varepsilon_h \to 0$ for $h \to 0$. In view of (2.30) this means that $E(f_*, S_{2n,m})_\infty \ge \|f_{n,r}(\omega)\|_C$. The opposite inequality is obvious.

For a concave ω proposition 7.2.4 gives the lower bound

$$E(W^rH^\omega, S_{2n,m})_q \ge \|f_{n,r}(\omega)\|_q, \qquad 1 \le q \le \infty, \tag{2.31}$$

and now (2.26), (2.28) and (2.31) together with (1.34) and (1.35) give the final result:

Theorem 7.2.5 For every concave modulus of continuity $\omega(\delta)$ and for all $m \ge r$ we have

$$E(W^rH^\omega, S_{2n,m})_C = \|f_{n,r}(\omega)\|_C = \frac{1}{n^{r+1}}\int_0^\pi R_{\pi,r}(t)\omega'\left(\frac{t}{n}\right)dt$$

$$= \frac{1}{2n^r}\int_0^\pi R_{\pi,r-1}(\pi - t)\omega\left(\frac{t}{n}\right)dt, \qquad n, r = 1, 2, \ldots; \tag{2.32}$$

$$E(W^rH^\omega, S_{2n,m})_1 = \|f_{n,r}(\omega)\|_1 = \frac{4}{n^{r+1}}\int_0^\pi R_{\pi,r+1}(t)\omega'\left(\frac{t}{n}\right)dt$$

$$= \frac{2}{n^r}\int_0^\pi R_{\pi,r}(\pi - t)\omega\left(\frac{t}{n}\right)dt,$$

$$n = 1, 2, \ldots, \qquad r = 0, 1, \ldots. \tag{2.33}$$

The case $r = 0$ in (2.32) will be considered in Section 7.2.5.

Comparing Theorems 7.2.2 and 7.2.5 we see that the supremums of the best approximations in class W^rH^ω ($r \ge 1$, concave ω) in metrics C and L_1 by subspaces $\mathcal{N}_{2n-1}^T$ and $S_{2n,m}$ ($m \ge r$) coincide. But one has to remember that $\dim \mathcal{N}_{2n-1}^T = 2n - 1$ while $\dim S_{2n,m} = 2n$.

We note that Corollary 7.2.3 and relations (2.11)–(2.13), (2.15) and

(2.16) remain valid if we replace the best approximation of class W^rH^ω by trigonometric polynomials with the best approximations by spline subspace $S_{2n,m}$ $(m \geq r)$.

7.2.3 Approximation by a convex set Here we consider the best uniform approximation of class W^rH^ω by the class $KW_\infty^{r+1} = W^rKH^1$ which is a convex closed set in C but not a subspace. In the case $r = 0$ this problem was solved in Section 6.2 by elementary methods. In view of Proposition 6.2.1 we have for every modulus of continuity ω and for every $f \in H^\omega$

$$E(f, KH^1)_C =: \inf_{\varphi \in KH^1} \|f - \varphi\|_C \leq \frac{1}{2} \max_{0 \leq t \leq \pi} [\omega(t) - Kt]. \tag{2.34}$$

This estimate is exact on the class H^ω because if f_0 is the even function coinciding with ω on $[0, \pi]$ then $\omega(f_0, \delta) = \omega(\delta)$ $(0 \leq \delta \leq \pi)$ and it follows from Proposition 6.2.1 that

$$E(f_0, KH^1)_C = \frac{1}{2} \max_{0 \leq t \leq \pi} [\omega(f_0, t) - Kt] = \frac{1}{2} \max_{0 \leq t \leq \pi} [\omega(t) - Kt].$$

For every positive integer r we can solve the formulated problem only for concave ω using the duality and the comparison theorems for Σ-rearrangements.

Fix a $K > 0$ and $r = 1, 2, \ldots,$. In view of Proposition 1.4.7 for every $f \in C \backslash KW_\infty^{r+1}$ there is a function g_0 defined on $[0, 2\pi]$ with a variation equal to 1 on $[0, 2\pi]$ such that

$$E(f, KW_\infty^{r+1})_C = \int_0^{2\pi} f(t)\, \mathrm{d}g_0(t) - \sup_{\varphi \in KW_\infty^{r+1}} \int_0^{2\pi} \varphi(t)\, \mathrm{d}g_0(t). \tag{2.35}$$

The function g_0 necessarily satisfies $g_0(0) = g_0(2\pi)$ because otherwise the supremum in (2.35) would be $+\infty$ (the class KW_∞^{r+1} contains all constants).

Continuing g_0 to the real line as a 2π-periodic function we get a function $g_0 \in V$ with variation 1 on $[0, 2\pi]$ satisfying (2.35).

Because KW_∞^{r+1} is locally compact and closed there is a function $\varphi_0 \in KW_\infty^{r+1}$ such that

$$\|f - \varphi_0\|_C = E(f, KW_\infty^{r+1})_C.$$

From Proposition 1.4.8 it follows that the functions f, g_0 and φ_0 are connected by the following relationships:

$$\|f - \varphi_0\|_C = \int_0^{2\pi} [f(t) - \varphi_0(t)]\, \mathrm{d}g_0(t), \tag{2.36}$$

$$\sup_{\varphi \in KW_\infty^{r+1}} \int_0^{2\pi} \varphi(t)\, \mathrm{d}g_0(t) = \int_0^{2\pi} \varphi_0(t)\, \mathrm{d}g_0(t). \tag{2.37}$$

In expressions (2.36) and (2.37) the function g_0 is defined up to an additive constant and we can choose this constant so that $g_0 \perp 1$. Denote by g_r the rth periodic integral of g_0 with a minimal L_1 norm. Therefore $g_r \in W_V^r$ and

$$\int_0^{2\pi} |g_r(t)| \, dt = \min_\lambda \int_0^{2\pi} |g_r(t) - \lambda| \, dt. \tag{2.38}$$

Moreover, the function $g_0 = g_r^{(r)}$ satisfies relations (2.35)–(2.37).

Because

$$\int_0^{2\pi} \varphi(t) \, dg_0(t) = (-1)^{r+1} \int_0^{2\pi} \varphi^{(r+1)}(t) g_r(t) \, dt$$

and because $\varphi^{(r+1)} \perp 1$ which is implied by Corollary 1.4.3 and relations (2.37) and (2.38) we obtain

$$\sup_{\varphi \in KW_\infty^{r+1}} \int_0^{2\pi} \varphi(t) \, dg_0(t) = K \int_0^{2\pi} |g_r(t)| \, dt$$

$$= (-1)^{r+1} \int_0^{2\pi} \varphi_0^{(r+1)}(t) g_r(t) \, dt. \tag{2.39}$$

The last equality, in particular, means that we have on the set of points t where $g_r(t) \neq 0$

$$\varphi_0^{(r+1)}(t) = (-1)^{r+1} K \operatorname{sgn} g_r(t).$$

Now we study the first integral on the right-hand side of (2.35). If $f \in C^r$ then

$$\int_0^{2\pi} f(t) \, dg_0(t) = (-1)^r \int_0^{2\pi} f^{(r)}(t) g_r'(t) \, dt,$$

where g_r is the same function from (2.39). Assuming that $f \in W^rH^\omega$ and ω is a concave modulus of continuity we obtain from Theorem 7.1.9

$$\int_0^{2\pi} f(t) \, dg_0(t) = \int_0^{2\pi} R(g_r, t) \omega'(t) \, dt. \tag{2.40}$$

If $R(g_r, t) > 0$ for $0 \leq t < \Delta$ and $R(g_r, \Delta) = 0$ then $\Delta \leq \pi$ because it is implied by (2.38) that the function g_r cannot preserve the sign on an interval with a length larger than π. From (2.39) we obtain

$$\sup_{\varphi \in KW_\infty^{r+1}} \int_0^{2\pi} \varphi(t) \, dg_0(t) = K \int_0^\Delta R(g_r, t) \, dt. \tag{2.41}$$

Now (2.35), (2.40) and (2.41) imply

$$E(f, KW_\infty^{r+1})_C \leq \int_0^\Delta [\omega'(t) - K] R(g_r, t) \, dt.$$

Choose $b > 0$ from the condition

$$\|R_{b,r}\|_1 = \|R(g_r)\|_1 = \|g_r\|_1 \tag{2.42}$$

Then Theorem 7.1.5 implies

$$R(g, 0) \leq R_{b,r}(0), \qquad |R'(g_r, t)| \leq |R'_{b,r}(t)|, \qquad 0 < t < b,$$

and in view of (2.42) $b \leq \Delta \leq \pi$ and if $\delta(t) = R_{b,r}(t) - R(g, t)$ then

$$\int_0^\pi \delta(t)\,\mathrm{d}t = 0.$$

Using this equality and Proposition 7.1.7 we obtain

$$\int_0^\pi [\omega'(t) - K]\delta(t) = \int_0^\pi \omega'(t)\delta(t)\,\mathrm{d}t \geq 0.$$

Thus for every $f \in W^r H^\omega$ we have established the estimate

$$E(f, KW_\infty^{r+1})_C \leq \int_0^b [\omega'(t) - K]R_{b,r}(t)\,\mathrm{d}t. \tag{2.43}$$

This, however, has the disadvantage that b depends on the approximating function f. We get a uniform estimate common for the whole class $W^r H^\omega$ if we take the maximum on $b \in [0, \pi]$ on the right-hand side of (2.43). Thus for a concave modulus of continuity ω and for every function $f \in W^r H^\omega$ $(r = 1, 2, \ldots,)$ we have

$$\begin{aligned} E(f, KW_\infty^{r+1})_C &\leq \max_{0 \leq a \leq \pi} \int_0^a R_{a,r}(t)[\omega'(t) - K]\,\mathrm{d}t \\ &= \frac{1}{2} \max_{0 \leq a \leq \pi} \int_0^a R_{a,r-1}(a - t)[\omega(t) - Kt]\,\mathrm{d}t. \end{aligned} \tag{2.44}$$

Is this estimate exact on class $W^r H^\omega$? In the general case, i.e. for every $K > 0$, it may not be. But if $K = K_{nr}$ is such that the maximum on a in (2.22) is attained for $a = \pi/n$ $(n = 1, 2, \ldots,)$ then the equality is realized by function $f_{n,r}(\omega, t)$. Indeed, in this case the right-hand side in (2.44) is equal (see (1.34) and (1.33)) to

$$\int_0^\pi R_{\pi/n,r}(t)\omega'(t)\,\mathrm{d}t - K\int_0^\pi R_{\pi/n,r}(t) = \|f_{n,r}(\omega)\|_C - K\|\varphi_{n,r+1}\|_C,$$

and hence for $f \in W^r H^\omega$

$$E(f, KW_\infty^{r+1})_C \leq \|f_{n,r}(\omega)\|_C - K\|\varphi_{n,r+1}\|_C. \tag{2.45}$$

On the other hand if $f \in C$ and $\varphi \in C$ then we have for the best approximations by trigonometric polynomials

$$E_n(f)_C \le E_n(f-\varphi)_C + E_n(\varphi)_C \le \|f-\varphi\|_C + E_n(\varphi)_C.$$

Hence we get for every function $\varphi \in KW_\infty^{r+1}$

$$\|f_{n,r}(\omega) - \varphi\|_C \ge E_n(f_{n,r}(\omega))_C - E_n(\varphi)_C = \|f_{n,r}(\omega)\|_C - E_n(\varphi)_C.$$

From (4.2.5) we have $E_n(\varphi)_C \le K\|\varphi_{n,r+1}\|_C$ and hence using the general relation (2.45) we obtain

$$E(f_{n,r}(\omega), KW_\infty^{r+1})_C = \|f_{n,r}(\omega)\|_C - K\|\varphi_{n,r+1}\|_C,$$

i.e. inequality (2.45) cannot be improved on the class W^rH^ω.

Let us state the result obtained.

Theorem 7.2.6 For every concave modulus of continuity ω and for all $r = 1, 2, \ldots,$ we have

$$E(W^rH^\omega, KW_\infty^{r+1})_C \le \max_{0\le a\le\pi} \int_0^a R_{a,r}(t)[\omega'(t) - K]\,\mathrm{d}t$$
$$= \frac{1}{2}\max_{0\le a\le\pi} \int_0^a R_{a,r-1}(a-t)[\omega(t) - Kt]\,\mathrm{d}t. \quad (2.46)$$

If $K = K_{nr}$ is such that the maximum on a is attained for $a = \pi/n$ $(n = 1, 2, \ldots,)$ then in (2.46) we have an equality. Moreover

$$E(W^rH^\omega, K_{nr}W_\infty^{r+1})_C = \|f_{n,r}(\omega)\|_C - K_{nr}\|\varphi_{n,r+1}\|_C$$
$$= \|f_{n,r}(\omega) - K_{nr}\varphi_{n,r+1}\|_C.$$

7.2.4 Best approximation on an interval Using the results from Sections 7.2.1 and 7.2.2 for the classes W^rH^ω of periodic functions we can, in some cases, find the exact asymptotic behaviour for the best approximations by algebraic polynomials or splines of classes $W^rH^\omega[a, b]$ of functions defined on the finite interval $[a, b]$.

We begin with approximating by the subspace $\mathcal{N}_n^A$ of algebraic polynomials of degree $n-1$. As in Chapter 4 we use the notation

$$E_n(f)_{C[a,b]} = E(f, \mathcal{N}_n^A)_{C[a,b]}, \qquad E_n(\mathcal{M})_{C[a,b]} = E(\mathcal{M}, \mathcal{N}_n^A)_{C[a,b]}.$$

From (2.17) and (2.18) and the asymptotic relation (4.4.2) for $n \to \infty$ we get

$$E_n(KH^\alpha[-1, 1])_{C[-1,1]} = \frac{K\pi^\alpha}{2n^\alpha} + O(n^{-\alpha}),$$

$$E_n(W^rKH^\alpha[-1, 1])_{C[-1,1]} = \frac{K}{2n^{r+\alpha}}\int_0^\pi t^\alpha R_{\pi,r-1}(\pi - t)\,\mathrm{d}t + O(n^{-r-\alpha}),$$
$$r = 1, 2, \ldots,.$$

Using other arguments one can also show that the supremums of the best uniform approximation by polynomials for the classes W^rH^ω and $W^rH^\omega[-1,1]$ are asymptotically equal, provided that ω is a concave modulus of continuity. This will be assumed in the following.

First we shall prove that we have for every fixed $r = 0, 1, \ldots$, and for every $f \in W^rH^\omega[-1,1]$

$$E_n(f)_{C[-1,1]} \le \|f_{n,r}(\omega)\|_C + O(n^{-r}\omega(1/n)), \tag{2.47}$$

where $f_{n,r}(\omega, t)$ is the 2π-periodic function defined in Section 7.1 and $\|\bullet\|_C = \|\bullet\|_{C[0,2\pi]}$.

This is obvious for $r = 0$ because if $\varphi(t) = f(\cos t)$ then we have from Proposition 2.1.6

$$E_n(f)_{C[-1,1]} = E_n(\varphi)_C =: E(\varphi, \mathcal{N}^{\mathrm{T}}_{2n-1})_C.$$

Using $\omega(\varphi, \delta) \le \omega(f, \delta) \le \omega(\delta)$ and (2.11) we conclude

$$E_n(f)_{C[-1,1]} \le \frac{1}{2}\omega\left(\frac{\pi}{n}\right) = \|f_{n,0}(\omega)\|_C$$

for $f \in H^\omega[-1,1]$.

For an arbitrary positive integer r we first consider the case

$$\lim_{\delta\to 0}\frac{\omega(\delta)}{\delta} = M < \infty \tag{2.48}$$

which implies $\omega(\delta) \le M\delta$ since ω is concave. Hence if $f \in W^rH^\omega[-1,1]$ then $f \in MW^{r+1}_\infty[-1,1]$ and the limit relation (4.4.4) gives

$$E_n(f)_{C[-1,1]} \le MK_{r+1}n^{-r-1} + o(n^{-r-1}) = M\|\varphi_{n,r+1}\|_C + o(n^{-r-1}). \tag{2.49}$$

From the definitions of standard functions $f_{n,r}(\omega, t)$ and $\varphi_{n,r+1}(t)$ and relation (2.48) for $n \to \infty$ it follows in our case that

$$\|f_{n,r}(\omega)\|_C = M\|\varphi_{n,r+1}\|_C\,(1 + o(1)),$$

Hence (2.49) can be written as (2.47) because (2.48) implies that $\omega(1/n) = O(1/n)$.

Now let $f \in W^rH^\omega[-1,1]$,

$$\lim_{\delta\to 0}\frac{\omega(\delta)}{\delta} = \infty \tag{2.50}$$

and $\varphi(t) = f(\cos t)$. The rth derivative of φ can be represented as (see (4.4.18) and (4.4.19))

$$\varphi^{(r)}(t) = (-1)^r f^{(r)}(\cos t)\sin^r t + \psi_r(t) = g(t) + \psi_r(t),$$

where $\psi_r \in M_rW^{r+1}_\infty$, M_r is a constant that depends on r. Let us estimate the modulus of continuity of g:

$$
\begin{aligned}
|g(t') - g(t'')| &\le |\sin^r t'|\,|f^{(r)}(\cos t') - f^{(r)}(\cos t'')| \\
&\quad + |f^{(r)}(\cos t'')|\,|\sin^r t' - sin^r t''| \\
&\le \omega(f^{(r)}, |\cos t' - \cos t''|) + \|f^{(r)}\|_C\,|t' - t''| \\
&\le \omega(f^{(r)}, |t' - t''|) + N|t' - t''|.
\end{aligned}
$$

and hence $\omega(g, \delta) \le \omega(f^{(r)}, \delta) + N\delta$. Therefore $\varphi \in W^rH^{\omega_*}$ where $\omega_*(\delta) = \omega(\delta) + \omega_0(\delta)$, where $\omega_0(\delta) = N\delta$ and where the constant $N = M_r + M$ does not depend on f.

Expression (2.3) implies that

$$E_n(f)_{C[-1,1]} = E_n(\varphi, \mathcal{N}^{\mathrm{T}}_{2n-1})_C \le \|f_{n,r}(\omega_*)\|_C$$

and since also $f_{n,r}(\omega_*, t) = f_{n,r}(\omega, t) + f_{n,r}(\omega_0, \mathrm{t})$ we obtain

$$E_n(f)_{C[-1,1]} = \|f_{n,r}(\omega)\|_C + \|f_{n,r}(\omega_0)\|_C.$$

Now using (2.50) it is easy to conclude that

$$\|f_{n,r}(\omega_0)\|_C = N\|\varphi_{n,r+1}\|_C = o(n^{-r}\omega(1/n))$$

for $n \to \infty$, i.e. inequality (2.47) holds. For every $r = 0, 1, \ldots,$ there is a sequence of functions $\{f_n\}$ in the class $W^rH^\omega[-1, 1]$ for which $E_n(f_n)_{C[-1,1]} = \|f_{n,r}(\omega)\|_C + o(n^{-r}\omega(1/n))$. The known procedure for constructing such a sequence (see references in the comments) is technically complicated and we shall not give it here. The final result is:

Theorem 7.2.7 For every concave modulus of continuity ω the following asymptotic relations (for $n \to \infty$) hold:

$$
\begin{aligned}
E_n(W^rH^\omega[-1, 1])_{C[-1,1]} &= \|f_{n,r}(\omega)\|_C + o(n^{-r}\omega(1/n)) \\
&= \frac{1}{n^{r+1}} \int_0^\pi R_{\pi,r}(t)\omega'\left(\frac{t}{n}\right) \mathrm{d}t + o(n^{-r}\omega(1/n)).
\end{aligned}
$$

Let us note two particular cases

$$E_n(H^\omega[-1, 1])_{C[-1,1]} = \frac{1}{2}\omega(\pi/n) + o(\omega(n^{-1})), \tag{2.51}$$

$$E_n(W^1H^\omega[-1, 1])_{C[-1,1]} = \frac{1}{4}\int_0^{\pi/n} \omega(t)\,\mathrm{d}t + o(n^{-1}\omega(n^{-1})). \tag{2.51}$$

It has been proved in Section 4.4 (Theorem 4.4.2 and (4.4.20)) that for every function $f \in W^r_\infty[-1, 1]$ there is a corresponding sequence $p_n(f, t)$ of algebraic polynomials which provides the asymptotically best uniform approximation on $[-1, 1]$ for the class and gives substantially smaller errors near the ends of the interval. Is there such a phenomenon for the functions from $W^rH^\omega[-1, 1]$? Although order relations, which give estimates depending on the position of the point, were established quite a long time ago (see e.g. [26A, p. 276]) without much difficulty, results supplying exact constants have only been obtained (for concave ω) for $r = 0$ and $r = 1$.

Theorem 7.2.8 For every concave modulus of continuity ω and for every $f \in H^\omega[-1, 1]$ and $g \in W^1H^\omega[-1, 1]$ there exist sequences $\{p_n(f, t)\}$ and $\{q_n(f, t)\}$ of algebraic polynomials of degree n such that we have uniformly on $t \in [-1, 1]$ for $n \to \infty$

$$|f(t) - p_{n-1}(f, t)| \leq \frac{1}{2}\,\omega\left(\frac{\pi}{n}(1 - t^2)^{1/2}\right) + o(\omega(n^{-1})), \tag{2.52}$$

$$|g(t) - q_{n-1}(f, t)| \leq \frac{1}{4}\int_0^{(\pi/n)(1-t^2)^{1/2}} \omega(u)\,du + o(n^{-1}\omega(n^{-1})). \tag{2.53}$$

The constants $\frac{1}{2}$ and $\frac{1}{4}$ on the right-hand sides of (2.52) and (2.53) cannot be decreased.

The proof is based on the idea of an intermediate approximation of functions $f \in H^\omega[-1, 1]$ and $g \in W^1H^\omega[-1, 1]$ by continuous functions possessing first and second derivatives respectively a.e. which are bounded by some majorant which depends on the position of the point $t \in [-1, 1]$. We shall illustrate the scheme of the proof for the class KH^α $(0 < \alpha < 1)$. We assume without loss of generality that $K = 1$.

Lemma 7.2.9 For every $f \in H^\alpha[-1, 1]$ $(0 < \alpha < 1)$ there exists a sequence $\{l_n\}$ of polygons on $[-1, 1]$ such that:

(1) for $n \geq n_\alpha$ we have at the points t where the derivative $l_n'(t)$ exists

$$|l_n'(t)| \leq \alpha\left(\frac{\pi}{n}(1 - t^2)^{1/2}\right)^{\alpha-1}, \qquad |t| < 1;$$

(2) for all $t \in [-1, 1]$ and $n = 1, 2, \ldots$ we have

$$|f(t) - l_n(t)| \leq \frac{1-\alpha}{2}\left(\frac{\pi}{n}(1 - t^2)^{1/2}\right)^{\alpha} + cn^{-3\alpha/2}, \qquad c = c(\alpha). \tag{2.54}$$

Clearly it suffices to prove the lemma for a polygonal function f. Let

$$-1 = t_0 < t_1 < \cdots < t_m = 1, \qquad |t_\nu - t_{\nu-1}| < \pi/n, \qquad \nu = 1, 2, \ldots, m$$

be a partition of the interval $[-1, 1]$ which contains the knots of the polygon

f. Denoting the right-hand side of (2.54) by $\mu_n(t)$ we define the continuous functions ψ_1 and ψ_2 by:

$$\psi_1(t_\nu) = f(t_\nu) + \mu_n(t_\nu); \qquad \psi_2(t_\nu) = f(t_\nu) - \mu_n(t_\nu); \qquad \mu = 0, 1, \ldots, m.$$

The functions ψ_1 and ψ_2 are linear on the intervals $[t_{\nu-1}, t_\nu]$ ($\nu = 1, 2, \ldots, m$). Functions ψ_1 and ψ_2 form a 'strip' and there is an algorithm providing a polygon l_n in this 'strip' which together with condition (2) satisfies condition (1) of the lemma.

One then considers the algebraic polynomials $V_{n-1}(f, \lambda^*, t)$ defined in (4.4.7) and establishes the estimate

$$|l_n(t) - V_{n-1}(l_n, \lambda^*, t)| \le \frac{\alpha}{2}\left(\frac{\pi}{n}(1-t^2)^{1/2}\right)^\alpha + O\left(\frac{\ln n}{n^{2\alpha}}\right).$$

which is uniform on $t \in [-1, 1]$. This estimate for an inequality (2.54) implies that we have for $f \in H^\alpha[-1, 1]$ and for $n \to \infty$

$$|f(t) - V_{n-1}(l_n, \lambda^*, t)| \le \frac{1}{2}\left(\frac{\pi}{n}(1-t^2)^{1/2}\right)^\alpha + O(n^{-3\alpha/2})$$

uniformly on $t \in [-1, 1]$.

The exactness of this estimate follows from (2.51). Let us note that, unlike the polynomial $p_n(f, t)$ which depends linearly on f in Theorem 4.4.2, polynomials $p_n(f, t)$ and $q_n(g, t)$ which are constructed for the proof of (2.52) and (2.53) depend non-linearly on f and g respectively.

Now let us turn to the approximation of classes $W^rH^\omega[a, b]$ by splines of minimal defect. Fix a partition

$$\Delta_N[a, b]\colon \mathrm{a} = \mathrm{t}_0 < \mathrm{t}_1 < \cdots < \mathrm{t}_N = \mathrm{b} \tag{2.55}$$

and the corresponding linear manifold of splines $S_m(\Delta_N[a, b])$. In view of the duality relations (1.5.18) and using (5.4.30) we can write for $f \in L_1^r[a, b]$

$$E(f, S_m(\Delta_N[a, b]))_q = \sup\left\{\int_a^b f^{(r)}(t)\psi^{(m-r+1)}(t)\,\mathrm{d}t\colon \psi \in W_{q'}^{m+1}(\Delta_N[a, b])_0\right\},$$

$$m \ge r, \tag{2.56}$$

where

$$W_p^{m+1}(\Delta_N[a, b])_0 = \{\psi\colon \psi \in W_p^{m+1}[a, b], \psi^{(\nu)}(a) = \psi^{(\nu)}(b) = 0,$$
$$\nu = 0, 1, \ldots, m; \qquad \psi(t_i) = 0, i = 1, 2, \ldots, N-1\}.$$

If $f \in W^rH^\omega[-1, 1]$ and if ω is a concave modulus of continuity then we get by an estimate of the integral on the right-hand side of (2.56) with the help of Theorem 7.1.9

$$E(f, S_m(\Delta_N[a,b]))_q = \sup\left\{\int_0^{b-a} R(\psi^{(m-r)}, t)\omega'(t)\,dt: \psi \in W_{q'}^{m+1}(\Delta_N[a,b])_0\right\},$$

$$m \geq r, \tag{2.57}$$

where $R(f,t) = R(f; a, b; t)$.

For fixed m, r, N, q and ω the right-hand side of (2.57) is determined only by the partition (2.55). It is interesting (at least for $\omega(\delta) = K\delta^\alpha$) to find the partition for which the supremum in (2.57) is minimal. We do not know the solution of this problem for arbitrary r and we restrict ourselves to upper bounds that follow directly from the results in the periodic case.

Let $[a,b] = [0, 2\pi]$, $\Delta_N[0, 2\pi] = \Delta_N$. If we consider the restrictions on $[0, 2\pi]$ of the functions from (2.22) then $W_p^{m+1}(\Delta_N)_0 \subset W_p^{m+1}(\Delta_N)$ and therefore

$$E(f, S_m(\Delta_N[0, 2\pi]))_q \leq \sup\left\{\int_0^{2\pi} R(\psi^{(m-r)}, t)\omega'(t)\,dt: \psi \in W_{q'}^{m+1}(\Delta_N)\right\}.$$

The supremum for $N = 2n$ and for the uniform partition $\bar{\Delta}_{2n}$ has been calculated in the cases $q' = 1, \infty$ in the proof of Theorem 7.2.5 of Section 7.2.2. It turns out that it is equal to $\|f_{n,r}(\omega)\|_q$, where $q = \infty, 1$. Thus we have for a concave ω for $X = C[0, 2\pi]$ or $X = L_1[0, 2\pi]$

$$E(W^r H^\omega[0, 2\pi], S_{N,m}[0, 2\pi])_X \leq \|f_{n,r}(\omega)\|_X, \qquad n = [N/2], \qquad m \geq r. \tag{2.57'}$$

In Section 8.2 it will be established that for every partition $\Delta_N\,[0, 2\pi]$ one has

$$E(W^r H^\omega[0, 2\pi], S_m(\Delta_N[0, 2\pi]))_X \geq \|f_{n_m,r}(\omega)\|_X,$$

$$n_m = [(N+m)/2] + 1, \qquad m \geq r, \tag{2.58}$$

where $X = C[0, 2\pi]$ or $X = L_1[0, 2\pi]$. Comparing inequalities (2.57′) and (2.58) and taking relations (*) and (**) into account which appear in Exercise 7.7 we conclude:

Theorem 7.2.10 For every concave modulus of continuity ω and for fixed natural m, r $(m \geq r)$ we have

$$E(W^r H^\omega[0, 2\pi], S_{N,m}[0, 2\pi])_X \leq \|f_{n,r}(\omega)\|_X(1 - \varepsilon_N), \qquad n = [N/2],$$

where $X = C[0, 2\pi]$ or $X = L_1[0, 2\pi]$ and $\varepsilon_N \to 0$ for $N \to \infty$.

7.2.5 Approximation by step functions Finally consider the best approximation of the class $H^\omega[a,b]$ by splines of zero degree, i.e. by step functions. Here we can obtain exact results in the L_p metric for all $1 \leq p \leq \infty$, but their real value will become clear in Section 8.2. Let us note that, in spite of the simplicity of the approximating tools, the way to this result for $1 \le p < \infty$ is far from being trivial. The path lies through solutions of extremal problems which are of independent importance.

Lemma 7.2.11 If $f \in C[a, b]$ then for all $p \ge 1$ we have the exact estimate

$$\inf_{\lambda\in\mathbb{R}} \int_a^b |f(t) - \lambda|^p \, dt \le 2^{-p} \int_0^{b-a} \omega^p(f, t) \, dt, \qquad \omega(f, t) = \omega(f, t)_{C[a,b]}. \quad (2.59)$$

Proof If $f \in C[a, b]$ then there exists for every $\varepsilon > 0$ a polygon l_ε without horizontal parts such that

$$\|f - l_\varepsilon\|_C < \varepsilon, \qquad |\omega(f, \delta) - \omega(l_\varepsilon, \delta)| < \varepsilon, \qquad 0 \le \delta \le b - a,$$

where the norm and moduli of continuity are taken on $[a, b]$. This fact can easily be established by interpolating the function f by polygons with respect to a sufficiently dense partition and after that changing its horizontal parts (if there are any) slightly.

Hence it is sufficient to prove (2.59) just for all such polygons $f \in C[a, b]$. The absence of horizontal parts guarantees the existence of a constant $\eta_p = \eta_p(f)$ for every $p \ge 1$ such that

$$\int_a^b |f(t) - \eta_p|^{p-1} \operatorname{sgn}[f(t) - \eta_p] \, dt = 0$$

and such that, in view of Proposition 1.4.6, realizing the best approximation to f in $L_p[a, b]$. Thus we have reduced the problem to proving

$$\int_a^b |f(t)|^p \, dt \le 2^{-p} \int_0^{b-a} \omega^p(f, t) \, dt \quad (2.60)$$

for every polygon f which does not have horizontal parts and which satisfies the condition

$$\int_a^b |f(t)|^{p-1} \operatorname{sgn} f(t) \, dt = 0. \quad (2.61)$$

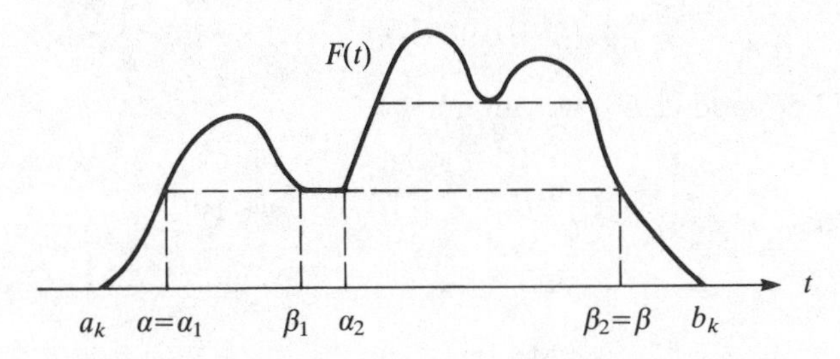

Fig. 7.5 To the proof of Lemma 7.2.11 (method of exhaustion).

A basic part of the proof is

Lemma 7.2.12 Let $f \in C[a, b]$ and let the function

$$F(t) = \int_a^t |f(u)|^{p-1} \operatorname{sgn} f(u)\, du, \qquad a < t \le b, \tag{2.62}$$

satisfy the conditions

$$F(b) = F(a) = 0, \qquad F(b') = F(a'), \qquad a < a' \le b' < b,$$

and let it be strictly monotone on every one of the intervals $[a, a']$ and $[b', b]$. Then

$$\int_e |f(t)|^p\, dt \le 2^{-p} \int_{b'-a'}^{b-a} \omega^p(f, t)\, dt, \tag{2.63}$$

where $e = [a, a'] \cup [b', b]$, $\omega(f, \delta) = \omega(f, \delta)_{C[a,b]}$.

The proof of Lemma 7.2.12 is similar to the proof of Lemma 7.1.2. Equation

$$F(t) = F(\rho(t)), \qquad a \le t \le a', \qquad b' \le \rho(t) \le b, \tag{2.64}$$

defines a strictly increasing absolutely continuous function ρ and its inverse ρ^{-1}. Moreover

$$F(\rho^{-1}(t)) = F(t), \qquad a \le \rho^{-1}(t) \le a', \qquad b' \le t \le b. \tag{2.65}$$

Changing the variables $t = \rho(u)$ and $t = \rho^{-1}(u)$ gives

$$\int_{b'}^{b} |f(t)|^p\, dt = -\int_a^{a'} |f(\rho(t))|^p\, \rho'(t)\, dt, \tag{2.66}$$

$$\int_a^{a'} |f(t)|^p\, dt = -\int_{b'}^{b} |f(\rho^{-1}(t))|^p (\rho^{-1}(t))'\, dt. \tag{2.66}$$

Differentiating equalities (2.64) and (2.65) and recalling the opposite signs of f on (a, a') and (b', b) we obtain

$$\begin{aligned} |f(t)|^{p-1} &= -|f(\rho(t))|^{p-1}\rho'(t), && a < t < a', \\ |f(t)|^{p-1} &= -|f(\rho^{-1}(t))|^{p-1}(\rho^{-1}(t))', && b' < t < b. \end{aligned} \tag{2.67}$$

From (2.66) and (2.67) we obtain

$$\int_e |f(t)|^p\, dt = \frac{1}{2}\int_e |f(t)|^{p-1}\psi(t)\, dt,$$

where

$$\psi(t) = \begin{cases} |f(t)| + |f(\rho(t))|, & a \le t \le a' \\ |f(t)| + |f(\rho^{-1}(t))|, & b' \le t \le b. \end{cases}$$

Hölder's inequality gives

$$2\int_e |f(t)|^p\,dt \le \left(\int_e |f(t)|^p\,dt\right)^{1-1/p}\left(\int_e |\psi(t)|^p\,dt\right)^{1/p}$$

and hence

$$2^p\int_e |f(t)|^p\,dt \le \int_e |\psi(t)|^p\,dt. \tag{2.68}$$

We estimate the right-hand side of this inequality. Reducing the range of integration to (a, a') by the change of variables $t = \rho(u)$ we obtain

$$\begin{aligned}\int_e |\psi(t)|^p\,dt &= \int_a^{a'} |\psi(t)|^p(1-\rho'(t))\,dt\\ &= \int_a^{a'} |f(t)-f(\rho(t))|^p(1-\rho'(t))\,dt\\ &\le \int_a^{a'} \omega^p(f, \rho(t)-t)(1-\rho'(t))\,dt = \int_{b'-a'}^{b-a} \omega^p(f, t)\,dt.\end{aligned}$$

This together with (2.68) proves (2.63).

Now using Lemma 7.2.12 as a basis we shall prove (2.60) (under the assumptions about f given there). We shall apply arguments which can be called the *exhaustion method*. Let f be a polygon without horizontal parts satisfying (2.61). If F is defined by (2.62) then $F(b) = 0$. The set of points $t \in (a, b)$ at which $F(t)$ is non-zero consists of a finite number of disjoint intervals (a_k, b_k) $(k = 1, 2, \ldots, n)$ where $F(a_k) = F(b_k) = 0$.

Take one of these intervals (a_k, b_k) and suppose $F(t) > 0$ for $a_k < t < b_k$. If the function F has only one local extremum on (a_k, b_k) then by Lemma 7.2.12 (with $a' = b'$) we obtain

$$\int_{a_k}^{b_k} |f(t)|p\,dt \le 2^{-p}\int_0^{b_k-a_k} \omega^p(f, t)\,dt. \tag{2.69}$$

In the opposite case there exists an interval $(\alpha, \beta) \subset (a_k, b_k)$ $(a_k < \alpha < \beta < b_k)$ such that

$$F(\alpha) = F(\beta), \qquad F(t) \ge F(\alpha)\ (\alpha < t < \beta), \qquad \max_{\alpha<t<\beta} F(t) > F(\alpha).$$

Moreover the difference $F(t) - F(\alpha)$ vanishes inside (α, β). Using Lemma 7.2.12 we can write

$$\int_{a_k}^{b_k} |f(t)|^p\,dt \le 2^{-p}\int_{\beta-\alpha}^{b_k-a_k} \omega^p(f, t)\,dt + \int_\alpha^\beta |f(t)|^p\,dt. \tag{2.70}$$

In order to evaluate the last integral we select all intervals (α_ν, β_ν) $(\nu = 1, 2, \ldots, m)$ in (α, β) such that

$$F(\alpha_\nu) = F(\beta_\nu) = F(\alpha), \qquad F(t) > F(\alpha), \qquad \alpha_\nu < t < \beta_\nu.$$

If we prove that

$$\int_{\alpha_\nu}^{\beta_\nu} |f(t)|^p \, dt \le 2^{-p} \int_0^{\beta_\nu - \alpha_\nu} \omega^p(f, t) \, dt, \qquad \nu = 1, 2, \ldots, m, \qquad (2.71)$$

then in view of the monotonicity of the modulus of continuity this will provide the estimate

$$\int_\alpha^\beta |f(t)|^p \, dt \le \sum_{\nu=1}^m 2^{-p} \int_0^{\beta_\nu - \alpha_\nu} \omega^p(f, t) \, dt \le 2^{-p} \int_0^{\beta - \alpha} \omega^p(f, t) \, dt,$$

which gives (2.69) when substituted in (2.70).

Thus we have to prove (2.71). But we can apply the arguments we have just applied on (a_k, b_k) to the difference $F(t) - F(\alpha)$ on the interval (α_ν, β_ν). As a result the proof of (2.71) will be reduced to proving similar inequalities on a finite number of intervals which are interior to (α_ν, β_ν). Continuing the reasoning in such a manner and applying Lemma 7.2.12 at every stage we shall exhaust the whole interval (a_k, b_k) after a finite number of steps. This will prove (2.69).

Assuming that (2.69) is proved for $k = 1, 2, \ldots, n$ we obtain

$$\int_a^b |f(t)|^p \, dt \sum_{k=1}^n 2^{-p} \int_0^{b_k - a_k} \omega^p(f, t) \, dt \le 2^{-p} \int_0^{b-a} \omega^p(f, t) \, dt.$$

This proves relation (2.60) and hence inequality (2.59) for every function $f \in C[a, b]$.

In order to show the exactness of estimate (2.59) for every concave modulus of continuity ω we consider the function

$$f_0(t) = \begin{cases} -\omega(a + b - 2t)/2, & a \le t \le (a + b)/2, \\ \omega(2t - a - b)/2, & (a + b)/2 \le t \le b, \end{cases} \qquad (2.72)$$

for which $\omega(f_0, \delta) = \omega(\delta)$ $(0 \le \delta \le b - a)$. Straightforward calculations show that one has equality in (2.59) for this function. The lemma is proved. □

Let $S_{N,0}[a, b]$ be the set of all step functions with possible jumps at the knots of the uniform partition $t_i = a + i(b - a)/N$ $(i = 1, 2, \ldots, N)$.

Theorem 7.2.13 For every modulus of continuity ω we have

$$E(H^\omega[a, b], S_{N,0}[a, b])_{L_p} \le \frac{1}{2}\left[N \int_0^{(b-a)/N} \omega^p(t) \, dt\right]^{1/p}, \qquad 1 \le p < \infty, \qquad (2.73)$$

$$E(H^\omega[a, b], S_{N,0}[a, b])_{L_\infty} = \frac{1}{2}\,\omega\left(\frac{b - a}{N}\right). \qquad (2.74)$$

For a concave ω inequality (2.73) becomes an equality.

In order to obtain (2.73) we apply Lemma 7.2.11 for every interval of the uniform partition. If ω is a concave modulus of continuity then equality is realized in (2.73) by a function $f \in H^\omega[a, b]$ constructed (up to a sign) on every interval of the partition in the same way as function (2.72) is constructed on $[a, b]$. The validity of (2.74) can be easily established by the reader.

7.2.6 Non-symmetric approximations The methods based on comparing Σ-rearrangements allow, in addition to the results on the best approximation of classes W^rH^ω by trigonometric polynomials and splines in the L_1 metric (see Theorems 7.2.2 and 7.2.5), the following theorem to be proved.

Theorem 7.2.14 Let ω be a concave modulus of continuity; let $n = 1, 2, \ldots;$ $r = 0, 2, 4, \ldots,$ $\alpha, \beta > 0$, and let $\mathcal{N}$ be either $\mathcal{N}^{\mathrm{T}}_{2n-1}$ or $S_{2n,m}$ $(m \geq r)$. Then

$$E(W^rH^\omega, \mathcal{N})_{1;\alpha,\beta} = \int_0^{\pi/n} R(\varphi_{n,r+1}(\alpha, \beta; \bullet), t)\omega'(t)\,\mathrm{d}t$$

$$= \frac{1}{n^r}\int_0^{\pi} |R'(\varphi_{1,r+1}(\alpha, \beta, \bullet), t)|\omega\left(\frac{t}{n}\right)\mathrm{d}t. \tag{2.75}$$

Letting α or β go to ∞ in (2.75) and taking Proposition 1.5.9 into account we get:

Theorem 7.2.15 Under the conditions of Theorem 7.2.14 we have

$$E^{\pm}(W^rH^\omega, \mathcal{N})_1 = \frac{2}{n^r}\int_0^{\pi} |R'(B_{r+1}, t)|\,\omega\left(\frac{t}{n}\right)\mathrm{d}t.$$

In particular, if $\omega(\delta) = \delta^\alpha$ $(0 < \alpha \leq 1)$ then

$$E^{\pm}(W^rH^\omega, \mathcal{N})_1 = \frac{2}{n^{r+\alpha}}\int_0^{\pi} |R'(B_{r+1}, t)|\,t^\alpha\,\mathrm{d}t.$$

Without going into details, we only give the scheme of the proof of Theorem 7.2.14. First, similarly to the proof for the symmetric case (Section 7.1.2), we reduce the evaluation of $E(W^rH^\omega\mathcal{N})_{1;\alpha,\beta}$ with the help of Proposition 1.4.9 to the evaluation of the supremum of the functional $F_\omega(g)$ on the set $W^r_{\infty;\alpha^{-1},\beta^{-1}}(\mathcal{N}) = \{g\colon g \in W^r_{\infty;\alpha^{-1},\beta^{-1}},\ g \perp 1,\ g^{(r)} \perp \mathcal{N}\}$. In view of (1.45) we have

$$F_\omega(g) \leq \int_0^{2\pi} R(G, t)\omega'(t)\,\mathrm{d}t, \tag{2.76}$$

where G is the periodic integral of function $g \in W^r_{\infty;\alpha^{-1},\beta^{-1}}(\mathcal{N})$ with zero mean value on the period. Now the basic role is played by Theorem 5.4.15 which we can employ in order to obtain for $g \in W^r_{\infty;\alpha^{-1},\beta^{-1}}(\mathcal{N})$, even r and all $x \in [0, 2\pi]$ that

$$\int_0^x R(G, t)\, dt \le \int_0^x R(\varphi_{n,r+1}(\alpha, \beta; \bullet), t)\, dt.$$

which together with (2.76) and Lemma 3.2.6 leads us to the inequality

$$E(W^r H^\omega, \mathcal{N})_{1;\alpha,\beta} = \int_0^{\pi/n} R(\varphi_{n,r+1}(\alpha, \beta; \bullet), t)\omega'(t)\, dt.$$

An inequality in the opposite direction can be easily obtained with the help of the extremal functions from Lemma 7.1.2.

7.3 Approximation by linear methods

7.3.1 Fourier means The approximation properties of the partial sums of the Fourier series

$$S_n(f, t) = \frac{a_0}{2} + \sum_{k=1}^{n} (a_k \cos kt + b_k \sin kt), \qquad n = 1, 2, \ldots, \tag{3.1}$$

with respect to classes of periodic functions were considered in Chapter 4, but the classes defined there were of functions f with a bounded rth derivative $f^{(r)}$ and the problem was reduced to calculating the L_1-norm of the corresponding integral operator (see (4.3.4)). In order to get an exact estimate of the approximation by sums (3.1) on $W^r H^\omega$ we have to use more detailed arguments.

The theorem stated below is due to Nikol'skii [4, 5, 9]. It is of particular importance both because it is the first asymptotically exact and non-trivial result on approximation of functional classes determined by a modulus of continuity and also because the concepts in its proof are at the basis of practically all further investigations of similar problems for Fourier means and other linear methods based on them.

We set

$$S_n(\mathcal{M}) = \sup_{f \in \mathcal{M}} \|f - S_n(f)\|_X, \qquad n = 1, 2, \ldots, \tag{3.2}$$

$W^r H^\omega$ ($r > 0$ is not necessarily an integer) is the class of all 2π-periodic functions f which can be represented as

$$f(t) = c + \frac{1}{\pi} \int_0^{2\pi} B_r(t - u)\varphi(u)\, du,$$

where $\varphi \in H^{\omega}$, $\varphi \perp 1$ and

$$B_r(t) = \sum_{k=1}^{\infty} \frac{\cos(kt - \pi r/2)}{k^r}, \qquad r > 0.$$

For integer $r = 1, 2, \ldots$, B_r are the known Bernoulli functions and the definition of classes $W^r H^{\omega}$ above coincides with the definition from Section 7.1.1. In general-type relations $W^0 H^{\omega}$ means, as usual, H^{ω}.

Notice that in view of the translation-invariance of the class $W^r H^{\omega}$ with respect to a translation of the argument for every t we have

$$\sup_{f \in W^r H^{\omega}} |f(t) - S_n(f, t)| = S_n(W^r H^{\omega})_C.$$

Theorem 7.3.1 For every modulus of continuity ω and for all $r \geq 0$ we have the asymptotic relation

$$S_n(W^r H^{\omega})_C = \theta_n(\omega) \frac{2 \ln n}{\pi^2 n^r} \int_0^{\pi/2} \omega\left(\frac{4t}{2n+1}\right) \sin t \, \mathrm{d}t + 0\left(\frac{\omega(1/n)}{n^r}\right), \quad (3.3)$$

where $\frac{1}{2} \leq \theta_n(\omega) \leq 1$. If ω is a concave modulus of continuity then $\theta_n(\omega) = 1$.

We give a detailed proof for $r = 0$ when the arguments are technically simpler. In view of Proposition 4.1.1

$$S_n(H^{\omega})_C = \sup_{f \in H_0^{\omega}} \frac{2}{\pi} \int_0^{\pi} D_n(t) f(t) \, \mathrm{d}t, \quad (3.4)$$

where $H_0^{\omega} = \{f \colon f \in H^{\omega}, f(0) = 0\}$ and

$$D_n(t) = \frac{1}{2} + \sum_{k=1}^{n} \cos kt = \frac{\sin(n + \frac{1}{2})t}{2 \sin(t/2)}. \quad (3.5)$$

Set

$$h = \pi/(2n+1), \qquad z_k = kh \ (k = 1, 2, \ldots, 2n+1),$$

$$\alpha_k = \left| \int_{z_k}^{z_{k+1}} D_n(t) \, \mathrm{d}t \right|, \qquad k = 1, 2, \ldots, 2n,$$

$$\Delta_\nu = \int_{z_{2\nu-1}}^{z_{2\nu+1}} D_n(t) \, \mathrm{d}t, \qquad \nu = 1, 2, \ldots, n.$$

The function (3.5) has zeros on $(0, \pi)$ with a change in sign only at points $z_{2\nu}$ ($\nu = 1, 2, \ldots, n$). Recalling the strong convexity and monotonicity on $(0, \pi)$ of the function $(\sin(t/2))^{-1}$ we easily see that

$$\alpha_1 > \alpha_2 > \cdots > \alpha_{2n};$$

$$|\Delta_\nu| > |\Delta_{\nu+1}|, \qquad \nu = 1, 2, \ldots, n-1, \tag{3.6}$$

$\operatorname{sgn} \Delta_\nu = (-1)^{\nu-1}$ and if we set $\gamma_\nu = \Delta_\nu + \Delta_{\nu+1} + \cdots + \Delta_n$ then

$$|\gamma_\nu| < |\Delta_\nu| = \alpha_{2\nu-1} - \alpha_{2\nu}. \tag{3.7}$$

Fix $f \in H_0^\omega$ and let

$$\psi(t) = f(t) - f(z_{2\nu}), \qquad z_{2\nu-1} < t < z_{2\nu+1}, \qquad \nu = 1, 2, \ldots, n. \tag{3.8}$$

Because $|D_n(t)| \le n + \frac{1}{2}$ we can write for $f \in H_0^\omega$

$$\left|\int_0^h f(t) D_n(t)\,\mathrm{d}t\right| \le \frac{\pi}{2}\,\omega(h)$$

and hence

$$\int_0^\pi f(t) D_n(t)\,\mathrm{d}t = \int_h^\pi \psi(t) D_n(t)\,\mathrm{d}t + \sum_{\nu=1}^n f(z_{2\nu})\Delta_\nu + O(\omega(h))$$

$$=: A(f) + \Sigma(f) + O(\omega(h)). \tag{3.9}$$

Applying Abel summation (4.3.7) to $\Sigma(f)$ and taking (3.6) and (3.7) into account we get

$$|\Sigma(f)| \le |\gamma_1 f(z_2)| + \sum_{\nu=2}^n |\gamma_\nu|\,|f(z_{2\nu-2}) - f(z_{2\nu})|$$

$$\le \omega(2h) \sum_{\nu=1}^n |\gamma_\nu| < \omega(2h)(\alpha_1 - \alpha_2 + \alpha_3 - \cdots - \alpha_{2n})$$

$$= O(\omega(h)), \tag{3.10}$$

We start by estimating the term $A(f)$ in (3.9). We have

$$|(\sin \nu h)^{-1} - (\sin(t/2))^{-1}| \le h^{-1} \sin^2(\nu - \tfrac{1}{2})h, \qquad z_{2\nu-1} < t < z_{2\nu+1}.$$

Setting

$$\mu(t) = (2 \sin \nu h)^{-1}, \qquad z_{2\nu-1} < t < z_{2\nu+1},$$

$$D_n^*(t) = \mu(t) \sin(n + \tfrac{1}{2})t, \qquad h \le t \le \pi,$$

we have (remember $\psi(z_{2\nu}) = 0$)

$$\sum_{\nu=1}^n \left|\int_{z_{2\nu-1}}^{z_{2\nu+1}} \psi(t)[D_n(t) - D_n^*(t)]\,\mathrm{d}t\right| \le \omega\left(\frac{h}{2}\right) \sum_{\nu=1}^n \frac{h^2}{2\sin^2(\nu - \frac{1}{2})h} = O(\omega(h)).$$

Therefore

$$A(f) = \int_h^\pi \psi(t) D_n^*(t)\,\mathrm{d}t + O(\omega(h))$$

$$= \sum_{\nu=1}^{n} \frac{1}{2\sin \nu h} \int_{z_{2\nu-1}}^{z_{2\nu+1}} \psi(t) \sin\left(n + \frac{1}{2}\right) t\,\mathrm{d}t + O(\omega(h)).$$

Let us estimate the integral in the last sum. Using (3.8) we have

$$\left|\int_{z_{2\nu-1}}^{z_{2\nu+1}} \psi(t) \sin\left(n + \frac{1}{2}\right) t\,\mathrm{d}t\right| = \left|\int_{z_{2\nu-1}}^{z_{2\nu}} [\psi(t) - \eta(2z_{2\nu} - t)] \sin\left(n + \frac{1}{2}\right) t\,\mathrm{d}t\right|$$

$$\leq \int_0^h \omega(2t) \sin\left(n + \frac{1}{2}\right) t\,\mathrm{d}t$$

$$\leq \frac{1}{n} \int_0^{\pi/2} \omega\left(\frac{4t}{2n+1}\right) \sin t\,\mathrm{d}t. \qquad (3.11)$$

Now using this estimate and

$$\sum_{\nu=1}^{n} \frac{1}{2\sin \nu h} = \frac{1}{2h} \sum_{\nu=1}^{n} \frac{1}{\nu} + O(n) = \frac{n}{\pi} \ln n + O(n)$$

we get

$$|A(f)| \leq \frac{\ln n}{\pi} \int_0^{\pi/2} \omega\left(\frac{4t}{2n+1}\right) \sin t\,\mathrm{d}t + O(\omega(h)) \qquad \forall f \in H_0^\omega. \quad (3.12)$$

Remark Some reduction in the calculations for the estimate of $A(f)$ will be provided by Theorem 7.1.9 which gives for $f \in H^\omega$

$$|A(f)| \leq \int_0^{2h} R'(\Psi_n, t)\omega(t)\,\mathrm{d}t + O(\omega(h)),$$

where

$$\Psi_n(t) = \int_h^t D_n^*(u)\,\mathrm{d}u, \qquad h \leq t \leq \pi,$$

and $R(\Psi_n, t)$ is the Σ-rearrangement of function Ψ_ν considered on the interval $[h, \pi]$. It is easy to see that we have for $0 \leq t \leq 2h$

$$R(\Psi_n, t) = \frac{\cos[(2n+1)t/4]}{2n+1} \sum_{\nu=1}^{n} \frac{1}{\sin \nu h} = \frac{\ln n}{\pi} \cos\left(\frac{2n+1}{4} t\right) + O(1).$$

For every $n = 2, 3, \ldots,$ we define a function f_n by the equalities

$$f_n(t) = \begin{cases} 0, & 0 \leq t \leq z_2, \\ (-1)^\nu \tfrac{1}{2}\omega(2t - 2z_{2\nu}), & z_{2\nu} \leq t \leq z_{2\nu+1}, \quad \nu = 1, 2, \ldots, n, \\ (-1)^\nu \tfrac{1}{2}\omega(2z_{2\nu+2} - 2t), & z_{2\nu+1} \leq t \leq z_{2\nu+1}, \quad \nu = 1, 2, \ldots, n-1, \end{cases}$$

$$f_n(t) = f_n(-t), \qquad -\pi \leq t \leq 0, \qquad f_n(t + 2\pi) = f_n(t).$$

If ω is a concave modulus of continuity then $f_n \in H_0^\omega$ – this can be checked as for the standard function $f_{n,r}(\omega)$ in Section 7.1.1. Following the calculations in estimating $A(f)$ for $f = f_n$ we easily see that the asymptotic relation (3.12) is fulfilled for $f = f_n$ as an equality. But then (3.4), (3.9), (3.10) and (3.12) give for a concave modulus of continuity ω

$$S_n(H^\omega)_C = \frac{2 \ln n}{\pi^2} \int_0^{\pi/2} \omega\left(\frac{4t}{2n+1}\right) \sin t \, dt + O\left(\omega\left(\frac{1}{n}\right)\right). \tag{3.13}$$

In the case of an arbitrary modulus of continuity ω the function f_n may not belong to H^ω but $f_n/2 \in H_0^\omega$ ($n = 2, 3, \ldots,$) and hence

$$|A(f)| \geq \frac{\ln n}{2\pi} \int_0^{\pi/2} \omega\left(\frac{4t}{2n+1}\right) \sin t \, dt + O(\omega(h)).$$

Therefore (3.13) remains correct if we multiply its right-hand side with a coefficient $\theta_n(\omega) \in [\tfrac{1}{2}, 1]$.

Now we give some indication for the proof of the theorem for $r > 0$. If $r = 1, 2, \ldots,$ then (see (4.3.3))

$$f(t) - S_n(f, t) = \frac{1}{\pi} \int_0^{2\pi} D_{n,r}(t - u) f^{(r)}(u) \, du, \tag{3.14}$$

where

$$D_{n,r}(t) = \sum_{k=n+1}^{\infty} \frac{\cos(kt - \pi r/2)}{k^r}.$$

Because $D_{n,r} \perp 1$ we have

$$S_n(W^r H^\omega)_C = \sup_{\varphi \in H_0^\omega, \varphi \perp 1} \frac{1}{\pi} \left| \int_0^{2\pi} D_{n,r}(t) \varphi(t) \, dt \right|, \tag{3.15}$$

and we have come to a problem similar to (3.4) but D_n is replaced by $D_{n,r}$.

As in Section 4.3 where we proved relation (4.3.1) with Abel summation we represent $D_{n,r}$ in the form

$$D_{n,r}(t) = -\frac{1}{(n+1)^r} \frac{\sin[(n + \tfrac{1}{2})t - \pi r/2]}{2 \sin(t/2)} + G_{n,r}(t). \tag{3.16}$$

After substituting (3.16) in (3.15) the integral corresponding to the first term on the right-hand side of (3.16) can be estimated following the scheme for

$r = 0$. This gives us the leading term in the asymptotic in (3.3). The proof that the integral which corresponds to the second term is $O(n^{-r}\omega(1/n))$ only involves overcoming some technical difficulties (see e.g. [26A, Chapter 8; 24A, Chapter 2]).

For a fractional $r > 0$ the error can also be represented in the form (3.14) where $f^{(r)}$ is replaced by φ, $\varphi \in H^\omega$. Representation (3.16) is used again, but the estimate for the terms of order $O(n^{-r}\omega(1/n))$ is more complicated.

Let us give a particular case of Theorem 7.3.1 when $\omega(\delta) = K\delta^\alpha$ $(0 < \alpha \le 1)$:

$$S_n(W^r KH^\alpha)_C = \frac{2^{\alpha+1}}{\pi^2}\frac{\ln n}{n^{r+\alpha}}\int_0^{\pi/2} t^\alpha \sin t \, \mathrm{d}t + O(n^{-r-\alpha}), \qquad r = 0, 1, \ldots.$$

For $\alpha = 1$ this is Kolmogorov's result (4.3.1).

In Section 7.2.1 and 4.2 the asymptotically best approximation by polynomials of individual functions of a given class was briefly considered. It is reasonable to ask this question for approximation by linear methods when the extremal functions depend on n. In particular, for the partial sums of Fourier series the quantities

$$S(\omega) = \sup_{f\in H^\omega} \varliminf_{n\to\infty} \frac{\|f - S_n(t)\|_C}{S_n(H^\omega)_C}$$

and $\bar{S}(\omega)$ for the upper limit respectively have been considered. It has been proved (see the references in the comments) that

$$\bar{S}(\omega) = 1; \qquad \omega(\delta) = \delta^\alpha; \qquad 0 < \alpha < 1; \qquad S(\omega) = \tfrac{1}{2} \tag{3.17}$$

7.3.2 λ-methods Consider the approximation of classes $W^r H^\omega$ by λ-means (2.2.20) and set

$$e_n(\mathcal{M}, \lambda)_X = \sup_{f\in\mathcal{M}} \|f - U_n(f, \lambda)\|_X.$$

In the case $r = 0$ we have in view of Proposition 4.1.1

$$e_n(H^\alpha, \lambda)_C = \sup_{f\in H_0^\omega} \frac{2}{\pi}\int_0^\pi f(t)K_n(\lambda, t)\,\mathrm{d}t, \tag{3.18}$$

where $K_n(\lambda, t) = \frac{1}{2} + \lambda_1^n \cos t + \lambda_2^n \cos 2t + \cdots + \lambda_n^n \cos nt$. For positive λ-methods, i.e. $K_n(\lambda, t) \ge 0$ we obviously have

$$e_n(H^\alpha, \lambda)_C = \frac{2}{\pi}\int_0^\pi \omega(t)K_n(\lambda, t)\,\mathrm{d}t,$$

and in some cases one can extract the leading term of the asymptotic of the right-hand side (see Section 4.3.5).

If the kernel $K_n(\lambda, t)$ oscillates on $(0, \pi)$ then even obtaining an asymptotically exact estimate of the integral in (3.18) is a difficult task. In the case of Fourier means $(\lambda_k^n = 1)$ we were able to solve it (Theorem 7.3.1 for $r = 0$)

using fine properties of the kernel D_n and the fact that the terms of order $O(\omega(1/n))$ go in the remainder. In order to realize the best order on class H^ω for λ-methods the remainder of the asymptotic exact estimate has to have order $o(\omega(1/n))$ and the problem becomes much more difficult. A rather general upper bound, can be obtained with the help of Theorem 7.1.9.

Let

$$\Psi_n(\lambda, t) = \int_t^\pi K_n(\lambda, u)\,du, \qquad 0 \le t \le \pi, \tag{3.19}$$

and let t_1 be the smallest positive zero of $\psi_n(\lambda, t)$ (in general, depending on n). For every function $f \in H^\omega$ we have in view of Theorem 7.1.9

$$\left|\int_{t_1}^\pi f(t) K_n(\lambda, t)\,dt\right| \le \int_0^{\pi - t_1} |R'(\Psi_n(\lambda), t)|\omega(t)\,dt, \tag{3.20}$$

and hence (see (3.18))

$$e_n(H^\omega, \lambda)_C \le \frac{2}{\pi}\int_0^{t_1} \omega(t) K_n(\lambda, t)\,dt + \frac{2}{\pi}\int_0^{\pi - t_1} |R'(\Psi_n(\lambda), t)|\omega(t)\,dt. \tag{3.21}$$

Here $R(\Psi_n(\lambda), t)$ is the Σ-rearrangement of the function $\Psi_n(\lambda, t)$ on the interval $[t_1, \pi]$.

Without touching the conditions under which one has equality in (3.21) we shall only give two particular cases of the λ-methods of best order which give equality.

Let $\lambda_k^n = \lambda_k^{n,*} = (k\pi/(n+1)) \operatorname{ctg}(k\pi/(n+1))$, i.e. these are Favard λ^*-means (see Section 4.1.3). If $t_k = k\pi/n$ then it is easy to see that

$$\Psi_{n-1}(\lambda^*, t_k) = 0, \qquad k = 1, 2, \ldots, n,$$

and on every interval (t_k, t_{k+1}) $(k = 1, 2, \ldots, n-1)$ the derivative $\Psi'_{n-1}(\lambda^*, t) = -K_{n-1}(\lambda^*, t)$ changes sign only once and hence satisfies the conditions of Lemma 7.1.2. According to this lemma one can, for a concave ω, easily construct extremal functions $f_k \in H^\omega[t_k, t_{k+1}]$ on every interval $[t_k, t_{k+1}]$ $(k = 1, 2, \ldots, n-1)$ which joined together give an even continuous function $f_* \in H_0^\omega$ ($f_*(t) = \omega(t)$ for $|t| \le t_1$) for which (3.20) is an equality.

Therefore we have for a concave ω

$$\begin{aligned} e_{n-1}(H^\omega, \lambda^*)_C &= \frac{2}{\pi}\int_0^\pi f_*(t) K_{n-1}(\lambda^*, t)\,dt \\ &= \frac{2}{\pi}\left[\int_0^{\pi/n} \omega(t) K_{n-1}(\lambda^*, t)\,dt \right. \\ &\qquad \left. + \int_0^{\pi/n} |R'(\Psi_{n-1}(\lambda^*), t)|\omega(t)\,dt\right]. \end{aligned} \tag{3.22}$$

It does not seem possible to extract in explicit form the leading term of the exact asymptotic of the right-hand side of (3.22) explicitly. On the other hand the obvious upper estimate

$$e_{n-1}(H^\omega, \lambda^*)_C \le \frac{2}{\pi}\,\omega\left(\frac{\pi}{n}\right)\int_0^{\pi/n}[k_{n-1}(\lambda^*, t) + |R'(\Psi_{n-1}(\lambda^*), t)|]\,dt.$$

It cannot be improved on the set of all classes H^ω.

The Rogosinski kernel

$$K\left(\frac{\pi}{2n}, t\right) = \frac{1}{2} + \sum_{k=1}^{n}\cos\frac{k\pi}{2n}\cos kt = \frac{\sin nt \sin[\pi/(2n)]}{2(\cos t - \cos[\pi/(2n)])}$$

with zeros $x_k = (2k+1)\pi/(2n)$ $(k = 1, 2, \ldots, n-1)$ also oscillates and the above arguments for Favard means are applicable to the sums (see Sections 2.2 and 4.3)

$$\begin{aligned}\mathcal{R}\left(f, \frac{\pi}{2n}, t\right) &= \frac{1}{2}\left[S_n\left(f, t - \frac{\pi}{2n}\right) + S_n\left(f, t + \frac{\pi}{2n}\right)\right]\\ &= \frac{1}{\pi}\int_{-\pi}^{\pi} f(t-u)K\left(\frac{\pi}{2n}, u\right)du. \end{aligned} \tag{3.23}$$

But the zeros of the integral (3.19) can in this case only be determined approximately.

Investigating the location of these zeros we can show (see e.g. [24A]) that for sums (3.23), where ω is concave, an equality holds in (3.20) and (3.21). Let us also give the following asymptotically exact result: for every modulus of continuity ω for $r = 2, 3, \ldots$, we have

$$\sup_{f\in W^rH^\omega}\left\|f - \mathcal{R}\left(f, \frac{\pi}{2n}\right)\right\|_C = \frac{\pi^2 M_{r-2}(\omega)}{8n^2} + \gamma_{nr}, \tag{3.24}$$

where

$$\gamma_{nr} = \begin{cases} O(n^{-2}\omega(1/n)), & r = 2,\\ O(n^{-3}), & r \ge 3,\end{cases}$$

$$M_k(\omega) = \sup\{\|f\|_C\colon f \in W^kH^\omega, f \perp 1\}, \qquad k = 0, 1, \ldots.$$

7.3.3 On L_1 polynomial approximation At the end of Section 4.3 cases of asymptotic coincidence of the approximation errors on classes W^r_∞ and W^r_1 in C and L_1 were given. Similar facts also hold for classes defined by the modulus of continuity ω.

Let H^ω_1 be the class of functions $f \in L_1$ such that $\omega(f, \delta)_1 \le \omega(\delta)$, where $\omega(f, \delta)_1$ is the integral modulus of continuity of the function f (see Section

6.1), and let $W^rH_1^\omega = \{f: f \in L_1^r, f^{(r)} \in H_1^\omega\}$. If the majorant $\omega(\delta)$ is concave then we have for the approximations by partial sums of Fourier means $S_n(f, t)$

$$S_n(W^rH_1^\omega)_{L_1} = S_n(W^rH^\omega)_C + O\left(\frac{\omega(1/n)}{n^r}\right)$$

$$= \frac{2\ln n}{\pi^2 n^r}\int_0^{\pi/2} \omega\left(\frac{4t}{2n+1}\right)\sin t\,\mathrm{d}t + O\left(\frac{\omega(1/n)}{n^r}\right). \quad (3.25)$$

It is also known that, for positive λ-methods, concave ω and all odd $r = 1, 3, \ldots$, the exact equality

$$e_n(W^rH_1^\omega, \lambda^+)_{L_1} = e_n(W^rH^\omega, \lambda^+)_C \quad (3.26)$$

holds.

7.3.4 Interpolating splines The results in Chapter 5 showed the excellent approximation properties of the interpolating splines of minimal defect on classes W_p^r. Here we give some exact results for spline interpolation on classes W^rH^ω, $W^rH^\omega[a, b]$.

We start with the case $r = 0$ and we recall the exact estimates (2.73) and (2.75) for concave ω for the best approximation of the class $H^\omega[a, b]$ by step functions. It can be shown (see Exercise 7.9) that a spline of an arbitrary degree of a fixed dimension N which interpolates $f \in H^\omega[a, b]$ at N points in $[a, b]$ cannot realize these estimates unless $\omega(\delta) = K\delta$. But the case is different for splines from $S_{N,0}[a, b]$ which interpolate the mean values of function f on the intervals of the partition.

Let $h_N = (b - a)/N$, $t_k = a + kh_N$ $(k = 0, 1, \ldots, N)$ and let $\psi_N(f, t)$ be the step function defined by

$$\psi_N(f, t) = \frac{1}{h_N}\int_{t_{k-1}}^{t_k} f(t)\,\mathrm{d}t, \qquad t_{k-1} < t < t_k, \quad (3.27)$$

$$k = 1, 2, \ldots, N, \qquad \psi_N(f, a) = f(a).$$

Theorem 7.3.2 For every modulus of continuity ω, for $0 < p \le 3$ and for $f \in H^\omega[a, b]$ we have the estimate

$$\int_a^b |f(t) - \psi_N(f, t)|^p\,\mathrm{d}t \le 2^{-p}N\int_0^{h_N} \omega^p(t)\,\mathrm{d}t, \quad (3.28)$$

which cannot be improved in the case of concave ω.

The assertion of the theorem follows immediately from:

Lemma 7.3.3 If $f \in C[a, b]$ and $\int_a^b f(t)\,dt = 0$ then for $0 < p \le 3$ we have the exact inequality

$$\int_a^b |f(t)|^p\,dt \le 2^{-p} \int_0^{b-a} \omega^p(t)\,dt, \tag{3.29}$$

where $\omega(f, t)$ is the modulus of continuity of f on $[a, b]$.

Proof Let us first consider the simplest case when $f(t) > 0$ ($f(t) < 0$) a.e. on (a, a'), $f(t) < 0$ ($f(t) > 0$) a.e. on (b', b) and $f(t) = 0$ for $t \in [a', b']$. Set

$$F(t) = \int_a^t f(u)\,du, \qquad a \le t \le b, \tag{3.30}$$

and as in the proof of Lemma 7.1.2 introduce the function ρ by

$$F(t) = F(\rho(t)), \qquad a \le t \le a' \le b' \le \rho(t) \le b.$$

Hence

$$f(t) = f(\rho(t))\rho'(t), \qquad a < t < a'. \tag{3.31}$$

The change in variable $t = \rho(z)$ on (b', b) gives

$$\int_a^b |f(t)|^p\,dt = \int_a^{a'} |f(\rho(t))|^p[|\rho'(t)|^p + |\rho'(t)|]\,dt, \qquad p > 0.$$

For $x \ge 0$ and $0 < p \le 3$ we have the inequality (Exercise 7.8)

$$2^p(x^p + x) \le (1 + x)^{p+1}.$$

Therefore using (3.31) and $\rho'(t) < 0$ we get for $0 < p \le 3$

$$\begin{aligned}
\int_a^b |f(t)|^p\,dt &\le 2^{-p} \int_a^{a'} |f(\rho(t))|^p\,|1 - \rho'(t)|^{p+1}\,dt \\
&= 2^{-p} \int_a^{a'} |f(\rho(t)) - f(t)|^p[1 - \rho'(t)]\,dt \\
&\le 2^{-p} \int_a^{a'} \omega^p(f, \rho(t) - t)[1 - \rho'(t)]\,dt \\
&= 2^{-p} \int_{b'-a'}^{b-a} \omega^p(f, t)\,dt.
\end{aligned}$$

Now considering the general case and assuming without loss of generality that the function f is a polygon without horizontal parts we can prove Lemma 7.3.3 repeating the corresponding arguments from the proof of Lemma 7.2.11. We introduce the function F by (3.30) and dividing (a, b) into disjoint intervals (a_k, b_k) $(k = 1, 2, \ldots, n)$ such that

$$F(a_k) = F(b_k) = 0, \qquad |F(t)| > 0, \qquad a_k < t < b_k.$$

Using the validity of the lemma in the simplest case we prove by following the exhaustion method that

$$\int_{a_k}^{b_k} |f(t)|^p \, dt \le 2^{-p} \int_0^{b_k - a_k} \omega^p(t) \, dt, \qquad k = 1, 2, \ldots, n.$$

In view of the monotonicity of ω this gives for $0 < p \le 3$

$$\int_a^b |f(t)|^p \, dt \le 2^{-p} \sum_{k=1}^{n} \int_0^{b_k - a_k} \omega^p(t) \, dt \le 2^{-p} \int_0^{b-a} \omega^p(t) \, dt.$$

The exactness of inequality (3.29) is easily checked for example on the function $f(t) = t - c$ $(a \le t \le b)$, where $c = (a + b)/2$. □

Let us deduce Theorem 7.3.2 from Lemma 7.3.3. Let $f \in C[a, b]$. From definition (3.27) of function $\psi_N(f, t)$ it is clear that the difference $f(t) - \psi_N(f, t)$ satisfies the conditions of the lemma on every interval (t_{k-1}, t_k). Hence we have for $0 < p \le 3$

$$\int_a^b |f(t) - \psi_N(f, t)|^p \, dt \le 2^{-p} N \int_0^{h_N} \omega^p(f, t) \, dt.$$

From here both the inequality (6.3.10) which was announced in Section 6.3 and the inequality (3.28) follow for $f \in H^\omega[a, b]$. □

One can directly check that (3.28) becomes an equality for $[a, b] = [0, \pi]$ and concave ω and for the functions $f_{N,0}$ defined in Section 7.1.1. By the way, the exactness of estimate (3.28) follows directly from Theorem 7.1.13 too. In Chapter 8 we shall see that (3.28) cannot be improved for $1 \le p \le 3$ even in a broader sense.

Let us turn our attention to interpolation by polygons with respect to the uniform partition, assuming for convenience $[a, b] = [0, \pi]$. Let $t_k = k\pi/N$ $(k = 0, 1, \ldots, N)$; let $\sigma_{N,1}(f, t)$ be the polygon (i.e. the spline from $S_{N,1}[0, \pi]$) interpolating f at the points $t_0, t_1, \ldots, t_N$ and let $f_{N,r}(\omega, t)$ be the standard functions constructed in Section 7.1.1.

Theorem 7.3.4 For every modulus of continuity ω and every $f \in W^1H^\omega[0, \pi]$ we have

$$\int_0^\pi |f(t) - \sigma_{N,1}(f, t)|^p \, dt \le \int_0^\pi |f_{N,1}(\omega, t)|^p \, dt, \qquad p > 0, \qquad (3.32)$$

$$\int_0^\pi |f'(t) - \sigma'_{N,1}(f, t)|^p \, dt \le \int_0^\pi |f_{N,0}(\omega, t)|^p \, dt, \qquad 0 < p \le 3, \quad (3.33)$$

which are exact for a concave ω.

Proof If we set

$$\eta(f,t) = f(t) - \sigma_{N,1}(f,t)$$

then

$$\eta'(f,t) = f'(t) - \psi_N(f',t), \qquad t \neq k\pi/N,$$

where $\psi_N(g,t)$ is defined in (3.27). The conditions of Lemma 6.3.1 are fulfilled on every interval (t_{k-1}, t_k) $(k = 1, 2, \ldots, N)$ for the function $\eta'(f,t)$. Hence we have in view of Lemma 6.3.2 (see the first inequality in (6.3.16))

$$\int_{t_{k-1}}^{t_k} |\eta(f,t)|^p \,\mathrm{d}t \leq \int_{t_{k-1}}^{t_k} |f_{N,1}(\omega,t)|^p \,\mathrm{d}t, \qquad k = 1, 2, \ldots, N,$$

i.e. (3.32) holds. Inequality (3.33) follows directly from Theorem 7.3.2 if we note that $\sigma'_{N,1}(f,t) = \psi_N(f',t)$ for $t \neq k\pi/N$ and that the right-hand sides of (3.28) and (3.33) coincide for $[a,b] = [0,\pi]$. The exactness of (3.32) and (3.33) for concave ω is obvious because in this case $f_{N,1} \in W^1H^\omega[0,\pi]$ and $\sigma_{N,1}(f_{\nu,1}(\omega,\bullet),t) \equiv 0$.

Taking $[a,b] = [0,2\pi]$, $N = 2n$, we see that in the periodic case for $f \in W^1H^\omega$ the inequalities

$$\begin{aligned} &\int_0^{2\pi} |f(t) - \sigma_{2n,1}(f,t)|^p \,\mathrm{d}t \leq \int_0^{2\pi} |f_{n,1}(\omega,t)|^p \,\mathrm{d}t, \qquad p > 0, \\ &\int_0^{2\pi} |f'(t) - \sigma'_{2n,1}(f,t)|^p \,\mathrm{d}t \leq \int_0^{2\pi} |f_{n,0}(\omega,t)|^p \,\mathrm{d}t, \qquad 0 < p \leq 3, \end{aligned} \tag{3.34}$$

hold and are exact for a concave modulus of continuity ω. Therefore, for concave ω we have

$$\sup_{f \in W^1H^\omega} \|f - \sigma_{2n,1}(f)\|_p = \|f_{n,1}(\omega)\|_p, \qquad 1 < p \leq \infty, \tag{3.35}$$

$$\sup_{f \in W^1H^\omega} \|f' - \sigma'_{2n,1}(f)\|_p = \|f_{n,0}(\omega)\|_p = 2^{-1}\left(2n \int_0^{\pi/n} \omega^p(t)\,\mathrm{d}t\right)^{1/p}, \qquad 1 \leq p \leq 3. \tag{3.36}$$

The comparison of relations (3.35) and (3.36) with Proposition 7.2.4 (with $r = 1$) and Theorem 7.2.13 shows that the interpolating polygons $\sigma_{2n,1}(f,t)$ realize for concave ω on classes W^1H^ω the best approximation $E(W^1H^\omega, S_{2n,1})_p$ for $1 \leq p \leq \infty$ and their derivatives $\sigma'_{2n,1}(f,t)$ simultaneously approximate the derivative f' in the best manner in the L_p $(1 \leq p \leq 3)$ metric. Stronger assertions for the approximation properties of the polygons $\sigma_{2n,1}(f,t)$ with respect to class W^1H^ω will follow from the results in Chapter 8.

We give without proof the following fact for concave ω:

$$\sup_{f\in H^\omega[0,\pi]} \|f-\sigma_{N,1}(f)\|_p = \left(2N\int_0^{\pi/(2N)} \omega^p(t)\,\mathrm{d}t\right)^{1/p}, \qquad p\geq 1. \quad (3.37)$$

Now let us consider the approximation of functions from classes W^rH^ω ($r = 2, 3, \ldots,$) by the splines $\sigma_{2n,r}(f, t) \in S_{2n,r}$ defined by the equalities

$$\sigma_{2n,r}(f, \tau_{ri}) = f(\tau_{ri}), \qquad i = 1, 2, \ldots, 2n,$$

where $\tau_{ri} = i\pi/n - \beta_r$, $\beta_r = [1 + (-1)^r]\pi/(4n)$. We make use of the integral representation of the error given in Section 5.2.4:

$$f(t) - \sigma_{2n,r}(f, t) = \frac{1}{\pi}\int_0^{2\pi} q_r(t, u) f^{(r)}(u + \beta_r)\,\mathrm{d}u, \qquad (3.38)$$

where

$$q_r(t, u) = B_r(t-u-\beta_r) - \sum_{i=1}^{2n} s_{r,i}(t) B_r(\tau_{ri} - u - \beta_r),$$

$s_{r,i}$ being the fundamental splines.

Theorem 7.1.9 gives the following estimate of the right-hand side of (3.38) for $f \in W^rH^\omega$ and for fixed $t \neq \tau_{ri}$: if $\psi \in H^\omega$ then

$$\left|\int_0^{2\pi} q_r(t, u)\psi(u)\,\mathrm{d}u\right| \leq \int_0^{2\pi} |R'(F_r(t, \bullet), u)|\omega(u)\,\mathrm{d}u, \qquad (3.39)$$

where $R'(F_r(t, \bullet), u)$ is the Σ-rearrangement (on variable u) of function

$$F_r(t, u) = \int_{-\beta_r}^{u} q_r(t, v)\,\mathrm{d}v.$$

Let us investigate the possibility for having equality in (3.39). In view of Lemma 5.2.10 $F_r(t, \tau_{ri}) = 0$ ($i = 1, 2, \ldots, 2n$), $q_r(t, u)$ (as a function of u) changes sign on $(\tau_{r,i-1}, \tau_i)$ only once and $F_r(t, u)$ has opposite signs on two consecutive intervals. This means that $F_r(t, u)$ is a sum of $2n$ simple functions on $(\tau_{r1}, \tau_{r1} + 2\pi)$. However, for concave ω the extremal functions from Lemma 7.1.2 combine to give a function $\psi_* \in H^\omega$ for which relation (3.39) holds as an equality.

Therefore we have for concave ω (assumed in the following too) at every point $t \neq \tau_{ri}$

$$\sup_{f\in W^rH^\omega} |f(t) - \sigma_{2n,r}(f, t)| = \int_0^{\pi/n} R(F_r(t, \bullet), u)\omega'(u)\,\mathrm{d}u, \qquad r = 2, 3, \ldots.$$

An investigation of the specific properties of functions $q_r(t, u)$ and $F_r(t, u)$ leads to the observation that we have for $\omega(\delta) \neq K\delta$ ($0 \leq \delta \leq \pi/n$)

$$\sup_{f\in W^rH^\omega} \|f-\sigma_{2n,r}(f)\|_C = \max_t \int_0^{\pi/n} R(F_r(t,\bullet),u)\omega'(u)\,du > \|f_{n,r}(\omega)\|_C$$

$$= E(W^rH^\omega, S_{2n,r})_C, \qquad r=2,3,\ldots, \tag{3.40}$$

i.e. the interpolating splines $\sigma_{2n,r}(f,t)$ for $r\geq 2$ do not realize, on class W^rH^ω, the supremum of the best approximations in the C metric.

The case is different for L_1. The following statement is valid. For every concave modulus of continuity ω and all $n, r = 1, 2, \ldots$, we have

$$\sup_{f\in W^rH^\omega} \|f-\sigma_{2n,r}(f)\|_1 = E(W^rH^\omega, S_{2n,r})_1 = \|f_{n,r}(\omega)\|_1. \tag{3.41}$$

The known proof of this fact is too complicated to be represented here. Let us only note that it is based on comparison theorems more fundamental than the ones proved in Chapter 3 and in Section 7.1.

Comments

Section 7.1 Classes H^ω and W^mH^ω were introduced by Nikol'skii [9] (see also [5]) who was the first to consider standard functions of the type $f_{n,m}(\omega,t)$. The statements of Lemma 7.1.2 without rearrangements were established for $f\in H^\alpha[a,b]$ $(0<\alpha\leq 1)$ by the author [1, 2] and (see the remark in [1]) Stechkin; in general form they were obtained in [5]. In the terms of rearrangements the statements of Lemma 7.1.2 are proved in [10, 11].

Σ-rearrangements were introduced by the author [10, 11]. All the results in Sections 7.1.3–7.1.5, Theorems 7.1.4–7.1.6, 7.1.9–7.1.11 and Propositions 7.1.7 and 7.1.8 are due to the author [10, 11, 13].

Section 7.2 Theorem 7.2.2 was proved by the author [10, 11]. The statements connected with relation (2.19) for the best approximation of an individual function $f\in W^rH^\omega$ are due to Temlyakov [1, 2]. Relations (2.32) and (2.33) were obtained by the author [14, 16]; in [14] relation (2.32) is proved (for $m=r$) with other arguments. Theorem 7.2.6 is also due to the author [13]. In connection with this result let us note that in the beginning relations (2.3) of Theorem 7.2.2 were obtained for $r=0,1,2$ by the author [3, 4, 7, 8] with the help of the intermediate approximation of class W^rH^ω by class KW_∞^{r+1} (see also [9A, Chapter 8]). A generalization of relations (2.3) and (2.32) to classes of convolutions of functions from H^ω with oscillation-diminishing kernels was obtained by V. F. Babenko [14]. Similar generalizations of relations (2.4) and (2.33) are given in the articles of Fang Gensun [1] and V. F. Babenko [10, 14]. A non-periodic variant of Theorem 7.2.6 is considered in [10A]. Using the method of an intermediate approximation Dzyadyk [5] extended (2.11) to the case of approximation of functions $f\in H^\alpha$ defined on the real line by entirely exponential-type functions. Results (including unimprovable ones) for the approximation of some functional classes by others are contained in the works of Korneichuk and Ligun [2], Taikov [3], Subbotin [2], Arestov and Gabushin [1].

The proof of inequality (2.47) given here was communicated to the author by Temlyakov, the sequence of functions from $W^rH^\omega[-1,1]$ realizing the equality in (2.47) was constructed by Polovina [1]. Theorem 7.2.8 was obtained by Korneichuk and Polovina [1–3]. The statements in Section 7.2.5 were proved by the author in [9, 12].

Exact results for the best one-sided (in the L_1 metric) approximation of functions from W^rH^ω by polynomials and splines were obtained by V. G. Doronin [1], V. G. Doronin and Ligun [3, 6]. Theorem 7.2.15 was proved in [6], the general Theorem 7.2.14 was established by V. F. Babenko [7]. Its generalization to classes of convolutions of functions from H^ω with oscillation-diminishing even kernels was obtained by him [10, 14]. V. G. Doronin and Ligun [6] have also considered the problem of one-sided approximation of one class by another (see also [11A]).

Section 7.3 After S. M. Nikol'skii's papers [5, 9] many articles devoted to estimates of the approximation error of the partial sums of Fourier series and their means on functional classes defined by a modulus of continuity have appeared; we note in particular the papers of Efimov [1–5] and Stepanets [1, 2]. A representation of results in this direction and more references are given in the monograph [24A]. Equalities (3.22) are actually contained in author's paper [2]; relation (3.25) for $\omega(\delta) = K\delta^\alpha$ $(0 < \alpha \le 1)$ was obtained by Gavrilyuk and Stepanets [1] and in the general case by Stepanets [24A]; quantity $M_k(\omega)$ was calculated in author's article [5]. Relation (3.25) was established in the papers of Berdyshev [1] $(r = 0)$ and Demchenko [1], equality (3.26) was obtained by Motornyi [2]. The first of equalities (3.17) is proved by G. Ya Doronin [1], the second one by K. I. Oskolkov [1, 2].

Lemma 7.3.3 was originally proved by the author [9] for $p = 1$, later Storchai [3, 4] proved it for $p = 2, 3$: the general result for $0 < p \le 3$ was established by the author [23], where Theorem 7.3.2, Theorem 7.3.4 and relations (3.34)–(3.36) are given. Equality (3.37) was proved by Storchai [1]. Inequality (3.40) for even r was proved by the author [14, 15] and for odd r by Zhensykbaev [4]; relations (3.41) are contained in the author's papers [18, 29, 30]. Some exact estimates for splines (in particular Hermitian) interpolation on classes W^mH^ω are given in the works of Cheney and Schurer [1], Velikin and Korneichuk [1] and Velikin [1, 2] $(m = 0, 1)$. Exact results on the uniform approximation by interpolation polygons are contained in the papers of Malozemov [1, 2] and Loginov [1]; some by-variable analogs can be found in Storchai's paper [2]. Asymptotically exact estimates for the error of interpolation of continuous mappings by partial linear ones were obtained by V. F. Babenko [2] and V. F. Babenko and Ligun [1].

Other results and exercises

1. For a concave modulus of continuity ω we have the inequality

$$(W^rH^\omega, NW_1^{r+\nu})_1 \le \sup_{b>0}\left\{b^{r+1}\int_0^\pi R(\varphi_{1,r+1}, t)\omega'(bt)\,dt - Nb^{r+\nu}K_{r+\nu}\right\},$$

$$r = 0, 1, \ldots; \qquad \nu = 2, 3, \ldots,$$

which becomes an equality for some increasing sequence N_m (Korneichuk and Ligun [2]).

2. For every concave ω we have

$$\sup\{\|f\|_1: f \in W^r H^\omega, f \perp 1\} = 4 \sum_{k=0}^{\infty} \frac{(-1)^{kr} b_{2k+1}}{(2k+1)^{r+1}}, \qquad r = 0, 1, \ldots,$$

where $b_m = 2/\pi \int_0^{\pi/2} \omega(2t) \sin mt \, dt$ (Korneichuk [9]).

3. If ω is concave modulus of continuity then

$$\sup_{f \in W^1 H^\omega} \|\tilde{f} - \tilde{\sigma}_{2n,1}(f)\|_1 = \|\tilde{f}_{n,1}(\omega)\|_1,$$

where $\tilde{g}$ is the conjugate of g (V. F. Babenko [9]).

4. For a concave ω find the quantities

$$M_k(\omega) = \sup\{\|f\|_C: f \in W^k H^\omega, f \perp 1\},$$

$$\Omega_k(h) = \sup\{\|f(\bullet + h) - f(\bullet)\|_C: f \in W^k H^\omega\}$$

and $\max_h \Omega_k(h)$. Hint. Use Lemma 7.1.2.

5. Prove that we have for every modulus of continuity

$$E^+(H^\omega, S_{2n,0})_p = \left\{2n \int_0^{\pi/n} \omega^p(t) \, dt\right\}^{1/p}, \qquad 1 \le p \le \infty, \qquad n = 1, 2, \ldots.$$

6. For concave ω prove

$$E^+(H^\omega, S_{2n,r})_1 = 2n \int_0^{\pi/n} \omega(t) \, dt, \qquad n, r = 1, 2, \ldots.$$

7. Using (1.34) for $1 \le p \le \infty$ prove

$$\|f_{n+j,r}(\omega)\|_p < \|f_{n,r}(\omega)\|_p, \qquad n, j = 1, 2, \ldots; r = 0, 1, \ldots, \qquad (*)$$

and for fixed j and r prove

$$\lim_{n \to \infty} \frac{\|f_{n+j,r}(\omega)\|_p}{\|f_{n,r}(\omega)\|_p} = 1. \qquad (**)$$

8. Prove the inequality

$$2^p(x^p + x) \le (1 + x)^{p+1}, \qquad x \ge 0, \qquad 0 < p \le 3$$

[10A, p. 225]; show that it is not valid for $p > 3$.

9. Prove that if $\omega(\delta) \ne K\delta$ $(0 \le \delta \le (b-a)/N)$ then the bounds (3.73) and (3.74) cannot be realized by the splines from $S_{N,0}[a, b]$ interpolating the values of the function at N points (cf. [9A, p. 213]).

10. Obtain (2.3) from Theorems 7.2.6 and 4.2.1.

11. For polynomials (2.2.20) prove

$$\sup_{f \in H^\omega} \|f - U_{n-1}(f, \lambda)\|_C \ge \frac{n}{\pi} \int_0^{\pi/n} \omega(t) \, dt.$$

8

N-widths of functional classes and closely related extremal problems

In Chapters 4 to 7 we considered approximation problems in which the approximating set (or the sequence of sets) is fixed. In particular, exact results for estimating the approximation error in functional classes by elements of finite dimensional polynomial or spline subspaces were obtained. Is it possible to improve these estimates by changing the approximating subspace to another of the same dimension? And which estimates cannot be improved on the whole set of N-dimensional approximating subspaces?

Here, we are referring to the problem stated in Section 1.2 of finding the N-widths of functional classes $\mathfrak{M}$ in a normed space X. The exact results from the previous chapters give upper bounds for the N-widths in the corresponding cases and now our attention will be concentrated on lower bounds for Kolmogorov N-widths. But a lower bound for the best approximations of the class $\mathfrak{M}$ which is simultaneously valid for all N-dimensional approximating subspaces can only be obtained by using some very general and deep result. Borsuk's topological Theorem 2.5.1 stated in Section 2.5 turns out to be a suitable result in many cases.

In order to make the application of this theorem both possible and effective, one has to identify an $(N+1)$ parametric set M_{N+1} in $\mathfrak{M}$ whose N-width is not less than the N-width of $\mathfrak{M}$. This depends on the metric of the space X and on the way of defining $\mathfrak{M}$. In the various cases the role of M_{N+1} may, for example, be played by a ball in subspaces of polynomials or splines, as some sets of perfect splines or their generalizations.

The main results in Chapter 8 are those in which exact or asymptotically exact solutions for N-widths problems for classes of functions are found. Some questions closely related with this problem are the minimization of the exact constant in Jackson inequalities, optimal coding and optimal recovery of functions and linear functionals with discrete information, which is treated in Section 8.3. We shall see that, in almost all cases when we are able to get an exact solution of the stated problems, it turns out that the polynomial or spline subspaces are extremal. By the way, polynomials and splines were widely used as approximation tools in theoretical and applied mathematics long before the facts mentioned above were discovered.

8.1 N-widths of classes of functions with bounded rth derivatives

8.1.1 The theorem for the N-width of the ball and its applications Recall the definition of a (Kolmogorov) N-width $d_N(\mathcal{M}, X)$ of a central symmetric set $\mathcal{M}$ in the linear normed space X (see Section 1.2):

$$d_0(\mathcal{M}, X) = \sup \{\|x\|: x \in \mathcal{M}\},$$
$$d_N(\mathcal{M}, X) = \inf_{\mathcal{N}_N} E(\mathcal{M}, \mathcal{N}_N)_X = \inf_{\mathcal{N}_N} \sup_{x\in\mathcal{M}} \inf_{u\in\mathcal{N}_N} \|x - u\|_X, \qquad N = 1, 2, \ldots,$$

where the infimum is taken on all subspaces $\mathcal{N}_N$ of dimension N.

An exact lower estimate for $d_N(\mathcal{M}, X)$ in many particular cases can be obtained using a statement known as the theorem for the N-width of the ball whose proof is based on Borsuk's Theorem 2.5.1.

Theorem 8.1.1 Let M_{N+1} be an $(N + 1)$-dimensional linear manifold in the normed space X and U_{N+1} be the closed unit ball in M_{N+1}, i.e.

$$U_{N+1} = \{x: x \in M_{N+1}, \|x\| \le 1\}.$$

Then $d_N(U_{N+1}, X) = 1$.

Proof Because $d_N(U_{N+1}, X) \le d_0(U_{N+1}, X) = 1$ we have to show that for every N-dimensional subspace $\mathcal{N}_N$ of X we have $E(U_{N+1}, \mathcal{N}_N)_X \ge 1$.

Fix $\mathcal{N}_N$ and consider the case when the norm of X is strictly normalized (see Section 1.1) first. Then, in view of Propositions 1.1.4 and 1.1.5, there exists for every $x \in X$ a unique element of best approximation $u = P(x)$ in the subspace $\mathcal{N}_N$.

The operator P is continuous and odd (see Proposition 1.1.6) and hence, if we consider it on the sphere $S^N = \{x: x \in M_{N+1}, \|x\| = 1\}$, we obtain from Theorem 2.5.1 that there exists $z \in S^N$ such that $P(z) = 0$. Because $S^N \subset U_{N+1}$ we have

$$E(U_{N+1}, \mathcal{N}_N)_X \ge E(z, \mathcal{N}_N)_X = \|z - P(z)\| = \|z\| = 1$$

and the theorem is proved.

If X is not strictly normalized, we can reduce our considerations to the case of a rotund norm. First we replace X by its finite dimensional subspace Y containing the linear manifolds M_{N+1} and $\mathcal{N}_N$. After that we approximate the norm $\|\bullet\|$ by a strict norm as follows. Let $\|\bullet\|'$ be a strictly normalized (e.g. Euclidian) norm in Y; then in view of the isomorphism of the finite dimensional spaces we have $\|x\|' \leq \beta\|x\|$ for some $\beta > 0$ and every $x \in Y$. For $\varepsilon > 0$ we set

$$\|x\|^0 = \|x\| + \varepsilon\|x\|'/\beta$$

and get a rotund norm $\|\bullet\|^0$ in Y such that

$$(1+\varepsilon)^{-1}\|x\|^0 \leq \|x\| \leq \|x\|^0, \qquad x \in Y. \tag{1.1}$$

In view of the above there is an element $z \in M_{N+1}$ such that

$$\inf_{u \in \mathcal{N}_N} \|z - u\|^0 = \|z\|^0 = 1,$$

and going back to the original norm $\|\bullet\|$ and using (1.1) we have

$$E(z, \mathcal{N}_N)_Y = \inf_{u \in \mathcal{N}_N} \|z - u\| \geq (1+\varepsilon)^{-1} > 1 - \varepsilon.$$

Moreover (1.1) also gives $\|z\| \leq 1$, i.e. $z \in U_{N+1}$, and hence

$$E(U_{N+1}, \mathcal{N}_N)_X \geq E(U_{N+1}, \mathcal{N}_N)_Y \geq E(z, \mathcal{N}_N)_Y > 1 - \varepsilon,$$

which implies $E(U_{N+1}, \mathcal{N}_N)_X \geq 1$ because ε was arbitrary positive. The theorem is proved. □

The application of Theorem 8.1.1 for obtaining lower estimates for the N-widths of particular sets is facilitated by the following immediate corollary:

Corollary 8.1.2 If the set $\mathcal{M}$ of the linear normed space X contains the ball γU_{N+1} of radius γ of some $(N+1)$-dimensional subspace then $d_N(\mathcal{M}, X) \geq \gamma$.

Indeed if $\varphi U_{N+1} \subset \mathcal{M}$ then using Theorem 8.1.1 and the definition of d_N we obtain

$$d_N(\mathcal{M}, X) \geq d_N(\gamma U_{N+1}, X) = \gamma d_N(U_{N+1}, X) = \gamma.$$

Before applying this, let us also recall the definitions of linear and projective N-widths of the central symmetric set $\mathcal{M}$ in the normed space X (see Section 1.2):

$$\lambda_N(\mathcal{M}, X) = \inf_{\mathcal{N}_N} \inf_{A \in \mathcal{L}(X, \mathcal{N}_N)} \sup_{x \in \mathcal{M}} \|x - Ax\|,$$

$$\pi_N(\mathcal{M}, X) = \inf_{\mathcal{N}_N} \inf_{A \in \mathcal{L}\perp(X, \mathcal{N}_N)} \sup_{x \in \mathcal{M}} \|x - Ax\|,$$

where $\mathcal{L}(X, \mathcal{N}_N)$ is the set of all linear bounded operators from X into $\mathcal{N}_N$, where $\mathscr{L}\perp(X,\mathscr{N}_N)$ is the subspace of projections in $\mathscr{L}(X,\mathscr{N}_N)$ and where the outer infimum is taken on all N-dimensional subspaces $\mathscr{N}_N$ of X. We shall make use of the monotonicity of the N-widths on N and also of the bands $d_N \le \lambda_N \le \pi_N$.

The first application of Theorem 8.1.1 is to the estimate for the N-widths of W_2^r $(r=1,2,\ldots,)$ in L_2. As before we denote

$$W_p^r = \{f: f\in L_p^r, \|f_p^{(r)}\|\le 1\}, \qquad r=1,2,\ldots; \qquad 1\le p\le\infty.$$

Let $n^{-r}U_{2n+1}^T$ be the intersection of the ball of radius n^{-r} in L_2 with the $(2n+1)$-dimensional subspace $\mathscr{N}_{2n+1}^T$ of trigonometric polynomials

$$f(t) = \frac{\alpha_0}{2} + \sum_{k=1}^{n} (\alpha_k \cos kt + \beta_k \sin kt),$$

i.e. $n^{-r}U_{2n+1}^T = \{f: f\in\mathscr{N}_{2n+1}^T, \|f\|_2 \le n^{-r}\}$. Thus for all $f\in n^{-r}U_{2n+1}^T$

$$\pi\left[\frac{\alpha_0^2}{2} + \sum_{k=1}^{n} (\alpha_k^2+\beta_k^2)\right] \le n^{-2r},$$

in view of Parseval's inequality. Applying this inequality we get for $f^{(r)}$

$$\|f^{(r)}\|_2^2 = \pi\sum_{k=1}^{n} k^{2r}(\alpha_k^2+\beta_k^2) \le \pi n^{2r}\sum_{k=1}^{n}(\alpha_k^2+\beta_k^2) \le 1.$$

This means that the ball $n^{-r}U_{2n+1}^T$ is contained in class W_2^r. Now Proposition 8.1.2 implies

$$d_{2n}(W_2^r, L_2) \ge n^{-r}.$$

On the other hand, for the approximation of the function $f\in W_2^r$ by the partial sums of Fourier series

$$S_{n-1}(f,t) = \frac{a_0}{2} + \sum_{k=1}^{n-1}(a_k\cos kt + b_k \sin kt), \qquad S_{n-1}(f,t)\in\mathscr{N}_{2n-1}^T,$$

we have (see Theorem 4.2.2) $\|f - S_{n-1}(f)\|_2 \le n^{-r}$. This gives an upper bound for the projective width: $\pi_{2n-1}(W_2^r, L_2)\le n^{-r}$. Hence:

Theorem 8.1.3 For all $n, r = 1, 2, \ldots,$ and for either $N=2n-1$ or $N=2n$ we have

$$d_N(W_2^r, L_2) = \lambda_N(W_2^r, L_2) = \pi_N(W_2^r, L_2) = n^{-r}. \tag{1.2}$$

All N-widths are realized by the trigonometric Fourier sums $S_{n-1}(f,t)$.

Let $H_2^{1/2}$ be the class of functions $f\in L_2$ such that

$$\omega(f,\delta)_2 := \sup_{|u|\le\delta} \|f(\bullet+u)-f(\bullet)\|_2 \le \delta^{1/2}.$$

For the subspace M_{N+1} from Theorem 8.1.1 we take the $2n$-dimensional linear manifold $S_{2n,0}$ of step functions with knots at the uniform partition $k\pi/n$ ($k = 0, 1, \ldots, 2n$), i.e. the set of functions $f \in L_\infty$ of the type

$$f(t) = \sum_{k=1}^{2n} c_k\, \psi_k(t), \tag{1.3}$$

where $\psi_k(t)$ is defined on $[0, 2\pi)$ by

$$\psi_k(t) = \begin{cases} 1, & (k-1)\pi/n \le t < k\pi/n, \\ 0, & 0 \le t < (k-1)\pi/n, \qquad k\pi/n \le t < 2\pi. \end{cases}$$

Consider the ball

$$\gamma_n U_{2n} = \{f: f \in S_{2n,0}, \|f\|_2 \le \gamma_n\},$$

where $\gamma_n = \frac{1}{2}(\pi/n)^{1/2}$. If $f \in \gamma_n U_{2n}$ then, employing representation (1.3), we obtain

$$\|f\|_2^2 = \frac{\pi}{n} \sum_{k=1}^{2n} c_k^2 \le \gamma_n^2 = \frac{1}{4}\frac{\pi}{n},$$

i.e. $c_1^2 + c_2^2 + \ldots + c_{2n}^2 \le \frac{1}{4}$. Also we get from the type of function (1.3) for $0 < \delta < \pi/n$

$$\omega^2(f,\delta)_2 = \|f(\bullet+\delta)-f(\bullet)\|_2^2 = \delta \sum_{k=1}^{2n} (c_{k+1}-c_k)^2, \qquad c_{2n+1} = c_1,$$

and because

$$\sum_{k=1}^{2n} (c_{k+1}-c_k)^2 \le 2\sum_{k=1}^{2n} (c_{k+1}^2 + c_k^2) = 4\sum_{k=1}^{2n} c_k^2$$

we finally obtain

$$\omega(f,\delta)_2 \le [(c_1^2 + c_2^2 + \ldots + c_{2n}^2)\delta]^{1/2} \le \delta^{1/2}.$$

If $\pi/n \le \delta \le \pi$ then we have for $f \in \gamma_n U_{2n}$

$$\omega(f,\delta)_2 \le 2\|f\|_2 \le 2\gamma_n = (\pi/n)^{1/2} \le \delta^{1/2}.$$

This proves $\gamma_n U_{2n} \subset H_2^{1/2}$ and also

$$d_{2n-1}(H_2^{1/2}, L_2) \ge \gamma_n = \frac{1}{2}\left(\frac{\pi}{n}\right)^{1/2}, \tag{1.4}$$

in view of Proposition 8.1.8.

Theorem 6.2.4 implies for $f \in H_2^{1/2}$

$$E_n(f)_2 = \|f - S_{n-1}(f)\|_2 \le \frac{1}{2}\left(\frac{\pi}{n}\right)^{1/2}, \qquad n = 1, 2, \ldots,$$

and hence $\pi_{2n-1}(H_2^{1/2}, L_2) \le 2^{-1}(\pi/n)^{1/2}$. Therefore:

Theorem 8.1.4 For all $n = 1, 2, \ldots$, we have

$$d_{2n-1}(H_2^{1/2}, L_2) = \lambda_{2n-1}(H_2^{1/2}, L_2) = \pi_{2n-1}(H_2^{1/2}, L_2) = \frac{1}{2}\left(\frac{\pi}{n}\right)^{1/2}.$$

The widths are realized by the partial sums of Fourier series $S_{n-1}(f, t)$.

With the help of Theorem 8.1.1 we can easily bound the widths d_{2n-1} of classes W_∞^r in C and W_1^r in L_1 from below. Here we construct balls in the $2n$-dimensional subspace $S_{2n,m}$ of splines of degree $m = r$ or $m = r - 1$ and defect 1 with respect to the uniform partition $k\pi/n$ ($k = 0, 1, \ldots, 2n$).

Let

$$K_r n^{-r} U_{2n,C} = \{f: f \in S_{2n,r}, \|f\|_C \le K_r n^{-r}\},$$

where K_r is the Favard constant. But if $f \in K_r n^{-r} U_{2n,C}$ then Proposition 3.4.1 implies $\|f^{(r)}\|_\infty \le 1$, i.e. the ball $K_r n^{-r} U_{2n,C}$ belongs to W_∞^r. Now from Proposition 8.1.2 we get

$$d_{2n-1}(W_\infty^r, C) \ge K_r n^{-r}, \qquad n, r = 1, 2, \ldots,. \tag{1.5}$$

We will show that a similar estimate is valid for the widths of W_1^r in L_1. But first we note that Proposition 1.5.1 implies

$$d_N(W_1^r, L_q) = d_N(W_V^{r-1}, L_q),$$

where

$$W_V^m = \{f: f \in V^m, \bigvee_0^{2\pi} f^{(m)} \le 1\}.$$

From Proposition 3.4.1 we obtain that the ball

$$K_r n^{-r} U_{2n,L_1} = \{f: f \in S_{2n,r}, \|f\|_1 \le K_r n^{-r}\}$$

is contained in class W_V^{r-1} and hence in view of Proposition 8.1.2 we have

$$d_{2n-1}(W_1^r, L_1) = d_{2n-1}(W_V^{r-1}, L_1) \ge K_r n^{-r}, \qquad n, r = 1, 2, \ldots. \tag{1.6}$$

The upper estimates of the odd linear widths λ_{2n-1} in the same cases and with the same right-hand sides as in (1.5) and (1.6) are given by Theorem 4.2.1 in which the results for approximation by Favard sums were formulated. Thus we have:

Theorem 8.1.5 For all $n, r = 1, 2, \ldots$, we have

$$d_{2n-1}(W^r_\infty, C) = \lambda_{2n-1}(W^r_\infty, C) = d_{2n-1}(W^r_1, L_1) = \lambda_{2n-1}(W^r_1, L_1)$$
$$= d_{2n-1}(W^{r-1}_V, L_1) = \lambda_{2n-1}(W^{r-1}_V, L_1) = K_r n^{-r}.$$

The widths are realized by Favard sums.

8.1.2 Minimization of the L_q norm of perfect splines In the cases considered in Section 8.1.1 we were able to embed in class $\mathcal{M}$ an $(N+1)$-dimensional ball with its radius coinciding with the N-width of $\mathcal{M}$. Usually this is not the case and we then have to try to identify another $(N+1)$ parametric set in $\mathcal{M}$ with the same N-width as $\mathcal{M}$.

When finding the widths of the classes W^r_∞ in L_q such a set is constructed from perfect splines and the way to the exact lower estimate for $d_N(W^r_\infty, L_q)$ is by means of the minimization of the L_q norm of the splines from this set. In Section 3.4 (see Corollary 3.4.10 and Proposition 3.4.12) we have established that the spline with the least norm in C or L_1 in the set $\Gamma_{2n,r}$ of 2π-periodic perfect splines of degree r with no more than $2n$ (free) knots on the period is the Euler spline $\varphi_{n,r}$. Now using Σ-rearrangements (see Section 7.1) we shall show that in $\Gamma_{2n,r}$ the spline $\varphi_{n,r}$ has the least L_q norm for all $1 \le q \le \infty$. This fact together with Borsuk's theorem will allow us to bound (exactly) the widths of classes W^r_∞ in L_q and W^r_p in L_1 from below.

Under the assumption $f \in D^1_0[a, b]$ (or, in particular, $f \in C^1[a, b]$ or $f \in V^1[a, b]$ and $f(a) = f(b) = 0$) it was established in Section 7.1 that the function f could be represented as a finite or countable sum of simple functions φ_k:

$$f(t) = \sum_k \varphi_k(t), \qquad a \le t \le b, \tag{1.7}$$

such that relations (7.1.21) and (7.1.22) hold.

Lemma 8.1.6 If $f \in D^1_0[a, b]$, $f'(t) = 0$ for a finite number of points and the number $\mu(f') = \mu(f'; (a, b))$ of essential sign changes in sign of f' on (a, b) is finite then there is a Σ-rearrangement (1.7) with no more than $\mu(f')$ simple functions φ_k.

Proof First assume that f preserves sign on (a, b), e.g. $f(t) > 0$ $(a < t < b)$. If $\mu(f') = 1$ then the assertion is trivial because f is a simple function. Assuming that the lemma holds for $\mu(f') = 1, 2, \ldots, k$ we shall prove its validity for $\mu(f') = k + 1$.

Let y_0 be the smallest of the numbers y for which the difference $f(t) - y$ has more than two zeros on (a, b). It is clear that $y_0 > 0$. Moreover, if c, d are outer zeros of the function $f_1(t) = f(t) - y_0$ on (a, b) then the function φ_1

coinciding with f on $[a, c]$ and $[d, b]$ and with y_0 on $[c, d]$ is simple (remember that $f(a) = f(b) = 0$) and

$$f(t) = \begin{cases} \varphi_1(t) + f_1(t), & c \leq t \leq d, \\ \varphi_1(t), & t \in [a, c] \cup [d, b]. \end{cases} \tag{1.8}$$

Then, let (α_j, β_j) $(j = 1, 2, \ldots, m)$ be the collection of all disjoint intervals from (c, d) such that $f_1(\alpha_j) = f_1(\beta_j) = 0, f_1(t) > 0$ $(\alpha_j < t < \beta_j)$. On each of these intervals function f_1 satisfies the condition of the lemma and the number k_j of essential changes of sign of f_1' on (α_j, β_j) is less than k. In view of the induction hypotheses the function f_1 is Σ-represented on (α_j, β_j) as a sum of no more than k_j simple functions. In fact, because $k_1 + k_2 + \cdots + k_m \leq k$ we see from (1.8) that f is Σ-represented on $[a, b]$ as a sum of no more than $(k + 1)$ simple functions.

If f vanishes in (a, b) then we separate the intervals $(a_i, b_i) \subset (a, b)$ $(i = 1, 2, \ldots, p)$ on which $|f(t)| > 0$ and $f(a_i) = f(b_i) = 0$. We then apply to each of them what we have proved above and use the fact that $\mu(f'; [a_1, b_1]) + \ldots + \mu(f'; [a_p, b_p]) \leq \mu(f')$. The lemma is proved. □

The lemma is also true for the periodic case when the function $f \in D_0^1$ considered on $[a, a + 2\pi]$, $f(a) = 0$, corresponds to the Σ-rearrangement

$$R(f, t) = \sum_k r(\varphi_k, t), \qquad 0 \leq t \leq 2\pi, \tag{1.9}$$

where $r(\varphi_k, t)$ is the non-increasing rearrangement of the simple function φ_k.

Proposition 8.1.7 Let $f \in D_0^1$ and f be Σ-representable (on the period) as a sum of m simple functions. Then

$$\int_0^t R(f, u)\,\mathrm{d}u \leq \int_0^{mt} r(f, u)\,\mathrm{d}u, \qquad 0 \leq t \leq \frac{2\pi}{m},$$

where $r(f, t)$ is the non-increasing rearrangement of $|f|$ on the period.

Proof For fixed $t \in (0, 2\pi/m)$ we obtain by integrating (1.9)

$$\int_0^t R(f, u)\,\mathrm{d}u = \sum_{k=1}^m \int_0^t r(\varphi_k, u)\,\mathrm{d}u.$$

From the definition of the simple function φ_k it follows that there exists an interval (a_k, b_k) such that $b_k - a_k = t$ and

$$\int_0^t r(\varphi_k, u)\,\mathrm{d}u = \int_{a_k}^{b_k} |\varphi_k(u)|\,\mathrm{d}u.$$

If $e = \cup_{k=1}^{m} (a_k, b_k)$ then meas $e \leq mt$ and, using Proposition 3.2.4, we get

$$\int_0^t R(f,u)\,du = \sum_{k=1}^{m} \int_{a_k}^{b_k} |\varphi_k(u)|\,du = \int_e |f(u)|\,du$$

$$\leq \int_0^{\text{meas}\, e} r(f,u)\,du \leq \int_0^{mt} r(f,u)\,du.$$

Theorem 8.1.8 For all $n = 1, 2, \ldots,; r = 0, 1, \ldots,$ we have

$$\inf_{f \in \Gamma_{2n,r}} \|f\|_q = \|\varphi_{n,r}\|_q, \qquad 1 \leq q \leq \infty. \tag{1.10}$$

Proof For $r = 0$ relation (1.10) is trivial; hence we assume $r \geq 1$. Because $\varphi_{n,r} \in \Gamma_{2n,r}$ we have to prove that we have for every $f \in \Gamma_{2n,r}$

$$\|f\|_q \geq \|\varphi_{n,r}\|_q, \tag{1.11}$$

where we may assume that f vanishes somewhere. Fix $f \in \Gamma_{2n,r}$ and let $f(a) = 0$. If $g(t) = f(t)/(4n)$ then $g \in W_V^r$ and the derivative g' ($g' \in V$) changes sign no more than $2n$ times. This can easily be proved by arguing by contradiction and by applying Rolle's theorem. In view of Lemma 8.1.6 there exists a Σ-representation

$$g(t) = \sum_{k=1}^{m} \varphi_k(t), \qquad a \leq t \leq a + 2\pi,$$

where $m \leq 2n$. Proposition 8.1.7 implies

$$\int_0^t r(g,u) \geq \int_0^{t/m} R(g,u)\,du \geq \int_0^{t/(2n)} R(g,u)\,du, \qquad 0 \leq t \leq 2\pi. \tag{1.12}$$

We estimate the last integral using the corresponding integral of the standard Σ-rearrangement $R_{\alpha,r}$ (see Section 7.1). We choose $\alpha > 0$ from the condition $R(g,0) = R_{\alpha,r}(0)$ which in view of (7.1.25) can be written as

$$\|g'\|_1 = 2\,R_{\alpha,r}(0).$$

By Proposition 3.4.12 and by (7.1.32) we have

$$\|g'\|_1 = \frac{1}{4n}\|f'\|_1 \geq \frac{1}{4n}\|\varphi_{n,r-1}\|_1 = \|g_{n,r-1}\|_1 = 2R_{\pi/n,r}(0).$$

Hence $\alpha \geq \pi/n$ and

$$R_{\alpha,r}(t) \geq R_{\pi/n,r}(t), \qquad 0 \leq t \leq 2\pi.$$

Now using Theorem 7.1.4 and relation (7.1.31) for $z \in [0, \pi/n]$ we have

$$\int_0^z R(g,u)\,\mathrm{d}u \geq \int_0^z R_{\alpha,r}(u)\,\mathrm{d}u \geq \int_0^z R_{\pi/n,r}(u)\,\mathrm{d}u = \int_0^{2nz} r(g_{n,r},t)\,\mathrm{d}t.$$

Comparing with (1.12) this gives

$$\int_0^t r(g,u)\,\mathrm{d}u \geq \int_0^t r(g_{n,r},t)\,\mathrm{d}t, \qquad 0 \leq t \leq 2\pi,$$

which implies $\|g\|_q \geq \|g_{n,r}\|_q$ $(1 \leq q \leq \infty)$ in view of Proposition 3.2.5. For the function $f(t) = 4ng(t)$ this gives (1.11) and the theorem is proved. □

8.1.3 Widths of classes W_∞^r in L_q Besides Theorem 8.1.8 in obtaining the lower bound of $d_{2n}(W_\infty^r, L_q)$ a key rôle is played by:

Lemma 8.1.9 Let $\mathcal{N}_{2n}$, $\dim \mathcal{N}_{2n} = 2n$, be a subspace of L_q $(q \geq 1)$ containing a constant. Then there is a $f_* \in \Gamma_{2n,r}$ $(r = 1, 2, \ldots,)$ such that

$$E(f_*, \mathcal{N}_{2n})_q = \|f_*\|_q.$$

If $r = 0$ then the above is true for every subspace $\mathcal{N}_{2n}$.

Proof In the $(N+1)$-dimensional space $\mathbb{R}^{N+1}$ of vectors $\boldsymbol{\xi} = \{\xi_1, \xi_2, \ldots, \xi_{N+1}\}$ with norm

$$\|\boldsymbol{\xi}\| = \sum_{i=1}^{N+1} |\xi_i|$$

consider the sphere

$$S^N = \{\boldsymbol{\xi}: \boldsymbol{\xi} \in \mathbb{R}^{N+1}, \|\boldsymbol{\xi}\| = 2\pi\}.$$

To every vector $\boldsymbol{\xi} \in S^N$ there is a system of corresponding non-negative numbers (we have also done this in Section 2.9)

$$\tau_0 = 0, \qquad \tau_k = \sum_{i=1}^{k} |\xi_i|, \qquad k = 1, 2, \ldots, N+1; \qquad \tau_{N+1} = 2\pi.$$

Now the function $g_0(\boldsymbol{\xi}, t)$ is defined on $[0, 2\pi]$ in the following way:

$$g_0(\boldsymbol{\xi}, \tau_k) = 0, \qquad k = 1, 2, \ldots, N, \tag{1.13}$$

and if $\xi_k \neq 0$ then

$$g_0(\boldsymbol{\xi}, t) = \operatorname{sgn} \xi_k, \qquad \tau_{k-1} < t < \tau_k, \qquad k = 2, 3, \ldots, N;$$

$$g_0(\boldsymbol{\xi}, t) = \operatorname{sgn} \xi_1, \qquad 0 \leq t < \tau_1;$$

$$g_0(\boldsymbol{\xi}, t) = \operatorname{sgn} \xi_{N+1}, \qquad \tau_N < t \leq 2\pi. \tag{1.14}$$

Thus $g_0(\boldsymbol{\xi}, t)$ is a perfect spline on $[0, 2\pi]$ of zero degree with no more than N knots on $(0, 2\pi)$. It is clear that we have for every $\boldsymbol{\xi} \in S^N$

$$g_0(-\boldsymbol{\xi}, t) = -g_0(\boldsymbol{\xi}, t), \qquad 0 \le t \le 2\pi.$$

Set

$$g_j(\boldsymbol{\xi}, t) = \int_0^t g_{j-1}(\boldsymbol{\xi}, u)\,du + c_j(\boldsymbol{\xi}), \qquad 0 \le t \le 2\pi, \qquad j = 1, 2, \ldots, r, \tag{1.15}$$

where

$$c_j(\boldsymbol{\xi}) = -\frac{1}{2\pi}\int_0^{2\pi}\left(\int_0^z g_{j-1}(\boldsymbol{\xi}, u)\,du\right)dz, \qquad j = 1, 2, \ldots, r-1, \tag{1.16}$$

(hence $g_j(\boldsymbol{\xi}, t) \perp 1$ for $j = 1, 2, \ldots, r-1$) and $c_r(\boldsymbol{\xi}) = 0$. Now for a fixed $q \ge 1$ set

$$f(\boldsymbol{\xi}, t) = g_r(\boldsymbol{\xi}, t) + c(\boldsymbol{\xi}),$$

where the constant $c(\boldsymbol{\xi})$ is uniquely determined from the condition

$$\int_0^{2\pi} |f(\boldsymbol{\xi}, t)|^{q-1}\operatorname{sgn} f(\boldsymbol{\xi}, t)\,dt = 0. \tag{1.17}$$

Hence in view of Proposition 1.4.6 we have

$$\|g_r(\boldsymbol{\xi}) + c(\boldsymbol{\xi})\|_q = \inf_{\lambda}\|g_r(\boldsymbol{\xi}) - \lambda\|_q. \tag{1.18}$$

From the definitions it is clear that $f(\boldsymbol{\xi}, t)$ is a perfect spline on $[0, 2\pi]$ of degree r and we have for every $\boldsymbol{\xi} \in S^N$

$$f(-\boldsymbol{\xi}, t) = -f(\boldsymbol{\xi}, t), \qquad 0 \le t \le 2\pi.$$

What is important for us is the continuous dependence of $f(\boldsymbol{\xi}, t)$ on $\boldsymbol{\xi} \in S^N$. Specifically, that if a sequence of vectors $\boldsymbol{\xi}^{(m)} = \{\xi_1^{(m)}, \xi_2^{(m)}, \ldots, \xi_{N+1}^{(m)}\}$ $(m = 1, 2, \ldots,)$ from S^N converges in the metric of $\mathbb{R}^{N+1}$ to $\boldsymbol{\xi} \in S^N$ $(\boldsymbol{\xi} = \{\xi_1, \xi_2, \ldots, \xi_{N+1}\})$ then

$$\lim_{m\to\infty}\|f(\boldsymbol{\xi}^{(m)}) - f(\boldsymbol{\xi})\|_C = 0. \tag{1.19}$$

This fact follows (we leave the details to the reader) from the definition of the function $f(\boldsymbol{\xi}, t)$ and from the relations

$$\lim_{m\to\infty}\operatorname{meas}\{t\colon t \in [0, 2\pi],\, g_0(\boldsymbol{\xi}^{(m)}, t) \ne g_0(\boldsymbol{\xi}, t)\} = 0, \qquad \lim_{m\to\infty} c(\boldsymbol{\xi}^{(m)}) = c(\boldsymbol{\xi}), \tag{1.20}$$

the last of which follows from (1.18) and from the continuity of the functional of best approximation.

Assuming $N = 2n$, $r \ge 1$, we fix the subspace $\mathcal{N}_{2n}$ of L_q which contains a constant and we let $\{e_1(t), e_2(t), \ldots, e_{2n}(t)\}$, where $e_1(t) \equiv 1$ be a basis.

Define a $2n$-dimensional vector field $\eta(\xi) = \{\eta_1(\xi), \eta_2(\xi), \ldots, \eta_{2n}(\xi)\}$ on S^{2n} by

$$\eta_1(\xi) = \int_0^{2\pi} g_0(\xi, t)\, dt,$$

$$\eta_k(\xi) = \int_0^{2\pi} e_k(t)\, |f(\xi, t)|^{q-1} \operatorname{sgn} f(\xi, t)\, dt, \qquad k = 2, 3, \ldots, 2n.$$

The field η is odd, i.e. $\eta(-\xi) = -\eta(\xi)$. This follows immediately from the definition of η and the oddness with respect to ξ of functions $g_0(\xi, t), f(\xi, t)$. The field η is also continuous on S^N, i.e. if $\xi^{(m)} \to \xi$ $(\xi^{(m)} \in S^N, \xi \in S^N)$ then

$$\lim_{m\to\infty} \eta_k(\xi^{(m)}) = \eta_k(\xi), \qquad k = 1, 2, \ldots, 2n.$$

For $k = 1$ this follows from (1.20); for $k = 2, 3, \ldots, 2m$ when estimating the difference $|\eta_k(\xi^{(m)}) - \eta_k(\xi)|$ we use (1.19) and the relation

$$\lim_{m\to\infty} \operatorname{meas} \{t: t \in [0, 2\pi], \operatorname{sgn} f(\xi^{(m)}, t) \neq \operatorname{sgn} f(\xi, t)\} = 0.$$

Thus on the sphere S^{2n} of the $(2n + 1)$-dimensional space $\mathbb{R}^{2n+1}$ we have an odd and continuous field $\eta(\xi) = \{\eta_1(\xi), \eta_2(\xi), \ldots, \eta_{2n}(\xi)\}$ and in view of Theorem 2.5.1 there is a vector $\xi^* \in S^{2n}$ such that $\eta(\xi^*) = 0$, i.e. $\eta_k(\xi^*) = 0$ $(k = 1, 2, \ldots, 2n)$. This implies

$$\eta_1(\xi^*) = \int_0^{2\pi} g_0(\xi^*, t)\, dt = 0$$

and hence, taking into account the choice of constants (1.16), the functions $g_j(\xi^*, t)$ satisfy the equalities

$$g_j(\xi^*, 0) = g_j(\xi^*, 2\pi), \qquad j = 1, 2, \ldots, r.$$

Therefore we can continue the function $f_*(t) = f(\xi^*, t)$ as 2π-periodic and so $f_* \in \Gamma_{2n,r}$.

From

$$\eta_k(\xi^*) = \int_0^{2\pi} e_k(t) |f(\xi^*, t)|^{q-1} \operatorname{sgn} f(\xi^*, t)\, dt = 0, \qquad k = 2, 3, \ldots, 2n,$$

and from equality (1.7) which is valid for every $\xi \in S^{2n}$ it follows (by Proposition 1.4.6) that the best approximant for f_* in $\mathcal{N}_{2n}$ is the constant zero, i.e. $E(f_*, \mathcal{N}_{2n})_q = \|f_*\|_q$.

In the case $r = 0$ we defined the vector field η by using the function $g_0(\xi, t)$ and the basis $\{e_k\}$ of an arbitrary subspace $\mathcal{N}_{2n}$ and by the equalities

$$\eta_k(\xi) = \int_0^{2\pi} e_k(t) |g_0(\xi, t)|^{q-1} \operatorname{sgn} g_0(\xi, t)\, dt, \qquad k = 1, 2, \ldots, 2n.$$

We then set $f_*(t) = g_0(\xi^*, t)$ where $\eta(\xi^*) = 0$. The lemma is proved. □

The lemma reduces the problem of bounding the widths of classes W^r_∞ from below to the one solved in Theorem 8.1.8, i.e. minimizing the norm of the perfect splines from $\Gamma_{2n,r}$.

Proposition 8.1.10 For every subspace $\mathcal{N}_{2n} \subset L_q$ $(1 \le q \le \infty, \dim \mathcal{N}_{2n} = 2n)$ we have

$$E(\Gamma_{2n,r}, \mathcal{N}_{2n})_q \ge \inf_{f \in \Gamma_{2n,r}} \|f\|_q = \|\varphi_{n,r}\|_q, \qquad r = 0, 1, \ldots, \tag{1.21}$$

Proof The set $\Gamma_{2n,r}$ $(r \ge 1)$ contains every function $f(t) + \lambda$ (λ is a constant) together with f. In view of the semi-additivity and positive homogeneity of the functional of best approximation (see Proposition 1.1.1) we have

$$E(f + \lambda, \mathcal{N}_{2n})_q \ge |\lambda|\, E(1, \mathcal{N}_{2n})_q - E(f, \mathcal{N}_{2n})_q.$$

If $\mathcal{N}_{2n}$ does not contain the constants then $E(1, \mathcal{N}_{2n})_q > 0$ and hence $E(\Gamma_{2n,r}, \mathcal{N}_{2n})_q = \infty$. Therefore we may assume $1 \in \mathcal{N}_{2n}$ when proving (1.21).

But then (1.21) for $1 \le q < \infty$ follows immediately from Lemma 8.1.9 and Theorem 8.1.8. The validity of (1.21) for $q = \infty$ can be easily established by passing to the limit. Indeed, for $g \in L_\infty$ $\|g\|_q \le (2\pi)^{1/q}\|g\|_\infty$ and (see e.g. [18A, p. 21]) $\lim_{q\to\infty} \|g\|_q = \|g\|_\infty$. Hence if $f \in \Gamma_{2n,r}$ and $E(f, \mathcal{N}_{2n})_\infty = \|f - \varphi\|_\infty$ $(\varphi \in \mathcal{N}_{2n})$ then

$$(2\pi)^{1/q} E(f, \mathcal{N}_{2n})_\infty \ge \|f - \varphi\|_q \ge E(f, \mathcal{N}_{2n})_q$$

and hence we have for large enough q

$$E(\Gamma_{2n,r}, \mathcal{N}_{2n})_\infty \ge (2\pi)^{-1/q} E(\Gamma_{2n,r}, \mathcal{N}_{2n})_q \ge (2\pi)^{-1/q}\|\varphi_{n,r}\|_q \ge \|\varphi_{n,r}\|_\infty - \varepsilon. \quad \square$$

From the obvious inclusion $\Gamma_{2n,r} \subset W^r_\infty$ and from Proposition 8.1.10 we get the lower bound

$$d_{2n}(W^r_\infty, L_q) \ge \|\varphi_{n,r}\|_q, \qquad 1 \le q \le \infty. \tag{1.22}$$

Now we establish the upper bounds. In view of Theorem 4.2.1

$$\sup_{f \in W^r_\infty} \|f - U_{n-1,r}(f, \lambda^*)\|_C = \|\varphi_{n,r}\|_C = K_r n^{-r}, \tag{1.23}$$

where $U_{n-1,r}(f, \lambda^*, t)$ are Favard sums (see Section 4.1.3) from the subspace $\mathcal{N}^T_{2n-1}$ representing a linear method of approximation. Because $\dim \mathcal{N}^T_{2n-1} = 2n - 1$ we obtain

$$\lambda_{2n-1}(W^r_\infty, C) \le K_r n^{-r}, \qquad n, r = 1, 2, \ldots. \tag{1.24}$$

Moreover it was established in Section 5.2 (Theorem 5.2.6) that

$$\sup_{f \in W^r_\infty} \|f - \sigma_{2n,r-1}(f)\|_q = \|\varphi_{n,r}\|_q, \qquad 1 \le q \le \infty, \tag{1.25}$$

where $\sigma_{2n,m}(f, t)$ are the splines from $S_{2n,m}$ interpolating the function f at the zeros of $\varphi_{n,m+1}$. Because $\dim S_{2n,m} = 2n$ and because $\sigma_{2,m}(f,t)$ is a linear projector (1.25) implies

$$\pi_{2n}(W^r_\infty, L_q) \le \|\varphi_{n,r}\|_q, \qquad 1 \le q \le \infty, \qquad n, r = 1, 2, \ldots. \quad (1.26)$$

Comparing relations (1.22)–(1.26) and recalling $E(W^r_\infty, S_{2n,m})_C = \|\varphi_{n,r}\|_C$ $(m \ge r-1)$ (see Theorem 5.4.8) we come to:

Theorem 8.1.11 For all $n, r = 1, 2, \ldots$, we have

$$d_{2n-1}(W^r_\infty, C) = d_{2n}(W^r_\infty, C) = \lambda_{2n-1}(W^r_\infty, C)$$
$$= \lambda_{2n}(W^r_\infty, C) = \pi_{2n}(W^r_\infty, C) = K_r n^{-r}, \qquad (1.27)$$
$$d_{2n}(W^r_\infty, L_q) = \lambda_{2n}(W^r_\infty, L_q) = \pi_{2n}(W^r_\infty, L_q) = \|\varphi_{n,r}\|_q, \qquad q \ge 1. \quad (1.28)$$

All even widths in (1.27) and (1.28) are realized by the interpolation splines $\sigma_{2n,r-1}(f,t)$; all widths in (1.27) except $\pi_{2n}(W^r_\infty, C)$ are realized by trigonometric Favard sums. Moreover the spline subspaces $S_{2n,m}$ (all $m \ge r-1$) are also extremal for $d_{2n}(W^r_\infty, C)$.

8.1.4 Widths of classes W^r_p in L_1 In this case we can also reduce the problem of binding the widths d_{2n} from below to that of minimizing the norm of perfect splines from $\Gamma_{2n,r}$. In order to do this we use duality arguments.

The class W^r_p $(r = 1, 2, \ldots,)$ contains all constants and hence (see Section 1.5.1) for establishing lower bounds of $d_{2n}(W^r_p, L_1)$ we only have to consider the subspaces $\mathcal{N}_{2n} \subset L_1$ which also contain the constants. In view of Proposition 1.5.4 we have for such subspaces

$$E(W^r_p, \mathcal{N}_{2n})_1 = \sup\{E_1(g)_{p'} : g \in W^r_\infty, g^{(r)} \perp \mathcal{N}_{2n}\}, \qquad p' = p/(p-1), \quad (1.29)$$

where $E_1(g)_q = \inf_\lambda \|g - \lambda\|_q$.

One way of giving a lower bound of the right-hand side of (1.29) is provided by:

Lemma 8.1.12 For every subspace $\mathcal{N}_{2n} \subset L_1$ $(\dim \mathcal{N}_{2n} = 2n)$ containing a constant and for every q $(1 \le q \le \infty)$ there exists a perfect spline $g_* \in \Gamma_{2n,r}$ such that $g^{(r)}_* \perp \mathcal{N}_{2n}$, $E_1(g_*)_q = \|g_*\|_q$.

The proof is based on applying Borsuk's theorem and we repeat the construction from the beginning of the proof of Lemma 8.1.9 including the definition of the function $g_0(\xi, t)$ by equalities (1.13) and (1.14). After that we fix a subspace $\mathcal{N}_{2n} \subset L_1$ with a basis $\{1, e_2(t), \ldots, e_{2n}(t)\}$. The vector field $\eta(\xi) = \{\eta_1(\xi), \ldots, \eta_{2n}(\xi)\}$ on S^{2n} is now defined by

$$\eta_k(\xi) = \int_0^{2\pi} e_k(t)\, g_0(\xi, t)\, dt, \qquad k = 1, 2, \ldots, 2n. \qquad (1.30)$$

It is clear that $\eta(-\xi) = -\eta(\xi)$. The continuity of the field η on S^{2n} can be established more easily than in Lemma 8.1.9: if $\xi^{(m)} \to \xi(\xi^{(m)} \in S^{2n}, \xi \in S^{2n})$ then

$$|\eta_k(\xi^{(m)}) - \eta_k(\xi)| \le \int_0^{2\pi} |e_k(t)|\,|g_0(\xi^{(m)}, t) - g_0(\xi, t)|\,\mathrm{d}t,$$

and in view of (1.20) the right-hand side goes to zero when $m \to \infty$.

From Theorem 2.5.1 if follows that there is a vector $\xi^* \in S^{2n}$ such that $\eta(\xi^*) = 0$. This means (see (1.30)) that $g_0(\xi^*, t) \perp \mathcal{N}_{2n}$ and in particular

$$\eta_1(\xi^*) = \int_0^{2\pi} g_0(\xi^*, t)\,\mathrm{d}t = 0.$$

The last equality allows us to continue $g_0(\xi^*,t)$ as a 2π-periodic function and to define g_* as the rth periodic integral of $g_0(\xi^*, t)$. This integral is defined up to an additive constant and we choose it so that $E_1(g_*)_q = \|g_*\|_q$. From the construction it is clear that $g_* \in \Gamma_{2n,r}$ and $g_*^{(r)} \perp \mathcal{N}_{2n}$. The lemma is proved. □

Because $\Gamma_{2n,r} \subset W_\infty^r$ from (1.29) and from Lemma 8.1.12 it follows that for every $2n$-dimensional subspace $\mathcal{N}_{2n} \subset L_1$ and for all $n,r = 1, 2, \ldots$, we have

$$E(W_p^r, \mathcal{N}_{2n})_1 \ge \inf_{f \in \Gamma_{2n,r}} \|f\|_p.$$

Hence (see Theorem 8.1.8)

$$d_{2n}(W_p^r, L_1) \ge \|\varphi_{n,r}\|_{p'}, \qquad 1 \le p \le \infty, \qquad 1/p + 1/p' = 1. \qquad (1.31)$$

As in the previous section upper bounds for $d_N(W_p^r, L_1), \lambda_N(W_p^r, L_1)$ are provided by the exact estimates of the approximations of classes W_p^r in L_1 by trigonometric polynomials and splines. Theorem 4.2.1 gives

$$\sup_{f \in W_1^r} \|f - U_{n-1}(f, \lambda^*)\|_1 = \|\varphi_{n,r}\|_\infty = K_r n^{-r}, \qquad n, r = 1, 2, \ldots, \qquad (1.32)$$

where, as in (1.23), $U_{n-1}(f, \lambda^*,t)$ are Favard means. Theorem 4.2.4 asserts that we have for the best approximation by subspace $\mathcal{N}_{2n-1}^T$ of trigonometric polynomials of degree $2n-1$ ($\dim \mathcal{N}_{2n-1}^T = 2n-1$)

$$E_n(W_p^r)_1 = \|\varphi_{n,r}\|_{p'}, \qquad p' = p/(p-1). \qquad (1.33)$$

Therefore

$$\lambda_{2n-1}(W_1^r, L_1) \le \|\varphi_{n,r}\|_\infty, \qquad d_{2n-1}(W_p^r, L_1) \le \|\varphi_{n,r}\|_{p'}. \qquad (1.34)$$

Further, in view of Theorems 5.2.8 and 5.4.8 we have for the approximation by spline subspaces $S_{2n,m}$ of dimension $2n$

$$E(W_p^r, S_{2n,m})_1 = \sup_{f \in W_p^r} \|f - \sigma_{2n,r-1}(f)\|_1 = \|\varphi_{n,r}\|_{p'}, \qquad m \ge r-1, \qquad (1.35)$$

where $\sigma_{2n,r-1}(f, t)$ are the interpolating splines from $S_{2n,r-1}$. Hence

$$\pi_{2n}(W_p^r, L_1) \le \|\varphi_{n,r}\|_{p'}. \tag{1.36}$$

Comparing relations (1.31) to (1.36) and recalling equality (1.5) we obtain:

Theorem 8.1.13 For all $n, r = 1, 2, \ldots,$ we have

$$\begin{aligned} d_N(W_1^r, L_1) &= \lambda_N(W_1^r, L_1) = d_N(W_V^{r-1}, L_1) \\ &= \lambda_N(W_V^{r-1}, L_1) = K_r n^{-r}, \quad N = 2n-1, 2n, \end{aligned} \tag{1.37}$$

$$\begin{aligned} d_{2n-1}(W_p^r, L_1) &= d_{2n}(W_p^r, L_1) = \lambda_{2n}(W_p^r, L_1) \\ &= \pi_{2n}(W_p^r, L_1) = \|\varphi_{n,r}\|_{p'}, \\ &1 \le p \le \infty, p' = p/(p-1). \end{aligned} \tag{1.38}$$

All even widths in (1.37) and (1.38) are realized by the interpolation splines $\sigma_{2n,r-1}(f, t)$; all widths in (1.37) are realized by Favard sums. The subspace of trigonometric polynomials of degree $(n-1)$ is extremal for $d_{2n-1}(W_p^r, L_1)$ and spline subspaces $S_{2n,m}$ (all $m \ge r-1$) are extremal for $d_{2n}(W_p^r, L_1)$.

8.1.5 N-widths of classes of functions on a finite interval Here arguments, connected with the application of Borsuk's theorem, allow the problem of finding the lower bounds of the N-widths of classes $W_p^r[a, b]$ to be reduced to that of minimizing the norms of perfect splines. But, unlike the periodic case, the spline of the minimal norm on a finite interval is not effectively constructed. Here we only sketch the proofs without going into technical detail because the scheme of reasoning is the same as in Sections 8.1.1–8.1.3.

In Section 2.5.3 we introduced the set $\Gamma_{N,m}[a, b]$ of perfect splines of degree m with no more than $(N-1)$ knots on (a, b). In Section 5.3.2 it was noticed that there is a spline $\Psi_{N,m}(t) = \Psi_{N,m}(q, t)$ in $\Gamma_{N,m}[a, b]$ with a minimal L_q norm:

$$\|\Psi_{N,m}\|_q = \inf\{\|f\|_q : f \in \Gamma_{N,m}[a, b]\}.$$

(We write $\|\bullet\|_q = \|\bullet\|_{L_q[a,b]}$ if there is no risk of misunderstanding.) It is important that the spline $\Psi_{N,m}(t)$ has, inside (a, b), exactly $(N-1)$ knots and $(N+m-1)$ distinct zeros; the distribution of the knots and zeros depends on q.

It has been established (see Relation 5.3.5) that if $\Delta_N^*[a, b]$ is the partition formed from the knots of $\Psi_{N,r}(q, t)$ then we have for the spline $s(f, t) \in S_{r-1}(\Delta_N^*[a, b])$ interpolating f at the zeros of $\Psi_{N,r}(q, t)$

$$\sup_{f \in W_\infty^r[a,b]} \|f - s(f)\|_q = \|\Psi_{N,r}(q)\|_q, \qquad 1 \le q \le \infty. \tag{1.39}$$

Because $\dim S_{r-1}(\Delta_N^*[a, b]) = N + r - 1$ we have

$$\pi_{N+r-1}(W_\infty^r[a,b], L_q[a,b]) \le \|\Psi_{N,r}(t)\|_q, \qquad 1 \le q \le \infty. \quad (1.40)$$

Let us now turn to the lower bound of Kolmogorov N-widths of $W_\infty^r[a,b]$ in $L_q[a,b]$.

Lemma 8.1.14 For every subspace $\mathcal{N}_{N+r-1} \subset L_q[a,b]$ ($\dim \mathcal{N}_{N+r-1} = N+r-1$) containing all algebraic polynomials of degree $(r-1)$ we have

$$E(\Gamma_{N,r}[a,b], \mathcal{N}_{N+r-1})_q \ge \inf\{\|f\|_q: f \in \Gamma_{N,r}[a,b]\}, \qquad 1 \le q \le \infty. \quad (1.41)$$

Proof First let $1 < q < \infty$. In the space $\mathbb{R}^N$ of vectors $\boldsymbol{\xi} = \{\xi_1, \xi_2, \ldots, \xi_N\}$ with norm $\|\boldsymbol{\xi}\| = |\xi_1| + |\xi_2| + \cdots + |\xi_N|$ we take the sphere $S^{N-1} = \{\boldsymbol{\xi}: \boldsymbol{\xi} \in \mathbb{R}^N, \|\boldsymbol{\xi}\| = b - a\}$ and set $\tau_0 = 0$, $\tau_k = |\xi_1| + |\xi_2| + \cdots + |\xi_k|$ $(k = 1, 2, \ldots, N\}$. To every $\boldsymbol{\xi} \in S^{N-1}$ there corresponds a function $g_0(\boldsymbol{\xi}, t)$ equal to $\operatorname{sgn} \xi_k$ on (τ_{k-1}, τ_k) and zero at points τ_k. The function

$$g(\boldsymbol{\xi}, t) = \frac{1}{(r-1)!} \int_a^b (t-u)_+^{r-1} g_0(\boldsymbol{\xi}, u)\, du, \qquad a \le t \le b, \quad (1.42)$$

belongs to $\Gamma_{N,r}[a,b]$; it is continuous and odd with respect to $\boldsymbol{\xi}$ on S^{N-1}.

Fix the subspace $\mathcal{N}_{N+r-1}$ with basis $\{1, t, \ldots, t^{r-1}, e_1(t), \ldots, e_{N-1}(t))$. For $1 < q < \infty$ $L_q[a,b]$ is strictly normalized and hence for $g(\boldsymbol{\xi}, t)$ there is a unique function in $\mathcal{N}_{N+r-1}$ of best approximation $\psi(\boldsymbol{\xi}, t) = \psi(g(\boldsymbol{\xi}), t)$; it can be written as

$$\psi(\boldsymbol{\xi}, t) = p(\boldsymbol{\xi}, t) + \sum_{k=1}^{N-1} \alpha_k(\boldsymbol{\xi}) e_k(t), \quad (1.43)$$

where $p(\boldsymbol{\xi}, t)$ is a polynomial of degree $(r-1)$ with coefficients depending on $\boldsymbol{\xi}$. The coefficients $\alpha_k(\boldsymbol{\xi})$ in (1.43) form an $(N-1)$-dimensional vector field $\alpha(\boldsymbol{\xi}) = \{\alpha_1(\boldsymbol{\xi}), \ldots, \alpha_{N-1}(\boldsymbol{\xi})\}$ which is continuous and odd on the sphere S^{N-1}. Hence Theorem 2.5.1 implies the existence of $\boldsymbol{\xi}^* \in S^{N-1}$ for which $\alpha_k(\boldsymbol{\xi}^*) = 0$ $(k = 1, 2, \ldots, N-1)$.

Then

$$E(g(\boldsymbol{\xi}^*), \mathcal{N}_{N+r-1})_q = \|g(\boldsymbol{\xi}^*) - \psi(\boldsymbol{\xi}^*)\|_q = \|g(\boldsymbol{\xi}^*) - p(\boldsymbol{\xi}^*)\|_q,$$

and because the difference $g(\boldsymbol{\xi}^*, t) - \psi(\boldsymbol{\xi}^*, t)$ belongs to $\Gamma_{N,r}[a,b]$ we obtain

$$E(\Gamma_{N,r}[a,b], \mathcal{N}_{N+r-1})_q \ge E(g(\boldsymbol{\xi}^*), \mathcal{N}_{N+r-1})_q \ge \inf_{f \in \Gamma_{N,r}[a,b]} \|f\|_q.$$

The validity of (1.41) for $q = 1, \infty$ can be established by passing to the limit. By the way, this is not entirely trivial [10A, p. 264] for $q = 1$. □

Notice that for the lower bound of $d_{N+r-1}(W_\infty^r[a,b], L_q[a,b])$ we can only consider the subspaces $\mathcal{N}_{N+r-1}$ containing the set P_{r-1} of algebraic polynomials of degree $(r-1)$ (see Section 1.5.1). Thus we have in view of Lemma 8.1.14 and the inclusion $\Gamma_{N,r}[a,b] \subset W_\infty^r[a,b]$

$$d_{N+r-1}(W_\infty^r[a,b], L_q[a,b]) \geq \inf_{f \in \Gamma_{N,r}[a,b]} \|f\|_q = \|\Psi_{N,r}(q)\|_q, \qquad 1 \leq q \leq \infty. \tag{1.44}$$

(1.39), (1.40) and (1.44) imply:

Theorem 8.1.15 For all $N = 2, 3, \ldots, r = 1, 2, \ldots$, we have

$$d_{N+r-1}(W_\infty^r[a,b], L_q[a,b]) = \lambda_{N+r-1}(W_\infty^r[a,b], L_q[a,b]) \\ = \pi_{N+r-1}(W_\infty^r[a,b], L_q[a,b]) = \|\Psi_{N,r}(q)\|_q. \tag{1.45}$$

The widths are realized by the splines $s(f,t)$ from $S_{r-1}(\Delta_N^*[a,b])$ interpolating the function f at the zeros of the perfect spline $\Psi_{N,r}(q,t)$.

Now we turn our attention to the widths of classes $W_p^r[a,b]$ in $L_1[a,b]$. Here we shall work with the set $\Gamma_{N,m}^0[a,b]$ (introduced in Section 5.3.2) of perfect splines ψ of degree m with no more than $(N+m-1)$ knots on (a,b) which satisfy the boundary conditions

$$\psi^{(\nu)}(a) = \psi^{(\nu)}(b) = 0, \qquad \nu = 0, 1, \ldots, m-1.$$

The spline $\Psi_{N,m}^0(t) = \Psi_{N,m}^0(q,t)$ from $\Gamma_{N,m}^0[a,b]$ with

$$\|\Psi_{N,m}^0\|_q = \inf\{\|f\|_q : f \in \Gamma_{N,m}^0[a,b]\},$$

has exactly $(N+m-1)$ knots and $(N-1)$ zeros inside (a,b). These zeros define the partition $\Delta_N^0[a,b]$ such that the subspace $S_{m-1}(\Delta_N^0[a,b])$ of splines interpolates at the $(N+m-1)$ knots of $\Psi_{N,m}^0$. If $\sigma(f,t)$ is the interpolating spline from $S_{r-1}(\Delta_N^0[a,b])$ then (see Proposition 5.3.7)

$$\sup_{f \in W_p^r[a,b]} \|f - \sigma(f)\|_1 = \|\Psi_{N,r}^0(p')\|_{p'}. \tag{1.46}$$

Because $\dim S_{r-1}(\Delta_N^0[a,b]) = N + r - 1$ we have

$$\pi_{N+r-1}(W_p^r[a,b], L_1[a,b]) \leq \|\Psi_{N,r}^0(p')\|_{p'}, \qquad p' = p/(p-1). \tag{1.47}$$

We obtain the lower bound (as in the similar instances of the periodic case) using duality arguments. In view of Proposition 1.5.3 we have for every subspace $\mathcal{M}$ in $L_1[a,b]$ containing P_{r-1}

$$E(W_p^r[a,b], \mathcal{N})_1 = \sup\{\|g\|_{p'} : g \in W_\infty^r[a,b], g^{(r)} \perp \mathcal{N}, g^{(\nu)}(a) \\ = g^{(\nu)}(b) = 0, \nu = 0, 1, \ldots, r-1\}. \tag{1.48}$$

We shall estimate the right-hand side assuming $\dim \mathcal{N} = N + r - 1$. In the space $\mathbb{R}^{N+r}$ of vectors $\boldsymbol{\xi} = \{\xi_1, \xi_2, \ldots, \xi_{N+r}\}$ with norm $\|\boldsymbol{\xi}\| = |\xi_1| +$

$|\xi_2| + \ldots + |\xi_{N+r}|$ we take the sphere $S^{N+r-1} = \{\boldsymbol{\xi}(t) \colon \boldsymbol{\xi} \in \mathbb{R}^{N+r}, \|\boldsymbol{\xi}\| = b - a\}$ and, as above, we define the function $g_0(\boldsymbol{\xi}, t)$ on it. We then define the function $g(\boldsymbol{\xi}, t)$ by equality (1.42) which is continuous and odd with respect to $\boldsymbol{\xi}$. It is clear that for every $\boldsymbol{\xi} \in S^{N+r-1}$ the function $g(\boldsymbol{\xi}, t)$ is a perfect spline with no more than $(N + r - 1)$ knots on (a, b) and $g^{(r)}(\boldsymbol{\xi}, t) = g_0(\boldsymbol{\xi}, t)$. Fix in L_1 a subspace $\mathcal{N}_{N+r-1}$ containing P_{r-1} with basis $\{e_1(t), \ldots, e_{N+r-1}(t)\}$, where $e_k(t) = t^{k-1}$ $(k = 1, 2, \ldots, r)$ and define on sphere S^{N+r-1} a continuous odd vector field $\beta(\boldsymbol{\xi}) = \{\beta_1(\boldsymbol{\xi}), \ldots, \beta_{N+r-1}(\boldsymbol{\xi})\}$ by

$$\beta_k(\boldsymbol{\xi}) = \int_a^b e_k(t)\, g_0(\boldsymbol{\xi}, t)\, \mathrm{d}t, \qquad k = 1, 2, \ldots, N + r - 1. \tag{1.49}$$

Theorem 2.5.1 gives a vector $\boldsymbol{\xi}^* \in S^{N+r-1}$ such that $\beta_k(\xi^*) = 0$. This means in view of (1.49), $g_0(\boldsymbol{\xi}^*, t) \perp \mathcal{N}_{N+r-1}$. In particular $g_0(\boldsymbol{\xi}^*, t) \perp P_{r-1}$ and, in view of Lemma 1.5.2, $g^{(\nu)}(\boldsymbol{\xi}^*, a) = g^{(\nu)}(\boldsymbol{\xi}^*, b) = 0$ $(\nu = 0, 1, \ldots, r - 1)$, i.e. $g(\boldsymbol{\xi}^*, t) \in \Gamma^0_{N,m}[a, b]$. But then the right-hand side of (1.48) is not less then $\|g(\boldsymbol{\xi}^*)\|_{p'}$ for $\mathcal{N} = \mathcal{N}_{N+r-1}$ and hence

$$E(W_p^r[a, b], \mathcal{N}_{N+r-1})_1 \geq \inf_{f \in \Gamma^0_{N,m}[a,b]} \|f\|_{p'} = \|\Psi^0_{N,r}(p')\|_{p'}.$$

This is true for every subspace of dimension $(N + r - 1)$ and therefore

$$d_{N+r-1}(W_p^r[a, b], L_1[a, b]) \geq \|\Psi^0_{N,r}(p')\|_{p'}.$$

Comparing this estimate with (1.46) and (1.47) we come to:

Theorem 8.1.16 For all $N = 2, 3, \ldots$; $r = 1, 2, \ldots$, we have

$$\begin{aligned} d_{N+r-1}(W_p^r[a, b], L_1[a, b]) &= \lambda_{N+r-1}(W_p^r[a, b], L_1[a, b]) \\ &= \pi_{N+r-1}(W_p^r[a, b], L_1[a, b]) = \|\Psi^0_{N,r}(p')\|_{p'}. \end{aligned} \tag{1.50}$$

The widths are realized by the splines from $S_{r-1}(\Delta^0_N[a, b])$ interpolating the function f from $W_p^r[a, b]$ at the zeros of the spline $\Psi^0_{N,r}(p', t)$.

Theorems 8.1.15 and 8.1.16 give the exact solution of the problem of the N-widths in the cases considered; a linear method providing the minimal possible error has been found. But the lack of explicit expressions both for the knots of the interpolating spline and the points of interpolation makes it difficult to apply this best method in practice. In this context it is of interest to study how much, in comparison with the N-widths, we lose by replacing the best method by interpolation with splines of equally spaced knots. It turns out that such splines asymptotically realize the N-widths when $N \to \infty$. The most convenient interval for these considerations is $[a, b] = [0, \pi]$.

In view of Proposition 5.3.3 we have

$$\sup_{f\in W^r_\infty[0,\pi]} \|f-\sigma_{r-1}(f)\|_{L_q[0,\pi]} = \|\varphi_{N,r}\|_{L_q[0,\pi]} = N^{-r}\|\varphi_{1,r}\|_{L_q[0,\pi]}, \quad (1.51)$$

where $\sigma_{r-1}(f,t)$ is the interpolating spline from $S_{r-1}(\bar{\Delta}_N[0,\pi])$ ($\dim S_{r-1}(\bar{\Delta}_N[0,f]) = N+r-1$) with Leedstone boundary conditions and the standard function $\varphi_{N,r}$ is defined as $\varphi_{n,r}$ (see Section 3.1). This gives an upper bound for $\pi_{N+r-1}(W^r_\infty[0,\pi], L_q[0,\pi])$.

In the same way the relation (see Proposition 5.3.4)

$$\sup_{f\in W^r_p[0,\pi]} \|f-\sigma_{r-1}(f)\|_{L_1[0,\pi]} = \|\varphi_{N,r}\|_{L_{p'}[0,\pi]} \quad (1.52)$$

gives an upper bound for the projective width $\pi_{N+r-1}(W^r_p[0,\pi], L_1[0,\pi])$.

On the other hand, as it can be easily seen, the length of the period does not play any rôle for the lower estimates of widths $d_N(W^r_\infty, L_q)$ of classes of 2π-periodic functions (see Sections 8.1.2 and 8.1.3) besides determining the interval on which the norm is taken. Therefore the equalities in Theorems 8.1.11 and 8.1.13 determining the values of the widths of classes W^r_∞ in L_q and W^r_p in L_1 can be directly transferred to the corresponding classes of functions of period $2l$ by replacing the standard function $\varphi_{n,r}$ with $\varphi_{ln/\pi,r}$.

Let $\hat{W}^r_p[0,\pi]$ denote the set of functions f from $W^r_p[0,\pi]$ satisfying the periodic boundary conditions

$$f^{(\nu)}(0) = f^{(\nu)}(\pi) = 0, \ \nu = 0, 1, \ldots, r-1.$$

Hence these functions can be continued as π-periodic functions while their smoothness on the real line is preserved. In view of the above we have for even $N = 2n$ (here $L_q = L_q[0,\pi]$)

$$d_N(\hat{W}^r_\infty[0,\pi], L_q) = \|\varphi_{N,r}\|_{L_q},$$
$$d_N(\hat{W}^r_p[0,\pi], L_1) = \|\varphi_{N,r}\|_{L_{p'}},$$

and because of $\hat{W}^r_p[0,\pi] \subset W^r_p[0,\pi]$ and the monotonicity of d_N with N we obtain the estimates

$$\begin{aligned} d_{N+r-1}(W^r_\infty[0,\pi], L_q) &\ge \|\varphi_{N+r,r}\|_{L_q}, \\ d_{N+r-1}(W^r_p[0,\pi], L_1) &= \|\varphi_{N+r,r}\|_{L_{p'}}. \end{aligned} \quad (1.53)$$

Comparing (1.51) and (1.52) with (1.53) and (1.54) we obtain:

Proposition 8.1.17 For all $N = 2, 3, \ldots,; r = 1, 2, \ldots,$ we have

$$\begin{aligned} \|\varphi_{N,r}\|_{L_q} &\le d_{N+r-1}(W^r_\infty[0,\pi], L_q) \le \lambda_{N+r-1}(W^r_\infty[0,\pi], L_q) \\ &\le \pi_{N+r-1}(W^r_\infty[0,\pi], L_q) \le (1+r/N)^r \|\varphi_{N,r}\|_q, \end{aligned}$$

$$1 \le q \le \infty;$$

$$\|\varphi_{N,r}\|_{L_{p'}} \le d_{N+r-1}(W_p^r[0,\pi], L_1) \le \lambda_{N+r-1}(W_p^r[0,\pi], L_1)$$
$$\le \pi_{N+r-1}(W_p^r[0,\pi], L_1) \le (1+r/N)^r \|\varphi_{N,r}\|_{p'},$$
$$1 \le p \le \infty, \qquad 1/p + 1/p' = 1,$$

where $L_q = L_q[0,\pi]$. The values of all widths are realized asymptotically (for $N \to \infty$) by the interpolating splines from $S_{r-1}(\bar{\Delta}_N[0,\pi])$ satisfying Leedstone boundary conditions.

8.1.6 Widths of non-symmetric classes The definition of Kolmogorov N-widths given at the beginning of the chapter also makes sense for sets which are not central symmetric. Let

$$W_{p;\gamma,\delta}^r =: \{f: f \in L_p^r, \|f^{(r)}\|_{p;\gamma,\delta} \le 1\},$$

where $r = 1, 2, \ldots; 1 \le p \le \infty; \gamma, \delta > 0$. Then:

Theorem 8.1.18 For all $n = 1, 2, \ldots$, we have

$$d_{2n-1}(W_{p;\gamma,\delta}^r, L_1) = d_{2n}(W_{p;\gamma,\delta}^r, L_1)$$
$$= n^{-r} \inf_{\lambda \in \mathbb{R}} \|\varphi_{1,r} - \lambda\|_{p';\gamma^{-1},\delta^{-1}}, \qquad p' = p/(p-1).$$

For d_{2n-1} the subspace $\mathcal{N}_{2n-1}^T$ of trigonometric polynomials of degree $(n-1)$ is external and for d_{2n} the subspaces of splines $S_{2n,m}$ $(m \ge r)$ is external.

The upper bounds for these widths follow from Theorem 5.4.13. For the lower bound of d_{2n} let us fix an arbitrary $2n$-dimensional subspace $\mathcal{N}_{2n} \subset L_1$ containing the constants. In view of Lemma 8.1.12 there is a perfect spline $f \in \Gamma_{2n,r}$ such that $f^{(r)} \perp \mathcal{N}_{2n}$. Taking Propositions 1.4.2 and 1.4.9 into account we obtain

$$E(W_{p;\gamma,\delta}^r, \mathcal{N}_{2n})_1 \ge \inf_{\lambda \in \mathbb{R}} \|f - \lambda\|_{p';\gamma^{-1},\delta^{-1}}$$

and it remains to show

$$\inf_{\lambda \in \mathbb{R}} \|f - \lambda\|_{p;\alpha,\beta} \ge \inf_{\lambda \in \mathbb{R}} \|\varphi_{n,r} - \lambda\|_{p;\alpha,\beta}, \qquad \alpha, \beta > 0.$$

This inequality follows from the following improvement of Theorem 8.1.8 and the inequalities for rearrangements obtained in its proof.

Theorem 8.1.19 Let $n, r = 1, 2, \ldots, f \in \Gamma_{2n,r}, f \perp 1$. Then we have for every $\lambda \in \mathbb{R}$ and $x \in [0, 2\pi]$

$$\int_0^x r((f-\lambda)_\pm, t)\, dt \ge \int_0^x r((\varphi_{n,r} - \lambda)_\pm, t)\, dt.$$

8.2 N-widths of classes defined by the modulus of continuity

8.2.1 ψ-splines Let us introduce a generalization for the polynomial splines which fit the specific features of the definition of classes (see Section 7.1)

$$W^rH^\omega = \{f: f \in C^r, |f^{(r)}(t') - f^{(r)}(t'')| \le \omega(|t' - t''|)\}, \qquad r = 0, 1, \ldots, .$$

Let ψ be a non-increasing function on $[0, \pi]$ with $\psi(0) = 0$ and $\psi \not\equiv 0$. For every fixed $n = 1, 2, \ldots,$ we define $2n$ 2π-periodic functions ψ_j by setting

$$\psi_1(t) = \begin{cases} \psi(t), & 0 \le t \le \pi/(2n), \\ \psi(\pi/n - t), & \pi/(2n) \le t \le \pi/n, \\ 0, & \pi/n \le t \le 2\pi; \end{cases}$$

$$\psi_j(t) = \psi_1(t - (j-1)\pi/n), \qquad j = 2, 3, \ldots, 2n.$$

Thus, every function ψ_j has one 'hat' on the jth interval of the uniform partition $\{j\pi/n\}_0^{2n}$ of the period and it is clear that this system of functions is linearly independent.

Assuming ψ and ψ_j are fixed we consider the set M_{2n}^0 of functions of the type

$$f(t) = \sum_{j=1}^{2n} c_j \psi_j(t), \tag{2.1}$$

which is a $2n$-dimensional linear manifold. By M_{2n}^r we denote the set of rth periodic integrals of functions (2.1) satisfying the additional condition

$$c_1 + c_2 + \cdots + c_{2n} = 0,$$

which guarantees the orthogonality of f to the unity: $f \perp 1$. Obviously M_{2n}^r is also a linear manifold and dim $M_{2n}^r = 2n$. If $\psi(t) = 1$ for $0 < t \le \pi$ then M_{2n}^r coincides with the subspace $S_{2n,r}$ of splines of degree r and defect 1 with respect to partition $\{j\pi/n\}$. The functions from M_{2n}^r will be called ψ-splines.

The standard functions $F_{n,r}$ in M_{2n}^r $\{r = 0, 1, \ldots,)$ are defined by

$$F_{n,0}(t) = F_{n,0}(\psi, t) = \sum_{j=1}^{2n} (-1)^{j-1} \psi_j(t),$$

$F_{n,r}(t) = F_{n,r}(\psi, t)$ – the rth periodic integral of $F_{n,0}$ with zero mean value on the period. If $\psi(t) = 1$ $(0 < t \le \pi)$ then $F_{n,r}$ coincides with the Euler spline $\varphi_{n,r}$. If $\psi(t) = \omega(2t)/2$, with ω some modulus of continuity, then $F_{n,r}(t) = f_{n,r}(\omega, t + \pi/(2n))$, where $f_{n,r}(\omega, t)$ is the standard function constructed in Section 7.1 and belonging to W^rH^ω for concave ω.

The next statement is a generalization of Lemma 2.4.10 for ψ-splines.

Lemma 8.2.1 For fixed $n = 1, 2, \ldots,$ let $\tau_k = \tau_{r,k}$ be the points of extremum for function $F_{n,r}$, i.e.

$$\tau_k = \tau_{r,k} = \begin{cases} k\pi/n, & k = 0, \pm 1, \pm 2, \ldots, \quad \text{for odd } r, \\ (2k-1)\pi/(2n), & k = 0, \pm 1, \pm 2, \ldots, \quad \text{for even } r. \end{cases} \tag{2.2}$$

If $f \in M_{2n}^r$ $(r = 0, 1, \ldots,)$ and

$$\max_{1 \le k \le 2n} |f(\tau_k)| \le \|F_{n,r}\|_\infty, \tag{2.3}$$

then the coefficients c_j in (2.1) for $r = 0$ and in the representation

$$f^{(r)}(t) = \sum_{j=1}^{2n} c_j \psi_j(t), \qquad c_1 + c_2 + \cdots + c_{2n} = 0, \tag{2.4}$$

for $r = 1, 2, \ldots,$ satisfy the inequalities

$$|c_j| \le 1, \qquad j = 1, 2, \ldots, 2n.$$

The idea of the proof is the same as in Lemma 2.4.10. For $r \ge 1$ (the assertion is trivial for $r = 0$) assume that for some $f \in M_{2n}^r$ (2.3) holds but that we have in representation (2.4)

$$\max_{1 \le j \le 2n} |c_j| = |c_\nu| > 1.$$

Set

$$f_*(t) = \varepsilon f(t)/|c_\nu|, \qquad \varepsilon = \pm 1, \tag{2.5}$$

where ε is chosen so that $\varepsilon \operatorname{sgn} c_\nu = (-1)^{\nu-1}$. But then from

$$f_*^{(r)}(t) = \sum_{j=1}^{2n} c_j^* \psi_j(t), \qquad c_j^* = \frac{\varepsilon c_j}{|c_\nu|},$$

we obtain

$$c_\nu^* = (-1)^{\nu-1}, \qquad |c_j^*| \le 1. \tag{2.6}$$

Consider the difference

$$\delta(t) = F_{n,r}(t) - f_*(t). \tag{2.7}$$

From (2.5), (2.3) and $|c_\nu| > 1$ it follows

$$\max_{1 \le j \le 2n} |f_*(\tau_j)| < \|F_{n,r}\|_C = |F_{n,r}(\tau_k)|,$$

and hence

$$\operatorname{sgn} \delta(\tau_k) = \operatorname{sgn} F_{n,r}(\tau_k),$$

i.e. the difference (2.7) changes sign on the period at least $2n$ times. Now Rolle's theorem (see also Proposition 2.4.2) implies that every derivative $\delta', \delta'', \ldots, \delta^{(r)}$ has on the period at least $2n$ essential changes of sign. But from

$$\delta^{(r)}(t) = F_{n,0}(t) - f_*^{(r)}(t) = \sum_{j=1}^{2n} [(-1)^{j-1} - c_j^*]\, \psi_j(t),$$

and (2.6) we see that the function $\delta^{(r)}$ is equal to zero on the νth interval of the partition $((\nu - 1)\pi/n, \nu\pi/n)$ and preserves sign on each of the remaining intervals $((j - 1)\pi/n, j\pi/n)$. Thus, the number of its essential changes of sign on the period does not exceed $(2n - 1)$. This contradiction proves the lemma. □

Together with the subspaces M_{2n}^r $(r = 0, 1, \ldots,)$ we shall need their translates

$$M_{2n,\beta}^r = \{f: f(t) = \varphi(t - \beta), \varphi \in M_{2n}^r\}, \tag{2.8}$$

which are of the same dimension $2n$ for every fixed β. Obviously the points

$$\tau_k^\beta = \tau_{rk}^\beta = \tau_{rk} + \beta, \qquad k = 1, 2, \ldots, 2n,$$

are the points of extremum for the translate $F_{n,r}(t - \beta)$ of the standard function $F_{n,r}$.

With every function $f \in M_{2n,\beta}^r$ we associate the vector

$$a_\beta(f) = \{f(\tau_1^\beta), f(\tau_2^\beta), \ldots, f(\tau_{2n}^\beta)\}$$

and we let

$$Q_{2n,\beta}^r = \{a_\beta(f): f \in M_{2n,\beta}^r\}. \tag{2.9}$$

The set $Q_{2n,\beta}^r$ is a linear manifold; one can say that it is 'cut' from subspace $M_{2n,\beta}^r$ by the points τ_k^β.

Lemma 8.2.2 For every β dim $Q_{2n,\beta}^r = 2n$ $(r = 0, 1, \ldots,)$.

Proof Let $f_1, \ldots, f_{2n}$ be a basis in $M_{2n,\beta}^r$. The lemma will be proved if we show that the vectors $a_\beta(f_i)$ $(i = 1, 2, \ldots, 2n)$ are linearly independent. Assume on the contrary that there are coefficients p_i $(i = 1, 2, \ldots, 2n)$ not all zero such that

$$p_1 a_\beta(f_1) + p_2 a_\beta(f_2) + \cdots + p_{2n} a_\beta(f_{2n}) = 0.$$

Then for the function

$$g(t) = p_1 f_1(t) + p_2 f_2(t) + \cdots + p_{2n} f_{2n}(t), \tag{2.10}$$

which obviously belongs to $M_{2n,\beta}^r$, we have $g(\tau_k^\beta) = 0$ $(k = 1, 2, \ldots, 2n)$. If we set $g_0(t) = g(t + \beta)$ then the functions Kg_0 will belong to M_{2n}^r for every $K \in \mathbb{R}$ and

$$Kg_0(\tau_k) = 0, \qquad k = 1, 2, \ldots, 2n.$$

Writing Kg_0 in the form of (2.1) ($r = 0$) or $Kg_0^{(r)}$ in the form of (2.4) ($r = 1, 2, \ldots,$) we have in view of Lemma 8.2.1

$$\|Kg_0^{(r)}\|_\infty = \max_j |c_j| \, \|\psi_j\|_\infty = \max_j \|\psi_j\|_\infty = \|F_{n,r}^{(r)}\|_\infty .$$

This is possible only when $g_0^{(r)}$ is identically zero (K was arbitrary) and hence $g^{(r)}$ and function (2.10) are also identically zero – a contradiction with the linear independence of $\{f_i\}$. The lemma is proved. □

A consequence is:

Proposition 8.2.3 The linear manifold $M_{2n,\beta}^r$ interpolates at the points τ_k^β, i.e. for every $2n$ numbers $a_1, a_2, \ldots, a_{2n}$ there exists a unique function $f \in M_{2n,\beta}^r$ such that

$$f(\tau_k^\beta) = a_k, \qquad k = 1, 2, \ldots, 2n.$$

Indeed in view of Lemma 8.2.2 every vector $\mathbf{a} = \{a_1, a_2, \ldots, a_{2n}\}$ with real coordinates belongs to $Q_{2n,\beta}^r$, i.e. there is a function $f \in M_{2n,\beta}^r$ such that $a_\beta(f) = a$. The uniqueness follows from the general fact that the linear mapping of the N-dimensional space X_N onto the N-dimensional space Y_N is a bijection.

8.2.2 Kolmogorov widths of classes W^rH^ω in C Concerning the odd $(2n - 1)$ widths we can easily estimate them from below using Theorem 8.1.1 for the width of a ball and using Lemma 8.2.1 from the previous section.

Denote by $M_{2n}^r(\omega)$ the linear manifold M_{2n}^r in the case when $\psi(t) = \omega(2t)/2$ and hence $F_{n,r}(t - \pi/(2n)) = f_{n,r}(\omega, t)$. From Lemma 8.2.1 we easily get:

Proposition 8.2.4 Let ω be a concave modulus of continuity, $f \in M_{2n}^r(\omega)$ and suppose that we have at the points $\tau_k = \tau_{rk}$ defined in (2.2) we have

$$|f(\tau_k)| \le \|f_{n,r}(\omega)\|_C, \qquad k = 1, 2, \ldots, 2n. \tag{2.11}$$

Then $f \in W^rH^\omega$.

Proof If $f \in M_{2n}^r(\omega)$ then

$$f^{(r)}(t) = \sum_{j=1}^{2n} c_j \, \psi_j(\omega, t), \tag{2.12}$$

where the function $\psi_j(\omega, t)$ coincides with $|f_{n,0}(\omega, t + \pi/(2n))|$ on the jth interval $\Delta_j = [(j - 1)\pi/n, j\pi/n]$ and is equal to zero on the complement

$[0, 2\pi]\backslash\Delta_j$. From inequality (2.11) and Lemma 8.2.1 we get that $|c_j| \le 1$ ($j = 1, 2, \ldots, 2n$) in representation (2.12) and this guarantees that $f^{(r)} \in H^\omega$ because we have for all points t' and t''

$$|f^{(r)}(t') - f^{(r)}(t'')| \le |f_{n,0}(\omega, t' + \pi/(2n)) - f_{n,0}(\omega, t'' + \pi/(2n))|$$

and for concave ω the function $f_{n,0}(\omega, t)$ belongs to H^ω. □

From Proposition 8.2.4 we get that $f \in M_{2n}^r(\omega)$ and $\|f\|_C \le \|f_{n,0}(\omega)\|_C$ implies $f \in W^rH^\omega$, i.e. the ball with radius $\|f_{n,0}(\omega)\|_C$ in subspace $M_{2n}^r(\omega)$ of C belongs to the class W^rH^ω provided ω is concave. But then Proposition 8.1.2 implies

$$d_{2n-1}(W^rH^\omega, C) \ge \|f_{n,0}(\omega)\|_C, \quad n = 1, 2, \ldots, \quad r = 0, 1, \ldots. \tag{2.13}$$

For the best approximation using trigonometric subspace $\mathcal{N}_{2n-1}^T$ of dimension $(2n - 1)$ we have, in view of Theorem 7.2.2,

$$E(W^rH^\omega, \mathcal{N}_{2n-1}^T)_C = \|f_{n,0}(\omega)\|_C, \quad n = 1, 2, \ldots; \quad r = 0, 1, \ldots, \tag{2.14}$$

and hence we have equality in (2.13) for concave ω.

Before we state the final result we need to bound the $2n$-widths from below. It turns out that we have an estimate with the same right-hand side as in (2.13) for them but we need more specific arguments. On the other hand these complications will allow us to by-pass Theorem 8.1.1 which is based on Borsuk's theorem.

The possibility of going to widths d_{2n} in estimate (2.13) gives, besides Lemma 8.2.2, one other interesting fact connected with periodicity. For every $\alpha \in \mathbb{R}$ we set

$$x_k^\alpha = k\pi/n + \alpha, \qquad k = 1, 2, \ldots, 2n, \tag{2.15}$$

and with every function $g \in C$ we associate the vector

$$\xi_\alpha(g) = \{g(x_1^\alpha), g(x_2^\alpha), \ldots, g(x_{2n}^\alpha)\}, \tag{2.16}$$

which obviously depends linearly on g. If $\mathcal{N}_{2n}$ is an arbitrary subspace $\mathcal{N}$ in C of dimension $2n$ then for a fixed α the set of vectors

$$N(\alpha) = N(\alpha, \mathcal{N}_{2n}) = \{\xi_\alpha(g) : g \in \mathcal{N}_{2n}\}, \tag{2.17}$$

which is 'cut' from $\mathcal{N}_{2n}$ by the system of points (2.15) is a linear manifold in $\mathbb{R}^{2n}$ of dimension not greater than $2n$. The following result is important for us:

Lemma 8.2.5 For every subspace $\mathcal{N}_{2n}$ of C there is a number $\alpha = \alpha_0$ ($0 \le \alpha_0 \le \pi/n$) such that $\dim N(\alpha_0) \le 2n - 1$.

Proof If $\dim N(0) \le 2n - 1$ then there is nothing to prove. So we assume that $\dim N(0) = 2n$ and hence the system of vectors $\boldsymbol{e}_1 = \{1, 0, \ldots, 0\}$,

$\ldots, \boldsymbol{e}_{2n} = \{0, \ldots 0, 1\}$ belongs to $N(0)$. According to definition (2.17) this means that there are $2n$ functions $g_i \in \mathcal{N}_{2n}$ $(i = 1, 2, \ldots, n)$ such that $\boldsymbol{e}_i = \xi_0(g_i)$ $(i = 1, 2, \ldots, 2n)$, i.e.

$$g_i(x_k^0) = \begin{cases} 0, & k \neq i \\ 1, & k = i, \qquad i, k = 1, 2, \ldots, 2n. \end{cases} \tag{2.18}$$

From the linear independence of the vectors $\boldsymbol{e}_i$ it follows that the functions g_i are linearly independent because if we assume that, e.g., $g_1(t) = \gamma_2 g_2(t) + \cdots + \gamma_{2n} g_{2n}(t)$ then

$$\boldsymbol{e}_1 = \xi_0(g_1) = \sum_{i=2}^{2n} \gamma_i \xi_0(g_i) = \sum_{i=2}^{2n} \gamma_i \boldsymbol{e}_i,$$

which is impossible. Thus the functions g_i $(i = 1, 2, \ldots, 2n)$ form a basis in $\mathcal{N}_{2n}$ and if $g(t) = \lambda_1 g_1(t) + \cdots + \lambda_{2n} g_{2n}(t)$ then we have for every α

$$\xi_\alpha(g) = \sum_{i=1}^{2n} \lambda_i \xi_\alpha(g_i).$$

Thus $N(\alpha)$ is the linear span of vectors $\xi_\alpha(g_i)$ $(i = 1, 2, \ldots, 2n)$ and the lemma will be proved if we show that for some $\alpha = \alpha_0$ $(\alpha_0 \in (0, \pi/n))$ vectors $\xi_{\alpha_0}(g_i)$ are linearly dependent. Consider the determinant

$$D(\alpha) = \begin{vmatrix} g_1(x_1^\alpha) \; g_1(x_2^\alpha) \ldots & g_1(x_{2n}^\alpha) \\ g_2(x_1^\alpha) \; g_2(x_2^\alpha) \ldots & g_2(x_{2n}^\alpha) \\ \ldots\ldots\ldots\ldots\ldots\ldots & \ldots \\ g_{2n}(x_1^\alpha) \; g_{2n}(x_2^\alpha) \ldots & g_{2n}(x_{2n}^\alpha) \end{vmatrix}, \tag{2.19}$$

which is a continuous function of α.

(2.18) implies that $D(0) = 1$. Let us calculate $D(\pi/n)$. Noticing that $x_k^{\pi/n} = x_{k+1}^0$ $(k = 1, 2, \ldots, 2n - 1)$ and in view of the periodicity $g_i(x_{2n}^{\pi/n}) = g_i(2\pi + \pi/n) = g_i(x_1^0)$ we get

$$D(\pi/n) = \begin{vmatrix} 0\,0 \ldots 0\,1 \\ 1\,0 \ldots 0\,0 \\ 0\,1 \ldots 0\,0 \\ \ldots\ldots \\ 0\,0 \ldots 1\,0 \end{vmatrix} = -1.$$

The continuous function $D(\alpha)$ takes opposite signs for $\alpha = 0$ and $\alpha = \pi/n$ and thus it has to vanish for some $\alpha_0 \in (0, \pi/n)$. This means that for $\alpha = \alpha_0$

the row vectors in (2.19) are linearly dependent and the lemma is proved because these vectors represent exactly $\xi_{\alpha_0}(g_i)$. □

Now we turn our attention to the lower bound for $d_{2n}(W^rH^\omega, C)$. Fix a subspace $\mathcal{N}_{2n}$ in C of dimension $2n$ and define the linear manifold $N(\alpha, \mathcal{N}_{2n})$ by equalities (2.15)–(2.17) assuming that α is chosen (according to Lemma 8.2.5) such that

$$\dim N(\alpha, \mathcal{N}_{2n}) = m \leq 2n - 1.$$

This choice fixes the system of points $\{x_k^\alpha\}$.

Let us go back to the sets $M_{2n,\beta}^r$ and $Q_{2n,\beta}^r$ defined in (2.8) and (2.9) and generated as M_{2n}^r by a fixed non-increasing function ψ. For $\psi(t) = \omega(2t)/2$, where ω is a modulus of continuity, we shall denote these sets by $M_{2n,\beta}^r(\omega)$, $Q_{2n,\beta}^r(\omega)$. Thus $Q_{2n,\beta}^r(\omega)$ is the linear manifold of vectors

$$a_\beta(f) = \{f(\tau_1^\beta), f(\tau_2^\beta), \ldots, f(\tau_{2n}^\beta)\},$$

of dimension $2n$ (Lemma 8.2.2), where $f \in M_{2n}^r(\omega)$. Define a norm in $Q_{2n,\beta}^r(\omega)$ by

$$\|a_\beta(f)\|_Q = \max_{1 \leq k \leq 2n} |f(\tau_k^\beta)|. \tag{2.20}$$

Choosing parameter β appropriately (see (2.8)) we make the system $\{\tau_k^\beta\}$ coincide with $\{x_k^\alpha\}$. This is possible because both systems are equally spaced with step-length π/n. Now the linear manifolds $N(\alpha, \mathcal{N}_{2n})$ and $Q_{2n,\beta}^r$ are 'cut' from $\mathcal{N}_{2n}$ and $M_{2n}^r(\omega)$ respectively by the same system of points for which we reserve the notation $\{\tau_k^\beta\}$. $N(\alpha, \mathcal{N}_{2n})$ can be considered as an m-dimensional ($m \leq 2\alpha - 1$) subspace of the normed space $Q_{2n,\beta}^r(\omega)$.

Lemma 8.2.6 If ω is a concave modulus of continuity then for every subspace $\mathcal{N}_{2n} \subset C$ ($\dim \mathcal{N}_{2n} = 2n$) there is a function $f_* \in W^rH^\omega$ such that

$$\inf_{g \in \mathcal{N}_{2n}} \max_{1 \leq k \leq 2n} |f_*(\tau_k^\beta) - g(\tau_k^\beta)| = \|f_{n,r}(\omega)\|_C, \qquad \tau_k^\beta = t_{rk}^\beta. \tag{2.21}$$

The lemma is an immediate consequence of the following general fact.

Proposition 8.2.7 Let X be a normed linear space of dimension greater than N and let $\mathcal{F}_N$ be its N-dimensional subspace. For every $\gamma > 0$ there is an element $y_\gamma \in X$ such that $E(y_\gamma, \mathcal{F}_N) = \|y_\gamma\| = \gamma$.

Indeed fix $x \in X \backslash \mathcal{F}_N$ and let u_* be the element nearest to x from $\mathcal{F}_N$, i.e.

$$\|x - u_*\| = \inf_{u \in \mathcal{F}_N} \|x - u\| =: d > 0.$$

Set $y_\gamma = \gamma(x - u_*)/d$. Hence $\|y_\gamma\| = \gamma$ and for every $u \in \mathcal{F}_N$ we have

$$\|y_\gamma - u\| = \frac{\gamma}{d}\left\| x - u_* - \frac{d}{\gamma}u \right\| \le \frac{\gamma}{d} E(x, \mathcal{F}_N) = \gamma,$$

i.e. $E(y_\gamma, \mathcal{F}_N) \ge \gamma$. But we actually have an equality here since $E(y_\gamma, \mathcal{F}_N) \le \|y_\gamma\| = \gamma$.

We now set in Proposition 8.2.7 $X = Q^r_{2n,\beta}(\omega)$, $\mathcal{F}_m = N(\alpha, \mathcal{N}_{2n})$, $\dim \mathcal{F}_m \le 2n-1$. Then there exists a vector $a_\beta(f_*) \in Q^r_{2n,\beta}(\omega)$ which is 'cut' by some function $f_* \in M^r_{2n,\beta}(\omega)$ such that

$$\|a_\beta(f_*)\|_Q = \|f_{n,r}(\omega)\|_C \tag{2.22}$$

and

$$\inf\{\|a_\beta(f_*) - \xi_\alpha(g)\|_Q : \xi_\alpha(g) \in N(\alpha, \mathcal{N}_{2n})\} = \|f_{n,r}(\omega)\|_C. \tag{2.23}$$

It follows from (2.22) and (2.20)

$$\max_{1\le k\le 2n} |f_*(\tau_k^\beta)| = \|f_{n,r}(\omega)\|_C$$

and hence $f_* \in W^rH^\omega$ in view of Proposition 8.2.4 and the invariance of the class W^rH^ω with respect to a translation of the argument. Finally we have to note that once more the definition of the norm (2.20) means (2.23) can be rewritten as (2.21). The lemma is proved. □

We deduce $E(f_*, \mathcal{N}_{2n})_C \ge \|f_{n,r}(\omega)\|_C$ from the lemma and hence

$$E(W^rH^\omega, \mathcal{N}_{2n})_C \ge \|f_{n,r}(\omega)\|_C.$$

Because of the validity of this inequality for every subspace $\mathcal{N}_{2n}$ (of dimension $2n$) of C we conclude

$$d_{2n}(W^rH^\omega, C) \ge \|f_{n,r}(\omega)\|_C, \qquad n = 1, 2, \ldots, \qquad r = 0, 1, \ldots \tag{2.24}$$

Estimate (2.24) and relation (2.14) allow us to formulate the final result also using Theorems 7.2.2 and 7.2.5.

Theorem 8.2.8 For every concave modulus of continuity $\omega(\delta)$ and all $n = 1, 2, \ldots, r = 0, 1, \ldots$, we have

$$d_{2n-1}(W^rH^\omega, C) = d_{2n}(W^rH^\omega, C) = \|f_{n,r}(\omega)\|_C. \tag{2.25}$$

All widths in (2.25) are realized by the subspace $\mathcal{N}^T_{2n-1}$ of trigonometric polynomials of degree $(n-1)$; the even widths are also realized by subspaces $S_{2n,m}$ of splines of degree $m \ge r$ and defect 1 with respect to the uniform partition $\{k\pi/n\}$.

Remembering the definition of function $f_{n,r}(\omega, t)$ (see Section 7.1.1) and also relations (7.1.34) which express $\|f_{n,r}(\omega)\|_C$ in the terms of standard Σ-rearrangements we can write (for concave ω)

$$d_{2n-1}(H^\omega, C) = d_{2n}(H^\omega, C) = \frac{1}{2}\,\omega\left(\frac{\pi}{n}\right), \qquad n = 1, 2, \ldots;$$

$$d_{2n-1}(W^1H^\omega, C) = d_{2n}(W^1H^\omega, C) = \frac{1}{4}\int_0^{\pi/n} \omega(t)\,\mathrm{d}t, \qquad n = 1, 2, \ldots;$$

$$d_{2n-1}(W^rH^\omega, C) = d_{2n}(W^rH^\omega, C) = \frac{1}{2n^r}\int_0^{\pi} R_{\pi,n-1}(\pi - t)\,\omega\left(\frac{t}{n}\right)\mathrm{d}t,$$

$$n = 1, 2, \ldots; r = 2, 3, \ldots.$$

Let us note that for a fixed n the subspace $\mathcal{N}^T_{2n-1}$ of trigonometric polynomials is extremal for W^rH^ω for all $r = 0, 1, \ldots$, while the subspace $S_{2n,m}$ of splines is extremal for W^rH^ω only for $m \geq r$.

The problem of the exact values of the linear widths of W^rH^ω in C is open if $\omega(\delta) \neq K\delta$ $(0 \leq \delta \leq \pi)$. (For $\omega(\delta) = K\delta$ the class W^rH^ω coincides with the class KW^{r+1}_∞ whose linear widths are known from Theorems 8.1.5 and 8.1.11.) The problem has not been solved even for the class H^α $(0 < \alpha < 1)$ of functions satisfying Hölder's condition. There is a conjecture that $\lambda_{2n-1}(H^\alpha, C) = [\pi^\alpha/(1+\alpha)]\,n^{-\alpha}$. This is, in particular, based on the fact that this is the quantity which bounds the approximation of class H^α from below by λ-means of Fourier sums (see Exercise 11, Chapter 7). Moreover it is known [9A, p. 213] that for a concave modulus of continuity $\omega(\delta)$ the supremum $E(H, \mathcal{N}^T_{2n-1})_C$ can be realized by a linear operator from C to $\mathcal{N}^T_{2n-1}$ if and only if function $\omega(\delta)$ is linear on $[0, \pi/n]$. A similar fact holds for approximation by splines.

It is interesting that the situation with the linear widths in L_p of classes defined by modulus of continuity is different. We study this problem in the next sections.

8.2.3 Widths of classes $H^\omega[a, b]$ in $L_p[a, b]$ We have (see Theorems 7.2.13 and 7.3.2) estimates for the best approximations in $L_p[a, b]$ of functions from $H^\omega[a, b]$ by the subspaces $S_{N,0}[a, b]$ of step functions with respect to the uniform partition

$$t_k = a + kh_N, \qquad h_N = (b-a)/N, \qquad k = 0, 1, \ldots, N,$$

and also (for $1 \leq p \leq 3$) by functions $\psi_N(f, t)$ from $S_{N,0}[a, b]$ interpolating the mean values of f on every interval (t_{k-1}, t_k) $(k = 1, 2, \ldots, N)$. This gives upper bounds for Kolmogorov and projective widths:

$$d_N(H^\omega[a, b], L_p) \leq E(H^\omega[a, b], S_{N,0}[a, b])_p \leq A_{N,p}(\omega), \qquad 1 \leq p \leq \infty, \tag{2.26}$$

$$\pi_N(H^\omega[a, b], L_p) \leq \sup_{f\in H^\omega} \|f - \psi_N(f)\|_p \leq A_{N,p}(\omega), \quad 1 \leq p \leq 3, \tag{2.27}$$

where $L_p = L_p[a, b]$ (the same in the following) and

$$A_{N,p}(\omega) = \begin{cases} \dfrac{1}{2}\left[N\displaystyle\int_0^{h_N} \omega^p(t)\,\mathrm{d}t\right]^{1/p}, & 1 \le p < \infty, \\ \dfrac{1}{2}\,\omega(h_N), & p = \infty. \end{cases} \tag{2.28}$$

Now we shall show that for concave ω the estimates (2.26) and (2.27) for the widths are exact, i.e. they cannot be improved if we change the approximation subspace to another of dimension N. Here we also supply Borsuk's theorem.

In the $(N+1)$-dimensional space $\mathbb{R}^{N+1}$ of vectors $\boldsymbol{\xi} = \{\xi_1, \xi_2, \ldots, \xi_{N+1}\}$ with norm $\|\boldsymbol{\xi}\| = |\xi_1| + \cdots + |\xi_{N+1}|$ we consider the sphere $S^N = \{\boldsymbol{\xi}\colon \boldsymbol{\xi} \in \mathbb{R}^{N+1}, \|\boldsymbol{\xi}\| = b - a\}$ and associate with $\boldsymbol{\xi} \in S^N$ the system of points $\tau_0 = a$, $\tau_k = |\xi_1| + \cdots + |\xi_k|$ $(k = 1, 2, \ldots, N+1)$ and the function (for a fixed ω)

$$g(\boldsymbol{\xi}, t) = \begin{cases} 2^{-1}\,\omega(2\tau_1 - 2t)\,\mathrm{sgn}\,\xi_1, & \tau_0 \le t \le \tau_1, \\ 2^{-1}\min\{\omega(2t - 2\tau_{k-1}), \omega(2\tau_k - 2t)\}\,\mathrm{sgn}\,\xi_k, & \\ \qquad\qquad \tau_{k-1} \le t \le \tau_k,\ k = 2, 3, \ldots, N, & \\ 2^{-1}\,\omega(2t - 2\tau_N)\,\mathrm{sgn}\,\xi_{N+1}, & \tau_N \le t \le \tau_{N+1} = b. \end{cases} \tag{2.29}$$

Of course, we assume that if $\xi_i = 0$ then the interval $[\tau_{i-1}, \tau_i]$ being a point does appear in the definition of $g(\boldsymbol{\xi}, t)$. From the definition we get $g(\boldsymbol{\xi}, \tau_k) = 0$, $g(-\boldsymbol{\xi}, t) = -g(\boldsymbol{\xi}, t)$ and the function $g(\boldsymbol{\xi}, t)$ depends on $\boldsymbol{\xi}$ continuously: if $\boldsymbol{\xi}^{(m)} \to \xi$ $(\boldsymbol{\xi}^{(m)}, \boldsymbol{\xi} \in S^N)$ then $\|g(\boldsymbol{\xi}^{(m)}) - g(\boldsymbol{\xi})\|_{C[a,b]} \to 0$. In the family of functions $g(\boldsymbol{\xi}, t)$ $(\boldsymbol{\xi} \in S^N)$ we identify a standard function $F_N(\omega, t)$ which is determined by the vector $\boldsymbol{\xi} = \bar{\boldsymbol{\xi}}$ with coordinates

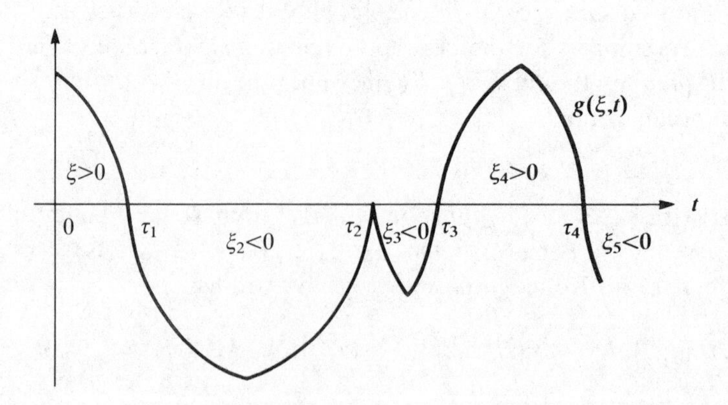

Fig. 8.1 Function $g(\boldsymbol{\xi}, t)$.

$\xi_1 = -h_N/2$, $\xi_i = (-1)^i h_N$ $(i = 2, 3, \ldots, N)$, $\xi_{N+1} = (-1)^{N+1} h_N/2$. If $[a, b] = [0, 2\pi]$, $N = 2n$, then $F_N(\omega, t)$ coincides on $[0, 2\pi]$ with the function $f_{n,0}(\omega, t)$ defined in Section 7.1.1. It is easy to check that the norm $\|F_N(\omega)\|_p$ coincides with quantity $A_{N,p}(\omega)$ defined in (2.28). The most important relation for us is:

$$\inf\{\|g(\boldsymbol{\xi})\|_p : \boldsymbol{\xi} \in S^N\} = \|F_N(\omega)\|_p, \qquad 1 \le p \le \infty, \qquad \|\bullet\|_p = \|\bullet\|_{L_p[a,b]}. \tag{2.30}$$

For $p = \infty$, i.e. for the norm in $C[a, b]$, it is obvious and for $1 \le p < \infty$ it follows from the following lemma given in two versions in view of later applications.

Lemma 8.2.9 Let ψ be a non-decreasing and non-negative function for $0 \le t \le b - a$. Let for a fixed $m = 1, 2, \ldots$, the vector $\boldsymbol{T} = \{\tau_1, \tau_2, \ldots, \tau_m\}$ $(a \le \tau_1 < \tau_2 < \cdots < \tau_m \le b)$ be associated with the functions

$$\Phi_1(\boldsymbol{T}, t) = \min_{1 \le k \le m} \psi(|t - \tau_k|), \qquad a \le t \le b,$$

$$\Phi_2(\boldsymbol{T}, t) = \min_{0 \le k \le m+1} \psi(|t - \tau_k|), \qquad \tau_0 = a, \qquad \tau_{m+1} = b, \qquad a \le t \le b.$$

Then

$$\int_a^b \Phi_1(\boldsymbol{T}, t)\, dt \ge \int_a^b \Phi_1(\boldsymbol{T}_1, t)\, dt, \qquad \int_a^b \Phi_2(\boldsymbol{T}, t)\, dt \ge \int_a^b \Phi_2(\boldsymbol{T}_2, t)\, dt, \tag{2.31}$$

where the vector $\boldsymbol{T}_1$ is determined by the coordinates $\tau_k = a + (2k - 1)(b - a)/(2m)$ and the vector $\boldsymbol{T}_2$ is determined by the coordinates $\tau_k = a + k(b - a)/(m + 1)$.

Indeed, setting

$$\Psi(t) = \int_0^t \psi(u)\, du, \qquad 0 \le t \le b - a,$$

it follows from the definitions of $\Phi_1(\boldsymbol{T}, t)$, $\Phi_2(\boldsymbol{T}, t)$ that

$$\int_a^b \Phi_1(\boldsymbol{T}, t)\, dt = \sum_{\nu=1}^{2m} \psi(\alpha_\nu), \qquad \int_a^b \Phi_2(\boldsymbol{T}, t)\, dt = 2 \sum_{\nu=1}^{m+1} \Psi(\beta_\nu), \tag{2.32}$$

where

$$\alpha_1 + \alpha_2 + \cdots + \alpha_{2m} = b - a, \qquad 2(\beta_1 + \beta_2 + \cdots + \beta_{m+1}) = b - a.$$

The function Ψ is convex and it follows from Jenssen's inequality (see e.g. [7B, p. 92]) that

$$\sum_{\nu=1}^{2m} \Psi(\alpha_\nu) \ge 2m\, \Psi\left(\frac{1}{2m} \sum_{\nu=1}^{2m} \alpha_\nu\right),$$

$$\sum_{\nu=1}^{m+1} \Psi(\beta_\nu) \geq (m+1)\,\Psi\left(\frac{1}{m+1}\sum_{\nu=1}^{m+1} \beta_\nu\right),$$

which in view of (2.32) is equivalent to (2.31).

In Lemma 8.2.9 setting $\psi(t) = 2^{-p}\,\omega^p(2t)$ $(p > 0)$ we obtain from its assertion about $\Phi_1(t, t)$ that

$$\inf\left\{\int_a^b |g(\boldsymbol{\xi}, t)|^p\, dt\colon \boldsymbol{\xi} \in S^N\right\} = \int_a^b |g(\bar{\boldsymbol{\xi}}, t)|^p\, dt = \int_a^b |F_N(\omega, t)|^p\, dt, \qquad p > 0.$$

Now we fix a subspace $\mathcal{N}_N \subset L_p[a, b]$ $(1 \leq p < \infty)$ with a basis $\{e_1(t), \ldots, e_N(t)\}$ and we define an N-dimensional vector field $\eta(\boldsymbol{\xi}) = \{\eta_1(\boldsymbol{\xi}), \ldots, \eta_N(\boldsymbol{\xi})\}$ on S^N by

$$\eta_k(\boldsymbol{\xi}) = \int_a^b e_k(t)\,|g(\boldsymbol{\xi}, t)|^{p-1}\,\mathrm{sgn}\, g(\boldsymbol{\xi}, t)\, dt, \qquad k = 1, 2, \ldots, N.$$

It is clear that $\eta(-\boldsymbol{\xi}) = -\eta(\boldsymbol{\xi})$. The continuity of the field η on S^N for $p > 1$ is obvious and for $p = 1$ the continuity follows from the fact that whenever $\boldsymbol{\xi}^{(m)} \to \boldsymbol{\xi}$ on S^N then

$$\mathrm{meas}\,\{t\colon t \in [a, b],\ \mathrm{sgn}\, g(\boldsymbol{\xi}^{(m)}, t) \neq \mathrm{sgn}\, g(\boldsymbol{\xi}, t)\} \to 0.$$

Thus we can apply Theorem 2.5.1 once more and it ensures the existence of $\boldsymbol{\xi}^* \in S^N$ such that $\eta_k(\boldsymbol{\xi}^*) = 0$, $(k = 1, 2, \ldots, N)$. Now having Proposition 1.4.6 and (2.30) in mind we get for $1 \leq p < \infty$

$$E(g(\boldsymbol{\xi}^*), \mathcal{N}_N)_p = \|g(\boldsymbol{\xi}^*)\|_p \geq \|F_N(\omega)\|_p.$$

Passing to the limit in these relations we see that they are valid for $p = \infty$ too. For a concave ω we have for every $\boldsymbol{\xi} \in S^N$ that $g(\boldsymbol{\xi}, t) \in H^\omega[a, b]$ (see Exercise 8). Hence the inequality $E(H^\omega[a, b], \mathcal{N}_N)_p \geq \|F_N(\omega)\|_p$ holds for every subspace $\mathcal{N}_N \subset L_p[a, b]$ ($\dim \mathcal{N}_N = N$). Thus

$$d_N(H^\omega[a, b], L_p[a, b]) \geq \|F_N(\omega)\|_p = A_{N,p}(\omega), \qquad 1 \leq p \leq \infty, \quad (2.33)$$

and a comparison of (2.26), (2.27) and (2.33) gives:

Theorem 8.2.10 For every concave modulus of continuity ω and for all $N = 1, 2, \ldots,$ we have

$$d_N(H^\omega[a, b], L_p[a, b]) = A_{N,p}(\omega), \qquad 1 \leq p \leq \infty, \qquad (2.34)$$

$$\lambda_N(H^\omega[a, b], L_p[a, b]) = \pi_N(H^\omega[a, b], L_p[a, b]) = A_{N,p}(\omega), \qquad 1 \leq p \leq 3. \qquad (2.35)$$

The widths are realized by subspace $S_{N,0}[a, b]$ and in (2.35) by the interpolating-in-mean operator $\psi_N(f, t)$.

We do not know whether the equalities (2.35) are true for $p > 3$.

8.2.4 Widths of classes H^ω, W^1H^ω in L_p Inequalities (2.26) and (2.27) give for $[a, b] = [0, 2\pi]$ upper bounds for the even widths $d_{2n}(H^\omega, L_p)$, $\pi_{2n}(H^\omega, L_p)$ of class H^ω of 2π-periodic functions. When we want to get the exact lower bound for $d_{2n}(H^\omega, L_p)$ using the standard arguments from Section 8.2.3 we have to ensure that the function $g(\boldsymbol{\xi}, t)$ satisfies the condition $g(\boldsymbol{\xi}, 0) = g(\boldsymbol{\xi}, 2\pi)$. This can be guaranteed by setting for $\boldsymbol{\xi} \in S^{2n}$

$$g(\boldsymbol{\xi}, t) = 2^{-1} \min \{\omega(2t - 2\tau_{k-1}), \omega(2\tau_k - 2t)\} \operatorname{sgn} \xi_k,$$
$$\tau_{k-1} \le t \le \tau_k, \qquad k = 1, 2, \ldots, 2n+1, \qquad \tau_0 = 0, \qquad \tau_{2n+1} = 2\pi. \tag{2.36}$$

The function $g(\boldsymbol{\xi}, t)$ depends continuously and oddly on $\boldsymbol{\xi}$ and vanishes at all points τ_k $(k = 1, 2, \ldots, 2n+1)$. In particular, $g(\boldsymbol{\xi}, 0) = g(\boldsymbol{\xi}, 2\pi) = 0$. Moreover on $[a, b]$ $g(\boldsymbol{\xi}, t)$ is joined together from m $(m \le 2n+1)$ positive or negative 'hats', where m is the number of the non-zero coordinates of $\boldsymbol{\xi}$. It follows from Lemma 8.2.9 (for Φ_2) that if one of the $(2n+1)$ components ξ_k of $\boldsymbol{\xi} \in S^{2n}$ is zero then

$$\int_a^b |g(\boldsymbol{\xi}, t)|^p \, \mathrm{d}t = \int_a^b |f_{n,0}(\omega, t)|^p \, \mathrm{d}t, \qquad p > 0, \tag{2.37}$$

and hence $\|g(\boldsymbol{\xi})\|_p \ge \|f_{n,0}(\omega)\|_p$ $(p \ge 1)$ and we get the same estimate for $d_{2n}(H^\omega, L_p)$ as on the final interval. If $\xi_k \ne 0$ $(k = 1, 2, \ldots, 2n+1)$ then unlike the non-periodic case inequality (2.37) may not be true, e.g. $\xi_k = 2\pi k/(2n+1)$ $(k = 1, 2, \ldots, 2n+1)$.

However we can save the case by using the periodicity. Without going into technical details, we shall only outline the idea of the construction. We assume that $g(\boldsymbol{\xi}, t)$ is continued as a 2π-periodic function on the real axis. If all ξ_k are non-zero then the number of 'hats' of $g(\boldsymbol{\xi}, t)$ on $[0, 2\pi]$ is $(2n+1)$, i.e. an odd number, and because every periodic function can only change sign an even number of times on the period we conclude that there are two consecutive 'hats' with the same sign. We can change the function $g(\boldsymbol{\xi}, t)$ slightly near the common point of the supports of these 'hats' preserving the continuous and odd dependence on $\boldsymbol{\xi}$ so that relation (2.37) will hold. The detailed computations are given in [10A, p. 273].

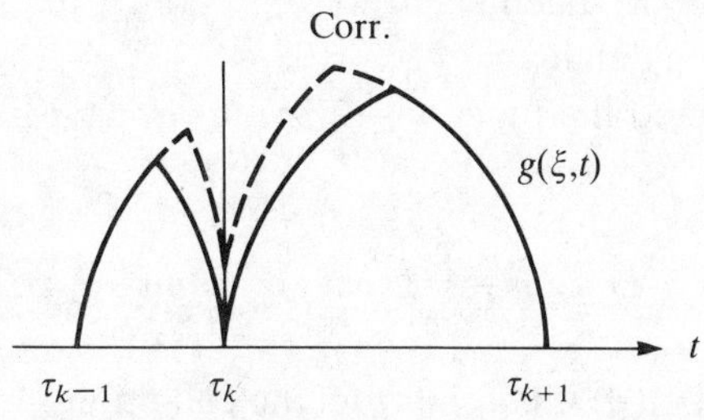

Fig. 8.2 Correction to function $g(\boldsymbol{\xi}, t)$.

Now we construct for a fixed subspace $\mathcal{N}_{2n}$, as on the interval, a continuous odd vector field $\eta(\boldsymbol{\xi})$ on S^{2n} and Theorem 2.5.1 gives the inequality

$$(H^\omega, \mathcal{N}_{2n})_p \geq \|f_{n,0}(\omega)\|_p, \qquad 1 \leq p \leq \infty,$$

valid for concave ω when $g(\boldsymbol{\xi}, t) \in H^\omega$.

Taking into account that we have $A_{2n,p}(\omega) = \|f_{n,0}(\omega)\|_p$ in relations (2.26) and (2.27) for $[a, b] = [0, 2\pi]$ we come to:

Theorem 8.2.11 For every concave modulus of continuity ω and for all $n = 1, 2, \ldots,$ we have

$$d_{2n}(H^\omega, L_p) = \|f_{n,0}(\omega)\|_p = A_{2n,p}(\omega), \qquad 1 \leq p \leq \infty, \tag{2.38}$$

$$\lambda_{2n}(H^\omega, L_p) = \pi_{2n}(H^\omega, L_p) = \|f_{n,0}(\omega)\|_p = A_{2n,p}(\omega), \qquad 1 \leq p \leq 3. \tag{2.39}$$

The widths (2.38) and (2.39) are realized by subspace $S_{2n,0}$ of 2π-periodic step functions with respect to the uniform partition $\{k\pi/n\}$ and in (2.39) by the interpolating-in-mean operator $\psi_{2n}(f, t)$.

The problem of the validity of (2.39) for $p > 3$ is open.

Now we shall consider the widths of W^1H^ω in L_p. In Section 7.3 (see (7.3.34)) it was established that

$$\sup_{f \in W^1H^\omega} \int_0^{2\pi} |f(t) - \sigma_{2n,1}(f, t)|^p \, dt \leq \int_0^{2\pi} |f_{n,1}(\omega, t)|^p \, dt, \qquad p > 0, \tag{2.40}$$

where $\sigma_{2n,1}(f, t)$ are the splines from $S_{2n,1}$ (i.e. polygons) which interpolate the function f at the points $k\pi/n$ $(k = 0, 1, \ldots, 2n)$. Moreover, $f_{n,1}(\omega, t)$ is the standard function constructed in Section 7.1. Because dim $S_{2n,1} = 2n$ we have

$$\pi_{2n}(W^1H^\omega, L_p) \leq \|f_{n,1}(\omega)\|_p, \qquad 1 \leq p \leq \infty. \tag{2.41}$$

Proving that we have for concave ω

$$d_{2n}(W^1H^\omega, L_p) \leq \|f_{n,1}(\omega)\|_p, \qquad 1 \leq p \leq \infty, \tag{2.42}$$

needs very detailed arguments and we only sketch the proof referring to [10A, p. 275] for the details.

Having constructed the functions (2.36) for every $\boldsymbol{\xi} \in S^{2n}$ and for every $p \geq 1$ we set

$$G(\boldsymbol{\xi}, t) = \int_0^t g(\boldsymbol{\xi}, u) \, du + c_1(\boldsymbol{\xi}),$$

where the constant $c_1(\boldsymbol{\xi})$ is chosen from the condition

$$\int_0^{2\pi} |G(\boldsymbol{\xi}, t)|^{p-1} \operatorname{sgn} G(\boldsymbol{\xi}, t)\, dt = 0. \tag{2.43}$$

The function $G(\boldsymbol{\xi}, t)$ depends continuously and oddly on $\boldsymbol{\xi}$.

We cannot guarantee that the inequality

$$\|G(\boldsymbol{\xi})\|_p \leq \|f_{n,1}(\omega)\|_p, \qquad 1 \leq p \leq \infty, \tag{2.44}$$

holds for every $\boldsymbol{\xi} \in S^{2n}$ and as in the lower bound of $d_N(H^\omega, L_p)$, we might have to make some corrections in the functions $g(\boldsymbol{\xi}, t)$ and $G(\boldsymbol{\xi}, t)$.

First we note (see Exercise 9) that if one of the components ξ_k of $\boldsymbol{\xi}$ is zero then (2.44) holds. Therefore only functions constructed with respect to vectors $\boldsymbol{\xi}$ with $\xi_k \neq 0$ $(k = 1, 2, \ldots, 2n+1)$ have to be corrected. But then at least at one of the points $\tau_1, \tau_2, \ldots, \tau_{2n+1}$ (see (2.36)) function $f(\boldsymbol{\xi}, t)$ does not change sign and we can correct it in such a way as to preserve the continuity and oddness with respect to $\boldsymbol{\xi}$, to keep $g(\boldsymbol{\xi}, t)$ in H^ω (for concave ω) and to let $G(\boldsymbol{\xi}, t)$ satisfy (2.43) and (2.44).

Assuming that this is done, we fix a basis $\{e_1(t), \ldots, e_{2n}(t)\}$ in subspace $\mathcal{N}_{2n}$ ($\dim \mathcal{N}_{2n} = 2n$), where we can set $e_1(t) \equiv 1$ because all constants are contained in W^rH^ω. The equalities

$$\eta_1(\boldsymbol{\xi}) = G(\boldsymbol{\xi}, 0) - G(\boldsymbol{\xi}, 2\pi),$$

$$\eta_k(\boldsymbol{\xi}) = \int_0^{2\pi} e_k(t)\, |G(\boldsymbol{\xi}, t)|^{p-1} \operatorname{sgn} G(\boldsymbol{\xi}, t)\, dt, \qquad k = 2, 3, \ldots, 2n,$$

define a continuous odd vector field $\eta(\boldsymbol{\xi})$ on sphere S^{2n} and for some $\boldsymbol{\xi}^* \in S^{2n}$ we have $\eta(\boldsymbol{\xi}^*) = 0$ because of Theorem 2.5.1. This implies that $G(\boldsymbol{\xi}^*, 0) = G(\boldsymbol{\xi}^*, 2\pi)$ and hence $g(\boldsymbol{\xi}, t) \in W^1H^\omega$ for concave ω. Further, the conditions $\eta_k(\boldsymbol{\xi}^*) = 0$ $(k = 2, 3, \ldots, 2n)$ together with (2.43) mean that for $1 \leq p < \infty$ we have $E(G(\boldsymbol{\xi}^*), \mathcal{N}_{2n})_p = \|G(\boldsymbol{\xi}^*)\|_p$. The validity of the last inequality for $p = \infty$ is established by passing to the limit.

Thus we have for concave ω

$$E(W^1H^\omega, \mathcal{N}_{2n})_p \leq \|f_{n,1}(\omega)\|_p, \qquad 1 \leq p \leq \infty,$$

and in view of the arbitrariness of subspace $\mathcal{N}_{2n}$ we come to (2.42). Therefore, in view of estimates (2.40)–(2.42) we have:

Theorem 8.2.12 For every concave modulus of continuity ω and for all $n = 1, 2, \ldots,$ we have

$$d_{2n}(W^1H^\omega, L_p) = \lambda_{2n}(W^1H^\omega, L_p) = \pi_{2n}(W^1H^\omega, L_p) = \|f_{n,1}(\omega)\|_p, \qquad 1 \leq p \leq \infty. \tag{2.45}$$

All widths are realized by the interpolating polygons $\sigma_{2n,1}(f, t)$ from subspaces $S_{2n,1}$.

Because $\sigma'_{2n,1}(f, t) = \psi_{2n}(f', t)$ for $f \in C^1$ in view of Theorem 8.2.11 we can assert that we have for concave ω and for every function $f \in W^1H^\omega$

$$\|f' - \sigma'_{2n,1}(f)\|_p \leq d_{2n}(H, L_p) = A_{2n,p}(\omega), \qquad 1 \leq p \leq 3. \quad (2.46)$$

Thus the interpolating splines $\sigma_{2n,1}(f, t)$ realize the best (in the sense of the width d_{2n}) approximation not only of the function $f \in W^1H^\omega$ in L_p $(p \geq 1)$ but simultaneously of its derivative in L_p $(1 \leq p \leq 3)$.

A few words about the widths of classes $W^1H^\omega[a, b]$ in connection with equalities (2.45). Theorem 7.3.4 gives

$$\pi_{N+1}(W^1H^\omega[0, 2\pi], L_p[0, \pi]) = \|f_{N,1}(\omega)\|_{L_p[0,\pi]}, \qquad 1 \leq p \leq \infty,$$

because subspace $S_{N,1}[0, \pi]$ containing the interpolating splines has dimension $(N + 1)$. On the other hand, standard arguments based on Theorem 2.5.1 establish (see Exercise 8.10) that we have for concave ω

$$d_N(W^1H^\omega[0, \pi], L_p[0, \pi]) \geq \|f_{N,1}(\omega)\|_{L_p[0,\pi]}, \qquad 1 \leq p \leq \infty,$$

and hence

$$\lim_{N\to\infty} d_N(W^1H^\omega[0, \pi], L_p[0, f])\, \|f_{N,1}(\omega)\|^{-1}_{L_p[0,\pi]}$$

$$= \lim_{N\to\infty} \pi_N(W^1H^\omega[0, \pi], L_p[0, \pi])\, \|f_{N,1}(\omega)\|^{-1}_{L_p[0,\pi]} = 1, \quad 1 \leq p \leq \infty.$$

8.2.5 On the widths of classes W^rH^ω $(r \geq 2)$ in L_1 Here we have restricted ourselves to a statement of the results only as we regard the presentation in this book of proofs as somewhat premature as they do not seem to be sufficiently simple and natural yet.

In this section ω is a concave modulus of continuity. It follows from Theorems 7.2.2 and 7.2.5

$$d_{2n-1}(W^rH^\omega, L_1) \leq E(W^rH^\omega, \mathcal{N}^T_{2n-1})_1 = E(W^rH^\omega, S_{2n,m})_1$$

$$= \|f_{n,r}(\omega)\|_1, \qquad m \geq r, \qquad n, r = 1, 2, \ldots, \quad (2.47)$$

and (7.3.41) also gives an estimate for the odd projective widths:

$$\pi_{2n}(W^rH^\omega, L_1) \leq \sup_{f\in W^rH^\omega} \|f - \sigma_{2n,r}(f)\|_1 = \|f_{n,r}(\omega)\|_1. \quad (2.48)$$

Obtaining exact lower estimates for the widths of class W^rH^ω in L_p, is a very complicated task as can be seen from the cases $r = 0, 1$ in Section 8.2.4. Nevertheless, the investigation of the extremal properties of ψ-splines (see Section 8.2.1) with the help of Σ-rearrangements allows us to prove (cf. references in the comments) that

$$d_{2n-1}(W^rH^\omega, L_1) \geq \|f_{n,r}(\omega)\|_1, \quad (2.49)$$

and the additional study of the topological properties of continuous mappings of a special type leads to the estimate

$$d_{2n}(W^r H^\omega, L_p) \geq \|f_{n,r}(\omega)\|_p, \qquad p \geq 1. \tag{2.50}$$

Relations (2.47)–(2.50) mean that we have for concave ω

$$d_{2n-1}(W^r H^\omega, L_1) = d_{2n}(W^r H^\omega, L_1) = \lambda_{2n}(W^r H^\omega, L_1)$$
$$= \pi_{2n}(W^r H^\omega, L_1) = \|f_{n,r}(\omega)\|_1, \tag{2.51}$$

where width d_{2n-1} is realized by trigonometric subspace $\mathcal{N}^T_{2n-1}$, d_{2n} is realized by subspaces $S_{2n,m}$ $(m \geq r)$ and all even widths by the interpolating splines $\sigma_{2n,r}(f,t)$ from $S_{2n,r}$.

8.3 Closely related extremal problems

There are other optimization problems connected with the N-widths of classes of functions.

8.3.1 Optimal recovery Many approximation problems of functions and linear functionals in which the approximating tools are constructed effectively can be interpreted as problems of their recovery from discrete information given, for example, by the values of the function and/or its derivatives at fixed points, its Fourier coefficients and so on. Considering the global recovery of the function f then a very general situation can be described as follows.

In the normed functional space X a set $M_N = \{\mu_1, \mu_2, \ldots, \mu_N\}$ of functionals μ_k defined on X can be considered as a coding method associating with $f \in X$ the vector

$$T(f, M_N) = \{\mu_1(f), \mu_2(f), \ldots, \mu_N(f)\}. \tag{3.1}$$

containing N bits of information for f. The problem of the recovery of f can be solved by corresponding the vector $T(f, M_N)$ to the function

$$A(f,t) = A(f; M_N, \Phi_N; t) = \sum_{k=1}^{N} \mu_k(f)\, \varphi_k(t), \tag{3.2}$$

where $\Phi_N = \{\varphi_1(t), \varphi_2(t), \ldots, \varphi_N(t)\}$ is some system of linearly independent functions from X. As a measure of the recovery error it is natural to take the quantity

$$\|f(\bullet) - A(f; M_N, \Phi_N; \bullet)\|_X, \tag{3.3}$$

and it is clear that the problem of estimating this error can be correctly stated under the existence of some *a priori* information about the function f which will guarantee at least the boundedness of the norm (3.3) on the set of all functions f with the same vector of information (3.1). This *a priori* information determines a class of functions $\mathcal{M}$ in X and

$$\rho(\mathcal{M}; M_N, \Phi_N)_X = \sup_{f \in M} \|f - A(f; M_N, \Phi_N)\|_X \tag{3.4}$$

is the recovery error of class $\mathcal{M}$.

Now given a fixed N one can consider the problem of minimizing (3.4) on all sets of functionals M_N for a fixed basis Φ_N or on all sets Φ_N of linearly independent functions from X for a fixed set of coding functionals M_N. We consider the problem closely related with N-widths, i.e. that of minimizing (3.4) simultaneously on M_N and Φ_N (N is fixed). Set

$$\rho_N(\mathcal{M}, X) = \inf_{M_N, \Phi_N} \rho(\mathcal{M}; M_N, \Phi_N)_X, \tag{3.5}$$

$$\rho'_N(\mathcal{M}, X) = \inf_{M'_N, \Phi_N} \rho(\mathcal{M}; M'_N, \Phi_N)_X, \tag{3.6}$$

where M'_N is a set of bounded linear functionals defined on X.

The quantities (3.5) and (3.6) characterize the minimal possible error in recovering the functions from $\mathcal{M}$ by methods (3.2) using N bits of information of the type (3.1), where in the case (3.6) the information vector (3.1) is defined by a set M'_N of N bounded linear functionals μ_k. We point out that in the problems of the optimal recovery (3.5) and (3.6) attention is paid not only to the determination of quantities $\rho_N(\mathcal{M}, X)$ and $\rho'_N(\mathcal{M}, X)$ but also to the effective construction of the best method, i.e. the pairs (M_N, Φ_N) or (M'_N, Φ_N) providing the minimal error on the class M.

The recovery method (3.2) can be made more flexible if we generalize it to

$$A(f, t) = \sum_{k=1}^{N} \gamma_k \mu_k(f)\, \varphi_k(t), \tag{3.7}$$

where γ_k are numerical coefficients which give the best method of fitting some class $\mathcal{M}$. This is exactly the form of Favard λ^*-means (see Section 4.1).

The determination of quantities (3.5) and (3.6) in almost all naturally arising cases can be reduced to the problems of N-widths and extremal subspaces and the particular linear methods which realize the N-widths provide the best methods for recovery.

Proposition 8.3.1 We have

$$\rho'_N(\mathcal{M}, X) = \lambda_N(\mathcal{M}, X), \tag{3.8}$$

$$\rho_N(\mathcal{M}, X) \geq d_N(\mathcal{M}, X). \tag{3.9}$$

If $\mathcal{M} = F \oplus \mathcal{N}$, where F is a compact set and $\mathcal{N}$ is finite dimensional subspace then an equality also holds in (3.9).

Indeed, the inequality $\rho'_N(\mathcal{M}, X) \geq \lambda_N(\mathcal{M}, X)$ follows by definition. On the other hand if $\{\varphi_1(t), \varphi_2(t), \ldots, \varphi_N(t)\}$ is a basis of $\mathcal{N}_N$ then every linear bounded operator $A: X \to \mathcal{N}_N$ can be represented (see Proposition 1.1.8) in the form (3.2), where μ_k are linear functionals. Therefore

$$\sup_{f \in \mathcal{M}} \|f - Af\|_X \geq \rho'_N(\mathcal{M}, X),$$

and because this is true for every N-dimensional subspace $\mathcal{N}_N$ and every linear operator $A: X \to \mathcal{N}_N$ we have $\lambda_N(\mathcal{M}, X) \geq \rho'_N(\mathcal{M}, X)$.

Inequality (3.9) follows by definition. If $\mathcal{M}$ is the sum of a compact and a finite dimensional subspace then the Kolmogorov width is defined by equality (1.2.11). Inequality $d_N(\mathcal{M}, X) \geq \rho_N(\mathcal{M}, X)$ is obtained as in the linear case using the fact that A can be represented in the form (3.2), where μ_k are only continuous functionals.

We consider some particular cases. If $\mathcal{M}$ is any of the classes W_p^{m+1}, $W^m H^\omega$ $(m = 0, 1, \ldots,)$ of 2π-periodic functions then the set $\mathcal{M}_0 = \{f: f \in \mathcal{M}, f(0) = 0\}$ is compact in C or L_p and hence $\mathcal{M}$ can be considered as the sum of a compact and the one-dimensional subspace of constants. In the same way the classes $W_p^{m+1}[a, b]$ and $W^m H^\omega[a, b]$ can be represented as a sum of a compact and the $(m + 1)$ dimensional subspace of the algebraic polynomials of degree m. In view of Proposition 8.3.1 we have in all these cases

$$\rho_N(\mathcal{M}, X) = d_N(\mathcal{M}, X), \qquad \rho'_N(\mathcal{M}, X) = \lambda_N(\mathcal{M}, X),$$

where X is one of the spaces C or L_q, $C[a, b]$ or $L_q[a, b]$ respectively. These equalities allow all the results considered in Sections 8.1 and 8.2 for the widths of the classes to be restated in terms of the optimal recovery. In this setting a high value is given to the result in which the widths are realized by the particular linear method which is effectively constructed on the basis of discrete information about the approximated function.

We shall not give the new formulations of the previous results (the reader can do this on his/her own). We only show how the approximation properties of some linear methods look from the new point of view.

The trigonometric Favard sums:

$$U_{n-1,r}(f, \lambda^*, t) = \frac{a_0}{2} + \sum_{k=1}^{n-1} \lambda_{rk}^{n-1,*} (a_k \cos kt + b_k \sin kt),$$

where a_k, b_k are Fourier coefficients of the function f and numbers $\lambda_{rk}^{n-1,*}$ are defined in (4.1.41), are obviously represented in the form (3.2) with $N = 2n - 1$. In view of Theorems 8.1.11 and 8.1.13 and Proposition 8.3.1

these sums are the best (in the sense of (3.5) and (3.6)) method of recovery of a function from classes W_∞^r in the C metric and a function from classes W_1^r, W_V^{r-1} in L_1 metric among all recovery methods of type (3.2) with $N \le 2n$.

Excellent properties in terms of optimal recovery are provided by the 2π-periodic interpolating splines $\sigma_{2n,m}(f, t)$ with respect to the uniform partition. They can be given as (see Section 2.4.4)

$$\sigma_{2n,m}(f,t) = \sum_{k=1}^{2n} f(\tau_{mk})\, s_k(t), \tag{3.10}$$

where s_k are the fundamental splines from $S_{2n,m}$ and τ_{mk} are the interpolation points ($\tau_{mk} = k\pi/n$ if m is odd and $\tau_{mk} = (2k-1)\pi/(2n)$ if m is even), and they define an operator of linear projection onto $S_{2n,m}$.

From Theorems 8.1.11, 8.1.13 and Proposition 8.3.1 it follows that among the methods of type (3.2) with $N = 2n$ the splines $\sigma_{2n,r-1}(f, t)$ provide the best method for recovery of functions $f \in W_\infty^r$ in L_q and functions $f \in W_p^r$ in L_1. Moreover Proposition 5.2.9 and Theorem 8.1.11 imply that splines $\sigma_{2n,r-1}(f, t)$ recover the minimal possible error in the L_q metric simultaneously with the derivative of a function $f \in W_\infty^r$. In view of (7.3.41) and (2.51) interpolating splines $\sigma_{2n,r}(f, t)$ are the best method for recovering functions f from class W^rH^ω $(r \ge 1)$ in the L_1 metric and for a concave ω; splines $\sigma_{2n,1}(f, t)$ recover in the best way a function $f \in W^1H^\omega$ in L_q $(1 \le q \le \infty)$ and simultaneously its derivative f' in L_q $(1 \le q \le 3)$ – this follows from the results in Section 8.2.4.

The results stated in Sections 8.1 and 8.2 for the widths of classes $W_p^r[a, b]$ and $W^rH^\omega[a, b]$ allow in many cases a sequence of linear methods for recovery which are asymptotically exact for some class of functions to be indicated. In particular, such properties are provided by the interpolating splines with respect to the uniform partition with Leedstone boundary conditions.

So far we have discussed the global recovery of a function by estimating the error in the norms of C or L_p. Another problem is the following: given the information vector (3.1) recover the value on f of some fixed linear functional of f, for example the value of f at a fixed point t. We consider this case now which gives us the opportunity of demonstrating another angle of the best approximating properties of the interpolating splines $\sigma_{2n,m}(f, t)$.

Given a fixed set of functionals $M_N = (\mu_1, \mu_2, \ldots, \mu_N\}$ we shall look for an approximate value of $f(t_*)$ represented in the form

$$\alpha(f, M_N, \mathbf{c}) = \sum_{k=1}^{N} c_k \mu_k(f), \tag{3.11}$$

where $\boldsymbol{c} = \{c_1, c_2, \ldots, c_N\}$ is the vector of the coefficients at our disposal. The problem is: find the quantity

$$\beta(\mathcal{M}, M_N, t_*) = \inf_{\boldsymbol{c}} \sup_{f \in \mathcal{M}} |f(t_*) - \alpha(f, M_N, \boldsymbol{c})| \qquad (3.12)$$

and indicate the coefficient vector $\tilde{\boldsymbol{c}} = \{\tilde{c}_1, \tilde{c}_2, \ldots, \tilde{c}_N\}$ which determines the best recovery method.

In some cases we can estimate quantity (3.12) from below (without any losses as we shall see later) using the following simple trick. Let $\mathcal{M}$ be a center-symmetric set of functions and

$$\mathcal{M}(M_N) = \{f : f \in \mathcal{M}, \mu_k(f) = 0, k = 1, 2, \ldots, N\}.$$

Then we have $\alpha(f, M_N, \boldsymbol{c}) = 0$ for $f \in \mathcal{M}(M_N)$ and hence

$$\beta(\mathcal{M}, M_N, t_*) \geq \beta(\mathcal{M}(M_N), M_N, t_*) \geq \sup_{f \in \mathcal{M}(M_N)} |f(t_*)|. \qquad (3.13)$$

We consider one particular case. Let $M_{2n,r}^{\tau}$ be the set of functionals $\mu_k(f) = f(\tau_{rk})$ $(k = 1, 2, \ldots, 2n)$, where τ_{rk} are the zeros of the perfect spline $\varphi_{n,r+1}$ at which the spline $\sigma_{2n,r}(f, t)$ interpolates the function f. As can be seen from (3.10) the spline $\sigma_{2n,r}(f, t)$ is constructed with respect to the information vector

$$T(f, M_{2n,r}^{\tau}) = \{f(\tau_{r1}), f(\tau_{r2}), \ldots, f(\tau_{r,2n})\}. \qquad (3.14)$$

From Proposition 5.2.4 we have for every function $f \in W_\infty^{r+1}$ and all t

$$|f(t) - \sigma_{2n,r}(f, t)| \leq |\varphi_{n,r+1}(t)|.$$

Thus, the spline $\sigma_{2n,r}(f, t)$ which uses the information (3.14) recovers the values of $f \in W_\infty^{r+1}$ at every point $t \neq \tau_{rk}$ with an error not exceeding $|\varphi_{n,r+1}(t)|$.

It turns out that no other method of recovery of type (3.11) using information (3.14) can guarantee a smaller error on the whole class W_∞^{r+1} for a fixed point t. Indeed, for a fixed point $t = t_*$ we get from (3.13) for every coefficient vector $\boldsymbol{c}$

$$\sup_{f \in W_\infty^{r+1}} |f(t_*) - \sum_{k=1}^{2n} c_k f(\tau_{rk})| \geq \sup \{|f(t_*)| : f \in W_\infty^{r+1},$$

$$f(\tau_{rk}) = 0, k = 1, 2, \ldots, 2n\} = |\varphi_{n,r+1}(t_*)|. \qquad (3.15)$$

The last equality follows from Corollary 5.1.2 of Theorem 5.1.1, by whose virtue we can, by the way, even replace class W_∞^{r+1} in (3.15) by the larger class $W^r H_{2n}^1$.

A remarkable fact is that the coefficients $\tilde{c}_k$ dependent on the point t_* in problem (3.12) have an uniform representation in the case considered:

$\tilde{c}_k(t_*) = s_k(t_*)$, where s_k are the fundamental splines from $S_{2n,r}$ for the system of interpolating points τ_{rk}.

8.3.2 Optimal coding of functions In Section 8.3.1 the image of the function f from the class $\mathcal{M}$ belongs to some N-dimensional subspace when recovered using vector (3.1). If we would want to recover f as a function in the same class $\mathcal{M}$ then the error of the optimal recovery is determined by the error with which function f is coded by vector $T(f, M_N)$; thus we come to the problem of optimal coding.

As a rule, the information vector (3.1) determines a whole bundle of functions in $\mathcal{M}$ which are the pre-images in class $\mathcal{M}$ of the vector $T(f, M_N)$ under the mapping $f \to T(f, M_N)$. This pre-image is the set

$$T^{-1}(f, M_N) = \{g: g \in \mathcal{M}, T(g, M_N) = T(f, M_N)\}$$

if we would like to 'remember' the function f from the vector $T(f, M_N)$ then it will be also correct to choose any one of the functions $g \in T^{-1}(f, M_N)$. In the case $\|f - g\|_X$ one can say that the error does not exceed the diameter of the set $T^{-1}(f, M_N)$ in the metric of X. In order to measure this error using the characteristics defining the class $\mathcal{M}$ we introduce the quantity

$$K(\mathcal{M}, M_N)_X = \sup_{f \in \mathcal{M}} \operatorname{diam}_X T^{-1}(f, M_N)$$

$$= \sup \{\|f_1 - f_2\|_X: f_1, f_2 \in \mathcal{M}, T(f_1, M_N) = T(f_2, M_N)\}, \quad (3.16)$$

which can be interpreted as the error on class $\mathcal{M}$ of the coding method given by a final set of functionals $M_N = \{\mu_1, \mu_2, \ldots, \mu_N\}$.

Restricting ourselves to the most important case $M_N = M'_N$ – a set of linear functionals – we formulate two naturally arising problems.

(1) For a fixed $N = 1, 2, \ldots$, find the quantity

$$\lambda^N(\mathcal{M}, X) = \inf_{M'_N} K(\mathcal{M}, M'_N)_X \quad (3.17)$$

and indicate a set of functionals M'_N realizing the infimum in (3.17) and, therefore, defining the best linear method for coding the functions in class $\mathcal{M}$.

(2) Indicate a way for the effective recovery of the pre-image $T^{-1}(f, M'_N) \subset \mathcal{M}$ for the vector $T(f, M'_N)$ of optimal coding.

Quantity (3.17) can easily be bounded from above using the linear width of class $\mathcal{M}$. Indeed, if $T(f_1, M_N) = T(f_2, M_N)$, i.e. $\mu_k(f_1) = \mu_k(f_2)$ $(k = 1, 2, \ldots, N)$, then (see (3.2)) $A(f_1; M_N, \Phi_N; t) \equiv A(f_2; M_N, \Phi_N; t)$ and hence

$$\|f_1 - f_2\|_X \leq \|f_1 - A(f_1; M_N, \Phi_N)\|_X + \|f_2 - A(f_2; M_N, \Phi_N)\|_X.$$

But then in view of definitions (3.4) and (3.16)

$$K(\mathcal{M}, M_N)_X \le 2\rho(\mathcal{M}, M_N, \Phi_N)_X.$$

Because this is true for every set of basis functions Φ_N we have for $M_N = M'_N$

$$\lambda^N(\mathcal{M}, X) \le 2\rho'_N(\mathcal{M}, X) = 2\lambda_N(\mathcal{M}, X). \tag{3.18}$$

Here, we have also used (3.8).

Obtaining an exact lower bound for $\lambda^N(\mathcal{M}, X)$ is essentially connected with the characteristics of the class $\mathcal{M}$. Sometimes this can be done using:

Proposition 8.3.2 If $\mathcal{M}$ is a convex central-symmetric set in a functional normed space X then

$$\lambda^N(\mathcal{M}, X) = 2 \inf_{M'_N} \sup \{\|f\|: f \in \mathcal{M}, T(f, M'_N) = 0\}. \tag{3.19}$$

Proof The set M contains function $-f$ together with f and because in our case $\mu_k(-f) = -\mu_k(f)$ it follows from definition (3.16) that

$$\begin{aligned} K(\mathcal{M}, M'_N)_X &\ge \sup \{\|f - (-f)\|_X: f \in \mathcal{M}, T(f, M'_N) = 0\} \\ &= 2 \sup \{\|f\|: f \in \mathcal{M}, T(f, M'_N) = 0\}. \end{aligned}$$

On the other hand, if we have the equality $T(f_1, M'_N) = T(f_2, M'_N)$ for functions $f_1, f_2 \in \mathcal{M}$, then we have $T(g, M'_N) = 0$ for function $g(t) = [f_1(t) - f_2(t)]/2$ which also belongs to $\mathcal{M}$, and using definition (3.16) again we get

$$K(\mathcal{M}, M'_N)_X \le 2 \sup \{\|g\|: g \in \mathcal{M}, T(g, M'_N) = 0\}.$$

Thus for every set M'_N of linear functionals μ_k determining an information vector $T(f, M'_N)$ we have

$$K(\mathcal{M}, M'_N)_X = 2 \sup \{\|f\|: f \in \mathcal{M}, T(f, M'_N) = 0\}, \tag{3.20}$$

and in order to get (3.19) we take the infimum in (3.20) on all sets M'_N. □

In particular cases of bounding the right-hand side of (3.19) from below we can use Borsuk's theorem. Let us illustrate this on class W^r_∞. We shall show that

$$\lambda^{2n}(W^r_\infty, L_q) \ge 2\,\|\varphi_{n,r}\|_q, \qquad 1 \le q \le \infty. \tag{3.21}$$

It is enough to prove that we have for every set M'_{2n} of linear functionals μ_k $(k = 1, 2, \ldots, 2n)$ given on L_q

$$K(W^r_\infty, M'_{2n})_{L_q} \ge 2\,\|\varphi_{n,r}\|_q.$$

Because the class W^r_∞ contains all constants we have to consider only those sets M'_N for which not all numbers $\mu_k(1)$ are zero – otherwise (3.20) implies $K(W^r_\infty, M'_{2n})_{L_q} = \infty$. Let $\mu_1(1) \neq 0$. In the proof of Lemma 8.1.9 on the sphere S^N of $(N+1)$-dimensional space we have defined a function $g_r(\xi, t)$ $(0 \le t \le 2\pi)$ (see (1.15)) which is continuous and odd with respect to $\xi \in S^r$ and which is such that $|g_r^{(r)}(\xi, t)| = 1$ a.e., such that $g_r^{(r)}(\xi, t)$ changes sign on $(0, 2\pi)$ no more than N times and finally such that

$$\int_0^{2\pi} g_r^{(\nu)}(\xi, t)\,dt = 0, \qquad \nu = 1, 2, \ldots, r-1. \tag{3.22}$$

Assuming $N = 2n$ we set

$$G(\xi, t) = g_r(\xi, t) - \mu_1(g_r(\xi))/\mu_1(1) \tag{3.23}$$

and define a $2n$-dimensional vector field η on sphere S^{2n} by:

$$\eta_1(\xi) = \int_0^{2\pi} g_r^{(r)}(\xi, t)\,dt, \qquad \eta_k(\xi) = \mu_k(G(\xi)), \qquad k = 2, 3, \ldots, 2n.$$

The field η is continuous and odd on S^{2n} and Theorem 2.5.1 implies the existence of $\xi^* \in S^{2n}$ such that $\eta_k(\xi^*) = 0$ $(k = 1, 2, \ldots, 2n)$. But then $g_r^{(r)}(\xi^*, t) \perp 1$ and hence (see (3.22)) $G(\xi^*, t)$ belongs to the set $\Gamma_{2n,r}$ of 2π-periodic perfect splines and hence $G(\xi^*, t) \in W^r_\infty$. Moreover, in view of (3.23) we have $\mu_1(G(\xi)) = 0$ for every $\xi \in S^{2n}$. Therefore $T(G(\xi^*), M'_N) = 0$. Thus, using (3.20) and Theorem 8.1.8 we have

$$K(W^r_\infty, M'_{2n})_{L_q} \ge 2\,\|G(\xi^*)\|_q \ge 2\,\|\varphi_{n,r}\|_q.$$

The proof of inequality is slightly more difficult

$$\lambda^{2n}(W^r_p, L_1) \ge 2\,\|\varphi_{n,r}\|_{p'}, \qquad 1 \le q \le \infty, \qquad p' = p/(p-1). \tag{3.24}$$

Here again it suffices to prove that we have for every set M'_{2n} of linear functionals μ_k $(k = 1, 2, \ldots, 2n)$, $\mu_1(1) \neq 0$,

$$\sup\{\|f\|_1 : f \in W^r_p,\ T(f, M'_{2n}) = 0\} \ge \|\varphi_{n,r}\|_{p'}. \tag{3.25}$$

We prove this relation for $1 < p \le \infty$; it can be extended to $p = 1$ by passing to the limit as $p \to 1$.

Let us go back to the function $g_r(\xi, t)$ which we constructed so that it is determined up to an additive constant. We now assume that

$$\int_0^{2\pi} |g_r(\xi, t)|^{p'-1} \operatorname{sgn} g_r(\xi, t)\,dt = 0.$$

Therefore there is a function $f \in L^r_p$, $f(t) = f(\xi, t)$, such that

$$f^{(r)}(\xi, t) = \|g_r(\xi)\|_{p'}^{1-p'}\,|g_r(\xi, t)|^{p'-1} \operatorname{sgn} g_r(\xi, t), \qquad 0 \le t \le 2\pi. \tag{3.26}$$

Since $\|f^{(r)}(\xi)\|_p = 1$ function $f(\xi, t)$ belongs to W^r_p.

It is easy to see that f is continuous and odd on $\xi \in S^{2n}$ and that the same is true for the function belonging to W_p^r:

$$F(\xi, t) = f(\xi, t) - \mu_1(f(\xi))/\mu_1(1), \tag{3.27}$$

where μ_1 is the first functional from the fixed set M'_{2n}, $\mu_1(1) \neq 0$.

Define the vector field η on sphere S^{2n} by:

$$\eta_1(\xi) = \int_0^{2\pi} g_r^{(r)}(\xi, t)\, dt,$$

$$\eta_k(\xi) = \mu_k(F(\xi)), \qquad k = 2, 3, \ldots, 2n.$$

which is also continuous and odd on S^{2n}. Theorem 2.5.1 implies the existence of $\xi^* \in S^{2n}$ such that $\eta_k(\xi^*) = 0$ $(k = 1, 2, \ldots, 2n)$. The first equality implies that $g_r(\xi^*, t) \in \Gamma_{2n,r}$ and the others together with the relation $\mu_1(F(\xi)) = 0$ which follows from (3.27) gives $T(F(\xi^*), M'_N) = 0$.

We now bound the norm $\|F(\xi^*)\|_1$ from below. Because $g_r^{(r)}(\xi^*) \perp 1$ and because of (3.26) we have

$$\|F(\xi^*)\|_1 \geq \int_0^{2\pi} F(\xi^*, t)\, g_r^{(r)}(\xi, t)\, dt = \left| \int_0^{2\pi} f^{(r)}(\xi^*, t)\, g_r(\xi, t)\, dt \right|$$

$$= \|g_r(\xi^*)\|_{p'} \geq \|\varphi_{n,r}\|_{p'},$$

where we have also used that $g_r(\xi^*, t) \in \Gamma_{2n,r}$ (see Theorem 8.1.8). Thus the estimate (3.25) is proved for every M'_{2n} and hence (3.24) too.

Inequalities (3.21), (3.24) and (3.18) together with Theorems 8.1.11 and 8.1.13 allow us to state:

Theorem 8.3.3 For all $n, r = 1, 2, \ldots,$ we have

$$\lambda^{2n-1}(W_\infty^r, C) = \lambda^{2n}(W_\infty^r, C) = 2K_r n^{-r}, \tag{3.28}$$

$$\lambda^{2n-1}(W_1^r, L_1) = \lambda^{2n}(W_1^r, L_1) = 2K_r n^{-r}, \tag{3.29}$$

$$\lambda^{2n}(W_\infty^r, L_q) = 2\|\varphi_{n,r}\|_q, \qquad \lambda^{2n}(W_p^r, L_1) = 2\|\varphi_{n,r}\|_{p'}, \tag{3.30}$$

$$1 \leq p \leq \infty, \qquad 1/p + 1/p' = 1.$$

The best method for coding the functions from the class W_∞^r in the L_q metric and from the class W_p^r in the L_1 metric is provided by the set $M_{2n,r-1}^{\tau}$ of functionals $\mu_k(f) = f(\tau_{r-1,k})$, where $\tau_{r-1,k}$ are the zeros of the perfect spline $\varphi_{n,r}$.

Now a few words about the second of the problems formulated at the beginning of the section. How can one 'remember' with possibly minimal error a function f from the class $\mathcal{M}$ provided an information vector $T(f, M_N)$

is given? Let $\mathcal{M} = W_\infty^{r+1}$ and suppose the function f from W_∞^{r+1} is coded by the vector $T(f, M_{2n,r}^\tau)$. One can guess that the recovery of the pre-image $T^{-1}(f, M_{2n,r}^\tau)$ with minimal error on W_∞^{r+1} is provided by the same spline $\sigma_{2n,r}(f,t)$ from $S_{2n,r}$ that coincides with f at the points τ_{rk}.

Indeed, we set (f is assumed fixed)

$$q(f) = \{g\colon g \in W_\infty^{r+1}, |g(t) - \sigma_{2n,r}(f,t)| \le |\varphi_{n,r+1}(t)|\}.$$

If $g \in W_\infty^{r+1}$ and $T(g, M_{2n,r}^\tau) = T(f, M_{2n,r}^\tau)$ then $\sigma_{2n,r}(g,t) \equiv \sigma_{2n,r}(f,t)$ and, in view of Proposition 5.2.4, we have $|g(t) - \sigma_{2n,r}(f,t)| \le |\varphi_{n,r+1}(t)|$, i.e. $g \in Q(f)$. This means that $T^{-1}(f, M_{2n,r}^\tau) \subset Q(f)$ and, as can be easily seen, for $f = \varphi_{n,r+1}$ these sets coincide. Thus, recovering a function f coded by vector $T(f, M_{2n,r}^\tau)$ we can take every function $g \in Q(f)$ and the error $|f(t) - g(t)|$ at every point $t \ne \tau_{rk}$ will be minimal on the whole class W_∞^{r+1}.

Considering the recovery error for the L_q norm we come to the estimate

$$\sup_{f \in W_\infty^{r+1}} \operatorname{diam}_{L_q} Q(f) = K(W_\infty^{r+1}, M_{2n,r}^\tau)_{L_q}$$

$$= 2\|\varphi_{n,r+1}\|_q = \lambda^{2n}(W_\infty^{r+1}, L_q), \qquad 1 \le q \le \infty,$$

which says that this error cannot be decreased on the class W_∞^{r+1} if we replace $M_{2n,r}^\tau$ by any other set M_{2n}' of linear functionals.

A similar fact is valid for the recovery of functions from class W_p^{r+1} in the L_1 metric (see Exercise 11).

8.3.3 Minimization of the exact constants in Jackson's inequalities In Chapter 6 we formulated problems (6.1.21), (6.1.22) and (6.1.24) as being equivalent to finding out the exact constants in Jackson's inequalities for a fixed approximating subspace $\mathcal{N}_N$. For some subspaces of polynomials and splines we obtained a solution for these problems. Here we shall look for the minimal constant with respect to the whole set of approximating subspaces of fixed dimension N. More exactly, we shall show that some of the results in Chapter 6 cannot be improved even if the approximating subspaces (of the same dimension) are changed.

Let X, Y be the spaces $C[a,b]$, $L_p[a,b]$ $(1 \le p < \infty)$ or their periodic analogs C, L_p, let X^r $(r = 1, 2, \ldots,)$ be the set of functions $f \in X$ which are rth integrals of functions $g \in X$, $X^r \subset Y$, $X^0 = X$. We set

$$\kappa_{N,\gamma}(X^r, Y) = \inf_{\mathcal{N}_N} \sup_{f \in X^r} \frac{E(f, \mathcal{N}_N)_Y}{\omega(f^{(r)}, \gamma)_X}, \tag{3.31}$$

$$\kappa_{N,\gamma}'(X^r, Y) = \inf_{\mathcal{N}_N} \inf_{A \in \mathcal{L}(Y, \mathcal{N}_N)} \sup_{f \in X^r} \frac{\|f - Af\|_Y}{\omega(f^{(r)}, \gamma)_X}, \tag{3.32}$$

where $\mathcal{L}(Y, \mathcal{N}_N)$ is the set of all bounded linear operators from Y to $\mathcal{N}_N$. The ratio 0/0 on the right-hand sides of (3.31) and (3.32) is to be assumed 0. In

(3.31) and (3.32) we have set the problem of minimizing quantities (6.1.21) and (6.1.24) on all subspaces of dimension N and a close connection with the N-widths is naturally expected.

Denote by X^r_γ the set of functions $f \in X^r$ such that $\omega(f^{(r)}, \gamma)_X \le 1$.

Proposition 8.3.4 We have

$$\kappa_{N,\gamma}(X^r, Y) = d_N(X^r_\gamma, Y), \tag{3.33}$$

$$\kappa'_{N,\gamma}(X^r, Y) = \lambda_N(X^r_\gamma, Y). \tag{3.34}$$

Proof Let $f \in X^r$ and $\omega(f^{(r)}, \gamma)_X = \beta > 0$. If $f_1(t) = f(t)/\beta$ then $\omega(f_1^{(r)}, \gamma)_X = 1$, i.e. $f_1 \in X^r_\gamma$. Taking the positive homogeneity of functionals $E(f, \mathcal{N}_N)_Y$ and (for fixed γ) $\omega(f^{(r)}, \gamma)_X$ into account we obtain

$$\sup_{f \in X^r} \frac{E(f, \mathcal{N}_N)_Y}{\omega(f^{(r)}, \gamma)_X} \le \sup_{f \in X^r_\gamma} E(f, \mathcal{N}_N)_Y.$$

Taking the infimum over all subspaces $\mathcal{N}_N \subset Y$ of dimension N we obtain $\kappa_{N,\gamma}(X^r, Y) \le d_N(X^r_\gamma, Y)$.

On the other hand we have for every function $f \in X^r_\gamma$ by the definition of the set X^r_γ

$$E(f, \mathcal{N}_N)_Y \le \frac{E(f, \mathcal{N}_N)_Y}{\omega(f^{(r)}, \gamma)_X}$$

and because of the validity of this inequality for every subspace $\mathcal{N}_N$ we get $d_N(X^r_\gamma, Y) \le \kappa_{N,\gamma}(X^r, Y)$. This proves (3.33) and the proof of (3.34) is similar. □

The calculation of quantities (3.31) and (3.32) in particular cases is based, besides on the results in Chapter 6, on some facts that were established when the lower bounds for Kolmogorov N-widths of classes of functions were proved.

In particular, in considering the periodic case we use the following corollary of Theorem 8.1.8 and Lemma 8.1.9.

Proposition 8.3.5 Let $\mathcal{N}_{2n}$ be a subspace in L_p $(p \ge 1)$ containing the constants and with $\dim \mathcal{N}_{2n} = 2n$. For every $m = 1, 2, \ldots,$ there exists a function $g \in L^m_\infty$ (depending on $\mathcal{N}_{2n}$) such that $|g^{(m)}(t)| = 1$ a.e., such that $g^{(m)}$ essentially changes sign no more than $2n$ times on the period and such that

$$E(g, \mathcal{N}_{2n})_p = \|g\|_p \ge \|\varphi_{n,m}\|_p. \tag{3.35}$$

For $m = 0$ the above holds for every subspace $\mathcal{N}_{2n}$.

With the help of this assertion it is not difficult to obtain an exact lower bound for the constant M when $X = C$ and $X = L_1$, $Y = L_p$ in the inequality

$$E(f, \mathcal{N}_{2n})_Y \leq M \, \|f^{(r)}\|_X,$$

which is weaker than the Jackson inequality

$$E(f, \mathcal{N}_{2n})_Y \leq M \, \omega(f^{(r)}, \gamma)_X.$$

Proposition 8.3.6 For every subspace $\mathcal{N}_{2n} \subset L_p$ $(1 \leq p \leq \infty)$ of dimension $2n$ we have

$$\sup_{f \in C^r} \frac{E(f, \mathcal{N}_{2n})_p}{\|f^{(r)}\|_C} \geq \|\varphi_{n,r}\|_p, \qquad r = 0, 1, \ldots, \tag{3.36}$$

$$\sup_{f \in L_1^r} \frac{E(f, \mathcal{N}_{2n})_p}{\|f^{(r)}\|_1} \geq \|g_{n,r-1}\|_1, \qquad r = 1, 2, \ldots, \tag{3.37}$$

where $\varphi_{n,m}$ is the perfect Euler spline, $g_{n,m}(t) = (4n)^{-1} \varphi_{n,m}(t)$.

Proof It is clear that for $r \geq 1$ we only have to consider the subspaces $\mathcal{N}_{2n}$ which contain the constants (otherwise the left-hand sides of (3.36) and (3.37) are infinity). We set $m = r$ and g_h to be the Stekhlov function of g in Proposition 8.3.5. Then $g_h \in C^r$ and $\|g_h^{(r)}\|_C = 1$ for small enough positive h. Because (see Section A2) $\|g_h - g\|_p \to 0$ as $h \to 0$ and because of the continuity of the functional of best approximation, we obtain from (3.35)

$$\lim_{h \to 0} E(g_h, \mathcal{N}_{2n})_p = E(g, \mathcal{N}_{2n})_p \geq \|\varphi_{n,r}\|_p. \tag{3.38}$$

Therefore

$$E(g_h, \mathcal{N}_{2n})_p > \|\varphi_{n,r}\|_p \, [1 - \varepsilon(h)],$$

where $\varepsilon(h) > 0$, $\varepsilon(h) \to 0$ as $h \to 0$. This proves (3.36) for $1 \leq p < \infty$. Taking the limit as $p \to \infty$ we also obtain it for $p = \infty$.

In order to prove (3.37) we once more consider the Stekhlov function g_h of g in Proposition 8.3.5 but we take now $m = r - 1$. In this case $g_h \in L_1^r$ and (see Proposition A2.3)

$$\|g_h^{(r)}\|_1 = \bigvee_0^{2\pi} g_h^{(r-1)} \leq \bigvee_0^{2\pi} g^{(r-1)} \leq 4n.$$

Relation (3.38) still holds with r replaced by $r - 1$. Hence

$$\frac{E(g_h, \mathcal{N}_{2n})_p}{\|g_h^{(r)}\|_1} > \|g_{n,r-1}\|_p \, [1 - \varepsilon_1(h)], \qquad 1 \leq p < \infty,$$

where $\varepsilon_1(h) \to 0$ if $h \to 0$. Here the case $p = \infty$ is also obtained by passing to the limit. □

Because $\omega(f,\delta)_X \le 2\|f\|_X$ we obtain from Proposition 8.3.6

$$\kappa_{2n,\gamma}(C^r, L_p) \ge \tfrac{1}{2}\|\varphi_{n,r}\|_p; \qquad \kappa_{2n,\gamma}(L_1^r, L_p) \ge \tfrac{1}{2}\|g_{n,r-1}\|_p,$$
$$1 \le p \le \infty; \qquad n = 1, 2, \ldots,. \tag{3.39}$$

Upper bounds for the quantities (3.31) and (3.32) in particular cases are provided by the exact results in Chapter 6 for approximation by polynomials and splines. Comparing (3.39) with Theorems 6.2.7 and 6.3.5 and with Proposition 6.3.7 (see also (6.3.31) and (6.3.32)) we obtain:

Theorem 8.3.7 We have

$$\kappa_{N,\pi/n}(C^{2\nu-1}, C) = \kappa'_{N,\pi/n}(C^{2\nu-1}, C) = \kappa_{N,\pi/n}(L_1^{2\nu-1}, L_1)$$
$$= \kappa'_{N,\pi/n}(L_1^{2\nu-1}, L_1) = K_{2\nu-1}/(2n^{2\nu-1}),$$
$$N = 2n-1, 2n;\ \nu = 1, 2, \ldots, \tag{3.40}$$

$$\kappa_{2n,\pi/n}(C^r, C) = K_r/(2n^r), \qquad n = 1, 2, \ldots, \qquad r = 0, 1, \ldots, \tag{3.41}$$

$$\kappa_{2n,\pi/n}(C^r, L_1) = \kappa'_{2n,\pi/n}(C^r, L_1) = 2\,K_{r+1}/n^r,$$
$$n = 1, 2, \ldots, \qquad r = 0, 1, \ldots. \tag{3.42}$$

The exact constants in (3.40) are realized by the subspace $\mathcal{N}_{2n-1}^T$ of trigonometric polynomials, in particular, by Favard λ^*-means, those in (3.41) and (3.42) – by the subspace $S_{2n,r}$ of splines and those in (3.42) – by interpolating splines $\sigma_{2n,r}$.

For $r = 0, 1$ relations (3.41), (3.42) can be extended by exact results for approximations in the L_p $(1 < p < \infty)$ metric. From (6.3.12), (6.3.13) and (6.3.23) and the first inequality in (3.39) we obtain

$$\kappa_{2n,\pi/n}(C, L_p) = (2\pi)^{1/p}/2, \qquad 1 \le p \le \infty, \tag{3.43}$$

$$\kappa'_{2n,\pi/n}(C, L_p) = (2\pi)^{1/p}/2, \qquad 1 \le p \le 3, \tag{3.44}$$

$$\kappa_{2n,\pi/n}(C^1, L_p) = \kappa'_{2n,\pi/n}(C^1, L_p) = \left(\frac{2\pi}{1+p}\right)^{1/p}\frac{\pi}{4n}, \qquad 1 \le p < \infty. \tag{3.45}$$

Moreover, the constants in (3.43) and (3.44) are realized by the subspace $S_{2n,0}$ (in (3.44) by step functions $\psi_{2n}(t,t)$ interpolating the mean values of f on the intervals from the partition); the constant in (3.45) is realized by subspace $S_{2n,1}$ interpolating by polygons $\sigma_{2n,1}(f,t)$.

Let us give another two results for the non-periodic case which follows from (6.3.7) and (6.3.11) and inequality (3.36) with $r = 0$ (when it holds without the assumption for periodicity). We have

$$\kappa_{2n,h_N}(C[a,b], L_p[a,b]) = (b-a)^{1/p}/2, \qquad 1 \le p \le \infty,$$

$$\kappa'_{2n,h_N}(C[a,b], L_p[a,b]) = (b-a)^{1/p}/2, \qquad 1 \le p \le 3,$$

where $h_N = (b-a)/N$. The constants are realized by step functions.

Comments

Section 8.1 Theorem 8.1.1 was established by Tikhomirov [2], the present proof follows the one in Lorentz's monograph [14A]. A series of relations for different widths of sets in Banach spaces are given in [25A]. Theorem 8.1.3 – the first exact result for N-widths – was obtained by Kolmogorov [2] and estimate (1.4) was established by Grigoryan [1]. The result in Theorem 8.1.3 for the C metric are due to Tikhomirov [1, 2] and the exact lower bounds for $d_{2n-1}(W_1^r, L_1)$ were obtained independently by Subbotin [3, 4] and Makovoz [1].

With respect to the results in Sections 8.1.2–8.1.4 and, in particular, Theorems 8.1.11 and 8.1.13 we note the following. The exact lower estimates for the even widths $d_{2n}(W_p^r, L_q)$ were obtained by Tikhomirov [5] for $p = q = \infty$ and by Ruban [9A, Ch. 10] for $p = \infty, q = 1$. In the cases $p = \infty, 1 < q < \infty$ and $1 < p < \infty, q = 1$, the problem was solved independently by different methods by Ligun [7], Makovoz [2] and Pinkus [1]. The proof in the book follows Ligun's scheme [7] who proved Lemma 8.1.6 and Theorem 8.1.8; by the way, for $q = 1$ relations (1.10) and (1.21) were previously known [9A, Section 10.5].

Concerning the widths of classes $W_p^r[a,b]$ defined on a finite interval, the exact result usually has an implicit form. The first result of this type was obtained by Kolmogorov [2] who expressed $d_N(W_2^r[0,1], L_2[0,1])$ by the numbers from the spectrum of an isoparametric problem (see also [25A]). The first expression via a norm of a perfect spline of an N-width $(d_N(W_\infty^r[a,b]), C[a,b])$ was due to Tikhomirov [5]. Theorems 8.1.15 (for $1 \le q < \infty$) and 8.1.16 were proved by Miccelli and Pinkus [1] by methods based on different ideas, although also using Borsuk's theorem. A series of additional results for widths of classes W_p^r, $W_p^r[a,b]$ are given in [25A, 19A]. The most general result expressing the widths $d_n(W_p^r, L_q)$ and $\lambda_n(W_p^r, L_p)$ for $p \ge q$ in terms of the spectral numbers of a non-linear differential operator (both in the periodic and the non-periodic cases) are contained in the paper of Buslaev and Tikhomirov [1], see also Tikhomirov [6]. The case $p = q$ was considered by Pinkus [2].

The results in Section 8.1.6 are due to V. F. Babenko [4, 5]. For one-sided widths see [11A].

Section 8.2 In the proof of Lemma 8.2.1 we use the scheme of Tikhomirov [5] who considered the case $\psi =$ constant. The results in Lemma 8.2.2 were communicated to the author by Ruban who indicated a method for obtaining exact lower estimates for $d_{2n}(W^r H^\omega, C)$ not based on Borsuk's theorem. Lemma 8.2.5 belongs to Tikhomirov [5]. Concerning Theorem 8.2.8, we note that relation (2.25) for odd widths was established by the author [10, 11] and for even widths by Ruban [4]. The exact lower estimates for $d_{2n-1}(W^rH^\omega, C)$ based on Theorem 8.1.1 are given in [14A].

The estimate (2.33) for $p = \infty$ was obtained by Tikhomirov [1, 2], the other results in Section 8.2.3 and, in particular, Theorem 8.2.10 were established by the author [12, 22, 23]. Theorems 8.2.11 and 8.2.12 for widths of classes of periodic functions are also due to the author [22, 23]. Estimate (2.49) was proved by Motornyi and Ruban [1], inequality (2.50) was contained in Ruban's paper [6, 7]. Extensions of the results on N-widths of classes of differentiable periodic functions from C or L_1 to convolution classes with oscillation-diminishing kernels and to classes defined with the help of linear differential operators can be found in the articles of Pinkus [1], Shevaldin [1, 2], Sun Yong-sheng and Hyang Daren [1], Fang Gensun [1] and V. F. Babenko [8, 14].

Section 8.3 The problems of optimal recovery in the form presented in Section 8.3.1 are stated in author's works [19, 22, 23] (see also [10A]), where exact results for $\rho_N(\mathcal{M}, X)$, $\rho'_N(\mathcal{M}, X)$ in particular cases are given. Estimate (3.15) was obtained by Velikin [4]; other results of similar nature for $f \in W^m_\infty$ with passage to the L_q norm are contained in Ligun's paper [6]. A series of exact results for the optimal recovery of differentiable functions were obtained by Boyanov [1–3]. Results on the optimal recovery of functions were also obtained by Sun Yong-sheng [4]. In different settings the optimal recovery of linear operators and functionals were considered by Stechkin [5], Bakhvalov [1], Subbotin [6], Arestov [2], Marchuk and Osipenko [1] and Grebennikov and Morozov [1].

The problem of optimal coding in general was formulated in Tikhomirov's book [25A], a comprehensive account can also be found in Vitushkin's monograph [16B]. Other than those in Section 8.3.2, exact results on optimal coding were contained in Korneichuk's papers [34, 37, 38]; the last two concern curve coding.

Quantities (3.31) and (3.32) were introduced in the author's papers [22, 23], problems connected with their determination or estimation were also considered in the author's papers [20, 21]. In these works the results given in Section 8.3.3 can also be found.

A look at the widths from the numerical analysis point of view together with new aspects to the optimization problems brought up here are contained in the survey paper of K. I. Babenko [2]. For the optimal recovery of the definite integral using the extremal properties of the splines see S. M. Nikol'skii book [18A] and Zhensykbaev's survey paper [5].

Other results and exercises

1. If for some $r = 0, 1, \ldots,$ the subspace $\mathcal{N}_N$ realizes the width $d_N(W^{r+1}_\infty, C)$ then the same subspace also realizes the width $d_N(W^rH^\infty, C)$ for every concave modulus of continuity ω (Korneichuk [14]).

2. If for some $r = 0, 1, \ldots; \nu = 2, 3, \ldots,$ subspace $\mathcal{N}_{2n-1}$ realizes the width $d_{2n-1}(W^{r+\nu}_1, L_1)$ then the same subspace also realizes the width $d_{2n-1}(W^rH^\omega, L_1)$ for every concave modulus of continuity ω (Korneichuk and Ligun [2]).

3. Let H_p be the Hardy space of analytic functions f in the unit disc width norm

$$\|f\|_{H_p} = \lim_{\rho\to 1-0}\left(\frac{1}{2\pi}\int_0^{2\pi} |f(\exp(it))|^p\, dt\right)^{1/p}, \qquad 1\le p<\infty,$$
$$\|f\|_{H_\infty} = \sup_{|z|<1} |f(z)|, \qquad p=\infty.$$

If $H_p^r = \{f: \|f^{(r)}\|_{H_p} \le 1\}$ then we have for $n>r$

$$d_n(H_p^r, H_p) = [n(n-1)\dots(n-r+1)]^{-r}, \qquad 1\le p\le\infty$$

(K. I. Babenko [1] ($p=\infty$, upper estimate), Tikhomirov [2] ($p=\infty$, lower estimate) and Taikov [3]).

4. Let $n = 1, 2, \dots,; r = 0, 1, \dots,; \alpha, \beta > 0$ and $\Gamma_{2n,r}(\alpha,\beta)$ be the set of functions $f \in L_\infty^r$ ($f \perp 1$) such that $\alpha^{-1}f_+^{(r)} + \beta^{-1}f_-^{(r)} \equiv 1$ and $f^{(r)}$ has no more than $2n$ changes of sign on the period. Then we have for every function $f \in \Gamma_{2n,r}(\alpha,\beta)$ and all $\lambda \in \mathbb{R}$

$$\int_0^x r((f-\lambda)_\pm, t)\, dt \ge \int_0^x r((\varphi_{n,r}(\alpha,\beta;\bullet)-\lambda)_\pm, t)\, dt, \qquad 0\le x\le 2\pi;$$

the inequality is exact (V. F. Babenko [5]).

5. If F is Π-invariant set in L_1 (Exercise 4.11) and $W^rF = \{f: f\in L_1^r, f^{(r)} \in F\}$ then for every $n = 1, 2, \dots$, and $m \ge r$ we have

$$d_{2n-1}(W^rF, L_1) = d_{2n}(W^rF, L_1) = E_n(W^rF)_1$$
$$= E(W^rF, S_{2n,m})_1 = \sup\left\{\int_0^{2\pi} \Pi(\varphi_{n,r}, t)\Pi(g,t)\, dt: g\in F, g\perp 1\right\}$$

(V. F. Babenko [5]).

6. If $\Gamma_{2n,r}$ is the set of 2π-periodic perfect splines of degree r with no more than $2n$ knots on the period then for $f\in\Gamma_{2n,r}, f\ge 0$, we have $R(f,t)\ge R(\varphi_{n,r}(\bullet)+K_rn^{-r}, t)$ and

$$\inf\{\|f\|_p: f\in\Gamma_{2n,r}, f\ge 0)\} = \|\varphi_{n,r}(\bullet)+K_rn^{-r}\|_p, \qquad 1\le p\le\infty$$

(Motornyi [1] ($p = 1, \infty$) and Ligun [14] ($1<p<\infty$)).

7. Let $W_p^{r,+} = \{f: f\in L_p^r, \|(f^{(r)})_+\|_p \le 1\}$. Then

$$d_{2n-1}(W_p^{r,+}, L_1) = d_{2n}(W_p^{r,+}, L_1) = \|\varphi_{n,r}(\bullet)+K_rn^{-r}\|_{p'}$$

(Ligun [14]).

8. For the class $\tilde W_p^1$ of functions conjugate with $f\in W_p^1$ for $p = 1, \infty$ we have

$$d_{2n-1}(\tilde W_\infty^1, C) = \lambda_{2n-1}(\tilde W_\infty^1, C) = d_{2n-1}(\tilde W_1^1, L_1) = \lambda_{2n-1}(\tilde W_1^1, L_1) = \|\tilde\varphi_{n,1}\|_\infty$$

(V. F. Babenko [6]).

9. Let $Q_{np}(t) = \xi_{0p} + \Sigma_{k=1}^n (\xi_{kp}\cos kt\,\eta_{kp}\sin kt)$ be the polynomial of best approximation of Bernoulli function B_r in the L_p ($1\le p\le\infty$) metric and $T(f) = \{a_0(f), \dots, a_n(f); b_1(f), \dots, b_n(f)\}$, where $a_k(f)$ and $b_k(f)$ are the Fourier coeffients of function $f\in L_1$. Among all the methods for the recovery of functions $f\in W_{p'}^r$ ($p' = p/(p-1)$) on information vector $T(f)$ the least error in the C metric is provided by the polynomial

$$S_f^*(t) = \frac{a_0(f)}{2} + \sum_{k=1}^{n} \mu_{kp}\,[a_k(f)\cos kt + b_k(f)\sin kt],$$

where $\mu_{kp} = (-1)^{r/2}k^r\xi_{kp}$ if r is even and $\mu_{kp} = (-1)^{(r-1)/2}k^r\xi_{kp}$ if r is odd. Moreover

$$\sup_{f\in W_{p'}^r} \|f - S_f^*\|_C = \frac{1}{\pi}\left\{\int_0^{2\pi} |B_r(t) - Q_{np}(t)|^p\, dt\right\}^{1/p}$$

(Boyanov [2]).

10. For quantities (3.17) and a concave modulus of continuity ω we have

$$\lambda^N(H^\omega[0, 2\pi], L_p[0, 2\pi]) = 2\|f_{N,0}(\omega)\|_p, \qquad 1 \le p \le 3;$$
$$\lambda^{2n}(W^1H^\omega, L_p) = 2\|f_{2n,1}(\omega)\|_p, \qquad 1 \le p \le \infty,$$

where the functions $f_{N,r}(\omega, t)$ are determined in Section 7.1 (Korneichuk [23, 10A]).

11. Prove that the function $g(\xi, t)$ defined by equalities (2.29) belongs to $H^\omega[a, b]$ if ω is a concave modulus of continuity.

12. Prove inequality (2.44) under the assumption that one of the coordinates of $\boldsymbol{\xi} = \{\xi_1, \ldots, \xi_{2n+1}\}$ is zero.

13. With the help of Theorem 2.5.1 prove that for concave ω

$$d_N(W^1H^\omega[0, \pi], L_p[0, \pi]) \ge \|f_{N1}(\omega)\|_{L_p[0,\pi]}, \qquad 1 \le p \le \infty.$$

14. Using Theorem 5.2.8 show (in the notation of Section 8.3.2) that if $P(f) = \{g\colon g \in W_p^{r+1}, \|g - \sigma_{,1}(f)\|_1 \ge \|\varphi_{n,r+1}\|_1\}$ then

$$\sup_{f\in W_p^{r+1}} \operatorname{diam}_{L_1} P(f) = K(W_p^{r+1}, M_{2n,r}^{\tau}) = 2\,\|\varphi_{n,r+1}\|_{p'}.$$

APPENDIX

A1 Hölder and Minkowski inequalities and some extremal relations

1. In this section we use the notation

$$\|f\|_p = \|f\|_{L_p[a,b]} = \left(\int_a^b |f(t)|^p \, dt\right)^{1/p}, \qquad 1 \le p < \infty,$$

$$\|f\|_\infty = \|f\|_{L_\infty[a,b]} = \operatorname*{ess\,sup}_{a \le t \le b} |f(t)|.$$

Here p and p' are related by $1/p + 1/p' = 1$, and hence $p' = \infty$ if $p = 1$, $p' = 1$ *if* $p = \infty$.

Proposition A1.1 If $f \in L_p[a, b]$ $(1 \le p \le \infty)$ and $\varphi \in L_{p'}[a, b]$ then

$$\int_a^b f(t)\varphi(t) \, dt \le \|f\|_p \, \|\varphi\|_{p'}. \tag{1}$$

Here we have an equality for $1 \le p < \infty$ if and only if

$$\varphi(t) = c|f(t)|^{p-1} \operatorname{sgn} f(t), \qquad c > 0 \tag{2}$$

a.e. on $[a, b]$. If $p = \infty$ then for every $f \in L_\infty[a, b]$, $\|f\|_\infty > 0$ and every $\varepsilon > 0$ there is $\varphi_\varepsilon \in L_1[a, b]$ such that

$$\int_a^b f(t)\varphi_\varepsilon(t) \, dt > \|f\|_\infty \, \|\varphi_\varepsilon\|_1 - \varepsilon. \tag{3}$$

Proof If $p = 1, \infty$ then (1) is trivial; for $1 < p < \infty$ relation (1) is known as the *Hölder inequality* which follows from the elementary inequality

$$ab \le a^p/p + b^{p'}/p', \qquad a, b \ge 0. \tag{4}$$

Indeed if $\|f\|_p > 0$ and $\|\varphi\|_{p'} > 0$ (otherwise the assertion is trivial) and if we set in (4)

$$a = |f(t)|/\|f\|_p, \qquad b = |\varphi(t)|/\|\varphi\|_{p'} \tag{5}$$

then we get by integration with respect to t from a to b

$$\int_a^b |f(t)\varphi(t)|\, dt \leq \|f\|_p \|\varphi\|_{p'}. \tag{6}$$

Inequality (4) can be proved by considering the function $\eta(x) = x^p/p + 1/p' - x$ $(x \geq 0)$ which has an absolute minimum only at point $x = 1$ and $\eta(1) = 0$. Hence

$$x^p/p + 1/p' \geq x, \qquad x \geq 0. \tag{7}$$

Setting $x = ab^{-p'/p}$ and multiplying both parts of the inequality by $b^{p'}$ we arrive at (4).

Because (7) is an equality only for $x = 1$ we see that (4) becomes an equality only when $a^p = b^{p'}$. Therefore (see (5)), an equality in (6) is possible only when $|f(t)|^p = \|f\|_p^p \|\varphi\|_{p'}^{-p'} |\varphi(t)|^{p'}$ a.e. on $[a, b]$. This is in turn equivalent to $|\varphi(t)| = c|f(t)|^{p-1}$, c being some positive constant.

In order to guarantee equality not only in (6) but also in (1), the signs of f and φ are required to coincide a.e. The fact that relation (2) is a necessary and sufficient condition for the equality in (1) can be directly verified.

Finally, let $p = \infty$, $f \in L_\infty[a, b]$, $\|f\|_\infty > 0$ and for a given $\varepsilon > 0$ $(\varepsilon < \|f\|_\infty)$

$$e = \{t: t \in [a, b], |f(t)| > \|f\|_\infty - \varepsilon\}.$$

Let $\varphi_\varepsilon(t) = (\text{meas } e)^{-1} \operatorname{sgn} f(t)$ for $t \in e$ and $\varphi_\varepsilon(t) = 0$ for $t \in [a, b]\backslash e$. Then $\|\varphi_\varepsilon\|_1 = 1$ and

$$\int_a^b f(t)\varphi_\varepsilon(t)\, dt = \frac{1}{\text{meas } e}\int_e |f(t)|\, dt > \|f\|_\infty - \varepsilon > \|f\|_\infty \|\varphi_\varepsilon\|_1 - \varepsilon.$$

The proposition is proved. □

An easy consequence is

Corollary A1.2 If $f \in L_p[a, b]$ $(1 \leq p \leq \infty)$ then

$$\sup_{\|\varphi\|_{p'} \leq 1} \int_a^b f(t)\varphi(t)\, dt = \|f\|_p. \tag{8}$$

Indeed, we get from (1) that the left-hand side does not exceed the right-hand side. On the other hand, if $1 \leq p < \infty$ and if we set

$$\varphi_0(t) = \|f\|_p^{1-p} |f(t)|^{p-1} \operatorname{sgn} f(t), \qquad a \leq t \leq b,$$

then $|\varphi_0(t)|^{p'} = \|f\|_p^{-p}|f(t)|^p$ and hence $\varphi_0 \in L_{p'}$, $\|\varphi_0\|_{p'} = 1$ and

$$\int_a^b f(t)\varphi_0(t)\,\mathrm{d}t = \|f\|_p.$$

In the case $p = \infty$ there may not be a function φ in $L_1[a, b]$ realizing the supremum in (8), but, as we have seen in the proof of Proposition A1.1, there is a sequence $\{\varphi_n\}$ ($n = 1, 2, \ldots$, $\varphi_n \in L_1[a, b]$) for every $f \in L_\infty[a, b]$ such that

$$\lim_{n\to\infty} \int_a^b f(t)\varphi_n(t)\,\mathrm{d}t = \|f\|_\infty.$$

2. If $f \in C$ then together with (8) (for $p = \infty$) one can write a similar relation for the Riemann–Stieltjes integral

$$\sup\left\{\int_a^b f(t)\,\mathrm{d}g(t)\colon g \in V[a, b],\ \bigvee_a^b g \le 1\right\} = \|f\|_C, \tag{9}$$

where $\|f\|_C = \|f\|_{C[a,b]}$.

Indeed, for every partition $a = t_0 < t_1 < \cdots < t_n = b$ we have

$$\sum_{k=1}^n f(t_k)[g(t_k) - g(t_{k-1})] \le \|f\|_C \sum_{k=1}^n |g(t_k) - g(t_{k-1})| \le \|f\|_C \bigvee_a^b g.$$

Passing to the limit for $\max_k |t_k - t_{k-1}| \to 0$ we get

$$\int_a^b f(t)\,\mathrm{d}g(t) \le \|f\|_C \bigvee_a^b g. \tag{10}$$

It remains to show that a function $g \in V[a, b]$ with a variation on $[a, b]$ equal to 1 exists for which (10) becomes equality. If $\|f\|_C = |f(t_*)|$ ($t_* \in [a, b]$) then we have to take $g(t) = 2^{-1}\,\mathrm{sgn}\,(t - t_*)\,\mathrm{sgn}\,f(t_*)$.

3. Let us now consider the non-symmetric case. Let $\alpha > 0$, $\beta > 0$, $f_\pm(t) = \max\{\pm f(t), 0\}$. Assuming $f \in L_p[a, b]$ ($1 \le p \le \infty$) we set

$$\begin{aligned}\|f\|_{p;\alpha,\beta} &= \|\alpha f_+ + \beta f_-\|_p \\ &= \left(\int_a^b |f(t)[\alpha\,\mathrm{sgn}\,f_+(t) + \beta\,\mathrm{sgn}\,f_-(t)]|^p\,\mathrm{d}t\right)^{1/p}.\end{aligned} \tag{11}$$

Because $(f+g)_\pm(t) \le f_\pm(t) + g_\pm(t)$, the functional $\|f\|_{p;\alpha,\beta}$ is semi-additive (see also Section A1.4):

$$\|f+g\|_{p;\alpha,\beta} \le \|f\|_{p;\alpha,\beta} + \|g\|_{p;\alpha,\beta}$$

and from the obvious inequalities

$$\min\{\alpha,\beta\}\|f\|_p \le \|f\|_{p;\alpha,\beta} \le \max\{\alpha,\beta\}\|f\|_p$$

we get its continuity. It is important that, although it is not homogeneous in the usual sense, the functional $\|f\|_{p;\alpha,\beta}$ is positive homogeneous: $\|\lambda f\|_{p;\alpha,\beta} = \lambda\|f\|_{p;\alpha,\beta}$ for every $\lambda > 0$. A non-symmetric analog of Proposition A1.1 is:

Proposition A1.3 If $f \in L_p[a, b]$ $(1 \le p < \infty)$ and $\varphi \in L_{p'}[a, b]$ then

$$\int_a^b f(t)\varphi(t)\,dt \le \|f\|_{p;\alpha,\beta}\|\varphi\|_{p';\alpha^{-1},\beta^{-1}}. \tag{11'}$$

Here, we have an equality if and only if a.e. on $[a, b]$

$$\varphi(t) = c|f(t)|^{p-1}[\alpha^p \operatorname{sgn} f_+(t) - \beta^p \operatorname{sgn} f_-(t)], \qquad c > 0. \tag{12}$$

Proof For brevity setting

$$\chi_{\alpha,\beta}(f, t) = \alpha \operatorname{sgn} f_+(t) - \beta \operatorname{sgn} f_-(t), \qquad a \le t \le b,$$

we first note that

$$\chi_{\alpha,\beta}(f, t)\chi_{\alpha^{-1},\beta^{-1}}(f, t) \equiv 1, \qquad f(t)\chi_{\alpha,\beta}(f, t) \ge 0. \tag{13}$$

For every function $f, g, \psi \in L_1[a, b]$ $(\psi(t) \ge 0)$ we have

$$\int_a^b \psi(t)f(t)\chi_{\alpha,\beta}(g, t)\,dt \le \int_a^b \psi(t)f(t)\chi_{\alpha,\beta}(f, t)\,dt. \tag{14}$$

Now if $f \in L_p[a, b]$ $(1 \le p < \infty)$, $\varphi \in L_{p'}[a, b]$ then consecutively applying (13), (14) and (1) we get

$$\begin{aligned}\int_a^b f(t)\varphi(t)\,dt &= \int_a^b f(t)\chi_{\alpha,\beta}(f, t)\varphi(t)\chi_{\alpha^{-1},\beta^{-1}}(f, t)\,dt \\ &\le \int_a^b [f(t)\chi_{\alpha,\beta}(f, t)][\varphi(t)\chi_{\alpha^{-1},\beta^{-1}}(\varphi, t)]\,dt \\ &\le \left(\int_a^b |f(t)|^p[\alpha^p \operatorname{sgn} f_+(t) + \beta^p \operatorname{sgn} f_-(t)]\,dt\right)^{1/p} \\ &\quad\times \left(\int_a^b |\varphi(t)|^{p'}[\alpha^{-p'} \operatorname{sgn} \varphi_+(t) + \beta^{-p'} \operatorname{sgn} \varphi_-(t)]\,dt\right)^{1/p'}\end{aligned} \tag{15}$$

and equality (11) is proved. It is easy to check that in the chain of inequalities (15) an equality is possible if and only if for a fixed $f \in L_p[a, b]$ function φ is given by (12).

If φ_0 is a function of the type (12) with $c = \|f(t)[\alpha \operatorname{sgn} f_+(t) - \beta \operatorname{sgn} f_-(t)\|_p^{1-p}$ then easy computations show $\|\varphi_0\|_{p';\alpha^{-1},\beta^{-1}} = 1$ and

$$\int_a^b f(t)\varphi_0(t)\,\mathrm{d}t \leq \|f\|_{p;\alpha,\beta},$$

Thus, we have

$$\sup_{\|\varphi\|_{p';\alpha,\beta} \leq 1} \int_a^b f(t)\varphi(t)\,\mathrm{d}t \leq \|f\|_{p;\alpha^{-1},\beta^{-1}}, \qquad 1 \leq p < \infty. \tag{16}$$

4. The third axiom for the norm in space $L_p[a, b]$:

$$\|f + \varphi\|_p \leq \|f\|_p + \|\varphi\|_p, \qquad f, g \in L_p[a, b], \tag{17}$$

which is trivial in the cases $p = 1, \infty$, for $1 < p < \infty$ is known as the *Minkowski inequality* for integrals. It can be easily derived from (1). Indeed

$$\int_a^b |f(t) + \varphi(t)|^p\,\mathrm{d}t \leq \int_a^b |f + \varphi|^{p-1}\,|f|\,\mathrm{d}t + \int_a^b |f + \varphi|^{p-1}\,|\varphi|\,\mathrm{d}t$$

$$\leq \|f\|_p \left(\int_a^b |f + \varphi|^p\,\mathrm{d}t\right)^{1/p'} + \|\varphi\|_p \left(\int_a^b |f + \varphi|^p\,\mathrm{d}t\right)^{1/p'}$$

and we get (17) dividing the last multiplier. Taking the conditions under which (1) is an equality into account, one can easily get that inequality (17) for $1 \leq p < \infty$ becomes an equality if and only if $\varphi(t) = cf(t)$ (c = constant > 0).

Inequality

$$\left\|\int_c^d f(\bullet, u)\,\mathrm{d}u\right\|_p \leq \int_c^d \|f(\bullet, u)\|_p\,\mathrm{d}u, \qquad 1 \leq p \leq \infty \tag{18}$$

(under the assumption that the integral on the right-hand side exists), can be considered as a generalization of inequality (17). For $p = 1, \infty$ the validity of relation (18) is obviously true and for $1 < p < \infty$ it is known as the *generalized Minkowski inequality*:

$$\left(\int_a^b \left|\int_c^d f(t, u)\,\mathrm{d}u\right|^p \mathrm{d}t\right)^{1/p} \leq \int_c^d \left(\int_a^b |f(t, u)|^p\,\mathrm{d}t\right)^{1/p} \mathrm{d}u. \tag{19}$$

In order to prove (19) we first assume that the function $J(t) = \int_c^d |f(t, u)|\,\mathrm{d}u$ belongs to $L_p[a, b]$. Then $J^{p-1} \in L_{p'}[a, b]$ and hence using Fubini's theorem [12B, p. 336] and inequality (1) consecutively we have

$$\int_a^b J^p(t)\,\mathrm{d}t = \int_a^b J^{p-1}(t)\left(\int_c^d |f(t, u)|\,\mathrm{d}u\right)\mathrm{d}t$$

$$= \int_c^d \left(\int_a^b J^{p-1}(t) |f(t, u)| \, dt \right) du$$

$$\leq \|J\|_p^{p/p'} \int_c^d \left(\int_a^b |f(t, u)|^p \, dt \right)^{1/p} du.$$

Dividing both parts by $\|J\|_p^{p/p'}$ we get (19).

If $J \notin L_p[a, b]$, i.e. $\|J\|_p^p = \infty$ then we set for $N > 0$

$$|f(t, u)|_N = \min \{|f(t, u)|, N\}, \qquad J_N(t) = \int_c^d |f(t, u)|_N \, du.$$

Then $J_N \in L_p[a, b]$ and in view of the result of the previous paragraph we have for every $N > 0$

$$\|J_N\|_p \leq \int_c^d \left(\int_a^b |f(t, u)|_N^p \, dt \right)^{1/p} du \leq \int_c^d \left(\int_a^b |f(t, u)|^p \, dt \right)^{1/p} du.$$

Taking the limit for $N \to \infty$ we obtain (19).

Remark All the arguments and results in Section A1 remain true if the functions are considered on arbitrary sets of finite measure instead of $[a, b]$.

A2 Steklov functions

Suppose the function f is defined on the real axis and summable on every finite interval (a, b). For every $h > 0$ we associate with f the function

$$f_h(t) = \frac{1}{2h} \int_{-h}^{h} f(t + u) \, du = \frac{1}{2h} \int_{t-h}^{t+h} f(u) \, du, \tag{1}$$

which is called the *Steklov function* for f (see e.g. [2A, p. 156]).

Thus, under the assumption that the function f is only integrable, the Steklov function f_h is absolutely continuous and a.e.

$$f_h'(t) =: [f_h(t)]' = \frac{1}{2h} [f(t + h) - f(t - h)] \tag{2}$$

and hence the derivative f_h' is integrable on every finite interval (a, b) as is f. It is clear that when f is locally absolutely continuous then the operations of differentiation and passing to the Steklov function can be interchanged and we have $f_h'(t) = [f'(t)]_h$.

Proposition A2.1 Let X be the space of 2π-periodic functions C or L_p $(1 \leq p \leq \infty)$. Then we have for $f \in X$

$$\|f_h\|_X \leq \|f\|_X, \tag{3}$$

and also, except the case $X = L_\infty$,

$$\lim_{h\to 0} \|f_h - f\|_X = 0. \tag{4}$$

Proof If $X = C$ or $X = L_\infty$ then inequality (3) is a trivial consequence of definition (1). If $X = L_p$ $(p \geq 1)$ then we apply the generalized Minkowski inequality (A1.19):

$$\|f_h\|_p = \left(\int_0^{2\pi} \left|\frac{1}{2h}\int_{-h}^{h} f(t+u)\,\mathrm{d}u\right|^p \mathrm{d}t\right)^{1/p}$$

$$\leq \frac{1}{2h}\int_{-h}^{h}\left(\int_0^{2\pi} |f(t+u)|^p\,\mathrm{d}t\right)^{1/p} \mathrm{d}u = \frac{1}{2h}\int_{-h}^{h} \|f(\bullet + u)\|_p\,\mathrm{d}u = \|f\|_p.$$

Further, inequality (A1.19) also gives

$$\|f_h - f\|_p = \left(\int_0^{2\pi} \left|\frac{1}{2h}\int_{-h}^{h} [f(t+u) - f(t)]\,\mathrm{d}u\right|^p \mathrm{d}t\right)^{1/p}$$

$$\leq \frac{1}{2h}\int_{-h}^{h}\left(\int_0^{2\pi} |f(t+u) - f(t)|^p\,\mathrm{d}t\right)^{1/p} \mathrm{d}u,$$

and hence we have for $X = L_p$ $(1 \leq p < \infty)$

$$\|f_h - f\|_X = \frac{1}{2h}\int_{-h}^{h} \|f(\bullet + u) - f(\bullet)\|_X\,\mathrm{d}u, \tag{5}$$

which also follows by trivial estimates for $X = L_\infty$ and $X = C$. It is clear that the right-hand side of (5) tends to zero with $h \to 0$ provided that

$$\lim_{h\to 0} \|f(\bullet + h) - f(\bullet)\|_X = 0, \qquad f \in X. \tag{6}$$

(6) *is obvious in the case* $X = C$, while relations (6) and (4) are no longer true for $X = L_\infty$ (as the example $f(t) = \operatorname{sgn} \sin t$ shows). Finally, relation (6) is the well-known Lebesgue theorem for $X = L_p$ $(1 \leq p < \infty)$ (see e.g. [15B, p. 407]). The salient fact that is used in its proof is that the functions from C, for which (6) holds, are dense in L_p. Indeed, if $f \in L_p$ $(1 \leq p < \infty)$ then there is a function $\varphi \in C$ for every $\varepsilon > 0$ such that $\|f - \varphi\|_p < \varepsilon$. But if $\|\varphi(\bullet + h) - \varphi(\bullet)\|_C \to 0$ as $h \to 0$ then $\|\varphi(\bullet + h) - \varphi(\bullet)\|_p \to 0$, i.e.

$$\|\varphi(\bullet + h) - \varphi(\bullet)\|_p < \varepsilon \qquad \text{for} \qquad |h| < \delta(\varepsilon).$$

Thus, we have for $|h| < \delta(\varepsilon)$

$$\|f(\bullet + h) - f(\bullet)\|_p \leq \|f(\bullet + h) - \varphi(\bullet + h)\|_p + \|\varphi(\bullet + h) - \varphi(\bullet)\|_p$$
$$+ \|\varphi(\bullet) - f(\bullet)\|_p < 3\varepsilon,$$

i.e. (6) and hence (4) holds for $X = L_p$ $(1 \leq p < \infty)$.

Note that it follows from (4) and the general inequality $|\|f\|_X - \|g\|_X| \leq \|f-g\|_X$ that in the cases $X = C$ and $X = L_p$ $(1 \leq p < \infty)$ that

$$\lim_{h\to 0} \|f_h\|_X = \|f\|_X, \qquad f \in X. \qquad (7) \qquad \square$$

Proposition A2.2 If $X = C$ or $X = L_p$ $(1 \leq p < \infty)$ then for $f \in X$ we have

$$\|f - f_h\|_X \leq \omega(f, h)_X, \qquad (8)$$

$$\|f_h'\|_X \leq h^{-1}\omega(f, h)_X, \qquad (9)$$

where $\omega(g, \delta)_X$ is the modulus of continuity of function $g \in X$ in X metric.

Inequality (8) follows immediately from (5) and the definition of a modulus of continuity (see §6.1). From (2) and the properties of the modulus of continuity we get

$$\|f_h'\|_X = \frac{1}{2h}\|f(\bullet + h) - f(\bullet - h)\|_X \leq \frac{1}{2h}\omega(f, 2h)_X \leq \frac{1}{h}\omega(f, h)_X.$$

Proposition A2.3 For $f \in L_1$ we have

$$\bigvee_0^{2\pi} f_h = \|f_h'\|_1 \leq \frac{1}{h}\omega(f, h)_1, \qquad (10)$$

and if moreover $f \in V$ then

$$\bigvee_0^{2\pi} f_h \leq \bigvee_0^{2\pi} f. \qquad (11)$$

The equality in (10) is a well-known fact for the variation of the indefinite Lebesgue integral (see e.g. [12B, p. 240]). The inequality in (10) follows from (9). Proving (11), for every partition $0 = t_0 < t_1 < \cdots < t_n = 2\pi$ we have

$$\sum_{k=1}^{n} |f_h(t_k) - f_h(t_{k-1})| = \sum_{k=1}^{n} \left| \frac{1}{2h}\int_{-h}^{h} [f(t_k + t) - f(t_{k-1} + t)]\,\mathrm{d}t \right|$$

$$\leq \frac{1}{2h}\int_{-h}^{h} \left(\sum_{k=1}^{n} |f(t_k + t) - f(t_{k-1} + t)| \right) \mathrm{d}t$$

$$\leq \frac{1}{2h}\int_{-h}^{h} \bigvee_t^{t+2\pi} f \,\mathrm{d}t = \bigvee_0^{2\pi} f.$$

A3 Proofs of comparison theorems for Σ-rearrangements

Here we give the proofs of the Theorems 7.1.4–7.1.6 stated in Section 7.1 which are at the base of the main results in Chapter 7 and Section 8.2. Together with the information for Σ-rearrangements given in Section 7.1.3 we need some additional properties.

Under the assumption that the 2π-periodic function f belongs to D_0^1 (i.e. $f(t) = \int_a^t \psi(u)\,du$, where $\psi \perp 1$ and ψ has one-sided limits $\psi(t \pm 0)$ at every point t), we consider the Σ-rearrangement of f, i.e. the function

$$R(f, t) = \sum_\nu r(\varphi_\nu, t), \qquad 0 \le t \le 2\pi, \tag{1}$$

where φ_ν are the simple functions from the Σ-representation

$$f(t) = \sum_\nu \varphi_\nu(t), \qquad |f(t)| = \sum_\nu |\varphi_\nu(t)|, \qquad a \le t \le a + 2\pi, \tag{2}$$

whose construction is described in Section 7.1.3.

Moreover we assume that

$$\psi(t) = \begin{cases} [\psi(t-0) + \psi(t+0)]/2, & \text{if } \psi(t-0)\psi(t+0) > 0; \\ 0, & \text{if } \psi(t-0)\psi(t+0) \le 0. \end{cases} \tag{3}$$

Condition (3) does not affect the values of function f, but it ensures that the derivative f' satisfies the following properties which are similar to those ones of continuous functions: (1) f' preserves sign in some neighbourhood of every point t where $f'(t) \neq 0$; moreover f' changes sign only when passing through zero; (2) the set of zeros of f' is closed.

1 Auxiliary results As we did in Section 7.1 with every simple function φ_ν from the Σ-representation (2) we associate its support $[\alpha_\nu, \beta_\nu] \subset [a, a + 2\pi]$ and the interval $[\alpha'_\nu, \beta'_\nu] \subset (\alpha_\nu, \beta_\nu)$ (which may degenerate into a point) where $|\varphi_\nu(t)| = \|\varphi_\nu\|_C$. We set

$$\delta_\nu = \beta_\nu - \alpha_\nu, \qquad \delta'_\nu = \beta'_\nu - \alpha'_\nu, \qquad \Delta = \max_\nu \delta_\nu.$$

In view of the definition of Σ-rearrangement $R(f, t)$, an interval $(t_\nu, \tau_\nu) \subset (\alpha_\nu, \beta_\nu)$ corresponds to every simple function φ_ν for a fixed $x \in (0, \Delta)$, provided $\delta'_\nu \le x < \delta_\nu$, such that

$$\tau_\nu - t_\nu = x, \qquad f(t_\nu) = f(\tau_\nu), \qquad |f(t)| > |f(t_\nu)|, \qquad t_\nu < t < \tau_\nu. \tag{4}$$

Moreover

$$r(\varphi_\nu, x) = |\varphi_\nu(t_\nu)| = |\varphi_\nu(\tau_\nu)| = |f(t_\nu) - f(\alpha_\nu)| = |f(\tau_\nu) - f(\beta_\nu)|. \tag{5}$$

It is clear that the intervals (t_ν, τ_ν) are disjoint and hence the number

$q = q(x)$ of simple functions φ_ν in (1) and (2) for which $\delta'_\nu \le x < \delta_\nu$ is finite: $q(x) < 2\pi/x$.

Lemma A3.1 For every $x \in (0, 2\pi)$ the following exact inequalities hold (the second one under the additional assumption that $R'(f, x)$ exists):

$$R(f, x) \le \frac{1}{2}\int_0^{2\pi - xq(x)} r(f', t)\,dt, \tag{6}$$

$$|R'(f, x)| \le \frac{1}{4}\sum_{k=1}^{q(x)} |f'(\tau_\nu) - f'(t_\nu)|. \tag{7}$$

Proof Fix x $(0 < x < \Delta)$ and take the family of simple functions φ_ν in (2) such that $\delta'_\nu \le x < \delta_\nu$. Let Q_x be the union of corresponding intervals (t_ν, τ_ν) and $F_x = [a, a + 2\pi]\backslash Q_x$, meas $F_x = 2\pi - xq(x)$. From (1) and (5) we get

$$R(f, x) \le \frac{1}{2}\int_{F_x} |f'(t)|\,dt,$$

and in order to obtain (6) we use the second inequality in (3.2.13).

We differentiate equality (1) for $t = x$ and we note that $r'(\varphi_\nu, x) = 0$ if $0 < x < \delta'_\nu$ or $x > \delta_\nu$. If $\delta'_\nu < x < \delta_\nu$ then using (4) and Proposition 7.1.3 we get

$$|r'(\varphi_\nu, x)| \le |f'(\tau_\nu) - f'(t_\nu)|/4,$$

which implies (7). (6) and (7) become equalities for $f(t) = \sin nt$ $(n = 1, 2, \ldots)$ or $f(t) = \varphi_{n,r}(t)$ $(r \ge 1)$, for example.

Remark It is easy to verify that if $f \in D_0^1$ then $f'_\pm(t) = f'(t \pm 0)$ and hence inequality (7) makes sense for the one-sided derivatives and it holds for every point $x \in [0, 2\pi]$.

Lemma A3.2 Let $f \in D_0^1$, $f(a) = 0$, $c \in (0, 2\pi)$ and S be a finite or countable system of disjoint intervals $(a_i, b_i) \subset [a, a + 2\pi]$ such that: (1) $b_i - a_i \le c$; (2) every two intervals from S are separated by a zero of f. Then

$$\sum_i \int_{a_i}^{b_i} |f(t)|\,dt \le \int_0^c R(f, t)\,dt.$$

Indeed, using Σ-representation (2) we can write

$$\sum_i \int_{a_i}^{b_i} |f(t)|\,dt = \sum_i \int_{a_i}^{b_i} \left(\sum_\nu |\varphi_\nu(t)|\right) dt = \sum_i \sum_\nu \int_{a_i}^{b_i} |\varphi_\nu(t)|\,dt.$$

In view of condition (2), the support $[\alpha_\nu, \beta_\nu]$ of every simple function φ_ν may have common points with no more than one being from the intervals from S. Taking condition (1) also into account and recalling Proposition 3.2.4 we obtain

$$\int_{a_i}^{b_i} |\varphi_\nu(t)|\, dt \le \int_0^c r(\varphi_\nu, t)\, dt$$

and

$$\sum_i \int_{a_i}^{b_i} |f(t)|\, dt \le \sum_\nu \int_0^c r(\varphi_\nu, t)\, dt = \int_0^c R(f, t)\, dt.$$

Lemma A3.3 Let $f \in C^2$ and $\min_t |f(t)| = f(a) = 0, f'' \in D_0^1$. If the derivative $R'(\varphi, x)$ exists at the point $x \in (0, 2\pi)$ then

$$|R'(f, x)| \le \frac{1}{4} \int_0^x R(f'', t)\, dt. \tag{8}$$

The inequality is exact for every $x \in (0, \Delta)$.

Proof Fix $x \in (0, \Delta)$ and for brevity set $q(x) = q$ in (7). Then (see (7))

$$4|R'(f, x) \le \sum_{\nu=1}^{q} |I_\nu|, \qquad I_\nu \le = \int_{t_\nu}^{\tau_\nu} f''(t)\, dt, \tag{9}$$

where the disjoint intervals (t_ν, τ_ν) $(\tau_\nu - t_\nu = x)$ are located in $(a, a + 2\pi)$ and we may assume

$$a \le t_1 < \tau_1 \le t_2 < \tau_2 \le \cdots \le t_q < \tau_q \le a + 2\pi.$$

Let $t'_\nu = t_\nu$ if $f''(t_\nu) = 0$ or $\operatorname{sgn} f''(t_\nu) = \operatorname{sgn} I_\nu$, t'_ν being the nearest from the right to the t_ν zero of f'' if $\operatorname{sgn} f''(t_\nu) = -\operatorname{sgn} I_\nu \ne 0$. Similarly $\tau'_\nu = \tau_\nu$ if $f''(\tau_\nu) = 0$ or $\operatorname{sgn} f''(\tau_\nu) = \operatorname{sgn} I_\nu$, τ'_ν being the nearest from the left to the τ_ν zero of f'' if $\operatorname{sgn} f''(\tau_\nu) = -\operatorname{sgn} I_\nu \ne 0$. Thus $\tau'_\nu - t'_\nu \le x$ and the intervals (t'_ν, τ'_ν) are disjoint and if

$$I'_\nu = \int_{t'_\nu}^{\tau'_\nu} f''(t)\, dt = f'(\tau'_\nu) - f'(t'_\nu), \tag{10}$$

then

$$|I'_\nu| \ge |I_\nu|, \qquad \operatorname{sgn} I'_\nu = \operatorname{sgn} I_\nu = -\operatorname{sgn}[f(t) - f(t_\nu)], \quad t_\nu < t < \tau_\nu. \tag{11}$$

Setting $t'_{q+1} = t'_1 + 2\pi$ and $\tau'_{q+1} = \tau'_1 + 2\pi$, we see that every two intervals (t'_ν, τ'_ν) and $(t'_{\nu+1}, \tau'_{\nu+1})$ $(\nu = 1, 2, \ldots, q)$ are separated by a zero of function f''. In view of the choice of the points $\tau'_\nu, t'_{\nu+1}$, this is not obvious only in the case $\operatorname{sgn} I'_\nu = \operatorname{sgn} I'_{\nu+1}$, $\tau'_\nu = \tau_\nu$, $t'_{\nu+1} = t_{\nu+1}$ and the values of f'' at points τ_ν, $t_{\nu+1}$ are different from zero and hence they have the signs of $I_\nu, I_{\nu+1}$. In this case we let for definiteness $I_\nu < 0$, $I_{\nu+1} < 0$. Then (see the last equality in (11)) $f'(\tau_\nu) \le 0$, $f'(\tau_{\nu+1}) \ge 0$ and therefore

$$\int_{\tau_\nu}^{t_{\nu+1}} f''(t)\,\mathrm{d}t = f'(t_{\nu+1}) - f'(\tau_\nu) \le 0.$$

Because $f''(\tau_\nu) < 0$, $f''(t_{\nu+1}) < 0$ we see that f'' has a zero at $(\tau_\nu, t_{\nu+1}) = (\tau'_\nu, t'_{\nu+1})$.

Going back to (9) we apply Lemma A3.2 and we obtain

$$4|R'(f,x)| \le \sum_{\nu=1}^{q} |I'_\nu| \le \int_0^x R(f'', t)\,\mathrm{d}t$$

and (8) follows. We have equality in (8) at $x \in (0, \pi/n)$ for $f(t) = \sin nt$ or $f(t) = \varphi_{n,r}(t)$ $(r \ge 2)$, for example.

Lemma A3.4 Let $f \in D_0^1$. If $f \in V^1$ then

$$|R'(f,x)| \le \frac{1}{4}\bigvee_0^{2\pi} f', \qquad 0 < x < 2\pi, \tag{12}$$

and if $f \in V^2$ then

$$|R'(f,x)| \le \frac{x}{8}\bigvee_0^{2\pi} f'', \qquad 0 < x < 2\pi, \tag{13}$$

The inequalities are exact.

Proof It is clear that (12) and (13) have to be shown only for $x \in (0, \Delta)$. Inequality (12) follows from (7) for $f \in V^1$ because the disjoint intervals (t_ν, τ_ν) are located in $(a, a + 2\pi)$.

Let $f \in V^2$, $x \in (0, \Delta)$ be fixed and $q = q(x)$ be the positive integer from Lemma A3.1. Without loss of generality, we can assume that function $\psi = f''$ from V satisfies condition (3) at the points of discontinuity. Hence f'' possesses the properties mentioned in connection with (3). Beginning with (7) and using these properties, as in the proof of Lemma A3.3, we come to

$$|R'(f,x)| \le \frac{1}{4}\sum_{\nu=1}^{q} |I'_\nu|, \tag{14}$$

where I'_ν are defined as in (10) and the disjoint intervals (t'_ν, τ'_ν) with length $\tau'_\nu - t'_\nu \le x$ are located in a period and are separated (also from the point $t'_{q+1} = t'_1 + 2\pi$) by the zeros of function f''. It is easy to see (e.g. arguing by the contrary) that there is a point η_ν such that $|I'_\nu| \le |f''(\eta_\nu)|(\tau'_\nu - t'_\nu)$ on every interval (t'_ν, τ'_ν).

Thus, we have found points

$$\gamma_0 < \eta_1 < \gamma_1 < \eta_2 < \gamma_2 < \cdots < \eta_q < \gamma_q = \gamma_0 + 2\pi,$$

such that $f''(\eta_\nu) = 0$ $(\nu = 0, 1, \ldots, q)$ and $|I'_\nu| \leq |f''(\eta_\nu)|x$ $(\nu = 1, 2, \ldots, q)$. Continuing estimate (14) we therefore obtain

$$|R'(f, x)| \leq \frac{x}{4} \sum_{\nu=1}^{q} |f''(\eta_\nu)|$$

$$= \frac{x}{8} \sum_{\nu=1}^{q} \{|f''(\eta_\nu) - f''(\gamma_{\nu-1})| + |f''(\gamma_\nu) - f''(\eta_\nu)|\}$$

$$\leq \frac{x}{8} \bigvee_{\gamma_0}^{\gamma_0+2\pi} f'' = \frac{x}{8} \bigvee_{0}^{2\pi} f''.$$

Relations (12) and (13) hold as equalities for every $x \in (0, \pi/n)$ for $f = g_{n,1}$ and for $f = g_{n,2}$ respectively.

2 Proofs of Theorems 7.1.4 and 7.1.5 It is convenient for us to prove the following assertion which is stronger than Theorem 7.1.4.

Theorem A3.5 Let $g \in W_V^m$ $(m = 1, 2, \ldots)$, $\min_t |g(t)| = 0$ and for some $a > 0$

$$R(g, 0) = R_{a,m}(0). \tag{15}$$

Then the following inequalities hold

$$|R'(g, t)| \leq |R'_{a,m}(t)|, \qquad 0 < t < a, \tag{16}$$

$$|R'(g, t)| \leq |R'_{a,m}(a - 0)|, \qquad a \leq t \leq 2\pi, \tag{17}$$

$$R(g, t) \geq R_{a,m}(t), \qquad 0 \leq t \leq 2\pi. \tag{18}$$

Proof Note that (18) follows immediately from (15) and (16). Hence we only have to prove (16) and (17). Recalling the definition of the standard Σ-rearrangement $R_{a,m}$ (see (7.1.28)) it is easy to see that we have for $0 < a < a_1$ and $m = 1, 2, \ldots,$

$$R_{a,m}(t) < R_{a_1,m}(t), \qquad 0 \leq t \leq a_1;$$

$$|R'_{a,m}(t)| \leq |R'_{a_1,m}(t)|, \qquad 0 < t < a_1. \tag{19}$$

From the properties of the Σ-rearrangement $R(f, t)$ (see Section 7.1.3) and from (15) for $m \geq 2$ we get

$$\|R(g')\|_1 = \|g'\|_1 = 2R(g, 0) = 2R_{a,m}(0) = \|R_{a,m-1}\|_1. \tag{20}$$

Because (see (7.1.28)) $R'_{a,1}(t) = -\frac{1}{4}$, $R'_{a,2}(t) = -t/8$ for $0 < t < a$ we see that (16) and (17) follow from Lemma A3.4 for $m = 1, 2$ and for $m = 1,$

respectively. The validity of (17) for $m = 2$ is established as follows: (20) implies $\|R(g')\|_1 \le \|R_{a,1}\|_1 = a^2/8$ and from $|R'(g,t)| \le \frac{1}{4}$ we get $R(g',0) \le a/4$ and

$$|R'(g,t)| \le \frac{1}{4}\|g''\|_1 = \frac{1}{2}R(g',0) \le \frac{a}{8} = |R'_{a,2}(a-0)|.$$

We now proceed by induction. Suppose Theorem A3.5 holds for $m = 1, 2, \ldots, k-1$ $(k \ge 3)$ and $g \in W_V^k$ satisfies its conditions for $m = k$. Then (see (20)) $\|R(g')\|_1 \le \|R_{a,k-1}\|_1$ and it is easy to see that

$$R(g',0) \le R_{a,k-1}(0). \tag{21}$$

Indeed, assuming that $R(g',0) > R_{a,k-1}(0)$ we choose $b > 0$ from the condition $R(g',0) = R_{b,k-1}(0)$ and because the theorem holds for $m = k-1$ by hypothesis we conclude that $\|R_{b,k-1}\|_1 \le \|R(g')\|_1 \le \|R_{a,k-1}\|_1$. This is impossible because $b > a$.

Applying the same arguments again, we get from (20) and (21)

$$\|R(g'')\|_1 \le \|R_{a,k-2}\|_1, \qquad R(g'',0) \le R_{a,k-2}(0). \tag{22}$$

If $R(g'',0) = R_{c,k-2}(0)$ $(c \le a)$ then our hypothesis for $m = k-2$ and (19) imply that

$$\begin{aligned}
|R'(g'',t)| &\le |R'_{c,k-2}(t)| \le |R'_{a,k-2}(t)|, \qquad 0 < t < c,\\
|R'(g'',t)| &\le |R'_{c,k-2}(c-0)| \le |R'_{a,k-2}(c)|\\
&\le |R'_{a,k-2}(t)|, \qquad c \le t \le a.
\end{aligned}$$

Thus

$$|R'(g'',t)| \le |R'_{a,k-2}(t)|, \qquad 0 < t < a. \tag{23}$$

If we set

$$\delta(t) = R_{a,k-2}(t) - R(g'',t), \qquad \delta_1(t) = \int_0^t \delta(u)\,\mathrm{d}u, \qquad 0 \le t \le 2\pi,$$

then it follows from (22) and (23) that the difference δ is either non-negative or changes sign only once – from '+' to '−', but we have always that $\delta_1(t) \ge 0$. This means that

$$\int_0^t R(g'',u)\,\mathrm{d}u \le \int_0^t R_{a,k-2}(u)\,\mathrm{d}u, \qquad 0 \le t \le 2\pi,$$

and we have in view of Lemma A3.3

$$4|R'(g,t)| \le \int_0^t R_{a,k-2}(u)\,\mathrm{d}u = \begin{cases} 4|R'_{a,k}(t)|, & 0 < t < a,\\ 4|R'_{a,k}(a-0)|, & a \le t \le 2\pi, \end{cases}$$

i.e. Theorem A3.5 is true for $m = k$.

Now we can easily prove Theorem 7.1.5. Let $g \in W_V^m$ $(m = 1, 2, \ldots,)$, $\min_t |g(t)| = 0$ and $\|g\|_1 \le \|R_{a,m}\|_1$. Choose $b > 0$ from the condition $R(g, 0) = R_{b,m}(0)$. Then we have from Theorem A3.5 (see (18))

$$\|R_{b,m}\|_1 \le \|R(g)\|_1 = \|g\|_1 \le \|R_{a,m}\|_1.$$

Hence $b \le a$ and $R(g, 0) = R_{b,m}(0) \le R_{a,m}(0)$. Applying Theorem A3.5 once more and taking (19) into account we obtain

$$|R'(g, t)| \le |R'_{b,m}(t)| \le |R'_{a,m}(t)|, \qquad 0 < t < b,$$

$$|R'(g, t)| \le |R'_{b,m}(b - 0)| \le |R'_{a,m}(t)|, \qquad b \le t \le a.$$

and Theorem 7.1.5 is proved.

3 Proof of theorem 7.1.6 Let $f \in W_\infty^m \cap D_0^1$ $(m = 1, 2, \ldots)$, $\min_t |f(t)| = 0$ and

$$\|f\|_C \le \|\varphi_{n,m}\|_C, \tag{24}$$

where we can assume that $f(0) = 0$ and that the Σ-rearrangement is constructed on $[0, 2\pi]$. Fixing $x \in (0, \pi/n)$ we start with the inequalities (6) and (7), recalling that the intervals (t_ν, τ_ν) contained in $(0, 2\pi)$ are disjoint, $\tau_\nu - t_\nu = x$ and $f(t_\nu) = f(\tau_\nu)$.

First we assume that we have $q = q(x) \le 2n$ in (6) and (7). If $m = 1$ then we get from (7)

$$|R'(f, x)| \le \frac{q}{2} \|f'\|_\infty \le \frac{q}{2} \le n = |R'(\varphi_{n,1}, x)|.$$

If $m \ge 2$ then, in view of (24), we can apply Proposition 3.3.3 to f and hence

$$4|R'(f, x)| \le \sum_{k=1}^{2n} |f'(\tau_\nu) - f'(t_\nu)| \le 2n\omega(f', x) \le 2n\omega(\varphi'_{n,m}, x),$$

where $\omega(g, \delta) = \omega(g, \delta)_C$ is the modulus of continuity of function $g \in C$. Taking the definitions of the standard function $\varphi_{n,m}$ and its Σ-rearrangement into account we easily see that $2|R'(\varphi_{n,m}, x)| = n\omega(\varphi'_{n,m}, x)$ for $x \in (0, \pi/n)$ and therefore

$$|R'(f, x)| \le |R'(\varphi_{n,m}, x)|, \qquad 0 < x < \pi/n. \tag{25}$$

We shall now estimate $R(f, x)$ in the case $q \ge \pi/n$. Inequality (24) implies

$$\max_{a,b} \left| \int_a^b f'(t)\, \mathrm{d}t \right| \le 2\|f\|_C \le 2\|\varphi_{n,m}\|_C$$

and Theorem 3.3.2 implies $\|f'\|_\infty \le \|\varphi_{n,m-1}\|_\infty$. We can therefore use Theorem 3.3.4. Applying (6) first and then theorem 3.3.4 for $q > 2n$ we obtain

$$R(f, x) \le \frac{1}{2}\int_0^{2\pi-2nx} r(f', t)\, \mathrm{d}t \le \frac{1}{2}\int_0^{2\pi-2nx} r(\varphi_{n,m-1}, t)\, \mathrm{d}t,$$

i.e.

$$R(f, x) \le R(\varphi_{n,m}, x), \qquad 0 \le x \le \pi/n.$$

It remains to show that inequality (25) is necessarily true for $m \ge 3$. Applying Theorems 3.3.2 and 3.3.4 once more we obtain

$$\|f^{(k)}\|_1 \le \|\varphi_{n,m-k}\|_1, \qquad \|f^{(k)}\|_C \le \|\varphi^{n,m-k}\|_C, \qquad k = 1, 2, \ldots, m,$$

and hence

$$R(f^{(k)}, 0) \le R(\varphi_{n,m-k}, 0), \qquad k = 0, 1, \ldots, m-1. \tag{26}$$

We can apply the above arguments for f to the derivatives $f^{(k)}$ $(k = 1, 2, \ldots, m-1)$. Thus, at every point $x \in (0, \pi/n)$ one of the following relations hold

$$|R'(f^{(k)}, x)| \le |R'(\varphi_{n,m-k}, x)| \qquad \text{or} \qquad R(f^{(k)}, x) \le R(\varphi_{n,m-k}, x).$$

Comparing this with (26) we see that the difference $R(\varphi_{n,m-k}, x) - R(f^{(k)}, x)$ is either non-negative or changes sign once from '+' to '−'. This fact, together with inequality $\|f^{(k)}\|_1 \le \|\varphi_{n,m-k}\|_1$, implies, as in the proof of Theorem A3.5, the inequality

$$\int_0^x R(f^{(k)}, t)\, \mathrm{d}t \le \int_0^x R(\varphi_{n,m-k}, t)\, \mathrm{d}t, \qquad 0 \le x \le 2\pi, \qquad 1 \le k \le m-1.$$

Now using Lemma A3.3 we obtain for $m \ge 3$, $k = 2$

$$|R'(f, x)| \le \frac{1}{4}\int_0^x R(f'', t)\, \mathrm{d}t \le \frac{1}{4}\int_0^x R(\varphi_{n,m-2}, t)\, \mathrm{d}t = |R'(\varphi_{n,m}, x)|.$$

Theorem 7.1.6 is proved. □.

We leave it to the reader to verify, that inequality (25) may not be true under the assumptions of Theorem 7.1.6 for $m = 1, 2$. For this, one should consider integrals of sgn sin $2nt$.

REFERENCES

I Monographs

A Monographs on approximation theory

1. Ahlberg, J. H., Nilson, E. N. and Walsh, J. L. (1972) *The Theory of Splines and Their Applications*. Mir, Moscow.
2. Akhiezer, N. I. (Ахиезер, Н. И.) (1965) *Lectures in the Theory of Approximation.* (Лекции по теории аппроксимации.) Nauka, Moscow.
3. Bernstein, S. N. (Бернштайн, С. Н.) (1952, 1954) *Collected Works. Constructive Theory of Functions.* (Собрание сочинения. Конструктивная теория функций.) vol. I, vol. II Akad. Nauk. USSR, Moscow.
4. De Boor, C. (1985) *A Practical Guide to Splines.* Radio i Svyaz, Moscow.
5. Davis, P. J. (1963) *Interpolation and approximation.* N.Y.
6. Dzyadyk, V. K. (1977) (Дзядык, В. К.) *Introduction to the Theory of Uniform Approximation of Functions by Polynomials. (*Введение в теорию равномерного приближения функций полиномами.) Nauka, Moscow.
7. Goncharov, V. L. (Гончаров, В. Л.) (1954) *Theory of Interpolation and Approximation of Functions.* (Теория интерполирования и приближения функций.) Gosudarstv. Tehn. Teor. Lit., Moscow.
8. Kiesewetter, H. (1973) *Vorlesungen über lineare Approximation.* Berlin, VEB.
9. Korneichuk, N. P. (Корнейчук, Н. П.) (1976) *Extremal Problems of Approximation Theory.* (Экстремальные задачи теории приближения.) Nauka, Moscow.
10. Korneichuk, N. P. (Корнейчук, Н. П.) (1984) *Splines in Approximation Theory* (Сплайны в теории приближения.) Nauka, Moscow.
11. Korneichuk, N. P., Ligun, A. A. and Doronin, V. G. (Корнейчук, Н. П., Лигун, А. А., Доронин, В. Г.) (1982) *Approximations with Constraints. (*Аппроксимация с ограничениями.) Naukova Dumka, Kiev.
12. Korovkin, P. P. (Коровкин, П. П.) (1959) *Linear Operators and Approximation Theory* (Линейные оператори и теория приближений.) Gosudarstv. Fiz.-Mat. Lit., Moscow.
13. Laurent, P.-J. (1975) *Approximation and Optimization.* Mir, Moscow.
14. Lorentz, G. G. (1966) *Approximation of Functions.* N.Y.
15. Lorentz, G. G., Jetter, K. and Riemenschneider, S. D. (1983) *Birkhoff Interpolation.* Addison-Wesley.

16. Natanson, I. P. (Натансон, И. П.) (1949) *Constructive Theory of Functions.* (Конструктивная теория функций.) Gosudarstv. Tehn.-Teor. Lit., Moscow.
17. Nikol'skii, S. M. (Никольский, С. М.) (1977) *Approximation of Functions of Several Variables and Imbedding Theorems.* (Приближение функций многих переменных и теоремы вложения.) Nauka, Moscow.
18. Nikol'skii, S. M. (Никольский, С. М.) (1979) *Quadrature Formulae.* (Квадратурные формули.) Nauka, Moscow.
19. Pinkus, A. (1985) *N-widths in Approximation Theory.* Berlin, Springer Verlag.
20. Schumaker, L. L. (1981) *Spline Functions: Basic Theory.* N.Y.
21. Sendov, Bl. (Сендов, Бл.) (1979) *Hausdorff Approximation.* (Хаусдорфовые приближения.) Sofia.
22. Singer, I. (1967) *Cea mai buna approximate in spatil vectoriale normate prin elemente din subspatil vectoriale.* Bucuresti.
23. Stechkin, S. B. and Subbotin, Yu. N. (Стечкин, С. Б., Субботин, Ю. Н.) (1976) *Splines in Numerical Mathematics.* (Сплайны в вычислительной математики.) Nauka, Moscow.
24. Stepanets, A. I. (Степанец, А. И.) (1981) *Uniform Approximation by Trigonometric Polynomials.* (Равномерные приближения тригонометрическими
полиномами.) Naukova Dumka, Kiev.
25. Tikhomirov, V. M. (Тихомиров, А. Ф.) (1976) *Some Questions in Approximation Theory.* (Некоторые вопросы теории приближений.) Moskow, Univ. Moscow.
26. Timan, A. F. (Тиман, А. Ф.) (1960) *Theory of Approximation of Functions of a Real Variable.* (Теория приближения функций действительного переменного.) Gosudarstv. Fiz.-Mat. Lit., Moscow.
27. Turetskii, A. H. (Турецкий, А. Х.) (1968) *Theory of Interpolating in Problem Form.* (Теория интерполирования в задачах.) Vyseisaja Skola, Minsk.
28. Zav'yalov, Yu. S., Kvasov, B. I. and Miroshnichenko, V. L. (1980) (Эавьялов, Ю. С., Квасов, Б. И., Мирошниченко, В. Л.) *Methods of Spline Functions.* (Методы сплайн-функций.) Nauka, Moscow.

B Other books

1. Bari, N. K. (Бари, Н. К.) (1961) *Trigonometric Series.* (Тригонометрические ряды) Gosudarstv. Fiz.-Mat. Lit., Moscow.
2. Chebyshev, P. L. (Чебышев, П. Л.) (1955) Selected papers. (Избраные труды.) Akad. Nauk SSSR, Moscow.
3. Danford, N. and Schwartz, J. T. (1962) *Linear Operators.* General theory. IL, Moscow.
4. Danilov, V. L. and others (Данилов, В. Л. и др.) (1961) *Calculus: Functions, Limits, Series, Continued Fractions.* (Математический анализ: функции, пределы, ряды, цепные дроби.) Gosudarstv. Izdat. Fiz.-Mat. Lit., Moscow.
5. Fikhtengol'ts, B. M. (Фихтенгольц, Г. М.) (1969) *A Course on Differential and Integral Calculus,* vol. II. Nauka, Moscow.
6. Gol'shtein, E. G. (Гольштейн, Е. Г.) (1871) *Duality Theory in Mathematical Programming and its Applications.* (Теория двойнствености в математическом программировании и ее приложения.) Nauka, Moscow.
7. Hardy, H., Littlwood, J. E. and Polya, G. (1948) *Inequalities.* IL, Moscow.

8. Ioffe, A. D. and Tikhomirov, V. M. (Иоффе, А. Д., Тихомиров, В. М.) (1974) *Theory of Extremal Problems* (Теория зкстремальных задач.) Nauka, Moscow.
9. Kolmogorov, A. N. and Fomin, S. V. (Колмогоров, А. Н., Фомин, С. В.) (1972) *Elements of the Theory of Functions and Functional Analysis.* (Элементы теории функций и функционального анализа.) Nauka, Moscow.
10. Lyusternik, L. A. and Sobolev, V. I. (Люстерник, Л. А., Соболев, В. И.) (1965) *Elements of Functional Analysis* (Элементы функционального анализа.) Nauka, Moscow.
11. Markov, A. A. (Марков, А. А.) (1948) *Selected papers.* (Избраные труды.) OGIZ, Moscow and Leningrad.
12. Natanson, I. P. (Натансон, И. П.) (1974) *Theory of Functions of a Real Variable.* (Теория функций вещественной переменной.) Nauka, Moscow.
13. Polya, G. and Szego, G. (1956) *Problems and Theorems in Analysis.* Gosudarstv. Tehn.-Teor. Lit., Moscow.
14. Suetin, P. K. (Суетни, П. К.) (1976) *Classical Orthogonal Polynomials.* (Классические ортогональные многочлены.) Nauka, Moscow.
15. Titchmarsh, E. C. (1980) *The Theory of Functions.* Nauka, Moscow, 1980.
16. Vitushkin, A. G. (Витушкин, А. Г.) (1959) *Estimation of the Complexity of the Tabulation Problem.* (Оценка сложности задачи табулирования.) Gosudarstv. Fiz.-Mat. Lit., Moscow.
17. Zygmund, A. (1965) *Trigonometric Series,* vol. II. Mir, Moscow.

II Articles

Ahlberg, J. H., Nilson, E. N. and Walsh, J. L.

1. Best approximation and convergence properties of higher order spline approximations. *J. Math. Mech.* **14** (1965), 231–44.

Akhiezer, N. I. (Ахиезер, Н. И.)

1. On best approximation of analytic functions. (О наилучшем приближении аналитических функций.) *Dokl. Akad. Nauk SSSR* **18** (1938), 4–5, 241–4.

Akhiezer, N. I. and Krein, M. G. (Ахиезер, Н. И., Крейн, М. Г.)

1. On best approximation of differentiable periodic functions by trigonometric sums. (О наилучшем приближении тригонометрическими суммами дифференцируемых перидических функций.) *Dokl. Akad. Nauk SSSR* **15** (1937), 107–12.

Arestov, V. V. (Арестов, В. В.)

1. Some extremal problems for differentiable functions of one variable (О некоторых зкстремальных задачах для дифференцируемых функций одной переменной.) *Trudy Math. Inst. Stekhlov* **138** (1975), 3–28.
2. Approximation of linear operators and related extremal problems. (Прнближение линейных операторов и родственные эхтремалрные задачи.) *Trudy Math. Inst. Stekhlov* **138** (1975), 29–42.
3. On Bernstein inequalities for algebraic and trigonometric polynomials. (О неравенствах С. Н. Бернщтейна для алгебраических и тригонометрических полиномов.) *Dokl. Akad. Nauk SSSR* **246** (1979), **6**, 1289–92.

Arestov, V. V. and Gabushin, V. N. (Арестов, В. В. и Габушин, Ф. Н.)

1. On approximation of classes of differentiable functions. (О приближении классов дифференцируемых функций.) *Math. Zametki* **9** (1971), 2, 105–12.

Avakyan, A. M. (Авакян, А. М.)

1. Exact error constants in local parabolic spline approximation. (Точные конс-

танты погрещностей при приближении локалрными параболическими сплайнами.) *Monogenic Functions and Mappings. Questions of the Theory of the Approximation of Functions.* Kiev, Akad. Nauk Ukrain. SSR Inst. Math., 1982, 114–21.

2. Approximation of functions of two variables by linear methods. (О приближении функций двух переменных линейными методами.) *Ukrain. Math. Zh.* **35** (1983), 4, 409–15.

Babenko, A. G. (Бабенко, А. Г.)

1. On the exact constant in Jackson inequality in L_1. (О точной константе в неравенстве Джексона в L_1.) *Math. Zametki* **39** (1986), 5, 651–64.

Babenko, K. I. (Бабенко, К. И.)

1. Best approximation to a class of analytic functions. (О наилучщих приближениях одного класса аналитических функций.) *Izv. Akad. Nauk. SSSR Ser. Math.* **22** (1958), 5, 631–40.
2. Some problems of the theory of approximation and numerical analysis. (О некоторых задачах теории приближений и численного анализа.) *Usp. Math. Nauk* **40** (1985), 1, 3–27.

Babenko, V. F. (Бабенко, В. Ф.)

1. Exact estimates for the norms of functions in conjugate classes in C and L metrics (Точные оценки для норм функций из сопряженных классов в метриках C и L.) *Studies in Current Problems of Summation and Approximation of Functions and Their Applications.* Dnepropetrovsk, Gos. Univ. Dnepropetrovsk, 1973, 3–5.
2. Interpolation of continuous mappings by piecewise linear ones. (Интерполяция непрерывных отображений кусочнолинейными.) *Math. Zametki* **24** (1978), 1, 43–52.
3. Non-symmetric approximations in spaces of summable functions. (Несимметричные приближения в пространствах суммируемых функций.) *Ukrain. Math. Zh.* **34** (1982), 4, 409–16.
4. Non-symmetric extremal problems in approximation theory. (Несимметричные экстремальные задачи теории приближения.) *Dokl. Akad. Nauk SSSR* **269** (1983), 3, 521–4.
5. Inequalities for rearrangements of differentiable periodic functions and problems of approximation and approximate integration. (Неравенства для перестановок дифференцируемых периодических функций, задачи приближения и приближенного интегрированиял.) *Dokl. Akad. Nauk SSSR* **272** (1983), 5, 1038–41.
6. The diameters of certain classes of convolutions. (О попречниках некоторых классов сверток.) *Ukrain. Math. Zh.* **35** (1983), 5, 603–7.
7. Asymmetric approximations of classes W^rH^ω by trigonometric polynomials and splines. (О несимметричных приближениях классов W^rH^ω тригонометрическими полиномами и сплайнами.) *Studies in Current Problems of Summation and Approximation of Functions and Their Applications.* Dnepropetrovsk, Gos. Univ. Dnepropetrovsk, 1983, 3–11.
8. Extremal problems in approximation theory and inequalities for rearrangements. (Экстремальные задачи теории приближения и неравенства для перестановок.) *Dokl. Akad. Nauk SSSR* **290** (1986), 5, 1033–6.
9. Exact inequalities for the norms of conjugate functions and their application. (Точные неравенства для норм сопряженных функций и их применения.) *Ukrain. Math. Zh.* **39** (1987), 2, 139–44.
10. Non-symmetric approximations and inequalities for rearrangements in extremal

problems of approximation theory. (Несимметричные приближения и неравенства для перестановок в экстремальных задачах теории приближения.) *Trudy Math. Inst. Stekhlov* **180** (1987), 33–5.

11. Some exact inequalities for approximation of periodic functions in space L_1. (Некоторые точные неравенства для приближения периодических функций в пространстве L_1.) *Studies in Current Problems of Summation and Approximation of Functions and Their Applications.* Dnepropetrovsk, Gos. Univ. Dnepropetrovsk, 1987, 4–6.
12. Comparison theorems and inequalities of Bernstein type. (Теоремы сравнения и неравенства типа неравенства Бернштейна.) *Teoriya priblizhenii i smezhnye voprosy analiza.* Kiev. Math. Inst. Akad. Nauk Ukrain. SSR (1987), 4–8.
13. Approximation of convolution classes. (Приближение классов сверток.) *Sibir. Math. Zh.* **28** (1987), 5, 6–21.
14. Approximation of classes of functions defined with the help of modulus of continuity. (Приближение классов функций, задаваемых с помощью модуля непрерывности.) *Dokl. Akad. Nauk SSSR* **298** (1988), 6, 1296–9.

Babenko, V. F. and Kofanov, V. A. (Бабенко В. Ф., Кофанов В. А.)

1. Non-symmetric approximations of classes of differentiable functions by algebraic polynomials. (Несимметричные приближения классов дифференцируемых функций алгебраическими полиномами.) *Dokl. Akad. Nauk Ukrain. SSR, Ser. A* (1984), 11, 3–5.

Babenko, V. F. and Ligun, A. A. (Бабенко, В. Ф., Лигун, А. А.)

1. Interpolation by polyhedral functions. (Об интерполяции многогранными функциями.) *Math. Zametki* **18** (1975), 6, 803–14.

Babenko, V. F. and Pichugov, S. A. (Бабенко, В. Ф., Пичугов, С. А.)

1. Best linear approximation of certain classes of differentiable periodic functions. (О наилучшем линейном приближении некоторых классов дифференцируемых периодических функций.) *Math. Zametki* **27** (1980), 683–9.
2. Asymptotic properties of best uniform approximation of periodic functions. (Асимптотические свойства наилучшего равномерного приближения периодических функций.) *Studies in Current Problems of Summation and Approximation of Functions and Their Applications.* Dnepropetrovsk, Gos. Univ. Dnepropetrovsk, 1980, 10–14.

Babenko, V. F. and Shalaev, V. V. (Бабенко, В. Ф., Шалаев, В. В.)

1. Inequalities of Jackson type for linear methods of approximation of functions. (О неравенствах типа неравенств Джексона для линейных методов приближения функций.) *Dokl. Akad. Nauk Ukrain. SSR Ser. A* (1980), 10, 31–4.

Bakhvalov, N. S. (Бахвалов, Н. С.)

1. The optimality of linear operator approximation methods on convex function classes. (Об оптимальности линейных методов приближений операторов на выпуклых классах функций.) *Zh. Vych. Math. and Math. Phys.* **11** (1971), 4, 1014–18.

Berdyshev, V. I. (Бердышев, В. И.)

1. Jackson's theorem in L_p. (О теореме Джексона в L_p.) *Trudy Math. Inst. Stekhlov* **88** (1967), 3–16.

Bernstein, S. N. (Бернштейн, С. Н.)

1. Sur la meilleure approximation de $|x|$ par des polynomes de degrées donnés. *Acta Sci. Math.* **37** (1913), 1–57.

2. Sur la limitation des dérivées des polynomes. *C.R. Acad. Sci. (Paris)* **190** (1930), 338–41.
3. Sur un procédé de sommation des series trigonometriques. *C.R. Acad. Sci. (Paris)* **191** (1930), 976–9.
4. On best approximation of $|x-c|^p$ (О наилучшем приближении $|x-c|^p$.) *Dokl. Akad. Nauk SSSR* **18** (1938), 379–84.
5. A generalization of a result of S. M. Nikol'skii (Обобщение одного результата С. М. Никольского.) *Dokl. Akad. Nauk SSSR* **53** (1946), 587–9.
6. On asymptotic relations between constants of best approximation theory. (О предельных зависимостях между константами теории наилучшего приближения.) *Dokl. Akad. Nauk SSSR* **57** (1947), 3–5.

Birkhoff, G., Schultz, M. H. and Varga, R. S.
1. Piecewise Hermite interpolation on one and two variables with applications to partial differential equations. *Num. Math.* **11** (1968), 3, 232–56.

Bohr, H.
1. Une théorème générale sur l'integration d'un polynome trigonometrique. *C.R. Acad. Sci. (Paris)* **200**, 1276–7.

De Boor, C.
1. On best interpolation. *J. Approx. Th.* **16** (1976), 28–42.

Borsuk, K.
1. Drie Satze über die *n*-dimensionale euklidische Spare. *Fund. Math.* **20** (1933), 177–91.

Boyanov, B. D. (Боянов, Б. Д.)
1. Best interpolation methods for certain classes of differentiable functions. (Наилучшие методы интерполирования для некоторых классов дифференцируемых функций.) *Math. Zametki* **17** (1975), 4, 511–24.
2. Best reconstruction of differentiable functions from their Fourier coefficients. (Наилучшее востоновление дифференцируемых периодических функций по их коефециентам Фурье.) *Serdica* **2** (1976), 4, 300–4.
3. A note on the optimal approximation of smooth periodic functions. *Dokl. Bul. Akad. Sci.* **30** (1977), 6, 809–19.

Bugaets, V. P. and Martynyuk, V. T. (Бугаец, В. П., Мартынюк, В. Т.)
1. Exact constants of approximation of continuous functions by Jackson integrals. (Точные константы приближения непрерывных периодических функций интегралами Джексона.) *Ukrain. Math. Zh.* **26** (1974), 4, 435–43.

Buslaev, A. P. and Tikhomirov, V. M. (Буслаев, А. П., Тихомиров, В. М.)
1. Some problems in non-linear analysis and approximation theory. (Некоторые вопросы нелинейного анализа и теории приближений.) *Dokl. Akad. Nauk SSSR* **283** (1985), 1, 13–18.

Chebyshev, P. L. (Чебышев, П. Л.)
1. Problems for the smallest quantities connected with approximate representation of functions. (Вопросы о наименьших величинах, связанных с приближенным представлением функций.) *Zap. Akad. Nauk* (1859) (See [2B, pp. 462–578]).

Cheney, E. M. and Schurer, F.
1. On interpolating by cubic splines with equally-spaced nodes. *Indag. Math.* **30** (1968), 517–24.

Chernykh, N. I. (Черных, Н. И.)
1. Jackson's inequality in L_2. (О неравенстве Джексона в L_2.) *Trudy Math. Inst. Stekhlov* **88** (1967), 71–4.
2. The best approximation of periodic functions by trigonometric polynomials in

L_2. (О наилучшем приближении периодических функций тригонометрическими полиномами в L_2.) *Math. Zametki* **2** (1967), 5, 513–22.

Ciarlet, P. G., Schultz, M. H. and Varga, R. S.

1. Numerical methods of high-order accuracy for nonlinear boundary problems. *Numer. Math.* **12** (1968), 394–430.

Davidchik, A. N. (Давидчик А. Н.)

1. Approximation of periodic functions by linear positive operators. (Приближение периодических функций линейными положителрными операторами.) *Studies in Current Problems of Summation and Approximation of Functions and Their Applications*. Dnepropetrovsk, Gos. Univ. Dnepropetrovsk, 1982, 187–93.
2. Extremal problems in approximation of periodic functions by linear operators. (Экстремальные задачи приближения периодических функций линейными операторами.) PhD Dissertation, Inst. Math. Akad. Nauk Ukrain. SSR, Kiev, 1983.

Davidchik, A. N. and Ligun, A. A. (Давидчик, А. Н., Лигун, А. А.)

1. About Jackson's theorem. (К теореме Джексона.) *Math. Zametki* **16** (1974), 5, 681–90.
2. Exact constants in Jackson-type inequalities. (О точных константах в неравенствах типа Джексона.) *Math. Zametki* **29** (1981), 5, 761–9.

Demchenko, A. G. (Демченко, А. Г.)

1. Approximation in the mean of functions of the classes $W^rH^{[\omega]}{}_L$ by Fourier sums. (Приближение в среднем функций класса $W^rH^{[\omega]}{}_L$ суммами Фурье.) *Ukrain. Math. Zh.* **25** (1973), 2, 267–77.

Doronin, V. G. (Доронин, В. Г.)

1. Best one-sided approximation on certain classes of functions. (Наилучшее одностороннее приближение на некоторых классах функций.) *Math. Zametki* **10** (1971), 6, 615–26.

Doronin, V. G. and Ligun, A. A. (Доронин, В. Г., Лигун, А. А.)

1. Upper bounds for best one-sided approximation of classes W^rL_1 by splines. (Верхние грани наилучших односторонних приближений сплайнами классов W^rL_1.) *Math. Zametki* **19** (1976), 1, 11–17.
2. The exact values of best one-sided approximation by splines. (О точных значениях наилучших односторонних приближений сплайнами.) *Math. Zametki* **20** (1976), 3, 417–24.
3. Best one-sided approximation of the classes W^rH^ω. (О наилучшем одностороннем приближении классов W^rH^ω.) *Math. Zametki* **21** (1977), 3, 313–27.
4. Upper bounds for the best one-sided approximation of the classes W^rL_ψ in the L metric. (Верхные грани наилучших односторонних приближений классов W^rL_ψ в метрике L.) *Math. Zametki* **22** (1977), 2, 257–68.
5. Best one-sided approximation of the classes $W^r_\alpha V$ $(r > -1)$ by trigonometric polynomials in the L_1 metric. (О наилучшем одностороннем приближении классов $W^r_\alpha V$ $(r > -1)$ тригонометрическими полиномами в метрике L_1.) *Math. Zametki* **22** (1977), 3, 357–70.
6. Best one-sided approximation of some classes of functions. (Наилучшее одностороннее приближение некоторых классов функции.) *Math. Zametki* **29** (1981), 3, 431–54.

Doronin, G. Ya. (Доронин, Г. Я.)

1. Some inequalities for approximation by trigonometric polynomials. (Некоторые неравенства для приближения тригонометрическими полиномами.) *Dokl. Akad. Nauk SSSR* **69** (1949), 3, 487–90.

Dzyadyk, V. K. (Дзядык В. К.)

1. On best mean approximation of periodic functions with singularities. (О наилучшем приближении в среднем периодических функции с особенностями.) *Dokl. Akad. Nauk SSSR* **77** (1951), 6, 949–52.
2. On best approximation of the class of periodic functions with bounded sth derivative ($0 < s < 1$). (О наилучшем приближении на классе периодических функций, имеющих ограниченую s-ю производную ($0 < s < 1$).) *Izv. Akad. Nauk SSSR Ser. Math.* (1953), 135–162.
3. Best approximation on classes of periodic functions defined by kernels which are integrals of absolutely monotonic functions. (О наилучшем приближении на классах периодических функций, определяемыых ядрами, являющимися интегралами от абсолютно монотонныых функций.) *Izv. Akad. Nauk SSSR Ser. Math.* **23** (1959), 933–50.
4. Best approximation on classes of periodic functions that are defined by integrals of a linear combination of absolutely monotonic kernels. (О наилучшем приближении на классах периодических функций, определяемых интегралами от линейной комбинации абсолютно монотонных ядер.) *Math. Zametki* **16** (1974), 5, 691–701.
5. The least upper bound of best approximation on certain classes of continuous functions that are defined on the real line. (О точных веьхных гранях наилучших приближений на некоторых классах функций, определенных на всей вещественной оси.) *Dopovidi Akad. Nauk Ukrain. RSR Ser. A* (1975), 7, 589–93.

Dzyadyk, V. K., Gavrilyuk, V. T. and Stepanets, A. I. (Дзядык В. К., Гаврилюк, В. Т., Степанец, А. И.)

1. The approximation of functions of Hölder classes by Rogosinski polynomials. (О приближении функций классов Гельдера полиномами Рогозинского.) *Dokl. Akad. Nauk Ukrain. RSR Ser. A* (1969), 3, 203–6.
2. The best upper bound of approximations on classes of differentiable periodic functions by means of Rogosinski polynomials. (О точной верхней грани приближений на классах дифференцируемых периодических функций при помощи полиномов Рогозинского.) *Ukrain. Math. Zh.* **22** (1970), 4, 481–93.

Dzyadyk, V. K. and Stepanets, A. I. (Дзядык, В. К., Степанец, А. И.)

1. Asymptotic equalities for least upper bounds of the approximations of functions of Hölder classes by means of Rogosinski polynomials. (Асимптотические равенства для точных верхних граней приближений функций классов Гельдера при помощи полиномов Рогозинского.) *Ukrain. Math. Zh.* **24** (1972), 4, 476–87.

Efimov, A. V. (Ефимов, А. В.)

1. Approximation of certain classes of continuous functions by Fourier sums and Fejér means. (О приближении некоторых классов неприрывных функций суммами Фурье и суммами Фейера.) *Izv. Akad. Nauk SSSR Ser. Math.* **22** (1958), 1, 81–116.
2. Approximation of functions with given modulus of continuity by Fourier sums. (Приближение функций с заданным модулем непрерывности суммами фурье.) *Izv. Akad. Nauk SSSR Ser. Math.* **23** (1959), 1, 115–34.
3. Approximation of continuous periodic functions by Fourier series. (Приближение непрерывных переодических функций суммами Фурье.) *Izv. Akad. Nauk SSSR Ser. Math.* **24** (1960), 2, 243–96.
4. Linear methods of approximating continuous periodic functions. (Линейные

методы приближения неприрывыых периодических функций.) *Math. Sb. (N.S.)* **54** (1961), 1, 51–90.
5. Best approximation of classes of periodic functions by trigonometric polynomials. (О наилучшем приближении классов периодических функций тригонометрическими полиномами.) *Izv. Akad. Nauk SSSR Ser. Math.* **30** (1966), 1163–78.

Egervary, E. V. and Szasz, O.
1. Einige Extremalprobleme in probleme in Bereich der trigonometrischen Polynome. *Math. Zeit.* **27** (1928), 641–52.

Fang Gensun
1. Asymptotic estimation of *N*-widths of periodic differentiable function class $\tilde{W}^r H_C^\omega$ and related extremal problems. *Northeastern Math. J.* **1** (1986), 2, 17–32.

Favard, J.
1. Application de la formule sommatoire d'Euler à la demonstration de quelques proprietes extremales des integrales des fonctions periodiques. *Math. Tidskrift.* **4** (1936), 81–4.
2. Sur l'approximation des fonctions periodiques par des polynomes trigonometriques. *C.R. Acad. Sci. (Paris)* **203** (1936), 1122–4.
3. Sur les meilleurs procédés d'approximation de certaines classes de fonctions par des polynomes trigonometriques. *Bull. Sci. Math. Ser. 2* **60** (1937), 209–24, 243–56.

Féjer, L.
1. Untersuchungen über Fouriersche Reihen. *Math. Ann.* **58** (1904), 501–69.

Freud, G.
1. Uber einseitige Approximation durch Polynome I. *Acta sci. Math. (Szeged)* **16** (1955), 12–18.

Gabushin, V. N. (Габушин, В. Н.)
1. Inequalities for norms of a function and its derivatives in L_p-metric. (Неравенства для норм функци и ее производных в метриках L_p.) *Math. Zametki* **1** (1967), 3, 291–8.
2. Precise constants in inequalities between the norms of the derivatives of a function. (Точные константы в неравенства между нормами производных функций.) *Math. Zametki* **4** (1968), 2, 221–32.

Galkin, P. V. (Галкин, П. В.)
1. Estimates for Lebesgue constants. (Оценки для констант Лебега.) *Trudy Math. Inst. Stekhlov* **109** (1971), 3–5.

Ganelius, T.
1. On one-sided approximation by trigonometrical polynomials. *Math. Scand.* **4** (1954), 247–58.

Garkavi, A. L. (Гаркави, А. Л.)
1. Duality theorems for approximation by elements of convex sets. (Теоремы двойнственности для прииближений посредством элементов выпуклых множеств.) *Usp. Math. Nauk* **16** (1961), 4, 141–5.
2. A criterion for the element of best approximation. (О критерии элемента наилучшего приближения.) *Sibirsk. Math. Zh.* **5** (1964), 472–6.
3. The theory of best approximation in normed linear spaces. (Теория наилучшего приближения в линейных нормированых пространствах.) *Mathematicheskii Analiz* (1969), 75–132.

Gavrilyuk, V. T. (Гаврилюк, В. Т.)
1. Approximation of continuous periodic functions by Rogosinski polynomials and Fourier sums. (Приближение непрерывных периодических функций суммами

Рогозинского и суммами Фурье.) *Questions in the Theory of Approximation of Functions and Its Applications.* Kiev, Inst. Math. Akad. Nauk Ukrain. SSR, 1976, 46–60.

2. Approximation of continuous periodic functions by trigonometric polynomials. (Приближение непрерывных периодических функций тригонометрическими полиномами.) *Approximation Functions Theory.* Nauka, Moscow, 1977, 101–3.

Gavrilyuk, V. T. and Stechkin, S. B. (Гаврилюк, В. Т., Стечкин, С. Б.)

1. Approximation of continuous periodic functions by Fourier series. (Приближение непрерывных периодических функций суммами Фурье.) *Dokl. Akad. Nauk SSSR* **241** (1978), 3, 525–7.

Gavrilyuk, V. T. and Stepanets, A. I. (Гаврилюк, В. Т. и Степанец, А. И.)

1. Approximation of differentiable functions by Rogosinski polynomials. (Приближение дифференцируемых функций полиномами Рогозинского.) *Ukrain. Math. Zh.* **25** (1973), 1, 3–13.

von Golitschek, M.

1. On *N*-widths and interpolation by polynomial splines. *J. Approx. Th.* **26** (1979), 133–41.

Grebennikov, A. I. and Morozov, V. A. (Гребенников, А. И., Морозов, В. А.)

1. Optimal approximation of operators. (Об оптимальном приближении операторов.) *Zh. Vyssh. Math. i Math. Fiz.* **17** (1977), 1, 3–14.

Grigoryan, Yu. I. (Григорян, Ю. И.)

1. Diameters of certain sets of function spaces. (Поперечники некоторых множеств в функциональных пространствов.) *Math. Zametki* **13** (1973), 5, 637–46.

Hall, Ch. A. and Meyer, W. W.

1. Optimal error bounds for cubic spline interpolation. *J. Approx. Th.* **16** (1976), 105–22.

Havinson, S. Ya. (Хавинсон, С. Я.)

1. Approximation by elements of convex sets. (Об апроксимации элементами выпуклых множеств.) *Dokl. Akad. Nauk SSSR* **172** (1967), 294–7.

Hörmander, L.

1. A new proof and generalization of inequality of Bohr. *Math. Scand.* **2** (1954), 33–45.

Ioffe, A. D. and Tikhomirov, V. M. (Иоффе, А. Д., Тихомиров, В. М.)

1. Duality of convex functions and extremal problems. (Двоинственость выпуклых функций и экстремальные задачи.) UMN **27** (1968), 6, 51–116.

Jackson, D.

1. Uber die Genauigkeit des Annaherung setetigen Funktionen durch ganze rationale Funktionen gegebenen Grades und trigonometrischen Summen dedebener. Ordnung. Diss., Göttingen, 1911.

Kofanov, V. A. (Кофанов, В. А.)

1. The best uniform approximation of differentiable functions by algebraic polynomials. (О наилучшем равномерном приближении дифференцируемых функций алгебраическими многочленами.) *Math. Zametki* **27** (1980), 3, 381–90.

2. Approximation of classes of differentiable functions by algebraic polynomials in the mean. (Приближения классов дифференцируемых функций алгебраи-

ческими многочленами в среднем.) *Dokl. Akad. Nauk SSSR* **262** (1982), 6, 1304–6.

Kolmogorov, A. N. (Колмогоров, А. Н.)

1. Zur Grossenordnung des restgliedes Fourierscher Reihen differenzierbarer Funktionen. *Ann. Math.* **36** (1935), 107–10.
2. Uber die besste Annaherung von Functionen einer gegebenen Funktionklassen. *Ann. Math.* **37** (1936), 107–10.
3. Inequalities between the upper bounds of consecutive derivatives of functions on the unbounded interval. (О неравенствах между верхними гранями последовательных производных функций на бесконечном интервале.) *Uchebnye Zap. MGU. Mathemathics* **30** (1939), 3, 3–13.

Kong-Ming Shong

1. Some extensions of a theorem of Hardy, Littlewood and Polya and their applications. *Canad. J. Math.* **26** (1947), 1321–40.

Konovalov, V. N. (Коновалов, В. Н.)

1. Sharp inequalities for the norm of functions and their third partial and second mixed or directional derivatives. (Точные неравенства для норм функций, третьих частных, вторых смешанных или косых производных.) *Math. Zametki* **23** (1978), 1, 67–78.

Korneichuk, N. P. (Корнейчук, Н. П.)

1. Approximation of periodic functions satisfying Lipschitz's condition by Bernstein–Rogosinski's sums. (О приближении функций, удовлетворяющих условию Липшица, суммами Бернштейна–Рогозинского.) *Dokl. Akad. Nauk SSSR* **125** (1959), 258–61.
2. On the degree of approximation of functions of class $H^{(\alpha)}$by means of trigonometric polynomials. (Об оценке приближений функций класса $H^{(\alpha)}$тригонометрическими многочленами.) *Studies of Modern Problems of Constructive Theory of Functions.* Moscow, Fizmathgiz, 1961, 148–54.
3. The best uniform approximation in certain classes of continuous functions. (О наилучшем равномерном приближении на некоторых классах непрерывных функций.) *Dokl. Akad. Nauk SSSR* **140** (1961), 748–51.
4. The best uniform approximation of differentiable functions. (О наилучшем равномерном приближении дифференцируемых функций.) *Dokl. Akad. Nauk SSSR* **141** (1961), 304–7.
5. Extremal properties of periodic functions. (Об экстремальных свойствах периодических функций.) *Dopovidi Akad. Nauk Ukrain. RSR* (1962), 993–8.
6. The exact constant in the theorem of D. Jackson on the best uniform approximation to continuous periodic functions. (Точная константа в теореме Джексона о наилучшем равномерном приближении непрерывных периодических функций.) *Dokl. Akad. Nauk SSSR* **145** (1962), 514–15.
7. On the best approximation of continuous functions. (О наилучшем приближении непрерывных функций.) *Izv. Akad. Nauk SSSR Ser. Math.* **27** (1963), 29–44.
8. The exact value of best approximations and widths of certain classes of function. (Точное значение наилучших приближений и поперечников некоторых классов функций.) *Dokl. Akad. Nauk SSSR* **150** (1963), 1218–20.
9. Precise estimates for the norms of differentiable periodic functions in the L metric. (Точные значения норм дифференцируемых периодических функций в метрике L.) *Math. Zametki* **2** (1967), 6, 569–76.

10. Upper bounds of best approximation on classes of differentiable periodic functions in the metrics C and L. (Верхные грани наилучших приближений на классах дифференцируемых периодических функций в метриках C и L.) *Dokl. Akad. Nauk SSSR* **190** (1970), 269–71.
11. Extremal values of functionals and the best approximation on classes of periodic functions. (Экстремальные значения функционалов и наилучшее приближение на классах периодических функций.) *Izv. Akad. Nauk SSSR Ser. Math.* **35** (1971), 1, 93–124.
12. The diameters of classes of continuous functions in the space L_p. (О поперечниках классов непрерывных функций в пространстве L_p.) *Math. Zametki* **10** (1971), 5, 493–500.
13. Inequalities for differentiable periodic functions and the best approximation of a certain class of functions by another. (Неравенства для дифференцируемых периодических функций и наилучшее приближение одного класа функций другим.) *Izv. Akad. Nauk SSSR Ser. Math.* **36** (1972), **2**, 423–34.
14. The uniform approximation of periodic functions by subspaces of finite dimension. (О равномерном приближении периодических функций подпространствами конечной размерности.) *Dokl. Akad. Nauk SSSR* **213** (1973), 3, 525–8.
15. On extremal subspaces and approximation of periodic functions by splines of minimal defect. *Analysis Math.* **1** (1975), 2, 91–101.
16. Best approximation by splines of classes of periodic functions in the L metric. (Наилучшее приближение сплайнами на классах периодических функций в метрике L.) *Math. Zametki* **20** (1976), 5, 655–64.
17. Exact error bound of approximation by interpolating splines on L-metric on the classes W_p^r ($1 \leqslant p < \infty$) of periodic functions. *Analysis Math.* **3** (1977), 2, 109–17.
18. Sharp inequalities for the best approximation by splines. (Точные неравенства для наилучшие приближения сплайнами.) *Dokl. Akad. Nauk SSSR* **242** (1978), 2, 280–3.
19. Optimal reconstruction of functions and their derivatives in the L_p metric. (Оптимальное восстановление функций и их производных в метрике L_p.) *Theory of cubature formulas and Numerical Mathematics*. Novosibirsk, Nauka, 1980, 152–7.
20. The best approximation on a segment of classes of functions with a bounded rth derivative via finite dimensional subspaces. (О наилучшем приближении на отрезке классов функций с ограниченной r-й производной конечномерными подпространствами.) *Ukrain. Math. Zh.* **31** (1979), 23–31.
21. Inequalities for the best approximation by splines of differentiable periodic functions. (Неравенсва для наилучшего приближения сплайнами дифференцируемых периодических функций.) *Ukrain. Math. Zh.* **31** (1979), 4, 380–8.
22. Diameters in L_p of classes of continuous and differentiable functions and the optimal reconstruction of functions and their derivatives. (Поперечники в L_p классов непрерывных и дифференцируемых функций и оптимальное востановление функций и их производных.) *Dokl. Akad. Nauk SSSR* **224** (1979), 6, 1317–21.
23. Diameters in L_p of classes of continuous and differentiable functions and optimal methods of coding and reconstructing functions and their derivatives. (Поперечники в L_p классов непрерывных и дифференцируемых функций и их производных.) *Izv. Akad. Nauk SSSR Ser. Math.* **45** (1981), 2, 266–90.

24. Existence and uniqueness of interpolation splines. (О существовании и единственности интерполяционных сплайнов.) *Geometric Theory of Functions and Topology*. Kiev, Akad. Nauk Ukrain. SSR Inst. Math., 1981, 42–55.
25. Approximation of functions and their derivatives by interpolation splines. (О приближении интерполяционными сплайнами функции и их производных.) *Dokl. Akad. Nauk SSSR* **264** (1982), 5, 1063–6.
26. Approximation by local splines of minimum defect. (О приближении локальными сплайнами минимального дефекта.) *Ukrain. Math. Zh.* **34** (1982), 5, 617–21.
27. The exact constant in the Jackson inequality for continuous periodic functions. (О точной константе в неравенстве Джексона для непрерывных периодических функций.) *Math. Zametki* **32** (1982), 5, 669–74.
28. Comparing of rearrangements and estimating the spline interpolation error. (О сравнении перестановок и оценке погрешности при интерполировании сплайнами.) *Dokl. Akad. Nauk Ukrain. SSR, Ser. A.* **4** (1983), 19–21.
29. Approximation by parabolic splines of differentiable functions and their derivatives. (О приближении параболическими сплайнами дифференцируемых функций и их производных.) *Ukrain. Math. Zh.* **35** (1983), 6, 702–10.
30. Some exact inequalities for differentiable functions and an estimate of the approximation of the functions and their derivatives by interpolation cubic splines. (Некоторые точные неравенства для дифференцируемых функций и оценка приближения функции и их производных интерполяционными кубическими сплайнами.) *Sibirsk. Math. Zh.* **34** (1983), 5, 94–108.
31. S. M. Nikol'skii and the development of research in the theory of approximation of functions in the USSR. (С. М. Никольский и развитие исследований по теории приближения функций в СССР.) *Usp. Math. Nauk* **40** (1985), 5, 71–131.
32. Optimal methods for coding and recovery of functions. (Оптимальные методы кодирования и восстановления функций.) *Optimal Algorithms,* Sofia, 1986, 157–71.
33. Rearrangement and extremal properties of functions. *Coll. math. soc. J.* Bolyai, Budapest (1986), 505–13.
34. On optimal coding of elements of a metric space. (Об оптимальном кодировании элементов метрического пространства.) *Ukrain. Math. Zh.* **39** (1987), 2, 168–73.
35. Exact constants in approximation theory. (Точные константы в теории аппроксимации.) *Teoriya priblizheniya funktsii.* Moscow, Nauka, 1987, 221–32.
36. On the estimate of the error of interpolation by splines of minimal defect. (Об оценке погрешности интерполирования силайнами минимального дефекта.) *Approximation Theory.* Warszawa, Center Publications, 1988, to appear.
37. On optimal coding of vector-functions. (Об оптимальном кодировании вектор-функций.) *Ukrain. Math. Zh.* **40** (1988), 6, 737–42.
38. Approximation and optimal coding of smooth plane curves. (Приближение и оптимальное кодирование гладких плоских кривых.) *Ukrain. Math. Zh.* **41** (1989), 2 (to appear).

Korneichuk, N. P. and Ligun, A. A. (Корнейчук, Н. П., Лигун, А. А.)
1. On the estimate of error of spline-interpolation in an integral metric. (Об оценке погрешности сплайн-интерполяции в интегральной метрике.) *Ukrain. Math. Zh.* **33** (1981), 3, 391–4.

2. On approximation of a class by another class and extremal subspaces in L_1. *Analysis Math.* **7** (1981), 2, 107–19.

Korneichuk, N. P. and Polovina, A. I. (Корнейчук, Н. П., Половина, А. И.)

1. Approximation of continuous and differentiable functions by algebraic polynomials on an interval. (О приближении непрерывных и дифференцируемых функций алгебраическими многочленами на отрезке.) *Dokl. Akad. Nauk SSSR* **166** (1966), 2, 281–3.
2. The approximation of functions that satisfy a Lipschitz condition by algebraic polynomials. (О приближении функций, удовлетворяющих условию Липшица алгебраическими многочленами.) *Math. Zametki* **9** (1971), 4, 441–7.
3. The approximation of continuous functions by algebraic polynomials. (О приближении непрерывных функций алгебраическими многочленами.) *Ukrain. Math. Zh.* **24** (1972), 3, 328–40.

Korovkin, P. P. (Коровкин, П. П.)

1. An asymptotic property of positive methods of summation of Fourier series and best approximation to functions of class Z_2 by linear positive polynomial operators. (Об одном асимптотическом свойстве положительных методов суммирования рядов Фурье и о наилучшем приближении финкции класса Z_2 линейными положительными полиномиальными операторами.) *Usp. Math. Nauk* **13** (1958), 6, 99–103.

Krein, M. G. (Крейн, М. Г.)

1. On the theory of best approximation of periodic functions. (К теории наилучшего приближения периодических функций.) *Dokl. Akad. Nauk SSSR* **18** (1938), 4–5, 245–51.

Kuptsov, N. P. (Купцов, Н. П.)

1. Kolmogorov estimates for derivatives in $L_2[0,\infty)$. (Колмогоровские оценки для производных в $L_2[0,\infty)$.) *Trudy Math. Inst. Stekhlov* **138** (1975), 94–117.

Ligun, A. A. (Лигун, А. А.)

1. Sharp inequalities for upper bounds of seminorms on classes of periodic functions. (Точные неравенства для верхных граней полунорм на классах периодических функций.) *Math. Zametki* **13** (1973), 5, 647–54.
2. The exact constants of approximation of differentiable periodic functions. (О точых константах приближения дифференцируемых периодических функций.) *Math. Zametki* **14** (1973), 1, 21–30.
3. Sharp inequalities for spline functions and best quadrature formulas for certain classes of functions. (Точные неравенства для сплайн-функций и наилучшие квадратурные формулы для некоторых классов функций.) *Math. Zametki* **19** (1976), 6, 913–26.
4. Inequalities for upper bounds of functions. *Analysis Math.* **2** (1976), 6, 913–62.
5. An inequality for a spline function with minimal defect. (Об одном неравенстве для сплайн-функций минимального дефекта.) *Math. Zametki* **24** (1978), 4, 547–52.
6. Optimal methods for the approximate calculation of functionals of classes W^rL_∞. *Analysis Math.* **5** (1979), 4, 269–86.
7. Diameters of certain classes of differentiable periodic functions. (О поперечниках некоторых классов дифференцируемых периодических функций.) *Math. Zametki* **27** (1980), 1, 61–75.
8. Approximation of periodic functions by splines of minimum defect. (О приближении периодических функций сплайнами минимального дефекта.) *Ukrain. Math. Zh.* **32** (1980), 3, 388–92.

9. Best approximation of differentiable functions by algebraic polynomials. (О наилучшем приближении дифференцируемых функций алгебраическими многочленами.) *Izv. Vissh. Uchebn. Zaved. Mathematika* (1980), 4, 53–60.
10. On inequalities between the norms of derivatives of periodic functions. (О неравенствах между нормами производных периодических функций.) *Math. Zametki* **33** (1983), 3, 385–91.
11. Constants in Jackson's theorem. (О констонтах в теореме Джексона.) *Math. Zametki* **37** (1985), 3, 326–36.
12. Exact constants in inequalities of Jackson type. (О точных константах в неравенствах типа Джексона.) *Math. Zametki* **38** (1985), 2, 248–56.
13. Sharp constants in inequalities of Jackson type. (О точных константах в неравенствах типа Джексона.) *Dokl. Akad. Nauk SSSR* **283** (1985), 1, 34–8.
14. Sharp inequalities for perfect splines and their applications. (Точные неравенства для совершенных сплайнов и их приложения.) *Izv. Vissh. Uchebn. Zaved. Mathematika* (1984), 5, 32–8.

Loginov, A. S. (Логинов, А. С.)
1. Approximation of continuous functions by polygons. (Приближение непрерывных функций ломаными.) *Math. Zametki* **6** (1969), 2, 149–60.

Lorch, L.
1. On approximation by Fejér means to periodic functions satisfying a Lipschitz condition. *Canad. Math. Bull.* **5** (1962), 1, 21–7.

Lorentz, G. G.
1. Convergence theorems for polynomials with many zeros. *Math. Zeit.* **186** (1984), 1, 117–23.

Makovoz, Yu. I. (Маковоз, Ю. И.)
1. Diameters of certain function classes in the space L. (Поперечники некоторых функциональных классов в пространстве L.) *Vesci Akad. Navuk BSSR Ser. Fiz.-Math. Navuk* (1969), 4, 19–28.
2. Diameters of Sobolev classes and splines deviating least from zero. (Поперечники Соболевских классов и сплайны, наименее уклоняющиеся от нуля.) *Math. Zametki* **26** (1979), 5, 805–12.

Malozemov, V. N. (Малоземов, В. Н.)
1. On the deviation of broken lines. (Об отклонении ломанных.) *Vestnik Leningrad. Univ.* **2** (1966), 7, 150–3.
2. On polygonal interpolation. (К полигональной интерполяции.) *Math. Zametki* **1** (1967), 5, 537–40.

Marchuk, A. G. and Osipenko, K. Yu. (Марчук, А. Г., Осипенко, К. Ю.)
1. Best approximation of functions defined with an error at a finite number of points. (Наилучшее приближение функций, заданных с погрешностью в конечном числе точек.) *Math. Zametki* **17** (1975), 3, 359–68.

Markov, A. A. (Марков, А. А.)
1. On limit quantities of integrals in connection with interpolation. (О предельных величинах интегралов в связи с интерполированием.) *Zap. Acad. Nauk Ser. VIII* **4** (1898) (See [11B, p. 146–230].)

Martynyuk, V. T. (Мартынюк, В. Т.)
1. On approximation by polygons defined by parametric equations. (О приближение ломаными кривых, заданных параметрическими уравнениями.) *Ukrain. Math. Zh.* **28** (1976), 1, 87–92.

2. Some questions for approximation of lines and surfaces. (Некоторые вопросы приближения линия и поверхностей.) *Teoriya priblizheniya funktsii,* Nauka, Moscow, 1987, 282–3.

Martynyuk, V. T. and Storchai, V. F. (Мартынюк, В. Т., Сторчай, В. Ф.)

1. Approximation by polyhedral functions in the Hausdorff metric. (Приближение многогранными функциями в хаусдорфовой метрике.) *Ukrain. Math. Zh.* **25** (1973), 1, 115–20.

Melkman, A. A.

1. Hermite–Birkhoff interpolation by splines. *J. Approx. Th.* **19** (1977), 3, 259–79.

Miccelli, Ch. A. and Pinkus, A.

1. Some problems on the approximation on functions of two variables and N-widths of integral operators. *J. Approx. Th.* **24** (1978), 51–77.

Motornyi, V. P. (Моторный, В. П.)

1. The best quadrature formula of the form $\Sigma_{k=1}^{n} p_k f(x_k)$ for certain classes of periodic functions. (О наилучией квадратурной формуле вида $\Sigma_{k=1}^{n} p_k f(x_k)$ для некоторых классов периодических функций.) *Izv. Akad. Nauk SSSR Ser. Math.* **38** (1974), 3, 538–614.
2. Approximation in the mean of periodic functions by trigonometric polynomials. (Приближение периодических функций тригонометрическими многочленами в среднем.) *Math. Zametki* **16** (1974), 1, 15–26.

Motornyi, V. P. and Ruban, V. I. (Моторный, В. П., Рубан, В. И.)

1. Diameters of certain classes of differentiable periodic functions in the space L. (Поперечники некоторых клоссов дифференцируемых периодических функций в пространстве L.) *Math. Zametki* **17** (1975), 4, 531–43.

Nazarenko, N. A. (Назаренко, Н. А.)

1. Approximation of plane curves by parametric Hermitian splines. (О приближении плоских кривых параметрическими эрмитовыми сплайнами.) *Ukrain. Math. Zh.* **31** (1979), 2, 201–5.
2. Approximation by splines with supplementary nodes of glueing in some classes of functions. (О приближении сплайнами с допольнительными узлами склейки на некоторых классах функций.) *Questions of the Theory of the Approximation of Functions.* Kiev, *Akad. Nauk Ukrain. SSR Inst. Math.,* 1980, 120–7.
3. Local reconstruction of curves by means of parametric splines. (О локальном восстановлении кривых с помощью параметрических сплайнов.) *Geometric Theory of Functions and Topology.* Kiev, *Akad. Nauk Ukrain. SSR, Inst. Math.,* 1981, 55–62.
4. Approximation of curves by means of parametric splines. (Приближения кривых с помощью параметрических сплайнов.) *Constructive Functions Theory '81 (Varna, 1981),* Sofia, Bulgar Akad. of Sci., 1983, 111–14.

Nazarenko, N. A. and Pereverzev, S. V. (Назаренко, Н. А., Переверзев, С. В.)

1. Exact values of an approximation by Hermite splines of even degree on classes of differentiable functions. (Точные значения приближения эрмитовыми сплайнами четной степени на классах дифференцируемых функций.) *Math. Zametki* **28** (1981), 1, 33–44.

Nguen Thi Hoa (Нгуен Тхи Хоа)

1. On an extremal problem for oscillation diminishing convolutions. (Об одной экстремальной задаче для сверток, не увеличивающих осцилляцию.) *Vestnik Moskow Univ. Math.-Mech.* (1982), 5, 3–7.

2. Rolle's theorem for differentiable operators and some extremal problems in approximation theory. (Теорема Ролля для дифференцируемых операторов и некторые экстремальные задачи теории приближений.) *Dokl. Akad. Nauk SSSR* **295** (1987), 6, 1313–18.

Nikol'skii, V. N. (Никольский, В. Н.)

1. Best approximation by elements of convex sets in normed linear spaces. (Наилучшее приближение элементами выпуклых миоожеств в линейныых нормированных пространствах.) *Kalinin. Gos. Ped. Uchen. Zap.* **29** (1963), 85–119.

Nikol'skii, S. M. (Никольский, С. М.)

1. On the asymptotic behaviour of the error of approximation of functions satisfying Lipschitz condition by Fejer means. (Об асимптотическом поведении остатка при приближении функций, удовлетворяющих условию Липшица, суммами Фейера.) *Izv. Akad. Nauk SSSR Ser. Math.* **4** (1940), 501–8.
2. On some methods for approximation by trigonometric sums. (О некоторых методах приближения тригонометрическими суммами.) *Izv. Akad. Nauk SSSR Ser. Math.* **4** (1940), 6, 509–20.
3. An asymptotic estimate of the trigonometric polynomials approximation error. (Асимптотическая оценка остатка при приближении интерполяционными тригонометрическими полиномами.) *Dokl. Akad. Nauk SSSR* **31** (1941), 3, 215–18.
4. Asymptotic estimate of the error of approximation by Fourier sums. (Асимптотическая оценка остатка при приближении суммами Фурье.) *Dokl. Akad. Nauk SSSR* **32** (1941), 6, 386–9.
5. Approximation of periodic functions by trigonometric polynomials. (Приближение периодических функций тригонометрическими многочленами.) *Trudy Math. Inst. Stekhlov* **15** (1945), 1–76.
6. Mean approximation of functions by trigonometric polynomials. (Приближение функций тригонометрическими многочленами в среднем.) *Izv. Akad. Nauk SSSR Ser. Math.* **10** (1946), 3, 207–56.
7. On the best approximation by polynomials of functions satisfying Lipschitz condition. (О наилучшем приближении многочленами функций, удовлетворяющих условию Липшица.) *Izv. Akad. Nauk SSSR Ser. Math.* **10** (1946), 4, 295–322.
8. On interpolation and best approximation of differentiable periodic functions by trigonometric polynomials. (Об интерполировании и наилучшем приближении дифференцируемых периодических функций тригонометрическими полиномами.) *Izv. Akad. Nauk SSSR Ser. Math.* **10** (1946), 5, 393–410.
9. Fourier series of functions with a given modulus of continuity. (Ряд Фурье функций с данным модулем непрерывности.) *Dokl. Akad. Nauk SSSR* **52** (1946), 3, 191–4.
10. On best approximation of functions whose sth derivative has discontinuities of the first kind. (О наилучшем приближении функций, s-я производная которых имеет разрывы первого рода.) *Dokl. Akad. Nauk SSSR* **55** (1947), 2, 99–102.
11. On the best linear method of approximation in mean of differentiable functions by polynomials. (О наилучшем линейном методе приближения многочле-

нами в среднем дифференцируемых функций.) *Dokl. Akad. Nauk SSSR* **58** (1947), 2, 185–8.
12. A generalization of an inequality of S. N. Bernstein. (Обобщение одного неравенства С. Н. Бернштейна.) *Dokl. Akad. Nauk SSSR* **60** (1948), 9, 1507–10.
13. On an asymptotically best linear method for approximation of differentiable functions by polynomials. (Об асимптотически наилучшем линйном методе приближения дифференцируемых функций многочленами.) *Dokl. Akad. Nauk SSSR* **69** (1949), 2, 129–32.

Oskolkov, K. I. (Осколков, К. И.)
1. Subsequences of Fourier sums of functions with a prescribed modulus of continuity. (Последовательности сум Фурье функций с заданым модулем непрерывности.) *Math. Sb. (N.S.)* **88** (1972), 3, 447–69.
2. An estimate for the approximation of continuous functions by sub-sequences of Fourier sums. (Оценка приближения непрерывных функций подпоследовательностями сумм Фурье.) *Trudy Math. Inst. Stekhlov* **134** (1975), 240–53.

Pereverzev, S. V. (Переверзев, С. В.)
1. Exact values of approximation by Hermite splines on a class of functions of two variables. (Точные значения приближения эрмитовыми сплайнами на одном классе функций двух переменных.) *Ukrain. Math. Zh.* **31** (1979), 5, 510–16.
2. An exact estimate of approximation by Hermite splines on a class of differentiable functions of two variables. (Точная оценка приближения эрмитовыми сплайнами на одном классе дифференцируемых функций двух переменных.) *Izv. Vissh. Uchebn. Zaved. Mathematika* (1981), 12, 58–66.

Petrov, I. M. (Петров, И. М.)
1. Order of approximation of functions of class Z by certain polynomial operators. (Порядок приближения функций класса Z некоторыми полиномиальными операторами.) *Usp. Math. Nauk* **133** (1958), 6, 127–32.

Pinkevich, V. T. (Пинкевич, В. Т.)
1. On the order of the remainder of Fourier series of functions differentiable in H. Weyl sense (О порядке остаточного члена ряда Фурье функций, дифференцируемых в смысле Вейля.) *Izv. Akad. Nauk SSSR Ser. Math.* **4** (1940), 6, 521–8.

Pinkus, A.
1. On *N*-widths of periodic functions. *J. Anal. Math.* **35** (1979), 209–35.
2. *N*-Widths of Sobolev spaces in L_p. *Constr. Approx.* **1** (1985), 1, 15–62.

Pokrovskii, A. V. (Покровский, А. В.)
1. On a theorem of A. F. Timan. (Об одной теореме А. Ф. Тимана.) *Funktional Anal. i Prilozhen.* **1** (1967), 3, 93–4.

Polovina, A. I. (Половина, А. И.)
1. The best uniform approximation by algebraic polynomials of differentiable functions. (О наилучшем равномерном приближении алгебраическими многочленами дифференцируемых функций.) *Izv. Vissh. Uchebn. Zaved. Mathematika* (1969), 12, 79–82.

Rogosinski, W.
1. Uber die Abschnitte trigonometrischer Reihen. *Math. Ann.* **95** (1925), 110–35.

Ruban, V. I. (Рубан, В. И.)
1. Diameters of class $W^1H_\omega^{2\pi}$ in L. (Поперечники класса $W^1H_\omega^{2\pi}$ в L.) *Studies in Current Problems of Summation and Approximation of Functions and Their Applications*. Dnepropetrovsk, Gos. Univ. Dnepropetrovsk, 1972, 59–61.

2. Diameters of some classes of continuously differentiable functions. (Поперечники некоторых классов непрерывно дифференцируемых функций.) *Studies in Current Problems of Summation and Approximation of Functions and Their Applications*. Dnepropetrovsk, Gos. Univ. Dnepropetrovsk, 1973, 78–81.
3. Diameters of some classes of functions in space $C_{2\pi}$. (Поперечники некоторых классов функций в пространстве $C_{2\pi}$.) *Studies in Current Problems of Summation and Approximation of Functions and Their Applications*. Dnepropetrovsk, Gos. Univ. Dnepropetrovsk, 1973, 75–7.
4. Even diameters of the classes W^rH_ω in the space $C_{2\pi}$. (Четные поперечники классов W^rH_ω в пространстве $C_{2\pi}$.) *Math. Zametki* **15** (1974), 3, 387–92.
5. Diameters of a class of 2π-periodic functions in space L_p $(1 < p < \infty)$. (Поперечники одного класса 2π-периодических функций в пространстве L_p $(1 < p < \infty)$.) *Teoriya priblizheniya funktsii i ee prilozheniya*. Kiev, Math. inst. Akad. Nauk Ukrain. SSR, 1974, 119–28.
6. Diameters of sets in spaces of periodic functions. (Поперечники множеств в пространствах периодических функций.) *Dokl. Akad. Nauk SSSR* **225** (1980), 1, 34–5.
7. Topological properties of spaces of perfect splines and N-widths of classes of periodic functions. (Топологические свойства пространств совершенных сплайнов и поперечники классов периодических функций.) *Teoriya priblizheniya funktsii,* Moskow, Nauka, 1987, 379–82.

Schoenberg, I. J.
1. Contributions to the problem of approximation of equidistant data by analytic functions. *Quart. Appl. Math.* **4** (1946), 45–99, 112–41.
2. On best approximation of linear operators. *Kon. Neder. Acad. Wetensch. Proceedings. Ser. A* **67** (1964), 155–63.
3. Cardinal interpolation and spline functions. *J. Approx. Th.* **2** (1969), 167–206.

Shabozov, M. (Шабозов, M.)
1. Estimates of approximation of differentiable periodic functions of two variables by interpolating mixed splines. (Оценки приближения дифференцируемых периодических функций двух переменных интерполяционными смешанными сплайнами.) *Questions of the Theory of the Approximation of Functions*. Kiev, Akad. Nauk Ukrain. SSR Inst. Math., 1980, 166–72.

Shalaev, V. V. (Шалаев, В. В.)
1. On the question of approximation of continuous periodic functions by trigonometric polynomials. (К вопросу о приближении непрерывных периодических функций тригонометрическими полиномами.) *Studies in Current Problems of Summation and Approximation of Functions and Their Applications*. Dnepropetrovsk, Gos. Univ. Dnepropetrovsk, 1977, 39–43.
2. On one estimate for approximation of differentiable periodic functions by trigonometric polynomials. (Об одной оценке приближения дифференцируемых периодических функций тригонометрическими полиномами.) *Studies in Current Problems of Summation and Approximation of Functions and Their Applications*. Dnepropetrovsk, Gos. Univ. Dnepropetrovsk, 1979, 31–5.

Sharma, A. and Tzimbalario, T.
1. Landau-type inequalities for some linear differential operators. *Illinois J. Math.* **20** (1976), 3, 443–55.

Shevaldin, V. T. (Шевалдин, В. Т.)
1. Some problems for extremal interpolation in mean for linear differential operators. (Некоторые задачи экстремальной интерполяции в среднем для

линейных дифференциальных операторов.) *Trudy Math. Inst. Stekhlov* **167** (1983), 203–40.
2. *L*-splines and *N*-widths. (*L*-сплайны и поперечники.) *Math. Zametki* **33** (1983), 5, 735–44.

Sokolov, I. G. (Соколов, И. Г.)
1. Remainder term of the Fourier series of differentiable functions. (Остаточны член ряда Фурье дифференцируемых фукций.) *Dokl. Akad. Nauk SSSR* **103** (1955), 1, 23–6.

Stechkin, S. B. (Стечкин, В. Б.)
1. Generalization of some inequalities of S. N. Bernstein. (Обобщение некоторых неравенств С. Н. Бернштейна.) *Dokl. Akad. Nauk SSSR* **60** (1948), 9, 1511–14.
2. On best approximation of certain classes of periodic functions by trigonometric polynomials. (О наилучшем приближении некоторых классов периодических функций тригонометрическими ролиномами.) *Izv. Akad. Nauk SSSR Ser. Math.* **20** (1956), 643–8.
3. Inequalities between norms of derivatives of arbitrary functions. (Неравенства между нормами производных произвольной функции.) *Acta Sci. Math.* **26** (1965), 225–30.
4. A remark on Jackson's theorem. (замечание к теореме Джексона.) *Trudy Math. Inst. Stekhlov.* **88** (1967), 17–19.
5. Best approximation of linear operators. (Наилучшее приблжение линейных операторов.) *Math. Zametki* **1** (1967), 2, 137–48.
6. The approximation of continuous periodic functions by Favard sums. (О приближении непрерывных периодических функций сумами Фавара.) *Trudy Math. Inst. Stekhlov* **109** (1971), 26–34.
7. Estimation of the remainder for the Fourier series for differentiable functions. (Оценка остатка ряда Фурье для дифференцируемых функций.) *Trudy Math. Inst. Stekhlov* **145** (1980), 126–51.

Stechkin, S. B. and Telyakovskii, S. A. (Стечкин, С. Б., Теляковский, С. А.)
1. The approximation of differentiable functions by trigonometric polynomials in the *L*-metric. (О приближении дифференцируемых функций тригонометрическими полиномами в метрике *L*.) *Trudy Math. Inst. Stekhlov* **88** (1967), 20–29.

Stein, E. M.
1. Functions of exponential type. *Ann. Math.* **65** (1957), 3, 582–92.

Stepanets, A. I. (Степанец, А. И.)
1. Classes of periodic functions and the approximation of their elements by Fourier sums. (Классы периодических функций и приближение их элементов суммами Фурье.) *Dokl. Akad. Nauk SSSR* **277** (1984), 5, 1074–7.
2. Simultaneous approximation of periodic functions and their derivatives. (Совместное приближение периодических функций и ис производных суммами Фурье.) *Math. Zametki* **36** (1984), 6, 873–82.

Storchai, V. F. (Сторчай, В. Ф.)
1. The deviation of polygonal lines in the L_p metric. (Об отклонении ломаных в метрике L_p.) *Math. Zametki* **5** (1969), 1, 31–7.
2. Approximation of functions of two variables by polyhedral functions in the uniform metric. (Прбличение функции переменных многогранными функциями в равномерной метрике.) *Izv. Vissh. Uchebn. Zaved. Mathematika* **8** (1973), 84–8.
3. Precise estimates for the norms of differentiable periodic functions in the L_2

metric. (Точные оценки для норм дифференцируемых периодических функций в метрике L_2.) *Ukrain. Math. Zh.* **25** (1973), 6, 835–41.
4. Exact bound for norms of differentiable periodic functions. (О точных оценках норм дифференцируемых периодических функций.) *Studies in Contemporary Problems of Summability and Approximation of Functions and Their Applications.* Dnepropetrovsk, Gos. Univ., Dnepropetrovsk, 1976, 50–4.

Subbotin, Yu. N. (Субботин, Ю. Н.)
1. Piecewise polynomial interpolation. (О кусочно-полиномиальной интерполяции.) *Math. Zametki* **1** (1967), 1, 63–70.
2. The best approximation of a class of functions by another class. (Наилучшее преближение класса функций другим классом.) *Math. Zametki* **2** (1967), 5, 495–504.
3. The diameter of the class W^rL in $L(0,2\pi)$ and approximation by spline functions. (Поперечники класса W^rL в $L(0,2\pi)$ и приближение сплайн функциями.) *Math. Zametki* **7** (1970), 1, 43–52.
4. Approximation by spline functions and estimates of widths. (Приближение сплайн-функциями и оценка поперечников.) *Trudy Math. Inst. Stekhlov* **109** (1971), 35–60.
5. Extremal problems of functional interpolation and mean interpolation splines. (Эхтремальные задачи функциональной интерполяции и интерполяционные в среднем сплайны.) *Trudy Math. Inst. Stekhlov* **138** (1975), 118–73.
6. Extremal problems of the theory of approximation of functions with incomplete information. (Экстремальные задачи теории приближения функций при неполной информации.) *Trudy Math. Inst. Stekhlov* **145** (1980), 152–68.

Sun Yong-sheng
1. On the best approximation of periodic differentiable functions by trigonometric polynomials. (О наилучшем приближении периодических дифференцируемых функций тригонометрическими полиномами.) *Izv. Akad. Nauk SSSR Ser. Math.* **22** (1959), 1, 67–92.
2. Some extremal problems on the class of differentiable functions. *J. Beiling Normal Univ. (Natur. Sci.)* (1984), 4, 11–26.
3. Inequalities of Landau–Kolmogorov type for some linear differential operators. *Kexue tongbao* **30** (1985), 8, 995–8.
4. On optimal interpolation for a differentiable function class. *Approx. Th. and Its Appl.* **2** (1986), 5, 49–54.

Sun Yong-sheng and Hyang Daren
1. On *N*-widths of the generalized Bernoulli kernel. *Approx. Th. and Its Appl.* **1** (1985), 2, 83–92.

Szokefalvy-Nagy, B.
1. Uber gewisse Extremalfragenbei transformierten trigonometrischen Entwicklungen 1. Periodischer Fall. *Berichte Math.-phys. Acad. d. Wiss. Leipzig* **90** (1938), 103–34.
2. Sur une class générale de procédés de sommation pour les series Fourier. *Hung. Acta Math.* **1** (1948), 3, 14–52.

Taikov, L. V. (Тайков, Л. В.)
1. A generalization of an inequality of S. N. Bernstein. (Одно обобщение неравенства Н. С. Бернщейна.) *Trudy Math. Inst. Stekhlov* **78** (1965), 43–7.
2. The approximation in the mean of certain classes of periodic functions. (О приближении в среднем некоторых классов периодических функций.) *Trudy Math. Inst. Stekhlov* **88** (1967), 61–70.

3. Best approximation in the mean of certain classes of analytic functions. (О наилучшем приближении в среднем некоторых классов аналитических функций.) *Math. Zametki* **1** (1967), 2, 155–62.

Telyakovskii, S. A. (Теляковский, С. А.)

1. On the norms of trigonometric polynomials and approximation of differentiable functions by linear means of their Fourier series: 1. (О нормах тригонометрических полиномах и приближение дифференцируемых функций линейными средными их рядов Фурье: 1.) *Trudy Math. Inst. Stekhlov* **62** (1961), 61–97.
2. Approximation of differentiable functions by partial sums of their Fourier series. (Приближение дифференцуруемых функций частнами сумами их рядов Фурье.) *Math. Zametki* **4** (1968), 3, 291–300.
3. The approximation by Féjer sums of functions which satisfy a Lipschitz condition. (О приближении функций, удовлетворяюших условию Липшица суммам Фейера.) *Ukrain. Math. Zh.* **21** (1969), 3, 334–43.

Temlyakov, V. N. (Темляков, В. Н.)

1. Asymptotic behaviour of the best approximation of continuous functions. (Асимптотическое поведение наилучших приближений непрерывных функциы.) *Izv. Akad. Nauk SSSR Ser. Math.* **41** (1977), 3, 587–606.
2. On the question of the asymptotic behaviour of the best approximations of continuous functions. (К вопросу об асимптотическом поведении наилучших приближений непрерывных функций.) *Math. Sb. (N.S.)* **110** (1979), 3, 399–413.

Tikhomirov, V. M. (Тихомиров, В. М.)

1. On n-dimensional diameters of certain functional classes. (Об n-мерных поперечниках некоторых функцональных классов.) *Dokl. Akad. Nauk SSSR* **130** (1960), 4, 734–7.
2. Diameters of sets in functional spaces and the theory of best approximations. (Поперечники множеств в функциональных пространствах и теория наилучших приближений.) *Usp. Math. Nauk* **15** (1960), 3, 81–120.
3. Some problems in approximation theory. (Некоторые вопросы теории приближений.) *Dokl. Akad. Nauk SSSR* **160** (1965), 4, 774–7.
4. A remark on n-dimensional diameters of sets in Banach spaces. (Одно замечание об n-мерных поперечниках множеств в Банаховых пространствах.) *Usp. Math. Nauk* **20** (1965), 1, 227–30.
5. Best methods of approximation and interpolation on classes of differentiable functions in the space $C[-1,1]$. (Наилучшие методы приближения и интерполирования дифференцируемых функций в пространстве $C[-1,1]$.) *Math. Sb. (N.S.)* **80** (1969), 2, 290–304.
6. Approximation theory. (Теория приближений.) *Itogi nauki i tekhniki. Sovremen. problemy matematiki* **14** (1987), 103–270.

Tikhomirov, V. M. and Boyanov, B. D. (Тихомиров, В. М., Бояноч, Б. Д.)

1. On some convex problems in approximation theory. (О некоторых выпуклых задачах теории приближений.) *Serdica,* Bulg. Math. Publications, vol. 5 (1979), 83–96.

Timan, A. F. (Тиман, А. Ф.)

1. On the best approximations of differentiable functions by algebraic polynomials on a finite interval of the real axis. (О наилучшем приближении дифференцируемых функций алгебраическими многочленами на конечном отрезке вещественной оси.) *Izv. Akad. Nauk SSSR Ser. Mat.* **22** (1958), 3, 335–60.

Turovets, S. P. (Туровец, С. П.)

1. Best approximation in the mean of differentiable functions. (О наилучшем приближении в среднем дифференцируемых функций.) *Dokl. Akad. Nauk Ukrain. RSR Ser. A* (1968), 5, 417–22.

Vakarchuk, S. V. (Вакарчук, С. Б.)

1. Approximation of parametric surfaces by bilinear ones. (О приближении параметрических поверхностей билинейными.) *Monogenic Functions and Mappings*. Kiev, Akad. Nauk Ukrain. SSR Inst. Math., 1982, 107–13.
2. Approximation of curves, specified in parametric form, by means of spline curves. (О приближении кривых, заданных в параметрическом виде, при помощи сплайн-кривых.) *Ukrain. Math. Zh.* **35** (1983), 3, 352–5.

De la Vallée–Poussin, Ch.-J.

1. Sur les polynomes d'approximation et la représentation approchée d'un angle. *Bull. Acad. Sci. Belgique* **2** (1910), 808–44.

Velikin, V. L. (Великин, В. Л.)

1. Best approximation by spline functions on classes of continuous functions. (О наилучшем приближении сплайн-функциями на классах непрерывных функций.) *Math. Zametki* **8** (1970), 1, 41–6.
2. Approximation by cubic splines on classes of continuously differentiable functions. (Приближение кубическими сплайнами на классах непрерывно дифференцируемых функций.) *Math. Zametki* **11** (1972), 2, 215–26.
3. Precise values of approximation by Hermitian splines on classes of differentiable functions. (Точные значения приближения эрмитовыми сплайнами на классах дифференцируемых функций.) *Izv. Akad. Nauk SSSR Ser. Math.* **37** (1973), 1, 165–85.
4. Optimal interpolation of periodic functions with a bounded higher derivative. (Оптимальная интерполяция пириодических дифференцируемых функций с ограниченной старшей производной.) *Math. Zametki* **22** (1977), 5, 663–70.
5. On the asymptotic relation between the approximations of continuous functions by splines and by trigonometric polynomials. (О предельной связи между приближениями периодических функций сплайнами и тригонометрическими полиномами.) *Dokl. Akad. Nauk SSSR* **258** (1984), 3, 525–9.

Velikin, V. L. and Korneichuk, N. P. (Великин, В. Л., Корнейчук, Н. П.)

1. Exact estimates of the approximation by spline functions on classes of differentiable functions. (Точные оценки приближения сплайн-функциями на классах дифференцируемых функций.) *Math. Zametki* **9** (1971), 5, 483–94.

Wang Hing-hua

1. The exact constant of approximation of continuous functions by the Jackson singular integral. *Chinese Math.* **5** (1964), 2, 254–60.

Zhensykbaev, A. A. (Женсыкбаев, А. А.)

1. Sharp estimates for the uniform approximation of continuous periodic functions by rth order splines. (Точные оценки равномерного приближения непрерывных периодических функций сплайнами r-того порядка.) *Math. Zametki* **13** (1973), 2, 217–28.
2. Approximation of differentiable periodic functions by splines on a uniform subdivision. (Приближение дифференцируемых периодических функций сплайнами по равномерному разбиению.) *Math. Zametki* **13** (1973), 6, 807–16.

3. Approximation of certain classes of differentiable periodic functions by interpolation splines on a uniform subdivision. (Приближение некоторых классов дифференцируемых периодических функций интерполяционными сплайнами по равномерному разбиению.) *Math. Zametki* **15** (1974), 6, 955–66.
4. Spline-interpolation and the best approximation by trigonometric polynomials. (Сплайн интерполаяция и наилучшее прибижение тригонометрическими многочленами.) *Math. Zametki* **26** (1979), 3, 355–66.
5. Monosplines of minimal norm and optimal quadrature formulas. (Моносплайны минимальной нормой и наилучшие квадратурные формулы.) *Usp. Math. Nauk* **36** (1981), 4, 107–59.

Zhuk, V. V. (Жук, В. В.)
1. Certain exact inequalities between best approximation and moduli of continuity. (О некоторых точных неравенствах между наилучшими приближениями и модулями непрерывности.) *Sibirsk. Math. Zh.* **12** (1971), 6, 1283–91.
2. Certain exact inequalities between uniform best approximation of periodic functions. (Некоторые точные неравенства между равномерными приближениями периодических функций.) *Dokl. Akad. Nauk SSSR* **201** (1971), 2, 263–5.

Zygmund, A.
1. A remark on conjugate series. *Proc. Am. Math. Soc.* **34** (1932), 435–46.

Added by the translator

Sendov, Bl. and Popov, V. A. (Сендов, Бл., Попов, В. А.)
1. *Averaged Moduli of Smoothness.* (Усреднени модули на гладкост.) Publ. house of the Bulg. Acad. of Sci., Sofia, 1983 (Russian translation – Mir, Moscow, 1988; English translation – John Wiley & Sons, NY, 1988.)

INDEX OF NOTATION

Approximation characteristics

Function classes

Approximating subspaces

Functions

Some sets

Constants and numerical characteristics

INDEX